Vegetable Crops: Quality Evaluation and Management

Editor: Laura Vivian

R CALLISTO REFERENCE

www.callistoreference.com

Callisto Reference,
118-35 Queens Blvd., Suite 400,
Forest Hills, NY 11375, USA

Visit us on the World Wide Web at:
www.callistoreference.com

ISBN: 978-1-63239-786-7 (Hardback)

The publisher's policy is to use permanent paper from mills that operate a sustainable forestry policy. Furthermore, the publisher ensures that the text paper and cover boards used have met acceptable environmental accreditation standards.

Trademark Notice: Registered trademark of products or corporate names are used only for explanation and identification without intent to infringe.

Printed in the United States of America.

Cataloging-in-publication Data

Vegetable crops : quality evaluation and management / edited by Laura Vivian.
    p. cm.
Includes bibliographical references and index.
ISBN 978-1-63239-786-7
1. Vegetables. 2. Vegetables--Quality. 3. Horticulture. 4. Vegetables--Diseases and pests. I. Vivian, Laura.
SB320.9 .V44 2017
635--dc23

Table of Contents

Preface

This book on vegetable crops focuses on the various cultivation practices that are implemented in the production of vegetables. Cultivation practices differ according to crops in question as well as the quantity that is being grown. Allied methods such as soil quality, soil preparation and monitoring are also important aspects of vegetable cultivation. With the prevalence of genetically modified crops, it is possible to reap greater quantities of improved quality vegetables. The book is compiled in such a manner that it will provide in-depth knowledge about the theory and practice of vegetable cultivation. The various advancements in this field are glanced at and their applications as well as ramifications are looked at in detail.

The purpose of the book is to provide a glimpse into the dynamics and to present opinions and studies of some of the scientists engaged in the development of new ideas in the field from very different standpoints. This book will prove useful to students and researchers owing to its high content quality.

At the end, I would like to appreciate all the efforts made by the authors in completing their chapters professionally. I express my deepest gratitude to all of them for contributing to this book by sharing their valuable works. A special thanks to my family and friends for their constant support in this journey.

Editor

Detection of Genome Donor Species of Neglected Tetraploid Crop *Vigna reflexo-pilosa* (Créole Bean), and Genetic Structure of Diploid Species Based on Newly Developed EST-SSR Markers from Azuki Bean (*Vigna angularis*)

Sompong Chankaew[1,9], Takehisa Isemura[2,9], Sachiko Isobe[3,9], Akito Kaga[2], Norihiko Tomooka[2*], Prakit Somta[4], Hideki Hirakawa[3], Kenta Shirasawa[3], Duncan A. Vaughan[2], Peerasak Srinives[4*]

1 Program in Plant Breeding, Faculty of Agriculture at Kamphaeng Saen, Kasetsart University, Kamphaeng Saen, Nakhon Pathom, Thailand, 2 Genetic Resources Center, National Institute of Agrobiological Sciences, Tsukuba, Ibaraki, Japan, 3 Kazusa DNA Research Institute, Kisarazu, Chiba, Japan, 4 Department of Agronomy, Faculty of Agriculture at Kamphaeng Saen, Kasetsart University, Kamphaeng Saen, Nakhon Pathom, Thailand

Abstract

Vigna reflexo-pilosa, which includes a neglected crop, is the only one tetraploid species in genus *Vigna*. The ancestral species that make up this allotetraploid species have not conclusively been identified, although previous studies suggested that a donor genome of *V. reflexo-pilosa* is *V. trinervia*. In this study, 1,429 azuki bean EST-SSR markers were developed of which 38 EST-SSR primer pairs that amplified one product in diploid species and two discrete products in tetraploid species were selected to analyze 268 accessions from eight taxa of seven Asian *Vigna* species including *V. reflexo-pilosa* var. *glabra*, *V. reflexo-pilosa* var. *reflexo-pilosa*, *V. exilis*, *V. hirtella*, *V. minima*, *V. radiata* var. *sublobata*, *V. tenuicaulis* and *V. trinervia* to identify genome donor of *V. reflexo-pilosa*. Since both diploid and tetraploid species were analyzed and each SSR primer pair detected two loci in the tetraploid species, we separated genomes of the tetraploid species into two different diploid types, viz. A and B. In total, 445 alleles were detected by 38 EST-SSR markers. The highest gene diversity was observed in *V. hirtella*. By assigning the discrete PCR products of *V. reflexo-pilosa* into two distinguished genomes, we were able to identify the two genome donor parents of créole bean. Phylogenetic and principal coordinate analyses suggested that *V. hirtella* is a species complex and may be composed of at least three distinct taxa. Both analyses also clearly demonstrated that *V. trinervia* and one taxon of *V. hirtella* are the genome donors of *V. reflexo-pilosa*. Gene diversity indicates that the evolution rate of EST-SSRs on genome B of créole bean might be faster than that on genome A. Species relationship among the *Vigna* species in relation to genetic data, morphology and geographical distribution are presented.

Editor: Manoj Prasad, National Institute of Plant Genome Research, India

Funding: The work was supported by the Thailand Research Fund (TRF). The funders had no role in study design, data collection and analysis, decision to publish, or preparation of the manuscript.

Competing Interests: The authors have declared that no competing interests exist.

* Email: tomooka@affrc.go.jp (NT); agrpss@yahoo.com (PS)

9 These authors contributed equally to this work.

Introduction

The Leguminosae genus *Vigna* comprises about 100 species. These species are morphologically diverse and geographically widespread and mainly found in Africa (African *Vigna*; subgenus *Vigna*) and Asia (Asian *Vigna*; subgenus *Ceratotropis*) [1]. The Asian *Vigna* comprises 21 species in which seven are domesticated and/or cultivated in various geographical and climatic regions, and cropping systems in Asia [2]. The seven domesticated/cultivated species include *V. aconitifolia* (moth bean), *V. angularis* (azuki bean), *V. mungo* (black gram), *V. radiata* (mungbean), *V. reflexo-pilosa* (créole bean), *V. stipulacea* (jungli bean) and *V.*

umbellata (rice bean). All Asian *Vigna* species are diploid having 11 haploid chromosomes ($2n = 2x = 22$) with the exception for créole bean which is a tetraploid species with number of haploid chromosome of 22 ($2n = 4x = 44$). In fact, créole bean is the only natural amphidiploid in the subtribe Phaseolinae [3]. Cytogenetic analyses of *V. reflexo-pilosa* showed that the species formed 22 bivalents without multivalent at meiosis and was, therefor considered to be an amphidiploid [4] [5].

Cultivated and wild forms of créole bean are classified as *V. reflexo-pilosa* var. *glabra* and *V. reflexo-pilosa* var. *reflexo-pilosa*, respectively [6]. Wild créole bean is widely distributed in East,

Southeast and South Asia, and across the islands from the west to the north Pacific islands. It is also found in Papua New Guinea and northern Australia [7] [8] [9]. The cultivated créole bean was formerly recognized as a glabrous variety of mungbean, *V. radiata* var. *glabra* [1]. Then, it was treated as a distinct species, *V. glabrescens* [3]. It is differentiated from its wild progenitor principally by thick glabrous stem and erect growth habit and reported to be cultivated as pulse in Vietnam and the Philippines or as forage in India, Mauritius and Tanzania [2]. This crop shows resistance to several insect pests and diseases such as bruchids, bean fly, powdery mildew, and cucumber mosaic virus [10], and is partially cross-compatible with mungbean [11]. Thus, créole bean has potential to be a gene source for breeding other *Vigna* crops. In addition, cultivated créole bean could be considered as a novel crop for the future and wild créole bean as a wider genepool to improve cultivated créole bean.

The origin and genome donors of créole bean have been the subject of debate. Based on the study on isozyme banding patterns and interspecific hybridization among the *Ceratotropis* species, Egawa et al [10] [12] proposed that *V. reflexo-pilosa* var. *reflexo-pilosa* is the descendant of natural interspecific hybridization between *V. trinervia* and *V. minima* followed by spontaneous chromosome doubling. An accession used as *V. minima* in Egawa et al. [10] [12] was correctly identified as *V. hirtella* by Konarev et al. [13], and was used in the present study (No. 21, JP108851, from Malaysia). Proteinase inhibitors polymorphism study in Asian *Vigna* by Konarev et al. [13] supported that one genome donor of créole bean is *V. trinervia* and suggested that the other genome donor is most likely *V. hirtella* or its closely related species. The phylogenetic studies based on plastid sequence data also supported that *V. trinervia* is a genome donor of créole bean [14] [15]. In contrast, rDNA-ITS sequence variation showed high similarity between créole bean and *V. exilis*, *V. hirtella* and *V. umbellata* [16], suggesting that one of these species is the genome donor of *V. reflexo-pilosa*.

The objective of this study was to determine the putative genome donor species of tetraploid créole bean using EST-SSR markers. To do so, we developed EST-SSR from azuki bean and used them to analyze accessions of créole bean and other *Vigna* species that are candidate genome donor species.

Materials and Methods

Plant Materials

A Japanese azuki bean cultivar 'Erimo-shouzu' (*V. angularis* var. *angularis*, accession no. JP37752) obtained from the Genebank, National Institute of Agrobiological Sciences (NIAS), Tsukuba, Japan was used for development of EST-SSR markers. Eight accessions consisting of four major *Vigna* crop species; azuki bean, rice bean, black gram and mungbean were used for assessing transferability and polymorphisms of the EST-SSR markers (Table 1). Two hundred and sixty-eight accessions from eight taxa of seven Asian *Vigna* species (Figure 1, Table S1) including 7 of *V. reflexo-pilosa* var. *glabra*, 51 of *V. reflexo-pilosa* var. *reflexo-pilosa*, 13 of *V. exilis*, 47 of *V. hirtella*, 49 of *V. minima*, 13 of *V. radiata* var. *sublobata*, 42 of *V. tenuicaulis*, and 46 of *V. trinervia* were used to analyze the genome origin of *V. reflexo-pilosa*.

Development of Azuki Bean EST-SSR Markers

Total RNA was extracted from seedlings and young pods of Erimo-shouzu using Plant RNA Purification Reagent (Invitrogen, CA, USA). Purification of polyadenylated RNA and conversion to cDNA were performed as described by [17]. Synthesized cDNA was resolved by 1% agarose gel electrophoresis, and fragments ranging from 1 to 3 kb were recovered. The recovered fragments were cloned into the *Eco* RI-*Xho* I site of the pBluescript II SK- plasmid vector (Stratagene, CA, USA) and introduced into the *E. coli* ElectroTen-Blue strain (Stratagene, CA, USA) by electroporation. To generate ESTs, plasmid DNAs were amplified from the colonies using TempliPhi (GE Healthcare UK Ltd, Buckinghamshire, England) and subjected to sequencing using a BigDye Terminator Cycle Sequencing Ready Reaction Kit (Applied Biosystems, CA, USA). The reaction mixtures were run on an automated ABI PRISM 3730 DNA Analyzer (Applied Biosystems).

Sequencing chromatograms were converted into nucleotide bases with Phred [18] [19], and the sequences derived from the vector and linkers were removed with CROSSMATCH [19]. The EST reads were quality-trimmed with TRIM2 [20] using the Phred quality score ≥ 20, and ambiguous regions including more than ten X or N bases were trimmed. Contiguous, high-quality reads ≥ 100 bp were submitted to the DDBJ/EMBL/GenBank databases under the accession numbers HX939204 to HX950377 (11,174 entries). The PHRAP program with default parameters was used to cluster and identify non-redundant azuki bean ESTs [19].

Simple sequence repeats (SSRs) ≥ 15 nucleotides in length, which contained all possible combinations of di-nucleotide (NN), tri-nucleotide (NNN), and tetra-nucleotide (NNNN) repeats, were identified from the non-redundant azuki bean ESTs using fuzznuc program in EMBOSS [21] for SSRs within two mismatches. Primer pairs for amplification of SSR-containing regions were designed based on the flanking sequences of each SSR with the aid of the Primer3 [22] so that the amplified fragment sizes were between 90 bp and 300 bp in length. The newly developed markers were designated as VES (Vigna EST-derived SSR) markers.

EST-SSR Marker Analysis

Total genomic DNA was extracted from young leaves of each *Vigna* accession using the method described by [23] with a slight modification. The DNA was quantified against a lambda DNA on 1.0% agarose gel stained with ethidium bromide and diluted to 5 ng/µl for PCR amplification.

Amplification, transferability and polymorphisms of all the EST-SSR markers were confirmed using eight accessions, two each from azuki bean complex, blackgram, rice bean and mungbean (Table 1). Five microliters of PCR reaction mixture including 0.5 ng of total genomic DNA, 0.02 U BIOTAQ DNA polymerase (BIOLINE, UK), 1x PCR buffer (BIOLINE, UK), 3 mM MgCl$_2$, 0.2 mM dNTPs and 2 pmol of the forward and reverse EST-SSR primers. The PCR thermal cycling was performed as follows: 94°C for 1 min; 3 cycles of 94°C for 30 s and 70°C for 30 s, followed by 3 rounds of the same program in which the annealing and extension temperatures were decreased by 2°C every 3 cycles; 3 cycles of 94°C for 30 s, 62°C for 30 s and 72°C for 30 s, followed by 2 rounds of the same program in which the annealing temperatures were decreased by 2°C every 2 cycles; 30 cycles of 94°C for 30 s, 55°C for 30 s and 72°C for 30 s, and a final cycle at 72°C for 10 min. The PCR products were separated by 10% polyacrylamide gel electrophoresis in tris-borate-ethylene diamine tetraacetic acid (TBE) buffer according to the standard protocol, and banding patterns in the gel were stored as pictures. The characteristics of PCR products such as intensity, banding pattern and size range were recorded for each accession and EST-SSR markers. Polymorphism between wild and cultivated accessions in each species was recorded.

Table 1. A summary of transferability and polymorphism of 1,429 azuki bean EST-SSR markers in the four Asian *Vigna* species.

Code	Species	Domestication status	Common name	Genebank acc. no	Origin	Linkage map	PCR products amplified	%	Simple banding pattern	%	Complex banding pattern	Exceeded of expected size	Number of markers compared between cultivated and wild	Polymorphic between cultivated and wild	%
VAC	V. angularis var. angularis	Cultivated	Azuki bean	JP81481	Japan	Han et al. (2005)	1327	92.9	1100	82.9	46	181	1307	236	18.1
VAW	V. nepalensis	Wild		JP107881	Nepal	Han et al. (2005)	1312	91.8	1090	83.1	45	177			
VMC	V. mungo var. mungo	Cultivated	Black gram	JP219132	Thailand	Chaitieng et al. (2006)	1216	85.1	1007	82.8	39	170	1196	164	13.7
VMW	V. mungo var. silvestris	Wild		JP107873	India	Chaitieng et al. (2006)	1194	83.6	998	83.6	36	160			
VUC	V. umbellata	Cultivated	Rice bean	JP217439	Myanmar	Isemura et al. (2010)	1324	92.7	1091	82.4	48	185	1296	187	14.4
VUW	V. umbellata	Wild		JP210639	Thailand	Isemura et al. (2010)	1304	91.3	1073	82.3	48	183			
VRC	V. radiata var. radiata	Cultivated	Mungbean	JP229096	Thailand	Isemura et al. (2012)	1197	83.8	987	82.5	44	166	1198	277	23.1
VRW	V. radiata var. sublobata	Wild		JP211874	Myanmar	Isemura et al. (2012)	1204	84.3	985	81.8	42	177			

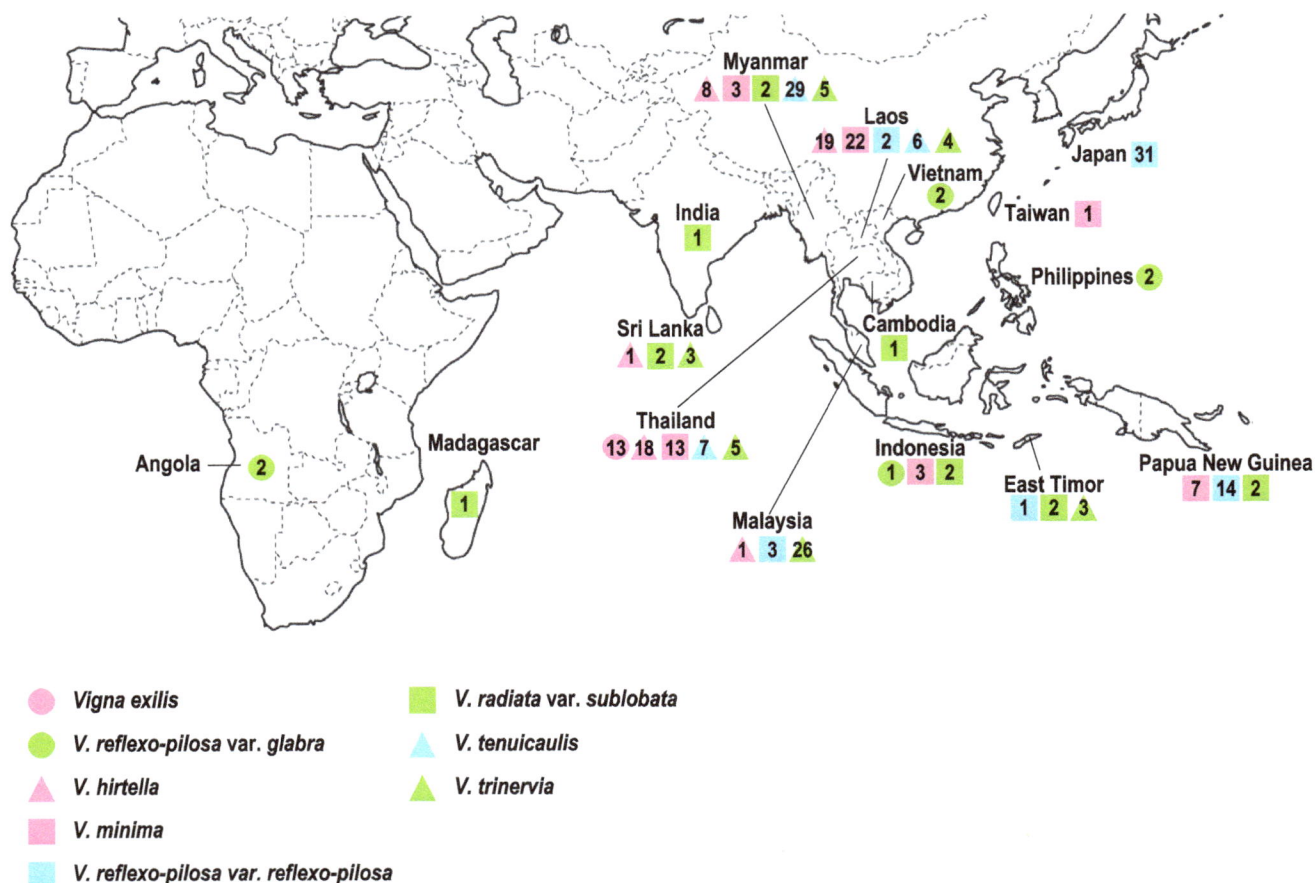

Figure 1. Distribution of 286 accessions of 8 taxa of the genus *Vigna* **subgenus** *Ceratotropis.* Numbers of accessions analyzed are shown within circles, triangles and squares.

Out of 1,429 azuki bean EST-SSR markers, 175 markers with good transferability in all the four major *Vigna* crop species were labeled with fluorescent dyes and further examined for polymorphism in a panel of eight taxa of Asian *Vigna* consisting of 16 accessions, two accessions for each species randomly chosen from the 268 accessions (Table S1). The 5′-end of the reverse primers were labeled with one of the following fluorescent dyes; 6-FAM (blue), HEX (green), and NED (yellow) (Applied Biosystems). Five microliters of PCR reaction mixture contained 5 ng of genomic DNA, 1× QIAGEN Multiplex PCR Master Mix, and 5 pmol of the forward and reverse primers. PCR reactions were performed in a GeneAmp PCR System 9700 (Applied Biosystems). The PCR thermal cycling was programmed as follows: 95°C for 15 min followed by 40 cycles of 94°C for 30 s, 55°C for 90 s, 72°C for 60 s, and a final cycle at 72°C for 10 min. One microliter of 10 times dilution PCR product was mixed with 8.5 µL of Hi-Di formamide and 0.125 µL of ROX size standard (Applied Biosystems). The mixture was denatured at 95°C for 5 min and then run on an ABI PRISM 3130xl DNA Analyzer (Applied Biosystems). Allele size for the highest stutter peak with the height ranging between 500 and 10,000 RFU were recorded. The genotyping was performed using GeneMapper 4.0 (Applied Biosystems).

Thirty-eight EST-SSR primer pairs amplified one product in diploid species and two discrete products in tetraploid species with height ranging RFU (Table S2) were selected for further analysis

in the 268 *Vigna* accessions. Three primers with different labels and size products were mixed into a single PCR reaction mixture and amplified as multiplex PCR. After genotyping, a single allele size was scored for each marker in each accession corresponding to the strongest peak.

Data Analysis

Since both diploid (1 genome set) and tetraploid (2 genome set) species were analyzed together in this study and each SSR primer pair detected 2 loci in the tetraploid species, we separated the genome of *V. reflexo-pilosa*, into 2 different diploid types, viz. A and B. Therefore 10 genomes were recognized from the 8 taxa of 7 *Vigna* species.

Genetic distance (D_A) [24] for all possible pairs of accessions was calculated using software POPULATIONS 1.2.28 (available at www.cnrs-gif.fr/pge/bioinfo/populations). D_A among the 10 genomes was also calculated using the same software. A phylogenic tree was constructed to reveal relationships among accessions based on D_A using neighbor-joining clustering method by software MEGA 5.05 [25] with bootstrap support (1,000 replicates) obtained by re-sampling the allelic frequency data. In order to confirm the results from neighbor-joining clustering, principal coordinate analysis (PCoA) was also performed using software PAST [26] to reveal the relationship among different accessions. In addition, based on sources of *V. hirtella*, Seehalak et al. (2006) divided *V. hirtella* in to 3 subgroups (a1, a2 and b), thus the relationship of individuals based on PCoA analysis enabled 12 sub-genome groups to be determined. Using average D_A genetic distance between 12 sub-genomes, an unrooted dendrogram showing relationships between these genomes was constructed by neighbor-joining method using software POPULATIONS 1.2.32. (available at www.cnrs-gif.fr/pge/bioinfo/populations) with bootstrap support (400 replicates) obtained by re-sampling the allelic frequency data. The trees were visualized with MEGA ver. 5.05 [25].

Results

Azuki Bean EST-SSR Markers

A total of 11,167 cDNA clones were sequenced consisting of 7,534 clones from a seedling library and 3,633 clones from a young pod library. After clustering, 4,896 potential non-redundant EST sequences, including 2,350 contigs and 2,546 singletons, were generated with a total of 4,284,693 qualified bases (Table S3). By using the fuzznuc program in EMBOSS, a total of 1,188 SSRs were identified in the 4,896 non-redundant EST sequences. Of the 1,188 SSRs, di-, tri-, and tetra-nucleotide SSRs accounted for 71.6%, 26.7%, and 1.7%, respectively (Table S4). Assuming that total length of the non-redundant azuki bean EST sequences is 4.3 Mbp, the frequency of occurrence of the SSRs in transcribed regions was estimated to be one in every 3.6 kb. Altogether105 primer pairs were initially designed on the flanking regions of the 132 perfect SSR motifs (Table S4). To increase the number of candidate EST-SSR markers, additional 1,324 primer pairs were designed on the flanking regions of 1,545 imperfect SSR motifs allowing one or two base mismatching. As a result, a total of 1,429 EST-SSR markers were designed (Table S4), of which 149, 28, 5 and 5 markers identified 2, 3, 4 and ≤5 SSRs in the regions between the primer pairs, respectively. Thus the total number of identified SSRs by the 1,429 EST-SSR markers was 1,677 that consisted of 137 (8.2%) di-nucleotide repeats, 1,400 (83.5%) tri-nucleotide repeats and 140 (8.3%) tetra-nucleotide repeats (Table S4). Among the di-nucleotide repeats, poly(AG)$_n$ (n = 88, 5.2% of total) were most frequently observed, followed by poly(AT)$_n$ (n = 35, 2.1%) and poly (AC)$_n$ (n = 14, 0.8%). Among the ten types of tri-nucleotide repeats observed, poly(AAG)$_n$ (n = 322, 19.2%) were the most abundant, followed by poly(GGA)$_n$ (n = 195, 11.6%) and poly(ATC)$_n$ (n = 180, 10.7%). Among the thirteen tetra-nucleotide repeats, poly(AAAG)$_n$ (n = 45, 2.7%) were the most

frequently observed, followed by poly(AAAT)$_n$ (n = 30, 1.8%), and poly(AAAC)$_n$ (n = 20, 1.2%). The details of the designed azuki bean EST-SSR primers, along with the corresponding SSR motif, product size, and primer sequence, are available at http://marker.kazusa.or.jp/Azuki and in Table S5.

Amplification and Transferability of Azuki Bean EST-SSR Markers

Transferability and polymorphism of the 1,429 azuki bean EST-SSR markers were initially examined by polyacrylamide gel electrophoresis using eight accessions consisting of four major *Vigna* crop species; azuki bean complex, black gram, rice bean and mungbean (Table 1, Table S1). These wild and cultivated parental accessions had been used for a linkage map construction in each species ([27] [28] [29] [30]). Amplification of azuki bean EST-SSR markers in the same species, *V. angularis*, and the closely related species, *V. umbellata,* were 91.3 and 92.9% which were higher than those in *V. radiata* and *V. mungo* (Table 1). However, the size and pattern of PCR products of EST-SSR markers did not always reveal simple banding pattern with expected size, even in the same species. Among the amplified markers, 81.8 to 83.6% of them revealed simple banding pattern suitable for further application. When polymorphism between cultivated and wild accessions was examined, *V. radiata* possessed the highest polymorphism in which 277 markers (23.1% of amplified markers) were polymorphic. Further, 175 markers with good transferability in the four *Vigna* species were tested for polymorphism in a panel of 16 accessions from 8 taxa of Asian *Vigna* (Table S1). The test revealed that 22 markers (12.6%) failed to amplify, 36 (20.6%) amplified some accessions, whereas 117 (66.8%) successfully amplified all the 16 accessions (Table 2). Among the 117 amplifiable markers, 2 (1.1%) gave product size larger than 500 bp, 4 (2.3%) amplified multiple products, 45 (25.7%) amplified one product in tetraploid species, 10 (5.7%) were monomorphic, and 56 (32%) were polymorphic, amplifying one product in diploid species and two discrete products in tetraploid species and were considered to be suitable for analyzing origin of genome donor of the tetraploid species.

EST-SSR Polymorphism and Genetic Diversity

Thirty-eight primer pairs of azuki bean EST-SSR were analyzed in 268 accessions of 10 genomes from eight taxa of Asian *Vigna*. Based on discrimination of the diploid species we

Table 2. Characteristics of 196 EST-SSR primers developed in this study.

Type	Description	Number of primers (%)
1	Not amplified in all accessions	**22 (12.6)**
	Amplified in some accessions	**36 (20.6)**
2–1)	1) Amplified in 1 to 8 accession(s)	5 (2.9)
2–2)	2) Amplified in 9 to 15 accessions	31 (17.7)
	Amplified in all 16 accessions	**117 (66.8)**
3–1)	1) PCR product of more than 500 bp	2 (1.1)
3–2)	2) Multiple PCR product	4 (2.3)
3–3)	3) Monomorphic PCR product	10 (5.7)
3–4)	4) Single PCR product in tetraploid	45 (25.7)
3–5)	5) Two PCR products in tetraploid	56 (32)
Total		**175 (100.0)**

distinguished two discrete products detected by each marker of *V. reflexo-pilosa* into two types, viz. A and B. The product size that is the same or very similar to *V. trinervia* was scored as type A, while the other product size was scored as type B. In total, 445 alleles were detected in 10 genomes by the 38 EST-SSR loci (Tables 3, 4). The number of alleles detected per locus ranged between 3 (VES0777 and VES1271) and 34 (VES1172) with a mean of 11.7 alleles per locus (Table 4). The *PIC* values ranged from 0.36 (VES0116) to 0.94 (VES1172) with a mean of 0.67. None of the EST-SSR markers had *PIC* values lower than 0.3, while 20 markers had *PIC* values higher than 0.7. The markers showed high *PIC* value in *V. hirtella* (0.44) and *V. minima* (0.42), but low value in *V. reflexo-pilosa* (0.04-0.09). In general, each marker showed a *PIC* value of 0 in both populations of *V. reflexo-pilosa*. The allelic richness was between 2.0 for marker VES0777 and 9.9 for marker VES1172 with a mean of 4.7. Fifteen primers had allelic richness higher than 5.0 (Table 4).

V. hirtella possessed the highest gene diversity (0.478), while *V. reflexo-pilosa* showed very low gene diversity (0.053 to 0.094) (Table 3). All genomes of *V. reflexo-pilosa* showed no observed heterozygosity, while the other genomes showed very low observed heterozygosity from 0.008 to 0.028 (Table 3).

Genetic Relationship among Genomes

Genetic distance (D_A) among the ten genomes is shown in Table 5a. In most cases D_A among genomes was high (>0.6). Nonetheless, D_A between *V. trinervia* and *V. reflexo-pilosa* var. *glabra* (A) or *V. reflexo-pilosa* var. *reflexo-pilosa* (A) was low being 0.333 and 0.313, respectively. While, both *V. reflexo-pilosa* var. *glabra* (B) and *V. reflexo-pilosa* var. *reflexo-pilosa* (B) showed lowest D_A with *V. hirtella* being 0.585 and 0.589 (Table 5a) or with *V. hirtella* (a1), being 0.429 and 0.422 (Table 6), respectively. Interestingly, *V. trinervia*, which belongs to section *Angulares* showed lower D_A with *V. radiata* var. *sublobata* (section *Ceratotropis*) than with those species in the section *Angulares*.

A neighbor-joining tree was constructed based on the genetic distances for all possible pairs of 268 accessions. The tree showed two major groups with 96.9% bootstrap value, namely azuki bean group and mungbean group (Figure 2). *V. exilis*, *V. minima*, *V. tenuicaulis*, *V. hirtella*, *V. reflexo-pilosa* var. *glabra* (B), and *V. reflexo-pilosa* var. *reflexo-pilosa* (B) were clustered in the azuki bean group. While *V. radiata* var. *sublobata*, *V. trinervia*, *V.*

reflexo-pilosa var. *glabra* (A) and *V. reflexo-pilosa* var. *reflexo-pilosa* (A) were clustered in the mungbean group. In the azuki bean group, 13 accessions of *V. exilis* and 49 accessions of *V. minima* were independent clusters with 95.3% and 83.5% bootstrap values, respectively. *V. hirtella* showed high divergence in which 9 accessions in the population of *V. hirtella* (b) clustered strongly with *V. tenuicaulis* with 91.5% bootstrap value, while the other 38 accessions in *V. hirtella* (a) clustered with *V. reflexo-pilosa* var. *glabra* (B) and *V. reflexo-pilosa* var. *reflexo-pilosa* (B) with 83.8% bootstrap value. Those 38 accessions of *V. hirtella* also showed two subclusters a1 and a2 (Figure 2 and 3). In the mungbean group, *V. trinervia* clustered with *V. reflexo-pilosa* var. *glabra* (A) and *V. reflexo-pilosa* var. *reflexo-pilosa* (A) with a 100% bootstrap support. Five accessions from Myanmar (40–44) formed a distinct branch with a bootstrap value of 87.8% and were separated from the other accessions (Figure 2). Three accessions from Laos (31–33) were differentiated from the other accessions and were clustered together with *V. reflexo-pilosa* (A) accessions.

The principle coordinate analysis (PCoA) was also conducted to confirm species relationship based on the neighbor-joining tree using the same genetic distance estimates. The first three PCs together accounted up to 56.6% of the total variation. The first, second and third PCs accounted for 38.1%, 10.6% and 7.9%, respectively (Figure 4). A three-dimensional PC plot (PC1, PC2 and PC3) of the twelve genome groups is shown in Figure 4, which clearly separated the genomes into 5 distinct groups. Group I comprised *V. reflexo-pilosa* var. *reflexo-pilosa* (A), *V. reflexo-pilosa* var. *glabra* (A), and *V. trinervia*. Group II comprised *V. minima* only. Group III comprised *V. reflexo-pilosa* var. *glabra* (B), *V. reflexo-pilosa* var. *reflexo-pilosa* (B), *V. exilis* and *V. hirtella* (a1, a2). Group IV comprised *V. hirtella* (b), and *V. tenuicaulis*. Group V comprised solely *V. radiata* var. *sublobata*. Distribution range of *V. reflexo-pilosa* var. *glabra* (A) and *V. reflexo-pilosa* var. *reflexo-pilosa* (A) in Group I was narrower than *V. reflexo-pilosa* var. *glabra* (B) and *V. reflexo-pilosa* var. *reflexo-pilosa* (B) in Group III. Of all species used in this study, only *V. hirtella* was unambiguously distinguished into two sub-groups, a and b. Subgroup a showed close relationship with populations of *V. reflexo-pilosa* var. *reflexo-pilosa* (B); *V. reflexo-pilosa* var. *glabra* (B) and *V. exilis*, while subgroup b showed close relationship with *V. tenuicaulis* (Figure 3).

Table 3. Genome number assigned in this study with number of alleles, gene diversity and observed heterozygosity analyzed by 38 EST-SSR markers.

Genome	No. of accessions	No. of loci typed	No. of alleles	Gene diversity	Observed heterozygosity
1 (*Vigna exilis*)	13	38	100	0.305	0.012
2 [*V. reflexo-pilosa* var. *glabra* (a)]	7	38	44	0.053	0.000
3 [*V. reflexo-pilosa* var. *glabra* (b)]	7	38	44	0.052	0.000
4 (*V. hirtella*)	47	38	171	0.478	0.019
5 (*V. minima*)	49	38	203	0.458	0.017
6 [*V. reflexo-pilosa* var. *reflexo-pilosa* (a)]	51	38	56	0.058	0.000
7 [*V. reflexo-pilosa* var. *reflexo-pilosa* (b)]	51	38	76	0.094	0.000
8 (*V. radiata* var. *sublobata*)	13	38	101	0.357	0.008
9 (*V. tenuicaulis*)	42	38	122	0.309	0.022
10 (*V. trinervia*)	46	38	98	0.306	0.028
Total	326	38	445	0.711	0.013

Table 4. EST-SSR primers used, number of alleles per locus, allele size range, polymorphic information content (PIC) and allelic richness for each genome.

| Primer | No. of alleles | Allele size range (bp)† | PIC Genome number | | | | | | | | | | | Allelic richness Genome number | | | | | | | | | | |
			1	2	3	4	5	6	7	8	9	10	Overall	1	2	3	4	5	6	7	8	9	10	Overall
VES0019	16	253–283 (30)	0.45	0.00	0.21	0.77	0.81	0.00	0.07	0.40	0.85	0.61	0.77	3.6	1.0	2.0	5.9	6.2	1.0	1.5	2.8	7.2	3.6	5.8
VES0021	18	217–289 (72)	0.46	0.00	0.00	0.51	0.59	0.00	0.00	0.13	0.37	0.04	0.75	3.0	1.0	1.0	3.6	4.1	1.0	1.0	1.8	3.0	1.3	5.3
VES0070	7	265–275 (10)	0.00	0.00	0.00	0.50	0.46	0.00	0.07	0.64	0.17	0.00	0.61	1.0	1.0	1.0	3.3	2.7	1.0	1.5	3.8	2.0	1.0	3.5
VES0093	11	183–199 (16)	0.00	0.00	0.00	0.51	0.15	0.00	0.00	0.66	0.05	0.40	0.75	1.0	1.0	1.0	3.3	1.9	1.0	1.0	3.9	1.3	2.3	5.0
VES0116	4	278–284 (6)	0.13	0.37	0.00	0.00	0.00	0.13	0.00	0.13	0.00	0.22	0.36	1.8	2.0	1.0	1.0	1.0	1.7	1.0	1.8	1.0	1.9	2.2
VES0120	16	245–293 (48)	0.13	0.00	0.00	0.16	0.52	0.00	0.10	0.73	0.00	0.00	0.51	1.8	1.0	1.0	2.0	4.2	1.0	1.6	4.8	1.0	1.0	3.5
VES0202	15	218–254 (36)	0.55	0.00	0.00	0.58	0.24	0.00	0.00	0.43	0.46	0.00	0.71	4.4	1.0	1.0	3.4	2.1	1.0	1.0	2.9	2.9	1.0	4.6
VES0204	20	310–366 (56)	0.61	0.00	0.00	0.69	0.70	0.10	0.24	0.44	0.21	0.31	0.71	4.5	1.0	1.0	4.7	5.4	1.6	2.6	2.8	2.4	2.5	5.0
VES0335	15	265–284 (19)	0.44	0.00	0.21	0.78	0.29	0.04	0.40	0.00	0.60	0.55	0.81	2.8	1.0	2.0	6.4	2.3	1.3	2.7	1.0	3.7	3.6	6.5
VES0427	9	316–329 (13)	0.50	0.32	0.00	0.26	0.47	0.00	0.00	0.34	0.16	0.31	0.71	3.0	2.0	1.0	2.1	3.3	1.0	1.0	2.8	1.8	2.4	4.5
VES0478	11	298–309 (11)	0.72	0.00	0.00	0.22	0.21	0.13	0.13	0.50	0.09	0.08	0.66	5.4	1.0	1.0	1.9	1.9	1.7	1.7	3.3	1.6	1.5	4.3
VES0546	26	431–476 (45)	0.63	0.00	0.00	0.32	0.84	0.00	0.37	0.48	0.58	0.48	0.85	4.4	1.0	1.0	2.9	7.2	1.0	3.2	3.0	3.0	2.9	7.5
VES0624	4	265–271 (6)	0.00	0.00	0.00	0.00	0.15	0.04	0.00	0.00	0.00	0.00	0.42	1.0	1.0	1.0	1.0	1.9	1.3	1.0	1.0	1.0	1.0	2.6
VES0665	9	195–213 (18)	0.26	0.00	0.00	0.31	0.35	0.00	0.00	0.67	0.13	0.24	0.68	2.6	1.0	1.0	2.7	2.8	1.0	1.0	5.0	1.9	2.3	4.2
VES0670	6	99–114 (15)	0.44	0.00	0.00	0.56	0.20	0.00	0.00	0.00	0.38	0.08	0.71	2.8	1.0	1.0	3.8	2.1	1.0	1.0	1.0	3.1	1.5	4.4
VES0678	7	293–318 (25)	0.45	0.00	0.00	0.40	0.54	0.00	0.00	0.36	0.28	0.41	0.70	2.9	1.0	1.0	2.3	3.3	1.0	1.0	2.0	2.0	2.5	4.5
VES0679	18	310–365 (55)	0.44	0.00	0.00	0.40	0.78	0.31	0.14	0.63	0.61	0.74	0.81	2.8	1.0	1.0	3.2	5.6	2.0	1.9	4.4	3.9	5.2	7.8
VES0749	15	208–236 (28)	0.29	0.00	0.00	0.60	0.27	0.00	0.00	0.62	0.09	0.43	0.79	2.0	1.0	1.0	3.8	2.3	1.0	1.0	4.4	1.6	2.5	5.8
VES0762	7	242–266 (24)	0.13	0.00	0.21	0.62	0.08	0.00	0.04	0.37	0.00	0.00	0.43	1.8	1.0	2.0	3.6	1.5	1.0	1.3	2.0	1.0	1.0	3.0
VES0777	3	166–175 (9)	0.00	0.00	0.00	0.36	0.40	0.07	0.10	0.18	0.14	0.00	0.37	1.0	1.0	1.0	2.0	2.3	1.5	1.6	1.9	1.7	1.0	2.0
VES0803	12	292–308 (16)	0.00	0.00	0.00	0.64	0.71	0.00	0.10	0.29	0.41	0.00	0.80	1.0	1.0	1.0	4.8	4.6	1.0	1.6	2.0	2.6	1.0	6.3
VES0868	12	213–288 (75)	0.13	0.00	0.00	0.53	0.27	0.00	0.00	0.26	0.35	0.00	0.62	1.8	1.0	1.0	4.1	2.2	1.0	1.0	2.6	2.0	1.0	4.2
VES0987	9	298–310 (12)	0.26	0.21	0.00	0.55	0.36	0.00	0.00	0.13	0.31	0.59	0.73	2.6	2.0	1.0	3.4	3.0	1.0	1.0	1.8	2.0	3.0	5.0
VES1001	10	227–250 (23)	0.29	0.00	0.00	0.39	0.55	0.00	0.00	0.64	0.00	0.00	0.65	2.5	1.0	1.0	2.7	3.6	1.0	1.0	3.9	1.0	1.0	3.9
VES1020	6	170–183 (13)	0.26	0.00	0.00	0.48	0.51	0.07	0.07	0.00	0.00	0.37	0.73	2.6	1.0	1.0	2.9	3.3	1.5	1.5	1.0	1.0	2.0	4.6
VES1023	7	140–157 (17)	0.00	0.00	0.00	0.28	0.56	0.00	0.10	0.00	0.37	0.00	0.59	1.0	1.0	1.0	2.2	3.7	1.0	1.6	1.0	2.2	1.0	3.4
VES1029	9	111–128 (17)	0.58	0.00	0.00	0.62	0.04	0.00	0.00	0.00	0.52	0.00	0.61	3.8	1.0	1.0	3.8	1.3	1.0	1.0	1.0	2.9	1.0	3.9
VES1067	14	447–477 (30)	0.13	0.00	0.00	0.68	0.60	0.00	0.32	0.36	0.17	0.37	0.83	1.8	1.0	2.0	4.5	3.9	1.0	2.8	2.0	2.1	2.0	6.7
VES1082	14	281–317 (36)	0.40	0.00	0.00	0.51	0.70	0.00	0.00	0.34	0.13	0.46	0.70	2.8	1.0	1.0	3.1	4.8	1.0	1.0	2.8	1.9	2.8	5.0
VES1085	13	390–403 (13)	0.00	0.00	0.00	0.33	0.60	0.31	0.13	0.00	0.50	0.74	0.82	1.0	1.0	1.0	2.6	4.1	2.5	1.7	1.0	2.8	4.7	6.6
VES1172	34	217–268 (51)	0.71	0.53	0.37	0.85	0.89	0.58	0.82	0.60	0.41	0.59	0.94	6.0	3.0	2.0	7.4	9.0	3.8	6.5	3.8	2.9	4.3	9.9
VES1196	5	203–215 (12)	0.00	0.00	0.00	0.30	0.15	0.00	0.00	0.09	0.09	0.00	0.38	1.0	1.0	1.0	2.5	1.9	1.0	1.0	1.0	1.5	1.0	2.1

Table 4. Cont.

Primer	No. of alleles	Allele size range (bp)[†]	PIC											Allelic richness										
			Genome number											Genome number										
			1	2	3	4	5	6	7	8	9	10	Overall	1	2	3	4	5	6	7	8	9	10	Overall
VES1231	6	120–139 (19)	0.23	0.00	0.00	0.53	0.08	0.00	0.00	0.00	0.34	0.00	0.52	2.0	1.0	1.0	2.9	1.5	1.0	1.0	1.0	2.6	1.0	3.5
VES1258	10	369–398 (29)	0.54	0.00	0.00	0.22	0.59	0.04	0.00	0.00	0.50	0.48	0.76	3.0	1.0	1.0	1.9	3.6	1.3	1.0	1.0	2.9	3.0	5.5
VES1263	8	305–320 (15)	0.13	0.00	0.00	0.00	0.43	0.07	0.00	0.56	0.00	0.30	0.68	1.8	1.0	1.0	1.0	2.7	1.5	1.0	3.6	1.0	2.0	4.4
VES1271	3	307–310 (3)	0.00	0.00	0.00	0.00	0.00	0.00	0.00	0.00	0.05	0.37	0.42	1.0	1.0	1.0	1.0	1.0	1.0	1.0	1.0	1.3	2.0	2.6
VES1310	7	288–309 (21)	0.23	0.00	0.00	0.37	0.25	0.00	0.10	0.60	0.60	0.30	0.55	2.0	1.0	1.0	3.0	2.5	1.0	1.6	3.8	3.6	2.0	3.3
VES1469	29	128–178 (50)	0.00	0.21	0.21	0.82	0.79	0.17	0.04	0.61	0.64	0.66	0.86	1.0	2.0	2.0	6.5	6.1	2.0	1.3	3.8	5.3	5.0	7.7
Total	445																							
Average	11.7		0.28	0.04	0.04	0.44	0.42	0.05	0.09	0.32	0.28	0.27	0.67	2.43	1.16	1.16	3.24	3.34	1.27	1.53	2.49	2.33	2.15	4.75

[†]Difference between the largest and smallest fragments amplified by each primer is shown in parentheses. See Table 3 for the abbreviations of genome number.

Table 5. Genetic distance(D_A) within and among 10 genomes with (*Vigna hirtella* is not divided).

Genome	1	2	3	4	5	6	7	8	9	10
1	**0.326**									
2	0.900	**0.061**								
3	0.695	0.985	**0.060**							
4	0.649	0.927	0.585	**0.482**						
5	0.740	0.881	0.740	0.693	**0.464**					
6	0.901	0.079	0.987	0.922	0.870	**0.059**				
7	0.690	0.989	0.098	0.589	0.740	0.990	**0.097**			
8	0.848	0.817	0.875	0.838	0.884	0.817	0.878	**0.385**		
9	0.712	0.895	0.716	0.652	0.716	0.894	0.720	0.829	**0.310**	
10	0.891	0.333	0.963	0.890	0.886	0.313	0.965	0.799	0.849	**0.304**

Table 6. Genetic distance(D_A) within and among 10 genomes (*Vigna hirtella* is divided into three genomes).

Genome	1	2	3	4 (a1)	4 (a2)	4 (b)	5	6	7	8	9	10
1	**0.326**											
2	0.900	**0.061**										
3	0.695	0.985	**0.060**									
4 (a1)	0.625	0.911	0.429	**0.377**								
4 (a2)	0.626	0.931	0.600	0.609	**0.298**							
4 (b)	0.749	0.926	0.672	0.672	0.646	**0.321**						
5	0.740	0.881	0.740	0.704	0.689	0.700	**0.464**					
6	0.901	0.079	0.987	0.908	0.925	0.922	0.870	**0.059**				
7	0.690	0.989	0.098	0.422	0.606	0.680	0.740	0.990	**0.097**			
8	0.848	0.817	0.875	0.899	0.824	0.831	0.884	0.817	0.878	**0.385**		
9	0.712	0.895	0.716	0.713	0.697	0.446	0.716	0.894	0.720	0.829	**0.310**	
10	0.891	0.333	0.963	0.892	0.895	0.874	0.886	0.313	0.965	0.799	0.849	**0.304**

See Table 3 for the abbreviations of genome number.
Genetic distances within and among genomes are shown on the diagonal and below the diagonal, respectively.
The accessions that belong to genome a1, a2 and b are shown in Figure 3.

Discussion

Transferability of Azuki Bean EST-SSRs

One of the advantages of SSRs derived from EST (EST-SSR markers) to other DNA marker systems is the high transferability across species and genera because of primer pairs are designed from conserved transcribed regions, although they are generally less polymorphic than genomic SSR markers (gSSR). In this study, more than 83% of the azuki bean EST-SSR markers were able to amplify DNA from 8 Asian *Vigna* species including four major crop species, viz. *V. angularis*, *V. umbellata*, *V. radiata* and *V. mungo* of the subgenus *Ceratotropis*. This result is in agreement with that of [31] who reported that more than 80% of mungbean EST-SSR markers were transferable to other species in the subgenus *Ceratotropis*. Transferability of the EST-SSR markers from azuki bean (in this study) and mungbean [31] is greater than that of gSSR markers from azuki bean (>67% transferable; [28]) and mungbean (>59% transferable; [32]). The high transferability of azuki bean EST-SSRs and of mungbean EST-SSRs to the related *Vigna* species indicates high genome conservation among the species in the subgenus *Ceratotropis*. Therefore, the azuki bean EST-SSR markers developed in this study will be useful for comparative genomic and genetic diversity studies in *Vigna* species. Nonetheless, as expected, polymorphism rate of the azuki bean EST-SSR markers detected between cultivated and wild accessions of azuki bean, rice bean, mungbean and black gram is low (14% in black gram and 23% in mungbean). In the same plant materials, Chaitieng et al. [28] reported that polymorphism rate of the genomic azuki bean SSRs was as high as 50% in black gram to 63% in mungbean. Thus low polymorphism rate of the azuki bean EST-SSRs can be an undesirable character for the use of these markers in comparative genome mapping in the genus *Vigna*. However, EST-SSRs represent true genetic diversity which may be more directly associated with traits of interest in breeding as compared to gSSRs [33], and they may reflect better relationships among the related species in genetic diversity study.

Evolution and Domestication of Tetraploid *V. reflexo-pilosa*

V. reflexo-pilosa is the only tetraploid species of the genus *Vigna*. Although previous studies on *Vigna* species using DNA markers and plastid sequences clearly demonstrated that *V. trinervia* is a genome donor of *V. reflexo-pilosa*, those studies were unable to identify the other genome donor. In the present study, five putative genome donor species of *V. reflexo-pilosa* belong to the section *Angulares* and one species representing the section *Ceratotropis* were analyzed with azuki bean EST-SSRs. By assigning the discrete PCR products of *V. reflexo-pilosa* into two distinguished genomes (designated as genome A and B), we were able to identify the two genome donor parents of créole bean. Phylogenetic tree (Figure 2) and PCoA plot (Figure 4) clearly demonstrated that genome A of *V. reflexo-pilosa* is received from *V. trinervia*. This result confirms the previous results reported by [7] [10] [13] [14] [15] that *V. trinervia* is a diploid genome donor. *V. trinervia* and *V. reflexo-pilosa* shares several similar morphological characters such as seed shape (rectangular), non-protruded hilum and large golden flower. The phylogenetic tree and the PCoA plot also unambiguously demonstrated that the other genome (genome B) of *V. reflexo-pilosa* is originated from *V. hirtella*. This result is in agreement with the result reported by Tateishi [7] [10] [13] [16] that *V. hirtella* is one of candidate genome donor of créole bean. Results from phylogenetic analysis of plastid DNA sequences of [14] [15] suggested that *V. trinervia* is the maternal genome donor of *V. reflexo-pilosa*. Since the

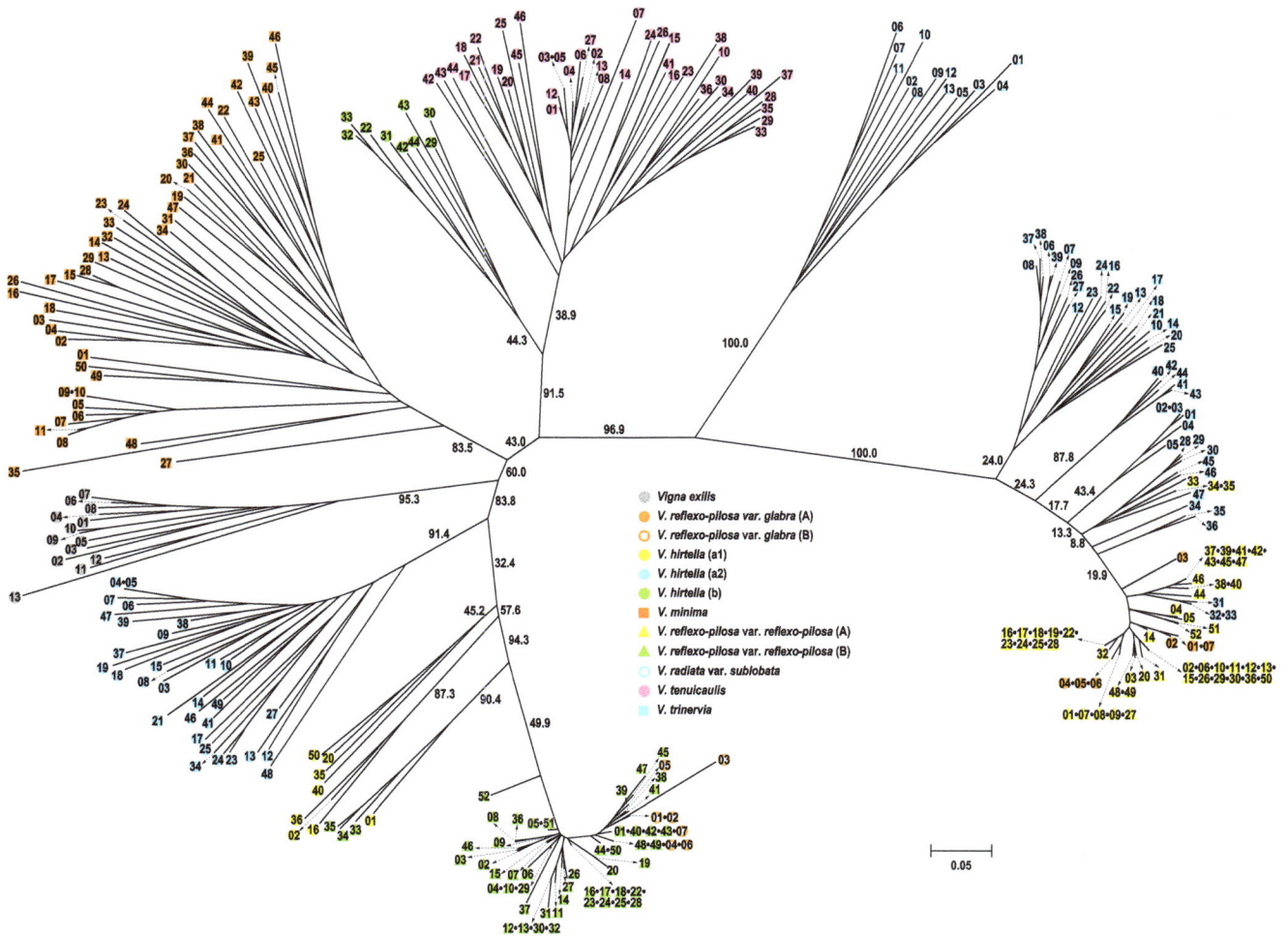

Figure 2. A phylogenetic tree showing relationship among 286 accessions in 12 sub-genome groups of the genus *Vigna* subgenus *Ceratotropis* based on variation at 38 EST-SSR loci.

previous and our results demonstrated close genetic relationship between tetraploid *V. reflexo-pilosa* and diploid *V. trinervia/V. hirtella*, we propose that créole bean evolved from interspecific hybridization between *V. trinervia* as female parent and *V. hirtella* as male parent, followed by genome duplication.

Phylogenetic tree (Figure 2) and PCoA plot (Figure 4) showed lower divergence of genome A of *V. reflexo-pilosa* compared to genome B. Very low gene diversity within both genomes A and B suggested that this tetraploid species has a monophyletic (single) origin and only evolved recently. Gene diversity within the genome A in both *V. reflexo-pilosa* var. *reflexo-pilosa* (wild form) and *V. reflexo-pilosa* var. *glabra* (cultivated form) is very similar (0.058 vs. 0.053), while gene diversity of the genome B in wild créole bean is about twice of that in the cultivated one (0.94 vs. 0.52) (Table 3). This indicates that the evolution rate of EST-SSRs on genome B of créole bean might be faster than that on genome A. The marked morphological differences between the cultivated and wild créole beans are thicker and erect glabrous stem in the former. Moreover, cultivated *V. reflexo-pilosa* still possesses a relatively high degree of pod shattering (Somta and Chankaew, personal observation), an important domestication trait for legume crops [34]. Therefore, the cultivated *V. reflexo-pilosa* can be treated as a semi-domesticated form. Based on cultivation,

utilization and distribution of diploid genome donor species, créole bean appears to have been domesticated in Southeast Asia. The crop is now very rarely cultivated, although it is produced in northern mountainous villages in Vietnam under the same name and consumed in the same way as mungbean [2].

Genetic Structure of Diploid Asian *Vigna* Species Detected by EST-SSR

We found that *V. hirtella* possessed greater genetic variation than *V. exilis*, *V. minima*, and *V. tenuicaulis*. A previous study based on AFLP suggested that *V. hirtella* germplasm collected from Thailand, Malaysia and Myanmar is a species complex consisting of two taxa (types), called *V. hirtella* (a) and *V. hirtella* (b) [35]. Although the study used only 3 accessions for *V. hirtella* (a) and 5 accessions for *V. hirtella* (b), all the 3 *V. hirtella* (a) accessions were included in *V. hirtella* (a1) group, and 5 *V. hirtella* (b) accessions were included in *V. hirtella* (a2) group in the present study. In addition, our EST-SSR results detected the existence of another genetically distinct type of materials designated as *V. hirtella* (b) which is closely related to *V. tenuicaulis* (Figure 2, 3, 4). These accessions should be re-examined morphologically whether they can be included within a morphological variation of *V. tenuicaulis*.

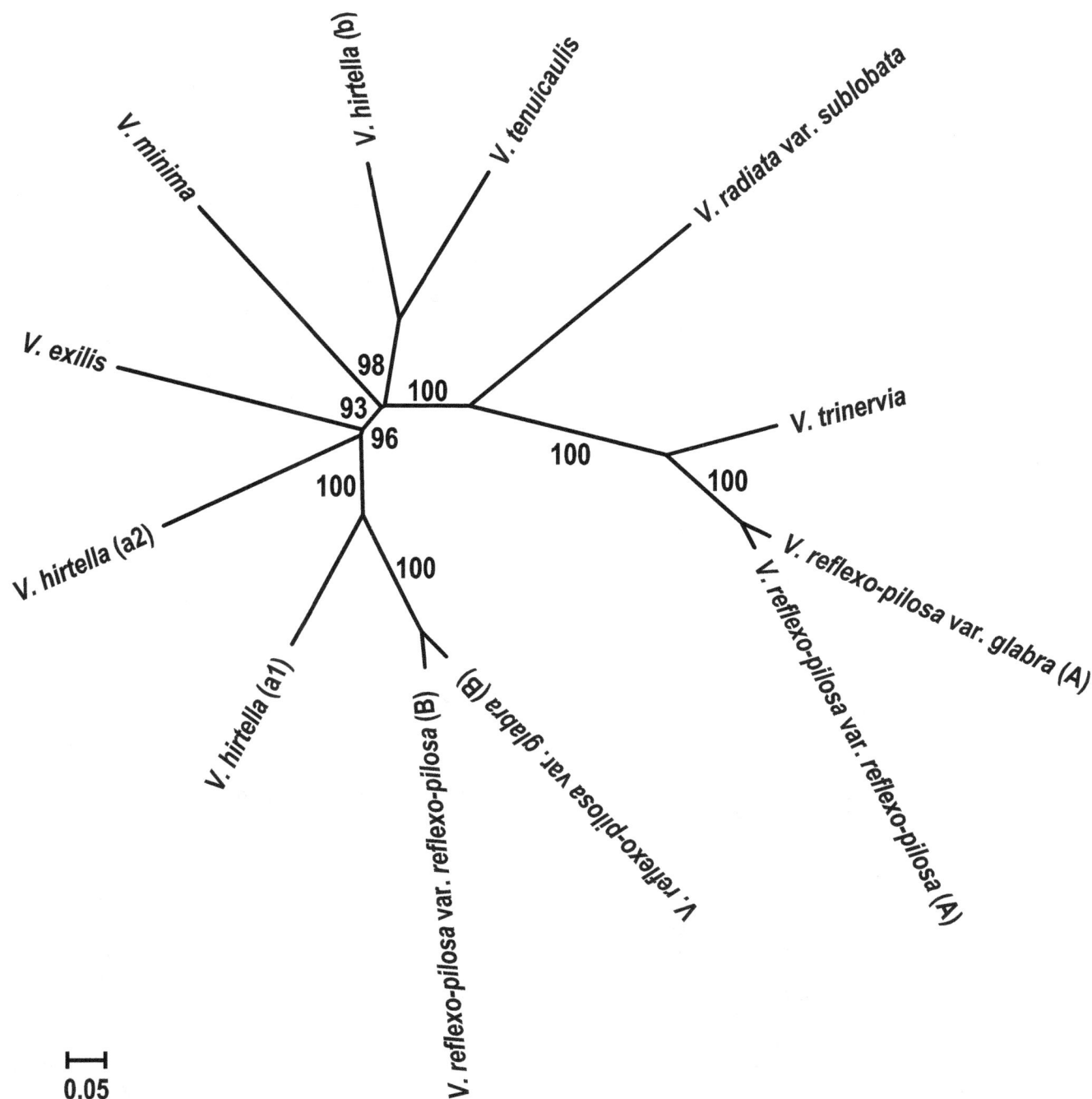

Figure 3. A phylogenetic tree showing relationship among 12 sub-genome groups of the genus *Vigna* subgenus *Ceratotropis* constructed based on the genetic distance shown in Table 6.

V. trinervia is widely distributed across Asia and also found in Papua New Guinea, Madagascar and East Africa. It is the second most widely distributed species in the subgenus *Ceratotropis* after *V. radiata* var. *sublobata* [2]. Nevertheless, gene diversity of *V. trinervia* was not high as compared to the other related species (Table 3). In Southeast Asia, especially in Thailand and the peninsular Malaysia, *V. trinervia* is often found on rural roadside habitats and in or near to rubber or oil palm plantations. It seems that some of the populations in those areas were introduced

recently as a cover plant in the plantations [2]. This may account for the low diversity of *V. trinervia*.

V. tenuicaulis distributes in open wet habitats in northern Southeast Asia, and appeared to be closely related to *V. hirtella* (b) and *V. angularis* [13] [35] [6] [36]. Tomooka et al. suggested that there are no major barriers to hybridization between *V. tenuicaulis* and *V. hirtella* (b) [2]. Our results confirmed the close relationship between these two species (Figure 2, 4). Previous studies based on molecular, biochemical and morphological variations all showed high level of distinctness and variation

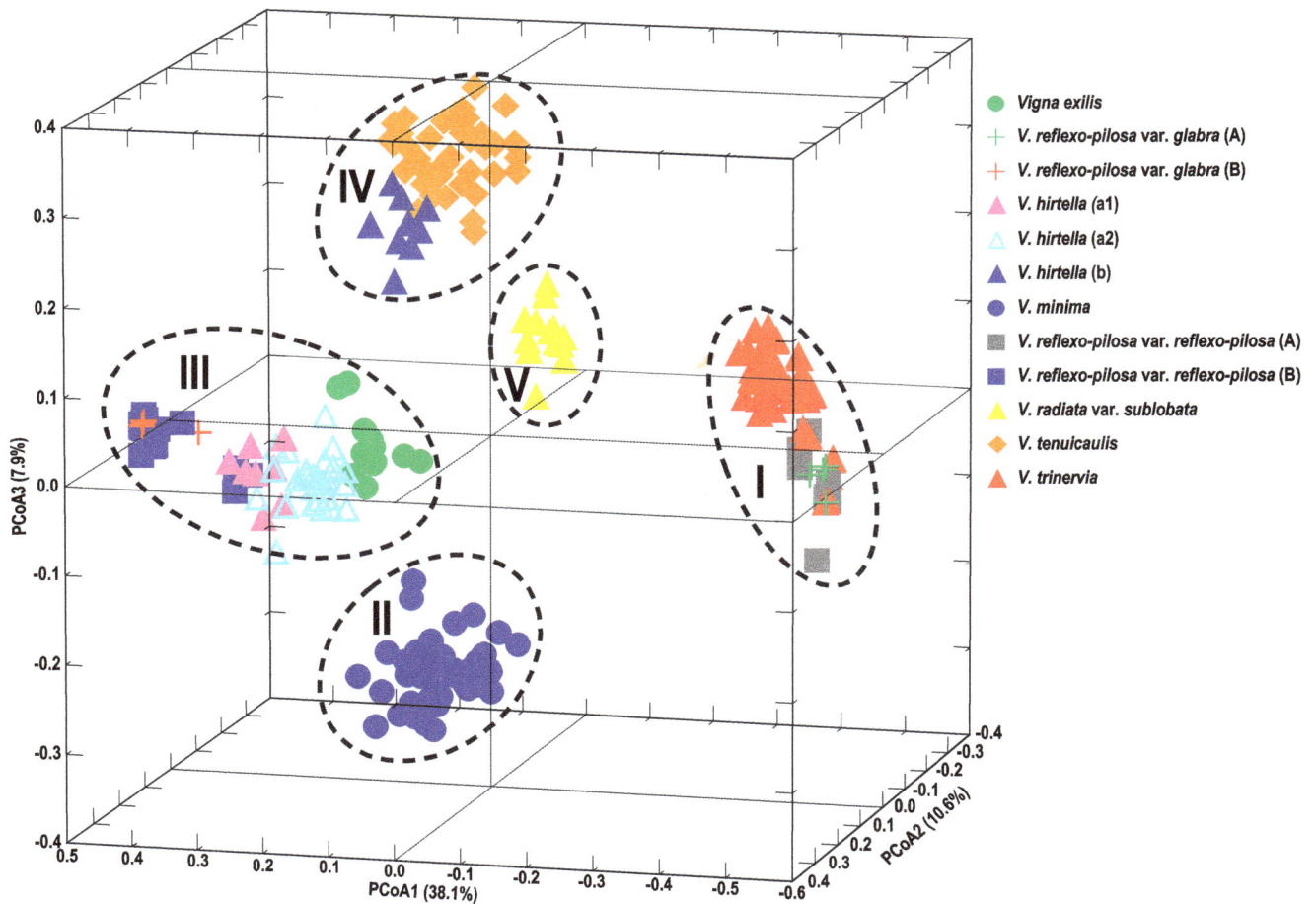

Figure 4. PCA scattered plot depicting relationship among 286 accessions in 12 sub-genome groups of the genus *Vigna* **subgenus** *Ceratotropis* **based on variation at 38 EST-SSR loci.**

within *V. tenuicaulis*. EST-SSR marker variation in our study revealed contrasting results. Although we used as many as 30 accessions of *V. tenuicaulis* from three countries, viz. Laos, Myanmar and Thailand. (Figure 1, Table S1), the species showed relatively low gene diversity (0.309).

V. exilis has been reported only in Thailand and Myanmar [2] [14]. In Thailand, this species is restricted to rocky limestone mountains. Its habitats suggest that it may be useful as gene sources for resistant to alkaline soil and drought conditions. SSR analysis revealed high level of intra-specific diversity in *V. exilis* [37]. Our results also supported this despite only 13 accessions from narrow geographical origin (west of Thailand) of this species were used (Figure 1, Table S1), the species showed similar level of gene diversity to *V. tenuicaulis* and *V. trinervia* (Table 3).

V. minima has broad environments adaptation and grows well in shaded deciduous forest floors and open-wet habitats in East and Southeast Asia [2]. It is the only species in section *Angulares* that is found on the forest floor. Among the Asian *Vigna* species analyzed in this study, *V. minima* is the second most diverse species after *V. hirtella* (Table 3). The high differentiation of *V. minima* is due to wide geographical distribution of the analyzed accessions. In Southeast Asia, isolation of *V. minima* in forests in different mountainous regions across Thailand and Myanmar and its sporadic occurrence in patches of forests in those regions may account for high level of population divergence [35]. Accessions of

V. minima in the present study were most closely related with *V. hirtella*, especially *V. hirtella* (b), followed by *V. exilis* (Figure 2, 4). Similar finding was reported by [35]. This agrees with their morphological appearance [2] that *V. minima* is sometimes confused with *V. hirtella*. *V. minima* can be distinguished from *V. hirtella* by smaller bracteole and more protruding hilum with well-developed rim-aril [2].

In summary, one of the advantages of EST-SSR markers to other DNA marker systems is their co-dominant nature and high transferability to closely related species. In this study, we used EST-SSR primer pairs, which amplified one PCR product in diploid species and two discrete PCR products in tetraploid species to separate genomes of *V. reflexo-pilosa* into 2 different diploid types, viz. A and B, that could successfully determine different sets of the donor genomes. Phylogenetic analyses revealed that *V. trinervia* and *V. hirtella* are the genome donors of *V. reflexo-pilosa*. Both genomes of cultivated and wild créole bean accessions showed low divergence suggesting that domestication of *V. reflexo-pilosa* is a relatively recent event.

Author Contributions

Conceived and designed the experiments: SC TI AK DAV NT P. Somta P. Srinives. Performed the experiments: SC TI AK DAV NT P. Somta P. Srinives. Analyzed the data: SC TI NT AK. Contributed reagents/materials/analysis tools: SI HH KS NT TI. Contributed to the writing of the manuscript: SC TI AK NT P. Somta P. Srinives.

References

1. Verdcourt B (1970) Studies in the Leguminosae–Papilionoideae for the flora of tropical East Africa, IV. Kew Bull 24: 507–569.
2. Tomooka N, Vaughan D A, Moss H, Maxted N (2002) The Asian *Vigna*: genus *Vigna* subgenus *Ceratotropis* genetic resources. Kluwer Academic Press. 270 pp.
3. Maréchal R, Mascherpa J M, Stainer F (1978) Etude taxonomique d'un groupe complexe d'espe'ces des genres *Phaseolus* et *Vigna* (Papilionaceae) sur la base de donne'es morphologiques et polliniques, traite'es par l'analyse informatique. Boissiera 28: 1–273.
4. Swindell RE, Watt EE, Evans GM (1973) A natural tetraploid mungbean of suspected amphidiploid origin. J Hered 64: 107.
5. Egawa Y, Siriwardhane D, Yagasaki K, Hayashi H, Takamatsu M, et al. (1990) Collection of millets and grain legume in Shimonai district of Nagano Prefecture, 1989. Annual report on exploration and introduction of plant genetic resources (NIAS, Tsukuba, Japan). 6: 1–22.
6. Tomooka N, Maxted N, Thavarasook C, Jayasuriya AHM (2002) Two new species, new species combinations and sectional designations in *Vigna* subgenus *Ceratotropis* (Piper) Verdcourt (*Leguminosae, Phaseoleae*). Kew Bull 57: 613–624.
7. Tateishi Y (1985) A revision of the Azuki bean group, the subgenus *Ceratotropis* of the genus *Vigna* (Leguminosae). Ph. D. Thesis, Tohoku University, Japan.
8. Tateishi Y, Ohashi H (1990) Systematics of the azuki bean group in the genus *Vigna*. In Bruchid and Legumes: Ecology and Coevolution. eds Fujii, K., *et al*. Kluwer Akademic Publishers, Netherlands, pp189–199.
9. Tomooka N, Kobayashi N, Kamuou RN, Risimeri J, Poafa J, et al. (2005) Ecological survey and conservation of legume symbiotic rhizobia genetic diversity in Papua New Guinea, 2004. Annual Report on Exploration and Introduction of Plant Genetic Resources. NIAS. Vol. 21: 135–143.
10. Egawa Y, Bujang IB, Chotechuen S, Tomooka N, TateishiY (1996) Phylogenetic differentiation of tetraploid *Vigna* species, *V. glabra* and *V. reflexo-pilosa*. JIRCAS J 3: 49–58.
11. Chen HK, Mok MC, Shanmugasundaram S, Mok DWS (1989) Interspecific hybridization between *Vigna radiata* L. Wilczek and *V. glabra*. Theor Appl Genet 78: 641–647.
12. Egawa Y, Chotechuen S, Tomooka N, Thavarasook C, Kitbamroong C (1996) Cross-compatability among the subgenus Ceratotropis of the genus *Vigna*. In Egawa Y, Chotechuen S (eds.) Phylogenetic differentiation of mungbean germplasm (subgenus *Ceratotropis* of the genus *Vigna*) and evaluation for breeding program. Japan International Research Center for Agricultural Science, Japan, pp. 19–30.
13. Konarev AV, Tomooka N, Vaughan DA (2002) Proteinase inhibitor polymorphism in the genus *Vigna* subgenus *Ceratotropis* and its biosystematic implications. Euphytica 123: 165–177.
14. Ye Tun Tun, Yamaguchi H (2007) Phylogenetic relationship of wild and cultivated *Vigna* (Subgenus *Ceratotropis*, Fabaceae) from Myanmar based on sequence variations in non-coding regions of *trnT-F*. Breed Sci 57: 271–280.
15. Javadi F, Ye Tun Tun, Kawase M, Guan K, Yamaguchi H (2011) Molecular phylogeny of the subgenus *Ceratotropis* (genus *Vigna*, Leguminosae) reveals three eco-geographical groups and Late Pliocene–Pleistocene diversification: evidence from four plastid DNA region sequences. Ann Bot 108: 367–380.
16. Doi K, Kaga A, Tomooka N, Vaughan DA (2002) Molecular phylogeny of genus *Vigna* subgenus Ceratotropis based on rDNA-ITS and *atpB-rbcL* intergenic spacer region of cpDNA sequences. Genetica 114: 129–145.
17. Asamizu E, Nakamura Y, Sato S, Fukuzawa H, Tabata S (1999) A large scale structural analysis of cDNAs in a unicellular green alga, *Chlamydomonas reinhardtii*. I. Generation of 3433 non-redundant expressed sequence tags. DNA Res 6: 369–373.
18. Ewing B, Hillier L, Wendl MC, Green P (1998) Base-calling of automated sequencer traces using phred. I. Accuracy assessment. Genome Res 8: 175–185.
19. Ewing B, Green P (1998) Base-calling of automated sequencer traces using phred. II. Error probabilities. Genome Res 8: 186–194.
20. Huang X, Wang J, Aluru S, Yang SP, Hillier L (2003) PCAP: a whole-genome assembly program. Genome Res13: 2164–2170.
21. Rice P, Longden I, Bleasby A (2000) EMBOSS: the European Molecular Biology Open Software Suite. Trends Genet 16: 276–277.
22. Rozen S, Skaletsky HJ (2000) Primer3 on the WWW for general users and for biologist programmers. In: Krawetz S, Misener S (eds) Bioinformatics Methods and Protocols: Methods in Molecular Biology. Humana Press, Totowa, NJ, pp 365–386.
23. Lodhi MA, Ye GN, Weeden NF, Reisch BI (1994) A simple and efficient method for DNA extraction from grapevine cultivars and *Vitis* species. Plant Mol Biol Rep 12: 6–13.
24. Nei M, Tajima F, TatenoY (1983) Accuracy of estimated phylogenetic trees from molecular data. J Mol Evol 19: 153–170.
25. Tamura K, Peterson D, Peterson N, Stecher G, Nei M, et al. (2011) MEGA5: Molecular evolutionary genetics analysis using maximum likelihood, evolutionary distance, and maximum parsimony methods. Mol Biol Evol 28: 2731–2739.
26. Hammer Ø, Harper DAT, Ryan PD (2001) PAST: Paleontological Statistics Software Package for Education and Data Analysis. Palaeontologia Electronica 4(1): 9 pp.
27. Han OK, Kaga A, Isemura T, Wang XW, Tomooka N, et al. (2005) A genetic linkage map for azuki bean [Vigna angularis (Willd.) Ohwi & Ohashi] Theor Appl Genet111: 1278–1287.
28. Chaitieng B, Kaga A, Tomooka N, Isemura T, Vaughan DA (2006) Development of a black gram [*Vigna mungo* (L.) Hepper] linkage map and its comparison with an azuki bean [*Vigna angularis* (Willd.) Ohwi and Ohashi] linkage map. Theor Appl Genet 113: 1261–1269.
29. Isemura T, Kaga A, Tomooka N, Shimizu T, Vaughan DA (2010) The genetics of domestication of rice bean, *Vigna umbellate*. Ann Bot 106: 927–944.
30. Isemura T, Kaga A, Tabata S, Somta P, Srinives P, et al. (2012) Construction of a genetic linkage map and genetic analysis of domestication related traits in mungbean (*Vigna radiata*) PLoS ONE 7(8): e41304.
31. Somta P, Seehalak W, Srinives P (2009) Development, characterization and cross-species amplification of mungbean (*Vigna radiata*) genic microsatellite markers. Conserv Genet 10: 1939–1943.
32. Tangphatsornruang S, Somta P, Uthaipaisanwong P, Chanprasert J, Sangsrakru D, et al. (2009) Characterization of microsatellites and gene contents from genome shotgun sequences of mungbean (*Vigna radiata* (L.) Wilczek). BMC Plant Biol 9: 137.
33. Choudhary S, Sethy NK, Shokeen B, Bhatia S (2009) Development of chickpea EST-SSR markers and analysis of allelic variation across related species. Theor Appl Genet 118: 591–608.
34. Harlan JR (1992) Crops and Man. 2ⁿᵈ ed. Am. Soc. Agronomy, Madison, WI.
35. Seehalak W, Tomooka N, Waranyuwat A, Thipyapong P, Laosuwan P, et al. (2006) Genetic diversity of the *Vigna* germplasm from Thailand and neighboring regions revealed by AFLP analysis. Genet Resour Crop Evol 53: 1043–1059.
36. Tomooka N, Yoon MS, Doi K, Kaga A, Vaughan DA (2002) AFLP analysis of a *Vigna* subgenus *Ceratotropis* core collection. Genet Resour Crop Evol 49: 521–530.
37. Kaewwongwal A, Jetsadu A, Somta P, Chankaew S, Srinives P (2013) Genetic diversity and population structure of *Vigna exilis* and *Vigna grandiflora* (Phaseoleae, Fabaceae) from Thailand based on microsatellite variation. Botany 91: 653–661.

2

Carrot yellow leaf virus Is Associated with Carrot Internal Necrosis

Ian P. Adams[1], Anna Skelton[1], Roy Macarthur[1], Tobias Hodges[1,3], Howard Hinds[2], Laura Flint[1], Palash Deb Nath[4], Neil Boonham[1], Adrian Fox[1]*

1 Centre for Crop Protection, Food and Environment Research Agency, Sand Hutton, York, United Kingdom, 2 RootCrop Ltd., Hoveringham, Nottinghamshire, United Kingdom, 3 University of York, York, United Kingdom, 4 Department of Plant Pathology, Assam Agricultural University, Jorhat, India

Abstract

Internal necrosis of carrot has been observed in UK carrots for at least 10 years, and has been anecdotally linked to virus infection. In the 2009 growing season some growers had up to 10% of yield with these symptoms. Traditional diagnostic methods are targeted towards specific pathogens. By using a metagenomic approach with high throughput sequencing technology, other, as yet unidentified causes of root necrosis were investigated. Additionally a statistical analysis has shown which viruses are most closely associated with disease symptoms. Carrot samples were collected from a crop exhibiting root necrosis (102 Affected: 99 Unaffected) and tested for the presence of the established carrot viruses: *Carrot red leaf virus (CtRLV)*, *Carrot mottle virus (CMoV)*, Carrot red leaf associated viral RNA (CtRLVaRNA) and *Parsnip yellow fleck virus (PYFV)*. The presence of these viruses was not associated with symptomatic carrot roots either as single viruses or in combinations. A sub-sample of carrots of mixed symptom status was subjected to MiSeq sequencing. The results from these tests suggested *Carrot yellow leaf virus* (CYLV) was associated with symptomatic roots. Additionally a novel Torradovirus, a novel Closterovirus and two novel Betaflexiviradae related plant viruses were detected. A specific diagnostic test was designed for CYLV. Of the 102 affected carrots, 98% were positive for CYLV compared to 22% of the unaffected carrots. From these data we conclude that although we have yet to practically demonstrate a causal link, CYLV appears to be strongly associated with the presence of necrosis of carrots.

Editor: Zhengguang Zhang, Nanjing Agricultural University, China

Funding: This work was funded by the Horticultural Development Company, a levy board of the Agricultural and Horticultural Development Board, as project FV382a. The funders had no role in study design, data collection and analysis, or preparation of the manuscript. Permission was required to allow submission of the manuscript.

Competing Interests: Howard Hinds is the sole owner and employee of RootCrop Ltd. Through the author HH, RootCrop Ltd. had a role in the study design, data collection and analysis, decision to publish, and preparation of the manuscript. The specific role of author, Howard Hinds, is articulated in the "author contributions" section. There are no patents, products in development, or marketed products to declare.

* Email: adrian.fox@fera.gsi.gov.uk

Introduction

For at least 10 years UK growers have reported carrot roots exhibiting internal necrosis around the root core extending from crown to tip, and these have been anecdotally associated with the presence of viruses. The 2009 growing season saw some growers with up to 10% of yield affected by these symptoms, although symptom development appeared to be locally significant, with many growers reporting no evidence of root symptoms in crops. It is difficult to grade out affected carrots because the symptoms tend to be internal. Results of a limited survey in 2010 [1] suggested a possible association between the presence of root necrosis symptoms and virus infection. However, a large proportion of the carrots tested in this earlier study were negative when tested for PYFV or the Carrot Motley Dwarf complex (CMD) of viruses. This finding raised the question of other viruses being a cause of the development of carrot root necrosis.

Globally more than 30 viruses are known to affect carrot [2]. The principal viruses known to affect commercial carrot crops in the UK are *Parsnip yellow fleck virus* (PYFV) and the CMD Complex consisting of *Carrot red leaf virus* (CtRLV), *Carrot mottle*

virus (CMoV) and Carrot red leaf associated RNA (CtRLVaRNA). The importance of PYFV and CtRLV as viruses causing economic damage have been recognised for over 20 years due to the foliar symptoms (CtRLV) and viral die-back of seedlings (PYFV) [3]. In the UK these viruses affect carrot crops only sporadically but when they do occur they can be devastating. Other carrot viruses are known to occur in the UK, however, their effects are not clear.

Carrot yellow leaf virus (CYLV) (Genus *Closterovirus*, Family *Closteroviridae*) was first isolated from carrot samples showing yellowing foliage from Japan [4] and described on the basis of particle morphology; measurement by Electron Microscopy (1,600×12 nm, 3.7 nm Helical pitch); being limited to phloem and having characteristics of closterovirus infection. Bem and Murant [5] later described a series of viruses found in the UK from hogweed (*Heracleum sphondylium*), among which were the filamentous viruses Hogweed 2 virus with a particle size of 700–750 nm, and Hogweed 6 virus (HV6) with a particle size of 1400 nm, and being transmissible by aphids these were tentatively assigned to the genus Closterovirus [5]. Hogweed 2 virus, with shorter particles, was characterised as *Heracleum latent virus*

(HLV) [6]. Hogweed 6 virus was subsequently shown to be transmitted by aphids including *Cavariella* spp. and to act as the helper virus for the transmission of HLV [7]. Murant again reported the length of HV6 as 1400 nm and that mechanical inoculation of HV6 was unsuccessful [8]. During a subsequent survey of umbellifers in the Netherlands [3] the virus previously reported by Murant [6,7,8] as HV6 was considered to be CYLV on the basis of host range similarity and the ability to facilitate co-transmission of HLV.

A virus isolated from carrot in the Netherlands was reported as 'resembling CYLV' due to host symptoms; the presence of closterovirus-like particles; and a host range that was different to *Beet yellows virus* [9]. Murant reported that sap inoculation of CYLV had been unsuccessful [8], van Dijk and Bos [9] reported poor sap transmissibility into *Nicotiana benthamiana*, but subsequent attempts to mechanically transmit this virus back into carrot were unsuccessful.

An isolate of an unknown closterovirus from a German carrot sample exhibiting foliar yellowing symptoms was shown to be CYLV through molecular characterisation [10]. There is no literature linking any carrot viruses with necrotic root symptoms.

Detection of carrot viruses is currently carried out using conventional PCR methods, which give efficient, specific detection of single targets. With the use of degenerate primer sets they can be used to detect a number of pathogens of the same genus [10]. However, such targeted testing will not reveal the presence of unexpected or unknown viruses. Even multi-target approaches such as micro-array based methods [11] are unlikely to reveal the presence of complete unknowns, unless cross-hybridisation to known close relatives occurs. A more efficient approach would be to use a 'non-targeted' method such as next generation (high throughput) sequencing for diagnosis of viral pathogens. These techniques have been successfully deployed in plant pathology for the detection of novel viruses [12,13,14] or for the diagnosis of unusual strains of plant viruses [15]. Such approaches are rapidly becoming more cost effective as the high throughput platforms develop. Previous reports utilising this technology have tended to identify the presence of a novel or unusual virus in single or pooled samples and then use the sequence generated to design targeted diagnostics to validate the finding from the original sample. Putting these findings into a broader context of field pathology is more challenging.

To definitively link a pathogenic cause to an observed symptom it is necessary to demonstrate Koch's postulates, the isolation from a diseased individual of a pure culture of a pathogen which is then used to induce symptoms in a previously healthy host. These requirements, first described in 1890 [16], were intended to set a standard methodology for proof of a causal relationship. As viruses are obligate pathogens, it is not possible to obtain a 'pure culture', in addition some viruses can be difficult to transmit and the specific transmission mechanism of a new virus may not be known. For diseases induced by a single virus species Koch's postulates may be satisfied in their broadest interpretation i.e. a pathogen is isolated from a symptomatic plant into an experimental host and then back-inoculated into the original host species to try and replicate the original symptom. Attempts have also been made to look at causation in light of developments in molecular detection [16]. However, where a complex of viruses may be affecting a host or where there may be environmental or agronomic influences on symptom development (temperature, moisture, time from exposure, time in ground/crop growth stage, etc) trying to link detection of pathogen/s with a symptom using a conventional cause-and-effect relation is often not possible. Therefore statistical approaches have been employed to demonstrate the possible influence of single or multiple pathogens on the expression of symptoms within a sampled population [17].

This paper describes a study of the potential causes of carrot internal necrosis using RT_PCR of common carrot viruses and next generation sequencing in carrots with and without symptoms of necrosis and a statistical approach to associate particular viruses with the incidence of necrotic symptoms.

Materials and Methods

Ethics statement
The carrots were obtained with the permission of Rodger Hobson, Hobson's Farming and no further permissions were required. The samples were taken post-harvest so there was no damage to endangered or protected species.

Source of carrot samples
The crop of carrots sampled were grown at Gothic Back Field, York, UK (OS Grid Ref: SE 65486 445297; Latitude 53.899711, Longitude -1.0048628). On the grading line carrots were cut and examined for symptoms, of these 3% of the 3300 individual carrots examined contained necrotic symptoms. Some of these carrots had surface necrosis which could have been graded out, others had only internal necrosis, and there were also carrots sampled with a combination of internal and external symptoms. From these samples 102 carrot roots were selected which were affected by disease (i.e. necrotic/symptomatic) and 99 carrot roots were selected which were un-affected (non-necrotic/asymptomatic). As the carrots were sampled on the grading line no assessment of foliar symptoms could be made.

Sap inoculations
Five plants of each of the standard indicators *Nicotiana benthamiana*, *N. debneyi*, *N. hesperis*, *N. tobaccum* (cv White Burley), *N. occidentalils* (P1), *Chenopodium quinoa*, and *C. amaranticolor*, Tomato and the umbelliferous plants Carrot, Chervil and Coriander were inoculated from carrot samples using the methods described in Hill [18]. Control and non-inoculated plants were maintained for all species. Plants were maintained in a green house with a mean temperature of 22°C with an 18 hr. photoperiod and assessed for symptoms weekly.

Reverse Transcription PCR
RNA was extracted from carrot roots by magnetic bead extraction using Invimag Virus DNA/RNA mini-kit (Invitek GMBH). Conventional RT-PCR was carried out for the presence of the four carrot viruses known to be common in the UK namely *Parsnip yellow fleck virus* [19] and the viruses of the Carrot Motley Dwarf complex, *Carrot red leaf virus*, *Carrot Mottle virus* [20] and Carrot red leaf associated viral RNA [21]. All RT-PCR reactions were carried out using Verso 1-Step RT-PCR ReddyMix Kit (Thermo Scientific) on a GeneAmp 9700 (Applied Biosystems) (Annealing Temperature 50°C).

High Throughput Sequencing
RNA was extracted from 12 affected and 12 unaffected carrots using an RNeasy kit (Qiagen, UK). TruSeq RNA Indexed sequencing libraries (Illumina) were then prepared following the manufacturers recommended protocols, before being sequenced on 2 500 cycle v2 flow cells using MiSeq Sequencer (Illumina). The resulting sequences were trimmed to remove low-quality nucleotides from the 3′ end, using a Phred score threshold of 30 and the *bwa* trimming approach implemented in SolexaQA [22], and assembled using Trinity [23]. Contigs produced were then compared to the GenBank protein database using BLASTx + [24] and viral reads extracted using MEGAN [25,26]. Open reading

frames were identified using Vecor NTi v11 (Invitrogen, UK) and alignments and phylogenetic trees produced using Mega5 with 500 bootstrapped replicates [27]. To determine the number of viral reads in the affected and unaffected samples reads were mapped back to the genomes of identified viruses using bwa *aln sampe* [28] and the numbers of matched read pairs extracted using Samtools [29]. To normalize mapped read counts against the length of genome and total number of reads, the values are reported as mapped reads per kilobase of viral genome per million reads (RPKM), an approach originally introduced for the comparison of mRNA abundance in differential expression analyses [30]. The fastq data produced during the project were submitted to the short read achive acc: SRP042501.

Real-time PCR

Real-time (TaqMan) primers and probes were designed using Primer Express 2 with sequences from GenBank and derived from the sequencing in this study when available (Applied Biosystems). Real-time RT-PCR was performed on previously extracted RNA in 96 well plates on an ABI 7900 instrument (Applied Biosystems). Reactions consisted of $1\times$ buffer A (Applied Biosystems), 0.2 mM of each dNTP, 5.5 mM $MgCl_2$, 0.025 U/µl AmpliTaq Gold (Applied Biosystems), 0.4 U/µl Revertaid (Fermentas), 300 nM of each primer, 100 nM of probe and 1 µl of extracted RNA (concentration as extracted) to give a final reaction volume of 25 µl. The cycling conditions used were: 30 min at 48°C, 10 min at 95°C, then 40 times, 15 sec at 95°C and 1 min at 60°C. Negative controls consisted of water replacing the template. Results were scored as positive or negative for CYLV based on presence or absence of amplification after 40 cycles.

Statistical analysis

The extent to which necrosis may be caused by viruses was assessed by counting the proportion of carrots that are necrotic and testing equal numbers of necrotic and non-necrotic carrots for the presence of viruses. Hence we observed three proportions:

$P(N)$: the proportion of carrots that were necrotic.

$P(V|N)$: the proportion of necrotic carrots that contained a virus,

$P(V|\sim N)$: the proportion of non-necrotic carrots that contained a virus.

Additionally estimates of three proportions were derived:

$P(V)$: the proportion of carrots that contained virus,

$P(N|V)$: the proportion of carrots with a virus that are necrotic

$P(N|\sim V)$: the proportion of carrots without a virus that are necrotic.

Estimates were derived from the law of total probability

$$P(V) = P(V|N)P(N) + P(V|\sim N)(1-P(N)) \quad \text{(Equation 1)}$$

and Bayes' Theorem

$$P(N|V) = \frac{P(V|N)P(N)}{P(V)}$$
$$P(N|\sim V) = \frac{P(\sim V|N)P(N)}{1-P(V)} \quad \text{(Equation 2)}$$

The size of the uncertainty associated with observed proportions was estimated using a Modified Jeffreys interval [1], where given x 'positives' out on n observations the probability p underlying the observed proportion is with confidence 1-α

$$B(\alpha/2, x+0.5, n-x+0.5) \leq p \leq B(1-\alpha/2, x+0.5, n-x+0.5)$$

if $0 < x < n$,

where $B(\alpha, b, c)$ is the α quantile of the *Beta*(*b*,*c*) distribution

$p \leq 1 - \alpha^{1/n}$

If $x = 0$, and

$p \leq \alpha^{1/n}$

If $x = n$.   (Equation 3)

The size of the predicted effect of removing a virus, on the prevalence of necrosis expressed as the proportional reduction in prevalence was estimated using

$$E = \frac{P(N) - P(N|\sim V)}{P(N)} \quad \text{(Equation 4)}$$

The uncertainty associated with derived estimates was estimated by generating independent random (uniform (0,1)) quantiles for each of the observed proportions (Equation 3) and calculating derived values using Equations 1 and 2. 95% confidence intervals were taken from the 2.5th and 97.5th percentiles of 10000 derived values.

Results

Sampling

A sample of 3300 carrots were examined for necrotic root symptoms. The prevalence of symptoms was estimated from this sample to be approximately 3%. The 102 necrotic/symptomatic carrot samples found within the sample were taken for analysis. These roots exhibited a range of symptoms which included cases of internal and external necrosis. (See Figure 1). 99 asymptomatic/non-necrotic carrots were also taken for analysis.

Reverse-Transcriptase PCR Testing

The sample of affected (symptomatic) and unaffected carrots were initially tested using conventional RT-PCR assays for the presence of the carrot viruses which are known to be common in the UK (PYFV, CMoV, CtRLV, CtRLVaRNA). Results from this testing are presented in Figure 2 which shows the percent of affected or unaffected carrots which contained the viruses. Approximately equal proportions of affected and unaffected carrots were positive for PYFV and CMD viruses (37% affected, 38% unaffected). The two groups also contained similar incidences of CtRLV (33% affected, 27% unaffected) and CMoV (9% affected, 14% unaffected). No CtRLVaRNA was detected from carrots of either symptom status. As RT-PCR results from the two groupings were broadly comparable no influence of the viruses upon the incidence of necrosis symptoms was detected. A subsample of carrots was subsequently tested using high through-put sequencing to investigate the presence of non-target pathogens.

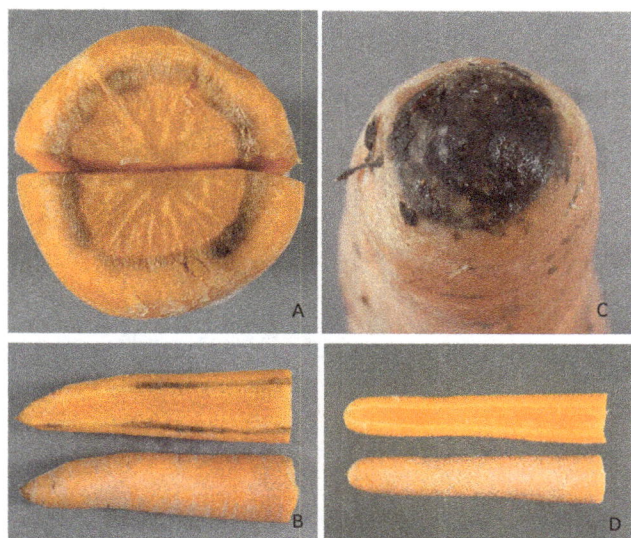

Figure 1. Examples of symptoms of 'affected' carrot samples (a) Cross section of carrot root showing internal necrosis around the root core. (b) Internal necrosis of carrot root along the root core. *(c)* External necrosis of the root tip. (d) Unaffected carrot.

High Throughput Sequencing

Indexed pair end reads (16225604 in total) with a sequence length of 2×250 bp were obtained from 12 affected and 12 unaffected carrot RNA extracts. A phylogram (figure 3), produced in MEGAN [25,26] details the viruses found in the affected and unaffected samples. *Carrot yellow leaf virus* (CYLV) was by far the most prevalent virus in the affected samples (RPKM = 221.3) but

it was not common in the unaffected samples (RPKM = 2.2). Comparison of the MEGAN pylograms for affected and unaffected carrots did not reveal any fungi or bacteria unique to the affected carrots.

Full genomes for CYLV were assembled from reads obtained from all but one of the affected samples, while only small (< 1000 bp) fragments were assembled from reads obtained from the unaffected samples. The genome sequences (KF533698–KF533708) were conserved between samples (>98% nucleotide identity) and closely related (95% nucleotide identity) to the reference genome for CYLV [31] (Acc NC013007.1). CYLV was not detected in one of the affected samples, but this sample was found to be infected with a novel virus related to CYLV. This virus, tentatively named Carrot closterovirus 1 (CtCV-1) for the purposes of this study, has a genome of 19923 nucleotides (KF533697). Open reading frame analysis identified 9 open reading frames analogous to 9 of the open reading frames found in CYLV. CtCV-1 lacks ORF 4 found in CYLV but a related virus *Mint virus 1* also lacks this open reading frame. Examination of the amino acid sequences of the putative coat, polymerase and HSP70h proteins suggest that CtCV-1 is a distinct member of the genus *Closterovirus*. When compared to its closest sequenced relative CYLV, the polymerase of CtCV-1 is 84% homologous, the HSP70h 64% homologous and the two coat proteins 69% and 46% homologous respectively. The species demarcation for closteroviruses specifies less than 75% homology in these values [32]. The coat proteins of CtCV-1 (24.7 kDa, 23.3 kDa) are also slightly larger than those of CYLV (24.5 kDa, 22.7 kDa). Figure 4 shows a phylogenetic tree constructed using the HSP70h sequences of CtCV-1, CYLV and related closteroviruses, again confirming CtCV-1 as a distinct member of the genus *Closterovirus*.

Almost complete genomes of CtRLV were recovered from 3 of the unaffected samples (KF533716–KF533718) but not from the

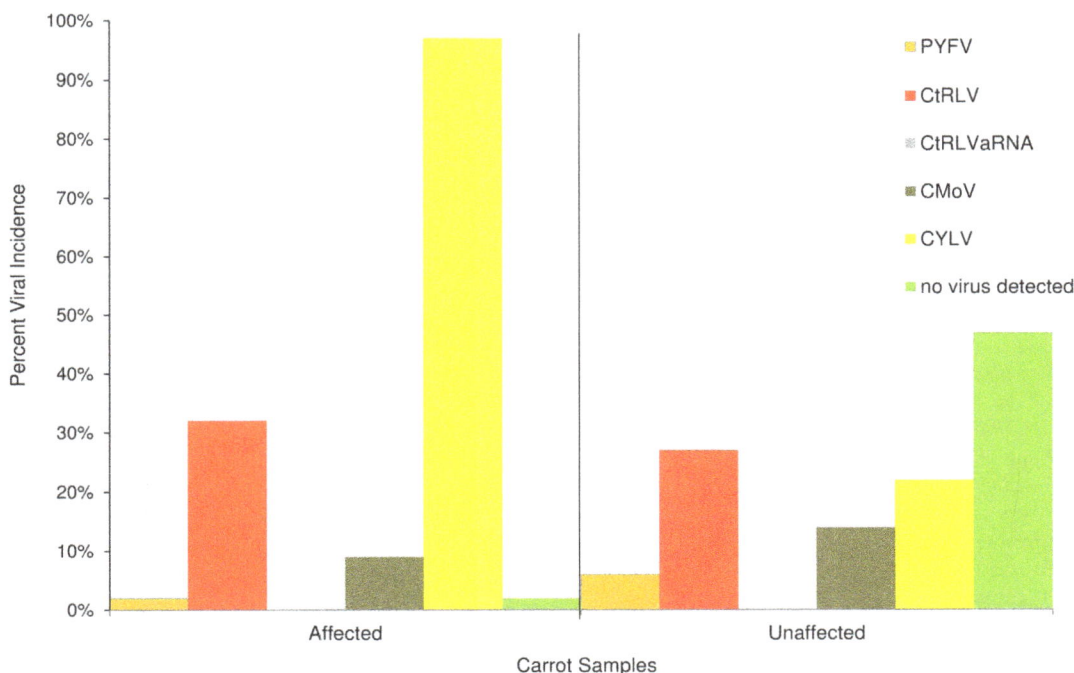

Figure 2. Percent virus incidence in carrot roots from the total field sample of 102 affected and 99 unaffected carrots for the presence of viruses including CYLV, presented as a percentage of carrots with necrosis symptoms (affected) and without necrosis symptoms (unaffected) where virus was detected by PCR or TaqMan.

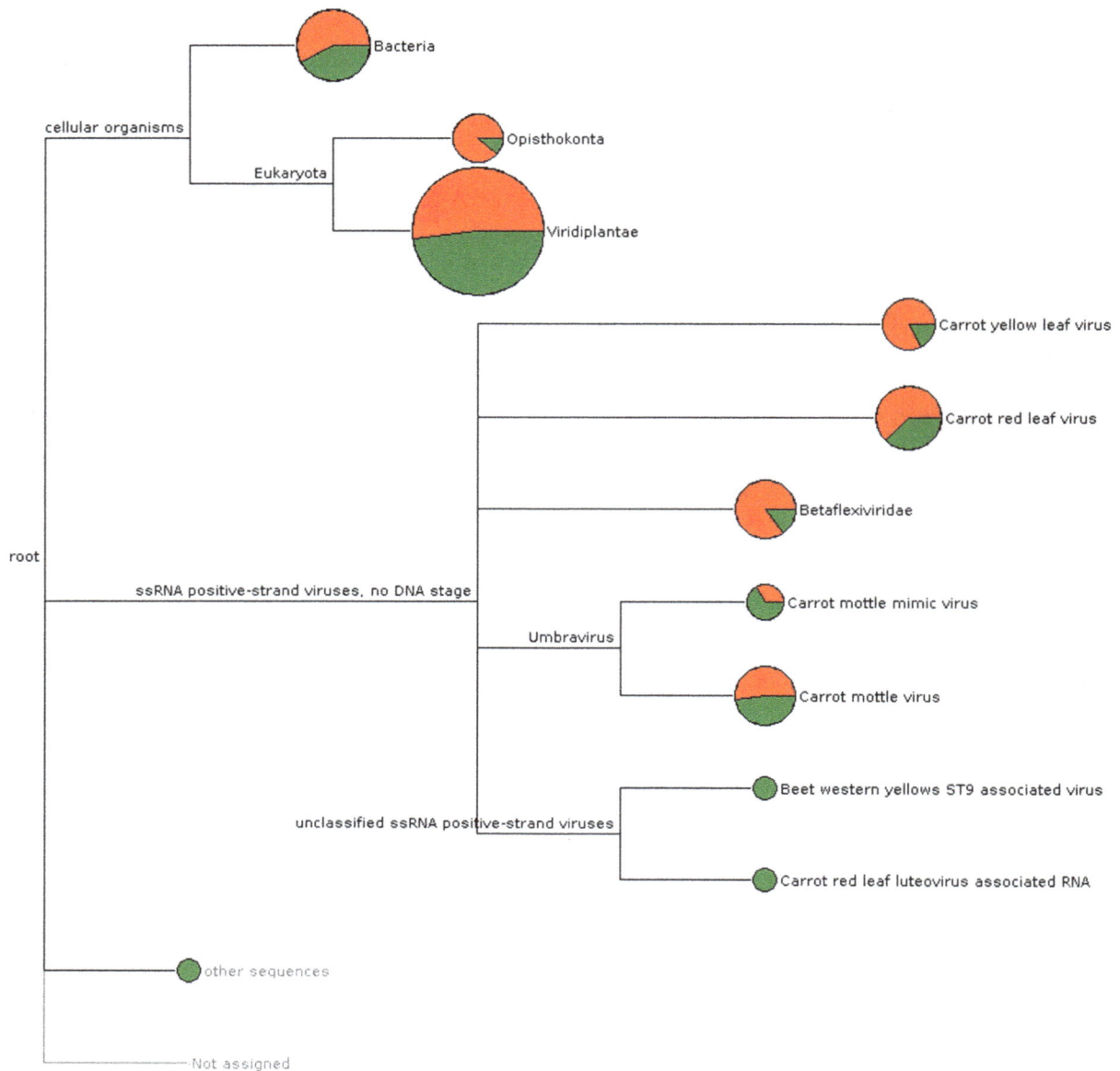

Figure 3. MEGAN derived Phylogram showing the putative identification of contigs in the Sequencing data. Circle size is derived from the number of contigs assigned to each taxa Red: affected, Green: unaffected.

affected samples. Small fragments of CtRLVaRNA were recovered from one of the unaffected samples (KF533715) which was also infected with CtRLV. A large fragment (2 kb, KF533709) was recovered from one unaffected sample, which was also infected with CtRLV. The analysis of this fragment showed 94% sequence identity to Beet western yellows virus associated RNA (BWY-VaRNA) and correspond to 75% of the complete genome.

Complete genomes of CMoV were recovered from 1 affected and 2 unaffected samples (KF533712–KF533714). These sequences have between 91–96% identity to the complete genome of CMoV (Acc: FJ88473).

Over 10,000 reads of a novel unclassified ssRNA positive strand virus were found in the unaffected and affected samples. Large fragments (6.9 k and 4.7 k nucleotides) of a bipartite viral genome were found in one affected and one unaffected sample and smaller fragments in another unaffected sample. The fragments from the

two different samples were >99% identical suggesting that they were infected with the same virus. Analysis suggests that the 6.9 k nucleotide fragment is the RNA1 genome of a novel *Torradovirus* tentatively named Carrot torradovirus 1 (CTV-1) (KF533719). It contains an open reading frame coding for a 2214 amino acid (249 kDa) polypeptide. BLAST analysis of this putative protein showed it to contain RNA helicase and RdRp domains and have 40% homology to the equivalent protein sequences from *Tomato marchitez virus* [33] and Tomato chocolate spot virus [34]. The 4.7 k nucleotide fragment appears to be the RNA2 genome of a novel torradovirus (KF533720). This contains two open reading frames ORF1 encoding a putative 202 amino acid (22 kDa) protein with 43% homology to the RNA2 ORF1 from Lettuce necrotic leaf curl virus, a recently reported torradovirus (KC855266). This rate of homology within ORF1 leads to a clear demarcation with the genus *Torradovirus*. The second ORF

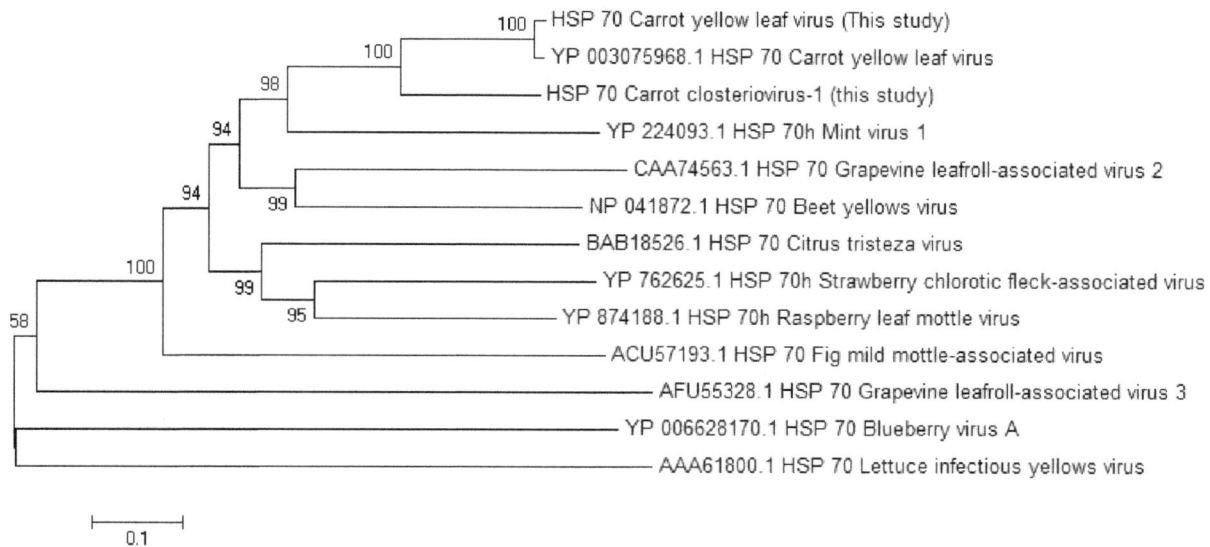

Figure 4. Bootstrapped neighbour joining tree of HSP70 proteins from viruses within the family *Closteroviridae* constructed with MEGA5 using 500 replicates.

encodes a putative 1167 amino acid (130 kDa) polyprotein. This polyprotein appears to contain movement and coat protein domains and have 35% homology to the RNA2 ORF2 from *Tomato torrado virus*. A phylogenetic tree (figure 5) constructed using the RNA2 polyprotein sequences from torradoviruses and other viruses from the *Secoviridiae* shows the closest related torradovirus to be Lettuce necrotic leaf curl virus.

Over 1000 reads with identity to the family *Betaflexiviridae* were found mainly in the affected samples. Table 1 details the relative abundance (expressed as RPKM) of the identified viruses in the affected and unaffected samples. Further examination revealed that 5 affected samples contained large fragments (>5 k nucleotides) and 3 other affected and 1 unaffected samples had fragments between 1000–5000 nucleotides. Analysis of the 5 larger fragments suggests that they are derived from 2 distinct viruses 1 from 4 samples tentatively named Carrot chordovirus-1 (CtChV-1) (KF533711), 1 from the fifth sample, tentatively named Carrot chordovirus-2 (CtChV-2) (KF533710). Both viruses have genomes of approximately 8.5 k nucleotides and encode 3 putative proteins expected in the *Betaflexiviridae*. The putative coat protein and polymerase nucleotide sequences have less than 45% identity to any previously sequenced virus suggesting they may constitute a new genus [32], within the *Betaflexiviridae*, tentatively named Chordovirus. Comparison of the putative coat protein and polymerase nucleotide and amino acid sequences suggests that the two viruses are in the same genus (>45% nucleotide identity) but distinct viruses 40% amino acid homology within the coat protein. Figure 6 shows a phylogenetic tree produced using the coat protein sequences of related members of the family *Betaflexiviridae* and further providing evidence that these viruses may constitute a novel genus.

Sequencing follow-up

Following the outcomes of sequencing, real-time RT-PCR (TaqMan) assays were designed to CYLV and CtCV-1 as follows: CYLV Forward: 5′-AAGATTCTCTTGTAACGAAGGTTTCC, reverse: 5′-GCCGCCTCCACGATCAC, Probe: 5′ Fam-AGA-CCTCACTATGCTAAACCCGAGCCGG-Tamra. CtCV-1 Forward: 5′-GCCTCCCGCTTGTTGGA reverse: 5′-AGCCGC-CAACGTCTATGAAG Probe 5′ Fam-AATAGGACCGTCGC-GAGTTTCTGCTCTG-Tamra.

These assays were then used to test the nucleic acid extracts of the 24 carrot sub-sample which had been analysed by sequencing. Of these only 1 of the affected carrots contained CtCV-1, with all 12 affected carrots testing positive for CYLV. Three of the unaffected carrot roots tested positive for CYLV, though in two of these cases the virus was detected at weak levels (>39 Ct).

On the basis of this finding the nucleic acid extracts from the field samples (affected and unaffected carrots) which had been previously tested using RT-PCR were tested for the presence of CYLV using the real-time RT-PCR assay. The results (Figure 2) show that of the carrots affected by necrotic symptoms 98% (99 of 102 carrots tested) were found to be positive, whilst in the unaffected sub-sample 22% (22 of 99 carrots tested) contained the virus. It was also found that the 3 affected samples which tested negative for CYLV tested positive for CtCV-1.

Attempts were made to sap inoculate CYLV from infected carrots into healthy carrots and other species of indicator plants. No infection was detected.

Statistical analysis

The crop of carrots sampled showing 3% of necrotic symptoms i.e. 3300 carrots were sliced to obtain 100 symptomatic samples. Some of these carrots had surface necrosis which could have been graded out, others had only internal necrosis, and there were also carrots sampled with a combination of internal and external symptoms with approximately even numbers of each within the sample set.

The statistical analysis is presented in Table 2. The results for carrot roots found to be positive for the established carrot viruses (PYFV, CtRLV, CtRLaVRNA and CMoV) show a similar proportion of carrots with and without necrosis indicating that necrosis is probably independent of infection with these viruses. With one exception (*Carrot yellow leaf virus*, CYLV) there is no association between virus status and the prevalence of necrosis. CYLV-positive carrots have an estimated prevalence of necrosis of 12.0% (8.4–17.1%) while the prevalence of necrosis in CYLV-negative carrots is estimated to be 0.1% (0.0–0.3%). The estimated

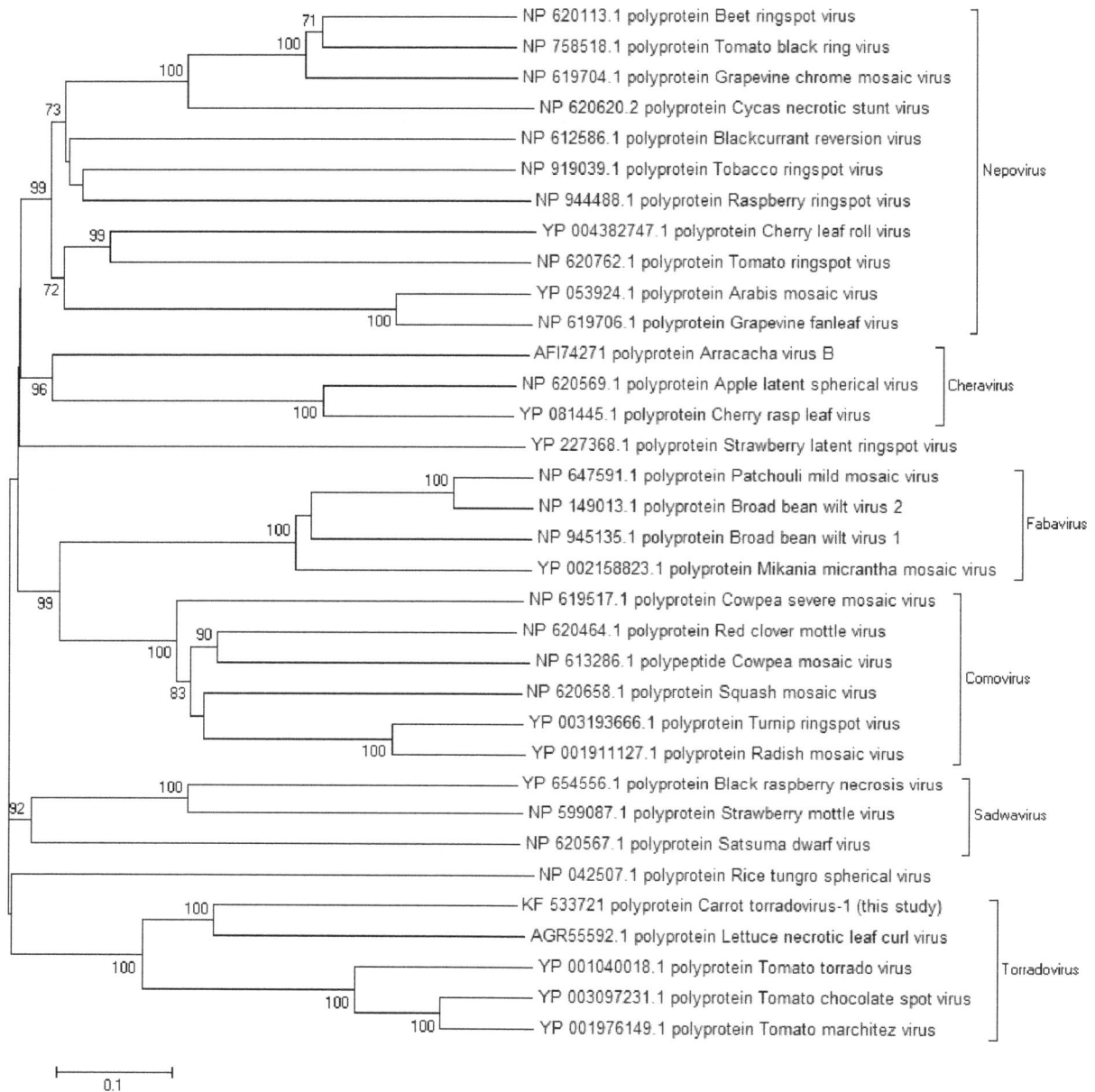

Figure 5. Bootstrapped neighbour joining tree of RNA2 polyproteins from viruses within the family *Secoviridae* constructed with MEGA5 using 500 replicates.

reduction in necrosis prevalence associated with the removal of a virus can be seen in table 2 (Column E%). Removing CYLV from the population is estimated to have a large potential effect, with an estimated reduction of 96% (89.6–98.8%). of necrosis. Because necrosis without CYLV, even in the presence of other viruses, is estimated to be rare (0.0–0.3%) the removal of CYLV alone may be sufficient to greatly reduce the prevalence of necrosis if CYLV is indeed causative.

Discussion

Carrot roots exhibiting internal necrosis have been a growing problem in the UK. Due to a lack of methods allowing the rapid screening of both symptomatic and asymptomatic roots for the

presence of a broad range of pathogens, progress on identifying the causal pathogen has been limited.

Of the thirty or so viruses known to have carrot as a host, at least twelve are known to be present in the UK. However, these viruses are not amongst those regularly tested for by diagnostic labs either due to unknown prevalence, poor symptomatic recognition, or more commonly, poor availability of targeted diagnostics. As a result this study has applied high throughput sequencing to screen carrots to help identify a putative causative agent for internal necrosis and a range of previously un-described viruses.

Testing for the four most common viruses (PYFV, CMoV, CtRLV, CtRLVaRNA) in affected and unaffected carrot samples

Table 1. Relative abundance (Reads per kilobase of viral genome per million sequenced reads-RPKM) of select viruses in sequenced affected and unaffected samples.

virus	unaffected	affected
CYLV	2.2	221.3
CtCV-1	0.0	22.0
CTV-1 RNA1	36.4	3.6
CTV-1 RNA2	15.9	5.1
CtRLV	16.2	1.4
CtChV-1	0.0	2.3
CtChV-2	0.0	2.1
CMoV	4.0	4.2

did not provide any evidence for a link between any single or group of viruses and necrosis.

The data obtained following high-throughput sequencing showed that *Carrot yellow leaf virus* was present in eleven of the affected samples and was by far the most common virus recovered warranting further study. The prevalence of this virus was much lower in the unaffected samples. The statistical analysis clearly indicates a link between this unexpected virus finding and the presence of necrosis in carrot roots at this site. Indeed if CYLV is the causal pathogen of carrot internal necrosis, removing CYLV from the sampled carrot population would give an estimated effect of reducing the incidence of necrosis by 96%. Demonstrating a mathematical statistical relationship does not show that there is biological causative relationship but it does point towards where further investigations should be carried out.

Attempts were made to sap inoculate CYLV into carrot or other indicator plants to carry out Koch's postulates but this proved unsuccessful. Previously Murant [8] reported that sap inoculation had been unsuccessful whereas van Dijk and Bos [9] reported poor sap transmissibility into *Nicotiana benthamiana*. Discussion with the authors of the 2009 paper on CLYV [31] also confirmed that they had been unable to sap inoculate CYLV. CYLV is aphid transmitted. In order to transmit CYLV using captive aphids live carrots with attached leaves would be required. In the current study the necrotic symptoms were determined by cutting the carrot root on the grading line when the carrot had already been harvested and the leaves removed. To date it has not been possible to obtain a live symptomatic carrot plant, therefore it has not been possible to further characterize the effects of CYLV on carrot root necrosis.

Due to the low incidence of expression (3%) and localised nature of these symptoms, it was decided that the strategy most likely to yield informative results was to focus in depth on an affected crop from a single site, sampled at the point where symptoms were evident (i.e. on the processing line). It is appreciated that caution should be applied in extrapolating from a single sampled site to other affected sites and future work will include a multi-site survey to confirm the applicability of these findings in a broader context. The statistical approach used in this study, based upon testing approximately equal numbers of affected and unaffected individuals for the presence of pathogens can be applied where infection and disease are often anecdotally linked but lack an empirically observed basis. This could be of particular use when applied to diseases thought to be caused by obligate pathogens such as viruses, viroids, phytoplasmas, or fungal obligates such as rusts or powdery and downy mildews.

Sequencing also identified sequences from 3 of the known carrot infecting viruses CtRLV, CtRLVaRNA and CMoV. CMoV was

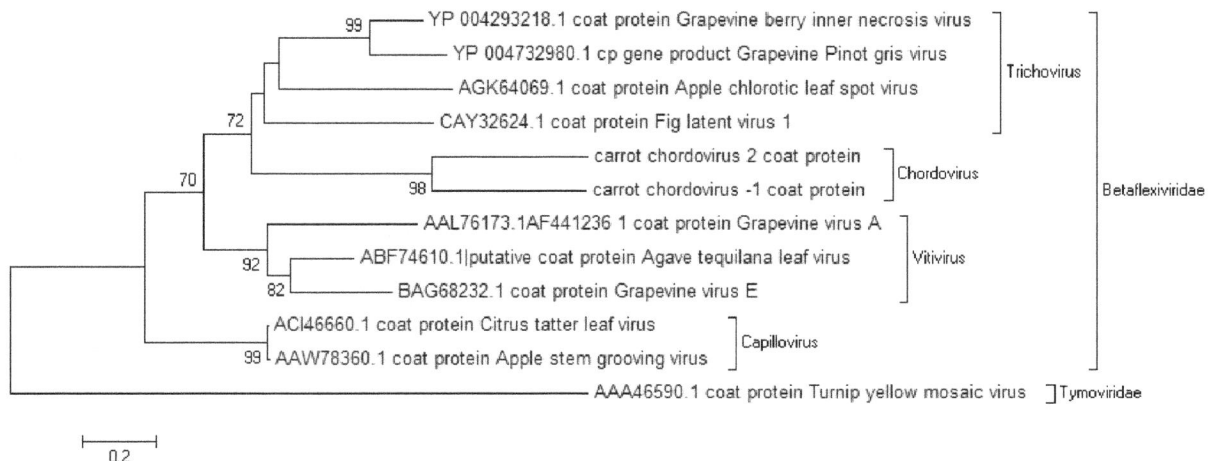

Figure 6. Bootstrapped neighbour joining tree of coat proteins from viruses within the family *Betaflexiviridae* constructed with MEGA5 using 500 replicates.

Table 2. Estimates of prevalence of virus and necrosis and the effect of virus removal on reducing necrosis (values in brackets are 95% confidence intervals).

Virus	P(V|N) (%)	P(V|~N) (%)	P(V) (%)	P(N|V) (%)	P(N|~V) (%)	E (%)
PYFV	2.0 (0.4–6.1)	6.1 (2.6–12.1)	5.9 (2.5–12.0)	1.0 (0.1–4.2)	3.1 (2.5–3.8)	−4.2 (−10.8–1.3)
CtRLV	32.4 (23.9–41.8)	27.3 (19.2–36.6)	27.4 (19.7–36.3)	3.5 (2.2–5.5)	2.8 (2.1–3.6)	6.8 (−11.3–21.8)
CtRLVaRNA	0.0 (0.0–2.9)	0.0 (0.0–3.0)	0.0 (0.0–3.5)	NE	NE	NE
CMoV	8.8 (4.5–15.5)	14.1(8.3–22.0)	14 (8.3–21.9)	1.9 (0.8–4.2)	3.2 (2.5–3.9)	−6.0 (−17–3.7)
CYLV	97.1 (92.4–99.2)	22.2 (14.9–31.1)	24.5 (17.1–33.1)	12.0 (8.4–17.1)	0.1 (0.0–0.3)	96.1 (89.6–98.8)
Any positive	98 (93.9–99.6)	52.5 (42.7–62.2)	53.9 (44.3–63.5)	5.5 (4.2–7.1)	0.1 (0.0–0.4)	95.7 (86.5–99.1)
CtRLV+CMoV	0.0 (0.0–2.9)	4.0 (1.4–9.3)	3.9 (1.4–9.1)	0.0 (0–17)	3.1 (2.5–3.7)	−4.1 (−9.2–0.4)
PYFV+CMoV	0.0 (0.0–2.9)	1.0 (0.1–4.6)	1.0 (0.1–4.6)	0.0 (0–17)	3.0 (2.4–3.6)	−1.0 (−4.2–2.5)
PYFV+CtRLV	0.0 (0.0–2.9)	1.0 (0.1–4.6)	1.0 (0.1–4.6)	0.0 (0–17)	3.0 (2.4–3.6)	−1.0 (−4.1–2.6)
PYFV+CYLV	1.0 (0.1–4.5)	0.0 (0.0–3.0)	0.0 (0–3.7)	100 (0.4–100)	3.0 (2.4–3.6)	1.0 (−2.5–3.7)
CtRLV+CYLV	26.5 (18.7–35.6)	7.1 (3.2–13.4)	7.7 (3.9–13.7)	10.4 (5.2–21.7)	2.4 (1.9–3.0)	20.4 (10.2–30.4)
CMoV+CYLV	2.9 (0.8–7.6)	0.0 (0.0–3.0)	0.1 (0.1–3.5)	100 (2.2–100)	2.9 (2.4–3.5)	2.9 (−1.3–6.5)
PYFV, CtRLV +CMoV	0.0 (0.0–2.9)	1.0 (0.1–4.6)	1.0 (0.1–4.5)	0.0 (0.0–24.3)	3.0 (2.4–3.6)	−1.0 (−4.2–2.3)
CtRLV, CMoV +CYLV	4.9 (1.9–10.4)	1.0 (0.1–4.6)	1.1 (0.3–4.5)	13.1 (2.4–61)	2.9 (2.3–3.5)	3.8 (−0.8–9.1)
PYFV, CMoV +CYLV	1.0 (0.1–4.5)	0.0 (0.0–3.0)	0.0 (0.0–3.6)	100 (0.4–100)	3.0 (2.4–3.6)	1.0 (−2.8–3.8)

Virus is presented both singly and in combinations. **P(V|N)**: proportion of necrotic carrots with the virus, **P(V|~N)**: proportion of non-necrotic carrots with the virus, **P(V)** prevalence of virus across all carrots, **P(N|V)**: proportion of carrots with the virus that are necrotic, **P(N|~V)**: proportion of carrots without the virus that are necrotic. **E**: estimated effect of removing the virus on the prevalence of necrosis expressed as a proportional reduction in the prevalence of necrosis, **NE**: not estimated.

detected in both affected and unaffected samples by both methods. CtRLV was found by PCR and sequencing to be present in both affected and unaffected samples. CtRLVaRNA was not detected by the conventional PCR [21] but small fragments were detected by sequencing. This might be due to specificity issues with the PCR assay. Therefore the new sequence may prove useful for improving the currently used primer sets. Conversely PYFV was detected by conventional PCR in 2% of affected and 6% of unaffected samples but none was found by sequencing, perhaps indicating the PCR approach is more sensitive than sequencing.

BWYVaRNA sequences were recovered from a sample also containing CtRLV. As far as we are aware, this is the first example of an association between these two viruses and deserves further examination as to whether BWYVaRNA is encapsidated by the CtRLV coat protein.

Four new viruses were identified in the sequencing data. A new closterovirus tentatively named CtCV-1, closely related to CYLV was found in necrotic carrots in the absence of CYLV. This suggests that CtCV-1 may also have a role in carrot necrosis and is worthy of further study. A novel torradovirus CTV-1 was found in affected and at a higher abundance in unaffected carrots whilst two novel betaflexiviruses CtChV-1 and CtChV-2 were found at low abundance predominantly in affected carrots. These viruses do not appear to be correlated to the necrotic root symptoms central to this study, but may be worthy of further investigation as they may be associated with other carrot diseases or be causing a reduction of crop yield. The novel viruses found within this small study demonstrate the limited knowledge of viral populations in carrots and it can be speculated in other crops also. The impact these viruses are having on crops are certainly not clear, however, these studies provide some of the knowledge (sequence data) and tools (specific tests) to enable us to investigate these effects in the future.

The overall aim of the project was to investigate viral causes of carrot root necrosis. The outcomes of this work show that, although potentially damaging to carrots in terms of lowering yield and causing growth defects, infection by the four established carrot viruses did not correlate with internal necrosis. There is a clear statistical association between the presence of internal necrosis and infection with CYLV. A closely related yet distinct virus tentatively named Carrot closterovirus-1 (CtCV-1) may yet be associated with necrosis, all be it with a lower incidence. This is the first report of a root necrosis symptom in carrot being associated with *Carrot yellow leaf virus*. On the basis of these findings, work is ongoing to demonstrate a biological causal relationship (Koch's Postulates) between CYLV and root necrosis.

Acknowledgments

This work was funded by the Horticultural Development Company, a levy board of the Agricultural and Horticultural Development Board, as project FV382a.

Author Contributions

Conceived and designed the experiments: IPA NB AF AS RM LF. Performed the experiments: IPA AS TH HH LF PDN. Analyzed the data: IPA AS RM TH NB AF LF PDN. Contributed reagents/materials/analysis tools: RM TH. Wrote the paper: IPA AS RM TH NB HH AF LF PDN.

References

1. Fox A (2011) Symptomatic survey of virus complexes of carrot. Horticultural Development Company. Available: http://www.hdc.org.uk/project/carrot-symptomatic-survey-virus-complexes-4. Accessed 2014 June 25.

2. Brunt AA, Crabtree K, Dallwitz MJ, Gibbs AJ, Watson L, et al. (1996) Plant Viruses Online: Descriptions and Lists from the VIDE Database. Available: http://pvo.bio-mirror.cn/refs.htm. Accessed 2014 June 25.

3. Dijk P, Bos L (1989) Survey and biological differentiation of viruses of wild and cultivated umbelliferae in the Netherlands. Netherlands Journal of Plant Pathology 95: 1–34.

4. Yamashita S, Ohki S, Doi Y, Yora K (1976) Identification of two viruses associated with the carrot yellow leaf syndrome. Annual Phytopathology Society of Japan 42: 382–383.

5. Bem F, Murant AF (1979) Transmission and differentiation of six viruses infecting hogweed (*Heracleum sphondylium*) in Scotland. Annals of Applied Biology 92: 237–242.

6. Bem F, Murant AF (1979) Host range, purification and serological properties of *heracleum latent virus*. Annals of Applied Biology 92: 243–256.

7. Murant AF (1982) Helper-dependent transmission of *Heracleum latent virus* (HLV) by aphids. Report of the scottish Crop Research Institute.191.

8. Murant AF (1983) *Heracleum latent virus* (HLV) and *Heracleum virus 6* (HV6). Report of the scottish Crop Research Institute189.

9. Van Dijk P, Bos L (1985) Viral dieback of carrot and other umbelliferae caused by the Anthriscus strain of parsnip yellow fleck virus, and its distinction from carrot motley dwarf. Netherlands Journal of Plant Pathology 91: 169–187.

10. Botermans M, van de Vossenberg BTLH, Verhoeven JTJ, Roenhorst JW, Hooftman M, et al. (2013) Development and validation of a real-time RT-PCR assay for generic detection of pospiviroids. Journal of Virological Methods 187: 43–50.

11. Boonham N, Walsh K, Smith P, Madagan K, Graham I, et al. (2003) Detection of potato viruses using microarray technology: towards a generic method for plant viral disease diagnosis. Journal of Virological Methods 108: 181–187.

12. Adams IP, Glover RH, Monger WA, Mumford R, Jackeviciene E, et al. (2009) Next-generation sequencing and metagenomic analysis: a universal diagnostic tool in plant virology. Molecular Plant Pathology 10: 537–545.

13. Harju V, Skelton A, Forde S, Bennett S, Glover RH, et al. (2011) New virus detected on *Nasturtium officinale*, Watercress New Disease Reports.

14. Kreuze JF, Perez A, Untiveros M, Quispe D, Fuentes S, et al. (2009) Complete viral genome sequence and discovery of novel viruses by deep sequencing of small RNAs: a generic method for diagnosis, discovery and sequencing of viruses. Virology 388: 1–7.

15. Adams IP, Miano DW, Kinyua ZM, Wangai A, Kimani E, et al. (2013) Use of next-generation sequencing for the identification and characterization of *Maize chlorotic mottle virus* and *Sugarcane mosaic virus* causing maize lethal necrosis in Kenya. Plant Pathology 62: 741–749.

16. Fredericks DN, Relman DA (1996) Sequence-based identification of microbial pathogens: a reconsideration of Koch's postulates. Clinical Microbiology Reviews 9: 18–33.

17. Carnegie SF, McCreath M (2010) Mosaic Virus Symptoms in Potato Crops and the Occurrence of Growth Cracking in Tubers. Potato Research 53: 17–24.

18. Hill SA (1984) Methods in Plant Virology. Oxford, UK: Blackwell Scienctific.

19. Morgan D (2004) Parsnip yellow fleck virus: Development of a disease management strategy. Horticultural Development Company. Available: http://www.hdc.org.uk/project/parsnip-yellow-fleck-virus-development-disease-management-strategy-4. Accessed 2014 June 25.

20. Vercruysse P, Gibbs M, Tirry L, Höfte M (2000) RT-PCR using redundant primers to detect the three viruses associated with carrot motley dwarf disease. Journal of Virological Methods 88: 153–161.

21. Morton A, Spence NJ, Boonham N, Barbara DJ (2003) Carrot red leaf associated RNA in carrots in the United Kingdom. Plant Pathology 52: 795–795.

22. Cox MP, Peterson DA, Biggs PJ (2010) SolexaQA: At-a-glance quality assessment of Illumina second-generation sequencing data. BMC Bioinformatics 11: 485.

23. Grabherr MG, Haas BJ, Yassour M, Levin JZ, Thompson DA, et al. (2011) Full-length transcriptome assembly from RNA-Seq data without a reference genome. Nature Biotechnology 29: 644–652.

24. Camacho C, Coulouris G, Avagyan V, Ma N, Papadopoulos J, et al. (2009) BLAST+: architecture and applications. BMC Bioinformatics 10: 421.

25. Huson DH, Auch AF, Qi J, Schuster SC (2007) MEGAN analysis of metagenomic data. Genome Research 17: 377–386.

26. Huson DH, Mitra S, Ruscheweyh HJ, Weber N, Schuster SC (2011) Integrative analysis of environmental sequences using MEGAN4. Genome Research 21: 1552–1560.

27. Tamura K, Peterson D, Peterson N, Stecher G, Nei M, et al. (2011) MEGA5: Molecular Evolutionary Genetics Analysis Using Maximum Likelihood, Evolutionary Distance, and Maximum Parsimony Methods. Molecular Biology and Evolution 28: 2731–2739.

28. Li H, Durbin R (2009) Fast and accurate short read alignment with Burrows-Wheeler transform. Bioinformatics 25: 1754–1760.

29. Li H, Handsaker B, Wysoker A, Fennell T, Ruan J, et al. (2009) The Sequence Alignment/Map format and SAMtools. Bioinformatics 25: 2078–2079.

30. Mortazavi A, Williams BA, McCue K, Schaeffer L, Wold B (2008) Mapping and quantifying mammalian transcriptomes by RNA-Seq. Nature Methods 5: 621–628.

31. Menzel W, Goetz R, Lesemann DE, Vetten HJ (2009) Molecular character-ization of a closterovirus from carrot and its identification as a German isolate of *Carrot yellow leaf virus*. Archives of Virology 154: 1343–1347.

32. King E, Kowitz E, Adams M, Carstens EB (2011) Virus Taxonomy: Ninth Report of the International Committee on Taxonomy of Viruses. London: Elsevier.

33. Verbeek M, Dullemans AM, van den Heuvel JF, Maris PC, van der Vlugt RA (2008) *Tomato marchitez virus*, a new plant picorna-like virus from tomato related to tomato torrado virus. Archives of Virology 153: 127–134.

34. Verbeek M, Dullemans A, van den Heuvel H, Maris P, van der Vlugt R (2010) *Tomato chocolate virus*: a new plant virus infecting tomato and a proposed member of the genus Torradovirus. Archives of Virology 155: 751–755.

3

Bank of Standardized Stimuli (BOSS) Phase II: 930 New Normative Photos

Mathieu B. Brodeur[1]*, **Katherine Guérard**[2], **Maria Bouras**[3]

1 Douglas Mental Health University Institute and Department of Psychiatry, McGill University, Montréal (Québec), Canada, **2** Department of Psychology, Université de Moncton, Moncton (New Brunswick), Canada, **3** Department of Education, University of Sheffield, Sheffield (South Yorkshire), United Kingdom

Abstract

Researchers have only recently started to take advantage of the developments in technology and communication for sharing data and documents. However, the exchange of experimental material has not taken advantage of this progress yet. In order to facilitate access to experimental material, the Bank of Standardized Stimuli (BOSS) project was created as a free standardized set of visual stimuli accessible to all researchers, through a normative database. The BOSS is currently the largest existing photo bank providing norms for more than 15 dimensions (e.g. familiarity, visual complexity, manipulability, etc.), making the BOSS an extremely useful research tool and a mean to homogenize scientific data worldwide. The first phase of the BOSS was completed in 2010, and contained 538 normative photos. The second phase of the BOSS project presented in this article, builds on the previous phase by adding 930 new normative photo stimuli. New categories of concepts were introduced, including animals, building infrastructures, body parts, and vehicles and the number of photos in other categories was increased. All new photos of the BOSS were normalized relative to their name, familiarity, visual complexity, object agreement, viewpoint agreement, and manipulability. The availability of these norms is a precious asset that should be considered for characterizing the stimuli as a function of the requirements of research and for controlling for potential confounding effects.

Editor: Kevin Paterson, University of Leicester, United Kingdom

Funding: This study was funded by the Natural Sciences and Engineering Research Council of Canada #388752-2012 (www.nserc-crsng.gc.ca/index_eng.asp). The funders had no role in study design, data collection and analysis, decision to publish, or preparation of the manuscript.

Competing Interests: The authors have declared that no competing interests exist.

* Email: mathieu.brodeur@douglas.mcgill.ca

Introduction

Stimuli are the key component of experiments. They must therefore be of outstanding quality and be selected meticulously as a function of specific criteria, which explains why they need to be normalized. Normalization is the process through which a representative sample of individuals evaluates images and their names according to specific variables. Normative data characterizes the images and provides a thorough description of their basic features. For instance, by discerning the name given to concepts depicted in images by the majority of individuals, it is possible to determine the level of consensus in naming the specific concepts. The name given by a majority of individuals is the modal name and the consensus is called the name agreement. Through an analysis of the different names given to each image, it is also possible to explain the variability in the given names and to determine how accurately concepts are identified [1]. Other norms commonly tested in sets of pictures include conceptual familiarity, visual complexity, and the typicality of the object. These variables are often normalized because they have a strong influence on many cognitive performances (e.g. object naming) and on the strategies used during image processing.

The need for normative sets of pictures in research is unequivocal and the number of normative studies has rapidly increased in the past years. Indeed, at least 12 new normative sets of pictures, including 2 sets intended to complement older sets [2,3], were developed between 2000 and 2009 [2–13] and 9 new normative sets of pictures were published since 2010 [14–22]. Each set is unique about the features of the visual stimuli it includes and the normative dimensions it provides. For example, in Viggiano and colleagues [13]'s dataset, stimuli were normalized in color and in greyscale tones. Op de Beeck and Wagemans [11]'s dataset includes multiple exemplars of each object. Adlington and colleagues [4]'s set includes concepts and images with a broad range of item difficulty and semantic subcategories. Finally, the sets of Barbarotto and colleagues [5] and Magnié and colleagues [10] present imaginary objects, created by combining different objects together. Some sets also offer stimuli normalized for specific visual attributes of the images (e.g. luminosity as opposed to familiarity or visual complexity). For example, the Amsterdam Library of Object Images (ALOI) is a color image set with a large number of images varying in angle, illumination and color. Other sets, such as that of Verfaille and Boutsen [23], use objects in 3-dimensional space instead of line drawings or 2-dimensional images.

The choice between sets of stimuli is made based on each set's distinctive character and stimulus type. Researchers must first decide whether line drawings or photos of objects are to be used as stimuli. An increasing number of researchers opt for photos of stimuli, highlighting the need for more ecological stimuli. Photos offer a more realistic depiction of everyday concepts. They provide great depth and richness, which potentially influences the way in which the stimulus is attended, memorized and acted upon [24–27]. Using photos as the experimental stimuli increases the chances of activating the same neuronal circuits that are activated in daily tasks. Line drawings, such as those created in 1980 by Snodgrass and Vanderwart [28], may also be privileged depending on the researchers' objective. Line drawings offer a simple and prototypal depiction of concepts, free of details (e.g. color, texture, or 3D cues) that could influence their naming and visual processing. Moreover, line drawings are easier to modify than photos of real objects in order to create additional experimental conditions. They can be made more difficult to recognize by fragmenting their line contours [29] and imaginary and impossible objects can easily be drawn [5,10].

Once researchers have chosen the type of stimuli they want to use, they have to decide which dimensions they want to control or manipulate in order to determine the set that best suits the needs of the experiment. The number of stimuli available is certainly an important feature that researchers must consider. Experiments often require hundreds of stimuli, especially those including multiple testing sessions, such as memory tasks and experiments involving recording of electrophysiological brain activities. The number of stimuli can be even more crucial for experiments requiring specific types of concepts, such as experiments including specific semantic categories. For example, if the selection of stimuli is limited to the category of fruits and vegetables, only 24 out of the 260 concepts from the Snodgrass and Vanderwart [28]'s set can be used. This issue is usually overcome by combining stimuli from different sets [2–3,30]. However, this practice increases the heterogeneity of the visual parameters and norms.

To our knowledge, the Bank of Standardized Stimuli (BOSS) [15] is the set offering the highest number of normative stimuli (see http://sites.google.com/site/bosstimuli/). It currently includes 538 normative photos of high quality color resolution. In Brodeur and colleagues (2010), stimuli that had a name agreement below 20% or were unrecognized by at least 20% of the participants were excluded from the analyses. Norms presented in this article were thus limited to 480 stimuli. These norms were for the name, familiarity, visual complexity, manipulability, object agreement and viewpoint agreement. Norms are described in more details below. The BOSS, however, does not include some categories that might be useful to researchers, such as animals, vehicles, and buildings. Moreover, a set of 538 images might still be insufficient for some experiments.

The present project further developed the BOSS by adding 930 normative photos. These photos increased the number of stimuli in the existing categories, and offer new categories including animals, building infrastructures, body parts, and vehicles. Differences of norms across categories as well as differences between males and females were also examined. Intrinsic (e.g. biological, neuropsychological, etc.) and extrinsic (e.g. social activities, exposure to specific stimuli, etc.) characteristics of men and women could indeed influence the way they name and rate the concepts. Surprisingly, this has not yet been examined in normative studies.

Materials and Methods

Participants

Participants, whose first language is English, were recruited through ads published in journals and newspapers, and via online classifieds such as Craigslist and Kijiji. A total of 141 participants between the ages of 18 and 55 participated in the project. They each participated in one of four normative studies. The subgroups participating in studies 1, 2, 3, and 4 respectively included 42 participants (22 female, mean age: 25.2, SD: 7.5), 33 participants (15 female, mean age: 30.7, SD: 9.3), 32 participants (17 female, mean age: 28.3, SD: 9.9), and 34 participants (15 female, mean age: 30.5, SD: 10.0).

Ethic Statements

This project was approved by the Research Ethics Board of the Douglas Institute and all participants gave their written consent. Their names were not written anywhere in order to secure confidentiality. Prior to the normative session, participants were told that they were free to interrupt their participation at any time and for any reason. Participants were compensated for their time.

Stimuli

The 930 new colored photos are all concepts that were not in the original BOSS, except for the cork, ice cube, kiwi, lollipop, mug, and recorder. These concepts were re-normalized by presenting new photos that were considered of better quality than those used in the original BOSS. The new photos depicted concepts of categories that were lacking in the original BOSS, including animals, building infrastructures, body parts, and vehicles. The number of concepts for other categories was significantly increased such as musical instruments, furniture, and weapons. The new set of 930 photo stimuli was created through a 5-step procedure, identical to the procedure used to generate the images for the first phase of the project [15]. Some objects were gathered, cleaned and digitally photographed one at a time in a box that uniformly diffused the light provided by two projectors. Other objects however, were photographed as part of a bigger scene and were then cut out of their backgrounds. These photos were taken in many locations. Consequently, the environmental conditions of the photos were not always uniform. The majority of animal photos were taken in museums and zoos. Few photos were taken from the internet and were generously donated to the project by their authors. Adobe Photoshop (Adobe Systems Inc., San Jose, U.S.A.) was used for image editing, including lighting adjustments and the cutting out of the objects. Examples of photos are presented in Figure 1.

General Procedure

Stimuli were presented using the software E-Prime 2.0. Participants were tested individually in a room equipped with one desktop computer and one laptop. The desktop was set up with E-Prime and the experiment's instructions. This computer was used for the stimuli presentation. The photos were presented in 500×500 pixels, centered on the computer screen. On the laptop screen, a blank response sheet was shown in which subjects recorded their responses by writing the name, selecting a category among a list, or entering a value between 1 and 5 on the keyboard. The response sheet was anonymous. The order of the stimuli in each study was random and differed across participants.

Study 1. The goal of this study was to normalize the new 930 photo stimuli for name, familiarity and visual complexity. Prior to the experiment, instructions were given orally and a written version was given to each participant. The first task was to "Identify the object as briefly and unambiguously as possible by writing only one name, the first name that comes to mind. The name can be composed of more than one word". Participants were told to write DKO (don't know object) if they had no idea what the object was. If they knew the object but not the name, they wrote DKN (don't know name) and if they knew the name but were

Figure 1. Examples of stimuli from the animal, food, body part, musical instrument, hand labour tool and accessory, vehicle, and weapon and war related item categories.

unable to retrieve it at that moment, they wrote TOT (tip-of-the-tongue).

For familiarity, participants were asked to "Rate the level to which you are familiar with the object". Responses were provided on a 5-point rating scale with 1 indicating very unfamiliar and 5 very familiar. Participants were asked to rate the concept itself and not the picture of the object. Responses were not required for the objects for which they responded DKO.

For visual complexity, participants were asked to "Subjectively rate the level to which the image appears to be complex in terms of the quantity of details and the intricacy of the lines". on a 5-point scale with value 1 indicating a very simple image and 5, a very complex image.

Images were presented one at a time and participants could change to the next image at their own pace, meaning that there was no set amount of time for the participants to see each image. Participants were unable to go back to previous images. For each concept, participants first wrote the name in one column and then provided their rating for familiarity and visual complexity rating in the two next columns of the response sheet. For both familiarity and visual complexity, participants were reminded to use the entire 5-point rating scale and not only its end points.

Study 2. The goal of this study was to normalize the photo stimuli for category agreement, which is the extent to which they are representative of their category. In the 2010 normative study (original BOSS) [15], participants classified each object within the most appropriate of 18 categories. This proved problematic when objects fell under more than one category heading. For example, a toy tank could be classified either within the weapon and war related category or within the games, toys and entertainment category. To avoid this problem, the participants in the present study had the possibility to classify the concept within two categories. Considering the change of instructions for this study, the categorization was performed for the 930 new photos as well as for the original 538 normative photos summing to 1468 categorizations.

Categories were created in a drop down box in an excel sheet in alphabetical order. The instructions read, "Determine to which category the concept belongs". Participants were asked to make a choice among the following five categories: animal, body part, building infrastructure, object, and vehicle. When they chose animal or object, participants were presented with a list of more specific categories allowing them to refine their selection. The list

of animals included bird, canine, crustacean, feline, fish, insect, mammal, reptile, and sea mammal. The list of objects included building material, clothing, decoration and gift accessory, electronic device and accessory, food, furniture, game toy and entertainment, hand labour tool and accessory, household article and cleaner, jewel and money, kitchen item and utensil, medical instrument and accessory, musical instrument, natural element, outdoor activity and sport item, skincare and bathroom item, stationary and school supply, weapon and war related item.

Study 3. The goal of this study was to normalize the photo stimuli for image agreement, which is the degree to which the mental image generated from the modal name (the name most commonly used), matched the object stimulus. Image agreement was separated into object and viewpoint agreement, meaning that participants had to decide to which extent the mentally generated concept was structurally similar to the photo concept (image agreement) and the extent the two concepts had comparable positions (viewpoint agreement).

For each concept, its name was first presented in black 14-point Times New Roman, centered on the computer screen. This name featured the modal name, which is the name that reached the greatest name agreement, as determined by the results from study 1. Only the 464 stimuli for which at least 21 participants (50%) gave the modal name in study 1 were normalized for object and viewpoint agreement. Following the appearance of the name, participants had to generate a mental image of the concept related to the name, after which, they pressed the space bar and the photo appeared. Participants were then asked to rate image agreement and viewpoint agreement. For object agreement, participants were asked "How closely does the picture of the BOSS resemble the mental image you had for the object name, independently from its position?" For viewpoint agreement, participants were asked to determine "How closely does the object of the BOSS match the position of the object you imagined?" In both tasks, participants had to provide a rating from 1 to 5, 1 corresponding to a low agreement and 5 corresponding to a high agreement. An example of low and high object and viewpoint agreements were presented before the session began.

Study 4. In the last study, all 930 stimuli were presented to participants at their own pace in order to rate the manipulability of the concept. Participants were instructed to determine "Could you easily mime the action usually associated with this object so that any person looking at you doing this action could decide which object is associated with this action?" Responses were provided on a 5-point rating scale where 1 was a definite "no" response and 5 was a definite "yes" response. Participants were instructed to use the entire scale and not only its end points.

Data analyses

Modal name and name agreement. For each image, the names provided by participants were analyzed after first excluding the data for which participants had responded DKN, DKO, or TOT. The name given by the highest percentage of participants was considered the modal name. The percentage of participants who agreed on the modal name is the name agreement. In the case where two names had the same percentage of responses, the most specific name for the object was used (e.g. plastic cup as opposed to cup). Composite names in which the order of the words was rearranged (e.g. ham slice or slice of ham) were considered to be the same name.

H value. The H value for each object was computed. The statistic H is a value sensitive to the number and weight of alternative names. It is computed with the following formula [28]:

$$H = \sum_{i=1}^{k} P_i \log_2(1/P_i)$$

Where k refers to the number of different names given to each picture and excludes the DKN, DKO, and TOT responses, and P_i is the proportion of participants that gave a name for each object. This proportion varies across pictures because of the exclusion of the DKN, DKO, and TOT responses. The H value of a picture with a unique name and no alternative is 0. The H value of a picture with two names provided with an equivalent frequency is 1.00. This value is smaller for an alternative that is provided to a lower frequency rate. On the other hand, the H value increases as a function of the number of alternatives. For instance, one picture with its modal name provided by 50% of participants and two alternative names each with a frequency of 25% would have an H value of 1.50.

Modal category and category agreement. The modal category and category agreement were computed following the same procedure used for the names. These statistics were computed on the first category selected by the participants. A second category was rarely selected by participants and was considered only when two or more categories were selected at the same frequency for a stimulus. However, the second response was not added to the percentage of agreement.

H$_{cat}$ value. An H value for the category, referred to as a H_{cat} value, was measured following the same procedure used for the names.

Variables rated on a 5-point scale. Familiarity, visual complexity, object agreement, viewpoint agreement, and manipulability were computed by averaging the scores on the 5-point rating scale and by calculating the standard deviations.

Statistical analyses. Means and standard deviations were analyzed using independent sample t-tests, with the stimuli as for the participants and the categories as for the between-"stimulus" variables. Comparisons of categories were limited to the most commonly used and studied categories in cognitive science. The categories included animal, food, tool, musical instrument, weapon and vehicle. Tool, musical instrument, weapon, and vehicle are typically used as non-living or man-made concepts and are generally opposed to food and animal that are used as living or natural concepts. Because many food items of the BOSS are non-living (e.g. bottle of wine), a second category of food was created

for the analyses which included only fruits, vegetables, and nuts. Categories analyzed thus consisted of animal (i.e. all animals collapsed together, except for the mussel, the seashell, and the fish skeleton), food, hand labour tool and accessory, musical instrument, vehicle, and weapon and war related item, as well as a seventh category including fruit/vegetable/nuts. Category comparisons were done for all norms except H value, H$_{cat}$ value, and TOT, in order to reduce the number of comparisons. Alpha threshold was Bonferroni corrected to .00003 for multiple (189) comparisons.

Comparisons between genders were also performed with independent sample t-tests. Samples opposed stimuli responded by males and by females. Gender differences were examined for the mean norms and within each of the seven categories retained for the analyses. Alpha threshold was Bonferroni corrected to .0042 for multiple (12) comparisons.

Results

Norms

Table 1 summarizes the agreement and ratings obtained for each normative dimension. The stimulus-specific norms are presented in supporting Tables S1 and S2. In these tables, photo stimuli are sorted as a function of their filename, which at times, differs from the modal name and is more precise. All norms except those related to category are listed in Table S1. Categories, category agreements and H$_{cat}$ for all stimuli, including the 538 photos of phase I, are presented in Table S2.

Norms per categories

The norms for each category, computed for all 1468 photos of the BOSS, are presented in Table 2. The first comparisons of categories, carried out on the categories in the upper part of Table 2, were conducted to determine whether some types of concepts were more difficult to recognize or name than others. DKO was significantly higher for tools (t(201) = 5.150, p<.00001) and weapons (t(201) = 5.150, p = .00003) than for animals. Tools were more difficult to name than all categories (all p<.00003) except musical instruments.

The next comparisons looked at differences of modal name and category agreement. Animals and fruits/vegetables/nuts were named with a relatively similar consensus and their modal name agreement was significantly higher (all p<.00003) than that for tools and vehicles, which yielded more inconsistent names. The modal name agreement for fruits/vegetables/nuts was also significantly higher than for foods (t(166) = 5.326, p<.00001). The lower name agreement for tools was contingent to the lowest category agreement. Tools were classified more inconsistently than animals, foods, fruits/vegetables/nuts, and musical instruments (all p<.00001). In contrast, fruits/vegetables/nuts were classified more consistently than all other categories (all p<.00001), except musical instruments.

The least familiar category was that of weapons. They were significantly different from vehicles, foods, and fruits/vegetables/nuts (all p<.00001). The most complex stimuli were animals which were rated significantly higher than all other categories (all p<.00001). Vehicles were also more visually complex than foods, fruits/vegetables/nuts, and weapons (all p<.00001). Finally, musical instruments were more complex than foods (t(211) = 5.907, p<.00001) and fruits/vegetables/nuts (t(117) = 5.127, p<.00001).

Object agreement was the highest for fruits/vegetables/nuts, meaning that the photos in this category matched the mental image evoked by the concepts to a larger extent than the other

Table 1. Norms.

| Normative dimension | Male (n = 20) | | | Female (n = 22) | | Gender |
	Mean	SD	SD	Mean	SD	comparison
Modal Name Agreement*	58%	25%	25%	61%	26%	t = 2.240***
H value*	1.89	1.06	0.97	1.53	0.09	t = 4.255***
DKO	3%	7%	6%	4%	10%	t = 4.903***
DKN	5%	8%	8%	6%	9%	t = 4.115***
TOT	3%	4%	3%	4%	7%	t = 12.458***
Category Agreement*	76%	21%	21%	77%	22%	t = 0.450
H_{cat} value*	0.97	0.75	0.72	0.87	0.75	t = 0.590
Familiarity	4.16	0.50	0.51	4.16	0.55	t = 0.147
Visual Complexity	2.43	0.54	0.52	2.49	0.59	t = 4.905***
Object Agreement**	3.69	0.52	0.54	3.57	0.57	t = 4.799***
Viewpoint Agreement**	3.60	0.44	0.45	3.45	0.51	t = 8.240***
Manipulability	2.57	0.78	0.84	2.67	0.76	t = 4.527***

*The modal name and category of males and females were not systematically the same, thus explaining why the norms of all subjects do not correspond to the averages of the two subgroups.
**Statistics for 464 stimuli.
***p<.0001.

stimuli. Object agreement for foods was significantly higher than all other categories except for musical instruments (all p<.00001). Viewpoint agreement was also the greatest for fruits/vegetables/nuts and for foods in general. These categories had a viewpoint agreement significantly higher than animals and vehicles (all p<.00001).

Finally, large differences were found with respect to manipulability. Foods, fruits/vegetables/nuts, and animals had manipulability ratings that were significantly smaller than all other categories (all p<.00001). In addition, musical instruments, which had the highest rating, were significantly more manipulable than vehicles (t(107) = 7.789, p<.00001).

Norms per sex

Norms of males and females and the statistics resulting from their comparisons are presented in Table 1. Modal name agreement, DKO, DKN, and TOT were all significantly higher in females than in males. Females also rated visual complexity and manipulability with higher scores than males. In contrast, males provided significantly higher scores for object and viewpoint agreement than females. No differences were denoted for category agreement and familiarity.

Table 3 presents the norms of males and females within seven categories. Although they were not systematically significant, differences between genders were consistent with those described in Table 1, except for the tool category. Tools were more familiar to males and named with a higher agreement. Tools were the items that females recognized and named with the greatest difficulty, compared to males. Those difficulties also occurred for weapons, despite a greater modal name agreement for females.

Correlations

As is generally done in normative studies, the relation between the different normative dimensions was examined using correlational analyses. The alpha threshold was Bonferroni corrected and lowered to .0014. Results, which are presented in Table 4, show that the strongest correlations were between the agreement (name

and category) and their respective H value. Name agreement correlated with all other norms except for visual complexity. In Brodeur and colleagues [15], modal name agreement did not correlate with category agreement however, in the present study there was a weak but significant correlation.

Object agreement and viewpoint agreement also exhibited a pattern of results very similar to that found in Brodeur and colleagues [15]. These ratings correlated with name dimensions and familiarity but not with category dimensions and visual complexity. The normative dimensions that differ the most between the present study and Brodeur and colleagues [15] are familiarity and visual complexity, which negatively correlated with each other in the present study. Moreover, familiarity no longer correlated with category agreement whereas visual complexity did. Finally, these two normative dimensions strongly correlated with manipulability.

Discussion

This project proposes 930 new normative photos of concepts from different categories to be added to the 538 photos that already compose the BOSS [15]. The norms for the new set are very similar to those collected for the initial set, except for name agreement, which is slightly lower than in the initial set. This difference is essentially due to the use of more stringent criteria for keeping stimuli in Brodeur and colleagues [15], where only photos with a DKO below 20% and a name agreement above 20% were included in the analyses.

Some norms also differ from those of other normative sets of photos. For instance, Moreno-Martinez and Montoro [20] and Adlington and colleagues [4] had name agreement of 72% and 67%, respectively. The lower name agreement of the BOSS mostly pertains to its high number of stimuli and the inclusion of concepts that are necessarily more difficult to name. As argued in Brodeur and colleagues [15], adding new stimuli is generally associated with a reduction of name agreement. Rating for familiarity was higher than in Moreno-Martinez and Montoro [20] and Adlington and colleagues [4] as well as in most normative

Table 2. Norms as a function of categories.

Category	Nb BOSS 2010	Nb BOSS 2014	Total Nb	DKO	DKN	TOT	NA	H	CA	Hcat	Fam	VC	OA	VA	Manip
Food	93	75	168	4%	4%	1%	63%	1.68	92%	0.38	4.24	2.25	4.05	3.89	1.96
- Fruit, vegetable & nut*	59	15	74	3%	5%	2%	74%	1.19	97%	0.20	4.28	2.25	4.27	3.99	1.66
Hand labour tool & accessory	49	46	95	7%	14%	4%	53%	2.15	70%	1.20	4.03	2.37	3.85	3.67	2.85
Musical instrument	5	40	45	4%	9%	4%	62%	1.65	89%	0.52	4.07	2.60	3.97	3.74	3.58
Vehicle	3	61	64	1%	5%	3%	56%	2.02	76%	0.97	4.24	2.74	3.60	3.40	2.59
Weapon & war related item	1	35	36	6%	6%	2%	59%	1.90	76%	1.04	3.73	2.24	3.65	3.56	2.90
Animal**	1	141	142	1%	5%	2%	71%	1.23	89%	0.49	4.02	3.09	3.77	3.50	1.89
- Bird	0	32	32	1%	5%	1%	72%	1.22	97%	0.18	4.01	3.16	3.74	3.46	1.85
- Canine	0	8	8	1%	6%	1%	63%	1.65	82%	0.70	4.20	3.00	3.46	3.45	2.38
- Crustacean	2	6	8	5%	7%	4%	54%	1.81	68%	1.32	3.88	3.13	3.57	3.69	1.88
- Feline	0	10	10	0%	2%	2%	62%	1.64	78%	0.93	4.27	3.10	3.83	3.58	2.06
- Fish	0	16	16	2%	7%	1%	75%	1.14	78%	0.94	3.71	3.02	3.44	3.49	1.74
- Insect	0	13	13	1%	1%	1%	73%	1.12	92%	0.40	4.18	3.06	3.86	3.38	1.82
- Mammal	0	42	42	1%	4%	2%	76%	1.00	93%	0.31	4.15	3.07	3.98	3.64	1.94
- Reptile	0	9	9	0%	3%	1%	60%	1.34	95%	0.29	3.98	3.21	3.31	3.34	1.98
- Sea mammal	0	7	7	2%	5%	2%	67%	1.49	76%	0.96	4.03	2.86	4.10	3.36	1.67
Body part	0	18	18	0%	0%	0%	69%	1.29	82%	0.78	4.67	2.57	4.01	3.64	3.14
Building infrastructure	0	96	96	3%	7%	4%	54%	2.17	67%	1.36	4.23	2.26	3.48	3.65	2.21
Building material	15	12	27	6%	11%	4%	49%	2.13	57%	1.64	4.35	2.18	3.81	3.60	2.19
Clothing	33	31	64	2%	3%	1%	63%	1.70	83%	0.70	4.13	2.18	3.65	3.56	2.93
Decoration & gift accessory	31	49	80	3%	6%	2%	56%	1.99	68%	1.24	4.05	2.52	3.45	3.61	2.21
Electronic device & accessory	44	47	91	4%	5%	2%	50%	2.30	68%	1.28	4.13	2.64	3.77	3.62	2.64
Furniture	3	32	35	1%	2%	2%	58%	1.87	78%	0.91	4.49	2.06	3.39	3.66	2.95
Game, toy & entertainment	27	40	67	3%	7%	2%	52%	2.15	76%	0.94	4.24	2.32	3.78	3.66	2.72
Household article & cleaner	38	21	59	2%	6%	2%	57%	2.02	55%	1.77	4.26	2.21	3.81	3.62	2.70
Jewel & money	7	1	8	2%	1%	1%	77%	1.07	60%	1.34	4.37	2.57	3.75	3.75	3.09
Kitchen & utensil	71	46	117	3%	8%	2%	53%	2.19	80%	0.88	4.23	2.23	3.86	3.78	2.62
Medical instrument & accessory	12	2	14	5%	9%	3%	57%	1.99	67%	1.38	4.24	2.41	3.94	3.73	3.11
Natural element	9	31	40	2%	3%	2%	65%	1.52	82%	0.74	4.14	2.61	3.56	3.66	2.02
Outdoor activity & sport item	19	75	94	4%	5%	3%	61%	1.73	72%	1.14	4.19	2.29	3.86	3.74	3.26
Skincare & bathroom item	34	11	45	2%	4%	2%	65%	1.60	74%	1.11	4.16	2.25	3.90	3.57	3.38
Stationary & school supply	42	18	60	1%	5%	3%	59%	1.87	75%	1.05	4.29	2.12	3.98	3.62	2.63

Nb = Number of stimuli, DKO = Don't know object, DKN = Don't know name, TOT = Tip-of-the-tongue, NA = Modal name agreement, H = H value, Fam = Familiarity, VC = Visual Complexity, CA = Category Agreement, Hcat = H value for category, OA = Object Agreement, VA = Viewpoint Agreement, Manip = Manipulability.
*Fruit, vegetable, and nut category was statistically compared with the other categories.
**All animals were collapsed together and compared statistically with the other categories. Subgroups of animals were not statistically compared.

Table 3. Norms as a function of gender.

Category	Gender	DKO	DKN	TOT	NA	H	CA	Hcat	Fam	VC	OA	VA	Manip
Food	Male	5%	3%	1%	54%	1.82	89%	0.46	4.29	2.23	3.81	3.86	2.08
(n = 75)	Female	4%	4%	2%	57%	1.63	92%	0.32	4.28	2.23	3.68	3.61	2.30
Fruit/vegetable/nut	Male	5%	5%	1%	52%	1.85	94%	0.28	4.27	2.21	3.91	4.18	1.86
(n = 15)	Female	4%	5%	4%	63%	1.43	98%	0.12	4.30	2.28	3.66	3.58	2.06
Tool	Male	3%**	8%**	1%**	62%*	1.67	74%	1.01	4.11*	2.24	3.84	3.84*	3.00
(n = 46)	Female	13%***	14%***	8%**	52%*	1.86	72%	1.07	3.72*	2.42	3.69	3.53*	2.83
Musical instrument	Male	4%	7%	2%**	64%	1.46	90%	0.46	4.01	2.47	4.09	3.89**	3.63
(n = 40)	Female	4%	10%	6%**	62%	1.41	91%	0.38	4.03	2.68	3.81	3.41**	3.62
Vehicle	Male	0%	5%	1%**	57%	1.82	75%	0.88	4.24	2.69	3.73*	3.59**	2.46
(n = 61)	Female	2%	7%	5%**	63%	1.52	79%	0.83	4.21	2.80	3.42*	3.21**	2.69
Weapon	Male	2%*	3%*	1%**	55%	1.87*	77%	0.98	3.86	2.18	3.88*	3.76*	3.05
(n = 35)	Female	10%*	9%*	4%**	65%	1.37*	77%	0.88	3.60	2.22	3.47*	3.34*	2.77
Animal	Male	1%	5%	1%*	70%	1.22	89%	0.45	3.94**	2.90**	3.88**	3.64**	1.70**
(n = 141)	Female	2%	4%	2%*	72%	1.06	89%	0.44	4.11**	3.24**	3.66**	3.37**	2.13**

DKO = Don't know object, DKN = Don't know name, TOT = Tip-of-the-tongue, NA = Modal name agreement, H = H value, Fam = Familiarity, VC = Visual Complexity, CA = Category Agreement, Hcat = H value for category, OA = Object Agreement, VA = Viewpoint Agreement, Manip = Manipulability.

*The agreement or rating is significantly different (.05< p <.0042) from the agreement or rating of the other gender.

**The agreement or rating is significantly different (p<.0042) from the agreement or rating of the other gender.

Table 4. Matrix of correlations.

	NA	H	Fam	VC	CA	Hcat	OA	VA
H	−.952*							
Fam	.340*	−.384*						
VC	.046	−.056	−.338*					
CA	.166*	−.194*	.081	.247*				
Hcat	−.172*	.203*	−.113	−.270*	−.960*			
OA	.288*	−.352*	.354*	−.034	.095	−.096		
VA	.176*	−.168*	.168*	−.097	.006	−.006	.323*	
Manip	.198*	−.223*	.445*	−.349*	.019	−.033	.183*	0.124

NA = Modal Name Agreement, Fam = Familiarity, VC = Visual Complexity, CA = Category Agreement, OA = Object Agreement, VA = Viewpoint Agreement, Manip = Manipulability.
*Significant correlation.

sets using line drawings. The BOSS includes a higher proportion of familiar everyday life objects (e.g. binder, pencil, toaster, etc.) than in these two other studies which, in contrast, offer a greater proportion of categories such as animals. Moreover, Adlington and colleagues [4] included concepts in their set that were intended to cover high, medium, and low familiarity ranges. Object and viewpoint agreements were very similar to the rating of typicality reported in Moreno-Martinez and Montoro [20] and visual complexity was only slightly smaller in the present study. Moreno-Martinez and Montero [20] also reported a higher rating for manipulability but the instructions were significantly different from those used in the present study.

The addition of animals, furniture, vehicles, weapons, musical instruments, and of many other types of concepts has not affected the mean ratings relative to Brodeur and colleagues [15] but it has slightly affected the pattern of correlations between norms. For instance, in contrast to Brodeur and colleagues [15], familiarity was negatively correlated with visual complexity. This negative correlation is consistent with most of the existing sets of images including a wide range of categories [28]. Moreover, manipulability was negatively correlated with visual complexity in the present study whereas this correlation was not significant in Brodeur and colleagues [15]. This new pattern of relationships is likely due to the addition of new categories in the present set. For instance, animals and vehicles, which were not in the original set, are amongst the most complex and the least manipulable concepts of the set. Moreover, the category of furniture was highly familiar but rated as visually simple, a pattern of correlation that contributes to the negative correlation found between familiarity and visual complexity. Overall, correlations found in this study are very similar to those reported in most previous studies using line drawings, likely because the present set includes animals, vehicles, furniture, and additional concepts also used in these other studies. For instance, like in other studies, name agreement correlated with familiarity [2,8,31–32] and norms of image agreement [2,31,33] but not with visual complexity [30–31,33–37]. Accordingly, correlations between norms must thus be examined cautiously as they highly depend on the categories included in the set of stimuli and they may be relatively independent from the stimulus format.

By adding new categories of concepts and by increasing the number of stimuli per category, the present study demonstrated how the norms vary across different categories of concepts. Overall, it was found that animals are easily recognized and named and that they are consistently categorized within their specific sub-categories (i.e. bird, reptile, mammal, etc). Animals are also the most visually complex, most likely due to furs and feathers that represent a rich texture. Most animals in the present set are common but there are also unfamiliar animals such as a fennec, a cuttlefish, and a horseshoe crab. This contributed to increase DKO responses and decrease the familiarity rating. Moreover, some animals were confounded with similar animals, such as the alligator which was recognized as a crocodile, the caribou as a moose, and the falcon as an eagle.

The food category also has distinctive features. Food, and more particularly fruits, vegetables, and nuts are among the concepts that are the easiest to recognize and name, along with the fact that they are also among the most familiar concepts. Fruits, vegetables, and nuts are also the least manipulable in the sense that they are not associated with specific manipulations that allow distinguishing among them. Finally, foods obtained the highest object agreement, which suggests that the BOSS pictures were very consistent with the way people imagined these concepts. This is probably due to the fact that most food items, including fruits, vegetables, and nuts are not man-made, and therefore, are less subject to various designs.

The tool category includes concepts with heterogeneous features which led to a large variability along the different dimensions. There are familiar tools (e.g. leaf rake) which are easily named, categorized, and associated to specific uses and there are unfamiliar tools, such as professional tools (e.g. flooring stapler) which are difficult to name and use. Moreover, category agreement was lowered because some tools can be used for multiple purposes that can be related to another category (e.g. ice scrapper, metal brush, etc.). This heterogeneity across tools calls for caution when interpreting norms and reminds that a mean is not warrant of the individual components of some categories.

For some categories, differences between genders were to be expected. For instance, previous studies showed a naming advantage for females with living things and a naming advantage for males with non-living things [38–39]. Comparisons of genders indicate that females had more difficulty recognizing and naming tools and weapons than males. Tools and weapons were also less familiar to females, although this difference was significant only for tools. This can be explained by a lower interest or use of these types of objects by females in general. Tools and weapons were the only categories with an atypical pattern of gender differences. The typical pattern, found in most categories, consisted in a greater use of the DKO, DKN, and TOT by females, in addition to a higher modal name agreement and a lower H value. Instead of reflecting a naming difficulty, the higher rate of DKO, DKN, and TOT in females could indicate that they tend to avoid giving a name when they think this name is incorrect. This tendency necessarily reduces the variability of names and increases the modal name agreement. Females also rated visual complexity and manipulability with higher scores. The two genders reach comparable agreement when categorizing concepts and rate familiarity similarly. On the other hand, object and viewpoint agreements were higher in males. This could simply be explained by the fact that most photos were selected and taken by a male (i.e. first author).

Norms are fundamental not only to characterize stimuli but also to measure variables that could introduce confounding effects. Confounding effects were demonstrated several times. For instance, Laws and Neve [40] compared living and non-living stimuli and showed that the disadvantage in naming living stimuli was reversed after controlling for familiarity, visual complexity, and name frequency. Similar findings were replicated with other categories and stimulus dimensions [12,41], which led Laws [42] to conclude that: "it is necessary to examine the performance of controls on sets of living and nonliving stimuli that are not confounded by these and other potential artefactual variables" (p. 842). In another study, Fillitier and colleagues [43] reported shorter response times for non-manipulable items compared to manipulable items. When they controlled for familiarity by including only familiar items in their analyses, they obtained the opposite effect. These confounding effects do not discard the existence of an effect inherent to the categories but they underline the importance of fully characterizing the stimuli before drawing conclusions on an effect.

Conclusion

Norms are a precious asset that should be considered when creating experimental conditions in order to control for potential confounding effects. The BOSS now includes 1,468 normative colored photos of various concepts from multiple categories. The BOSS also offers 1,179 non-normative photos depicting other exemplars of the normative concepts, and the normative concepts photographed from different viewpoints. In addition, 275 photos are also available in a black and white line drawing version. Norms collected thus far for the BOSS include those described in the present study as well as norms related to symmetry [44]), color diagnosticity (unpublished), and different actions afforded by the objects including those for grasping, using, and moving the object [45–47]. There are yet no norms on the names of the concepts, such as frequency and age-of-acquisition, but they may be collected in the future. Finally, norms were collected from English native speakers [15] and French native speakers [48]. More information about the BOSS can be found at http://sites.google. com/site/bosstimuli/.

Acknowledgments

We are grateful to Lara Berliner, Sébastien Lagacé, Frédéric Downing-Doucet, and Samantha Burns for having edited a number of stimuli.

Author Contributions

Conceived and designed the experiments: MBB KG. Performed the experiments: MB. Analyzed the data: MBB KG MB. Contributed reagents/materials/analysis tools: MBB KG MB. Contributed to the writing of the manuscript: MBB KG MB.

References

1. O'Sullivan M, Lepage M, Bouras M, Montreuil T, Brodeur MB (2012) North-american norms for name disagreement: pictorial stimuli naming discrepancies. PLoS One 7: e47802.

2. Bonin P, Peereman R, Malardier N, Meot A, Chalard M (2003) A new set of 299 pictures for psycholinguistic studies: French norms for name agreement, image agreement, conceptual familiarity, visual complexity, image variability, age of acquisition, and naming latencies. Behav Res Methods InstrumComput 35: 158–167.

3. Nishimoto T, Miyawaki K, Ueda T, Une Y, Takahashi M (2005) Japanese normative set of 359 pictures. Behav Res Methods 37: 398–416.

4. Adlington RL, Laws KR, Gale TM (2009) The Hatfield Image Test (HIT): a new picture test and norms for experimental and clinical use. J Clin ExpNeuropsychol 31: 731–753.

5. Barbarotto R, Laiacona M, Macchi V, Capitani E (2002) Picture reality decision, semantic categories and gender. A new set of pictures, with norms and an experimental study. Neuropsychologia 40: 1637–1653.

6. Cuetos F, Alija M (2003) Normative data and naming times for action pictures. Behav Res Methods InstrumComput 35: 168–177.

7. De Winter J, Wagemans J (2004) Contour-based object identification and segmentation: stimuli, norms and data, and software tools. Behav Res Methods InstrumComput 36: 604–624.

8. Dell'Acqua R, Lotto L, Job R (2000) Naming times and standardized norms for the Italian PD/DPSS set of 266 pictures: direct comparisons with American, English, French, and Spanish published databases. Behav Res Methods Instrum Comput 32: 588–615.

9. Kremin H, Akhutina T, Basso A, Davidoff J, De Wilde M, et al. (2003) A cross-linguistic data bank for oral picture naming in Dutch, English, German, French, Italian, Russian, Spanish, and Swedish (PEDOI) Brain Cogn 53: 243–246.

10. Magnié MN, Besson M, Poncet M, Dolisi C (2003) The Snodgrass and Vanderwart set revisited: norms for object manipulability and for pictorial ambiguity of objects, chimeric objects, and nonobjects. J Clin Exp Neuropsychol 25: 521–560.

11. Op De Beeck H, Wagemans J (2001) Visual object categorisation at distinct levels of abstraction: A new stimulus set. Perception 30: 1337–1361.

12. Rossion B, Pourtois G (2004) Revisiting Snodgrass and Vanderwart's object pictorial set: the role of surface detail in basic-level object recognition. Perception 33: 217–236.

13. Viggiano MP, Vannucci M, Righi S (2004) A new standardized set of ecological pictures for experimental and clinical research on visual object processing. Cortex 40: 491–509.

14. Brielmann AA, Stolarova M (2014) A New Standardized Stimulus Set for Studying Need-of-Help Recognition (NeoHelp) PLoS ONE. DOI:10.1371/journal.pone.0084373.

15. Brodeur MB, Dionne-Dostie E, Montreuil T, Lepage M (2010) The Bank of Standardized Stimuli (BOSS), a new set of 480 normative photos of objects to be used as visual stimuli in cognitive research. PLoS One 5: e10773.

16. Dan-Glauser ES, Scherer KR (2011) The Geneva affective picture database (GAPED): a new 730-picture database focusing on valence and normative significance. Behav Res 43: 468–477.

17. Denkinger B, Koutstaal W (2014) A set of 265 pictures standardized for studies of the cognitive processing of temporal and causal order information. Behav Res Methods, DOI:10.3758/s13428-013-0338-x.

18. Janssen N, Pajtas PE, Caramazza A (2011) A set of 150 pictures with morphologically complex English compound names: norms for name agreement, familiarity, image agreement, and visual complexity. Behav Res Methods 43: 478–490.

19. Migo EM, Montaldi D, Mayes AR (2013) A visual object stimulus database with standardized similarity information. Behav Res Methods 45: 344–354.

20. Moreno-Martinez FJ, Montoro PR (2012) An ecological alternative to Snodgrass &Vanderwart: 360 high quality colour images with norms for seven psycholinguistic variables. PLoS One 7: e37527.

21. Nishimoto T, Ueda T, Miyawaki K, Une Y, Takahashi M (2010) A normative set of 98 pairs of nonsensical pictures (droodles) Behav Res Methods 42: 685–691.

22. Salmon JP, McMullen PA, Filliter JH (2010) Norms for two types of manipulability (graspability and functional usage), familiarity, and age of acquisition for 320 photographs of objects. Behav Res Methods 42: 82–95.

23. Verfaillie K, Boutsen L (1995) A corpus of 714 full-color images of depth-rotated objects. Percept Psychophys 57: 925–961.

24. Biederman I, Ju G (1988) Surface versus edge-based determinants of visual recognition. Cogn Psychol 20: 38–64.

25. Brodie EE, Wallace AM, Sharrat B (1991) Effect of surface characteristics and style of production on naming and verification of pictorial stimuli. Am J Psychol 104: 517–545.

26. Leder H (1999) Matching person identity from facial line drawings. Perception 28: 1171–1175.

27. Rhodes G, Brennan S, Carey S (1987) Identification and Ratings of Caricatures: Implications for Mental Representations of Faces. Cogn Psychol 19: 473–497.

28. Snodgrass JG, Vanderwart M (1980) A standardized set of 260 pictures: norms for name agreement, image agreement, familiarity, and visual complexity. J Exp Psychol Hum Learn 6: 174–215.

29. Snodgrass JG, Corwin J (1988) Perceptual identification thresholds for 150 fragmented pictures from the Snodgrass and Vanderwart picture set. Percept Mot Skills 67: 3–36.

30. Cycowicz YM, Friedman D, Rothstein M, Snodgrass JG (1997) Picture naming by young children: norms for name agreement, familiarity, and visual complexity. J Exp Child Psychol 65: 171–237.

31. Alario FX, Ferrand L, Laganaro M, New B, Frauenfelder UH, et al. (2004) Predictors of picture naming speed. Behav Res Methods Instrum Comput 36: 140–155.

32. Pompeia S, Miranda MC, Bueno OF (2001) A set of 400 pictures standardised for Portuguese: norms for name agreement, familiarity and visual complexity for children and adults. Arq Neuropsiquiatr 59: 330–337.

33. Sanfeliu MC, Fernandez A (1996) A set of 254 Snodgrass Vanderwart pictures standardized for Spanish: Norms for name agreement, image agreement, familiarity, and visual complexity. Behav Res Methods Instrum Comput 28: 537–555.

34. Barry C, Morrison CM, Ellis AW (1997) Naming the Snodgrass and Vanderwart pictures: Effect of age of acquisition, frequancy, and name agreement. Q J Exp Psychol 50A: 560–585.

35. Berman S, Friedman D, Hamberger M, Snodgrass JG (1989) Developmental picture norms: Relationships between name agreement, familiarity, and visual complexity for child and adult ratings of two sets of line drawings. Behav Res Methods Instrum Comput 21: 371–382.

36. Sirois M, Kremin H, Cohen H (2006) Picture-naming norms for Canadian French: name agreement, familiarity, visual complexity, and age of acquisition. Behav Res Methods 38: 300–306.

37. Weekes BS, Shu H, Hao M, Liu Y, Tan LH (2007) Predictors of timed picture naming in Chinese. Behav Res Methods 39: 335–342.

38. Laws KR (2000) Category-specific naming errors in normal subjects: the influence of evolution and experience. Brain Lang 75: 123–133.

39. Laws KR (2004) Sex differences in lexical size across semantic categories. Pers Individ Dif 36: 23–32.

40. Laws KR, Neve C (1999) A 'normal' category-specific advantage for naming living things. Neuropsychologia 37: 1263–1269.

41. Laws KR (2001) What is structural similarity and is it greater in living things? Behav Brain Sci 24: 486–487.

42. Laws KR (2005) "Illusions of Normality": a Methodological Critique of Category-Specific Naming. Cortex 41: 842–851.

43. Filliter JH, McMullen PA, Westwood D (2005) Manipulability and living/non-living category effects on object identification (2005) Brain Cogn 57: 61–65.

44. Brodeur MB, Chauret M, Dion-Lessard G, Lepage M (2011) Symmetry brings an impression of familiarity but does not improve recognition memory. Acta Psychol 137: 359–370.

45. Lagacé S, Downing-Doucet F, Guérard K (2013) Norms for grip agreement for 296 photographs of objects. Behav Res Methods 45: 772–781.

46. Guérard K, Lagacé S, Brodeur MB (2014) Four types of manipulability ratings and naming latencies for a set of 560 photographs of objects. Behav Res Methods. DOI:10.3758/s13428-014-0488-5.

47. Guérard K, Brodeur MB (2014) Manipulability agreement as a predictor of action initiation latency. Behav Res Methods. DOI:10.3758/s13428-014-0495-6.

48. Brodeur MB, Kehayia E, Dion-Lessard G, Chauret M, Montreuil T, et al. (2012) The bank of standardized stimuli (BOSS): comparison between French and English norms. Behav Res Methods 44: 961–970.

4

Colonization of Onions by Endophytic Fungi and Their Impacts on the Biology of *Thrips tabaci*

Alexander M. Muvea[1,2], Rainer Meyhöfer[1], Sevgan Subramanian[2]*, Hans-Michael Poehling[1], Sunday Ekesi[2], Nguya K. Maniania[2]

1 Institute of Horticultural Production Systems, Section Phytomedicine, Leibniz Universität Hannover, Hannover, Germany, **2** Plant Health Division, IPM cluster, International Centre of Insect Physiology and Ecology, Nairobi, Kenya

Abstract

Endophytic fungi, which live within host plant tissues without causing any visible symptom of infection, are important mutualists that mediate plant–herbivore interactions. *Thrips tabaci* (Lindeman) is one of the key pests of onion, *Allium cepa* L., an economically important agricultural crop cultivated worldwide. However, information on endophyte colonization of onions, and their impacts on the biology of thrips feeding on them, is lacking. We tested the colonization of onion plants by selected fungal endophyte isolates using two inoculation methods. The effects of inoculated endophytes on *T. tabaci* infesting onion were also examined. Seven fungal endophytes used in our study were able to colonize onion plants either by the seed or seedling inoculation methods. Seed inoculation resulted in 1.47 times higher mean percentage post-inoculation recovery of all the endophytes tested as compared to seedling inoculation. Fewer thrips were observed on plants inoculated with *Clonostachys rosea* ICIPE 707, *Trichoderma asperellum* M2RT4, *Trichoderma atroviride* ICIPE 710, *Trichoderma harzianum* 709, *Hypocrea lixii* F3ST1 and *Fusarium* sp. ICIPE 712 isolates as compared to those inoculated with *Fusarium* sp. ICIPE 717 and the control treatments. Onion plants colonized by *C. rosea* ICIPE 707, *T. asperellum* M2RT4, *T. atroviride* ICIPE 710 and *H. lixii* F3ST1 had significantly lower feeding punctures as compared to the other treatments. Among the isolates tested, the lowest numbers of eggs were laid by *T. tabaci* on *H. lixii* F3ST1 and *C. rosea* ICIPE 707 inoculated plants. These results extend the knowledge on colonization of onions by fungal endophytes and their effects on *Thrips tabaci*.

Editor: Ren-Sen Zeng, South China Agricultural University, China

Funding: This study was funded by the BMZ (The German Federal Ministry for Economic Cooperation and Development) through GIZ (Deutsche Gesellschaft für Internationale Zusammenarbeit) through a project grant entitled "Implementation of integrated thrips and tospovirus management strategies in small-holder vegetable cropping systems of Eastern Africa" (Project number: 11.7860.7-001.00, Contract number: 81141840) to which we are grateful. The funders had no role in study design, data collection and analysis, decision to publish, or preparation of the manuscript.

Competing Interests: The authors have declared that no competing interests exist.

* Email: ssubramania@icipe.org

Introduction

In Kenya, onions *Allium cepa* L. (Asparagales: Amaryllidaceae), are grown in all regions by both large- and small-scale farmers, where they have a ready domestic and regional market [1]. Onion thrips, *Thrips tabaci* Lindeman (Thysanoptera: Thripidae), is considered the most economically important pest of onion worldwide [2], [3]. In Kenya, it is present in all onion growing areas and can cause up to 59% loss in yield [4]. Currently, growers manage thrips by applying insecticides which are ineffective due to the cryptic feeding behavior of thrips, overlapping generations and insecticide resistance [5], [6]. Therefore, an integrated approach that includes the use of entomopathogens, cultural practices, host plant resistance and judicious use of insecticides is needed [7], [8].

Entomopathogenic fungi (EPF) are considered as important biocontrol agents (BCAs). They are traditionally applied in an inundative approach [9], but recent studies have shown that EPF play diverse roles in nature including as endophytes [10]. Indeed, the endophytic niche in a plant is a rich source of microorganisms that can directly and indirectly promote plant growth and development through plant defence against herbivorous insects [11] and plant pathogens [12], [13] due to their ability to produce secondary metabolites with biocidal activity [14], [15]. On a wide variety of crops, fungal endophytes have been reported to deter feeding, oviposition and performance of stem boring, sap sucking, chewing, and leaf mining insects [11], [16], [17], [18]. For example, endophytic colonization of banana by *Beauveria bassiana* significantly reduced larval survivorship of banana weevil, *Cosmopolites sordidus* (Coleoptera: Curculionidae), resulting in 42–87% reduction in plant damage [19]. Reduction in feeding and reproduction by *Aphis gossypii* (Hemiptera: Aphididae) has also been reported on cotton endophytically colonized by either *B. bassiana* or *Lecanicillium lecanii* (Hypocreales: Clavicipitaceae) [20].

Advantages of the application of endophytes over conventional foliar application of fungal entomopathogens [7] [21] are the ability to colonize plants systemically, thereby offering continuous protection and enhanced persistence [22]. Moreover, considerably low inoculum is required when applied as seed treatment [23]. However, colonization of a host plant by an endophyte is influenced by the inoculation method, species of fungal endophytes and the host plant species itself. Based on the inoculation technique, the endophytes differ in their ability to colonize different plant parts and to persist over a crop growth cycle [24], [25], [26]. Akello et al. [25] reported a higher colonization of tissue-cultured banana by *B. bassiana* through dipping roots and rhizome in a conidial suspension as compared to injecting a conidial suspension into the plant rhizome and by growing the plants in sterile soil mixed with *B. bassiana*-colonized rice substrate. Bing and Lewis [24] reported improved colonization of corn plants through foliar spray of conidia as compared to injection of conidia suspension. They also demonstrated the ability of *B. bassiana* to invade maize plants through the epidermis, thereafter persisting in the plant through the entire growing season, which conferred crop resistance against damage by European corn borer. However, information on endophyte colonization of onions and impact of endophytes on thrips infesting onions is not available. Hence, this study aimed to evaluate the efficacy of two procedures, seed and seedling inoculation methods, on colonization of onion plants by fungal endophytes and further assess their impact on infestation by onion thrips, and on thrips feeding and oviposition. Post-inoculation recovery of all the endophytes tested wase highest with seed inoculation method compared to seedling inoculation method, and this method was selected for the additional impact studies with thrips.

Materials and Methods

Ethics statement

The study was not undertaken in a national park or any other protected areas of land. The plants (onion), endophytes and the insect pest (thrips) involved in the study are not endangered or protected species. No specific permits were required to undertake the field studies in the locations mentioned. However, we obtained prior permission from the farmers in whose fields the sampling was undertaken.

Biological material

Fungal isolates. Five fungal isolates (*Clonostachys rosea* ICIPE 707, *Trichoderma atroviride* ICIPE 710, *Trichoderma harzianum* ICIPE 709, *Fusarium* sp. ICIPE 712, and *Fusarium* sp. ICIPE 717 with GenBank Accession Nos: KJ619987, KJ619990, KJ619989, KJ619992 and KJ619993, respectively) were used in this study. The endophytes were isolated from onion plants asymptomatic of any pathogenic infection, collected during a field survey conducted in different altitudinal gradients of Kenya, namely Nakuru (00.01 N 36.26 E, 2000 m.a.s.l.), Loitokitok (02.71 S 37.53 E, 1200 m.a.s.l.) and Kibwezi (02.25 S 38.08 E, 825 m.a.s.l.) as detailed in the GenBank Accessions mentioned above. Two fungal isolates (*Hypocrea lixii* F3ST1 and *Trichoderma asperellum* M2RT4) isolated from the aboveground parts of maize and sorghum, and previously reported endophytic on maize and bean seedlings [27], were also included. Conidia were obtained from two-week-old cultures grown on potato dextrose agar (PDA) plates. The conidia were harvested by scraping the surface of sporulating cultures with a sterile scalpel. The harvested conidia were then placed in universal bottles with 10 ml sterile distilled water containing 0.05% Triton X-100 and vortexed for 5 min to produce homogenous conidial suspensions. The conidial concentration was determined using Neubauer hemocytometer.

The conidial concentration was adjusted to 1×10^8 conidia mL^{-1} through dilution prior to inoculation of seeds and seedlings.

To assess the viability of the conidia, 100 µL of conidial suspension was inoculated to the surface of two fresh plates of PDA for each isolate. A sterile microscope cover slip (2×2 cm) was placed on top of the agar in each plate before incubation. The inoculated plates were incubated for 24 h at 20°C. The percentage conidial germination was assessed by counting the number of germinated conidia out of 100 in one randomly selected field. Conidia were considered as germinated when germ tubes exceeded half of the diameter of the conidium. The percent germination of the different isolates exceeded 90%, which is recommended by Parsa et al. [28].

Insects. Initial cultures of *T. tabaci* were field-collected from onion plants at the International Centre of Insect Physiology and Ecology (icipe) organic farm. Thrips were reared on snow peas, *Pisum sativum* L. (Fabales: Fabaceae), for over 30 generations in ventilated plastic jars at the icipe's insectary at 25±1°C, 50–60% relative humidity (RH), 12 h L: 12 h D photoperiod.

Onion seeds. Onion can be established using either direct seed sowing or seedling transplanting [29]. Seeds of onion (var. Red Creole) were surface-sterilized in 70% ethanol and then immersed in 2% NaOCl (bleach) for 2 and 3 min, respectively. The seeds were finally rinsed three times using sterile distilled water to ensure epiphytes were not carried on the seed surface. To confirm the efficiency of the surface sterilization methods, 100 µl of the last rinse water [28], [30] was spread onto potato dextrose agar and plates were incubated at 20°C for 14 days. The absence of fungal growth on the medium confirmed the reliability of the sterilization procedure. The seeds were then placed on sterile filter paper to dry for 20 min before being divided into two portions, one for the seed and the other for the seedling inoculation.

Seed and seedling colonization of onion plants by fungal endophytes

Seed inoculation of fungal endophytes. For seed inoculation, 10 g of surface-sterilized seeds were subdivided into eight equal portions whereby seven portions were individually soaked in a conidial suspension of 1×10^8 conidia ml^{-1} of each isolate for 10 hours. In the control, the eighth portion was soaked in sterile distilled water containing 0.05% Triton X-100. The inoculated seeds were air dried on a sterile paper towel for 20 min and then transferred in plastic pots (8 cm diameter×7.5 cm height) containing sterile planting substrate. The substrate was a mixture of red soil and livestock manure in a 5:1 ratio and was sterilized in an autoclave for 2 hr at 121°C and allowed to cool up to ambient temperature before being used. Seeds were sown 1 cm below the surface of the substrate and maintained at room temperature (~25°C and 60% RH) in the screen house. After germination, seedlings were thinned to one per pot for all the eight treatments and the four replicates. The plants were watered once per day in the evening. No additional fertilizer was added to the planting substrate.

Seedling inoculation of fungal endophytes. For seedling inoculation, surface-sterilized seeds, as described earlier, were raised in a plastic bucket (30 cm diameter×28 cm height) with sterile planting substrate and maintained in a screenhouse at room temperature (~25°C and 60% RH) for one month before transplanting. Before transplanting, seedlings (height 7–8 cm) were watered and uprooted carefully to minimize damage to roots. After uprooting, the plants were shaken gently to dislodge excess soil on the roots, which were further washed with running tap water. Roots of four seedlings with well-developed shoots were dipped in each of the seven endophyte conidial suspensions of

1×10^8 conidia ml^{-1} for 10 hours. Control plants were dipped in sterile distilled water containing 0.05% Triton X-100. The inoculated seedlings were transplanted in pots containing sterile soil as described earlier. The experimental design was a completely randomized block design (CRBD) with four replicates. The plants were maintained under similar conditions as those inoculated through the seeds.

Assessment of colonization. To determine colonization by inoculated fungal isolates, onion plants were carefully uprooted from the pots after 50 and 70 days for inoculated seeds and seedlings, respectively. Plants were then washed gently with water. Leaves, stems and roots were separated from each plant. Sections of leaves were sampled from the middle and outer leaves of the plant while the whole lengths of stems and roots were used for sampling. The sampled plant parts were then surface-sterilized by dipping them in 70% ethanol and then immersing in 2% NaOCl for 2 and 3 min, respectively, and rinsed three times using sterile distilled water. The final rinse water was plated on PDA to confirm elimination of epiphytic microorganisms as described earlier. The surface-sterilized plant parts were then aseptically cut into 1 cm lengths under a laminar flow hood. Five randomly selected pieces were placed in uniform distribution on PDA plates amended with antibiotics (tetracycline and streptomycin sulfate salt at 0.05%) [31] and incubated in the dark at 25°C for 10 days, after which the presence of fungal growth was observed. Positive colonization was scored by counting the number of pieces of the different plant parts with growth of inoculated endophyte. To confirm whether the growing endophytes were the ones initially inoculated; slides prepared from the mother plates were used for comparison and morphological identification.

Effect of endophytically colonized onion plants on proportion of thrips observed on plants, feeding punctures and oviposition

Seed inoculation technique was found to be effective for colonization and was therefore adopted for this study. Seeds inoculated with all fungal isolates and a control were transplanted in smaller pots (diameter 8 cm) with one plant per pot until 3- to 5-leaf stage before being used in the experiment. Plants with four fully grown leaves were exposed to one-day-old (presumably mated) adult female thrips (10 individuals) for 72 h in Plexiglas cages (30×30×25 cm) and were maintained at 26±1°C, 50–70% RH and 12L: 12D photoperiod. A total of four cages were used for each treatment. After 72 h, all adult thrips observed on the plants were recorded. The individual plants were cut and placed in labeled polythene paper bags for later quantification of thrips feeding and oviposition activities. Two leaves from each plant were cut into three sections of 4 cm each, from the base, middle and tip of the leaf. The number of feeding punctures was counted under a stereomicroscope and recorded. The sections were stained in boiling lactophenol-acid fuchsin solution [32] for 30–40 mins. After staining, the leaves were placed in 90 mm Petri dishes for 1 h before being destained. Destaining was done by immersing the leaves in warm water for three minutes after which the eggs were counted under a stereomicroscope. Treatments were randomized in complete block design and the experiment replicated four times. Verification of colonization of onions by the endophytes was performed at the end of the experiment.

Data analysis

Binary data on colonization (presence or absence) were fitted in a generalized linear mixed model assuming binomial distribution error and logit using package *lme*4 [33] in R 2.15.2 statistical

software [34]. Treatments were considered as fixed effects and the plant pieces nested within the plant as random effects. The extent of fungal colonization (%) of host plant parts was calculated as detailed below.

$$\text{Colonization}(\%) = \left(\frac{\text{PF}}{\text{TP}} \right) \times 100$$

where – PF – Number of pieces exhibiting fungal growth, TP – Total number of pieces plated out.

The numbers of thrips observed on the onion plants were recorded for all treatments and replicates. Analysis was performed using logistic regression model which was fitted to the data on proportion of thrips recovered 72 h post-exposure using package HSAUR [35] in R 2.15.2. The number of feeding punctures on each leaf section were determined and summed up per plant before staining the leaves for eggs count. All count data on feeding and oviposition of *T. tabaci* were checked for normality and homogeneity of variance using Shapiro-Wilk and Levene tests, respectively, before analysis by negative binomial regression using R 2.15.2 [34] with package MASS [36]. The negative binomial distribution was chosen, based on its biological appropriateness in handling overdispersion in count data. P-values of <0.05 were considered as significant.

Results

Seed and seedling colonization of onion plants by fungal endophytes

The viability tests yielded >90% germination of conidia, for all the isolates. Since the final rinse water did not show any sign of fungal growth on the media, it was concluded that the surface sterilization technique used was effective. All the tested fungal isolates were able to colonize onion plants following seed or seedling inoculation (Figures 1, 2). However, the extent of colonization of the different plant parts depended on the inoculation method and the fungal isolate. Seed inoculation resulted in 1.47 times higher mean percentage post-inoculation recovery of all the endophytes tested as compared to seedling inoculation (F = 11.13; df = 1, 3; p = 0.002). For example, mean colonization of roots by *C. rosea* ICIPE 707 isolate was 75.00±9.7% through seed inoculation and 29.85±3.7% through seedling inoculation. Seed inoculation method resulted in higher mean post-inoculation recovery of all the endophytes tested for roots, stems and leaves (76.06±4.1%, 44.24±3.6% and 44.73±5.4%), respectively (Figure 1). On the other hand, seedling inoculation recorded 55.62±4.5%, 31.75±5.8% and 24.65±6.8% for roots, stems and leaves, respectively (Figure 2).

Effect of endophytically colonized onion plants on proportion of thrips observed on plants, feeding punctures and oviposition

The treatments had a significant effect on the proportion of thrips observed on the onion plants 72 h post-exposure ($\chi^2 = 87.79$, df = 7, p<0.001) (Figure 3). Overall *Hypocrea lixii* outperformed all the other treatments in affecting the proportion of thrips on the plants. Fewer thrips were observed on plants inoculated with *C. rosea* ICIPE 707, *T. asperellum* M2RT4, *T. atroviride* ICIPE 710, *T. harzianum* ICIPE 709, *H. lixii* F3ST1 and *Fusarium sp.* ICIPE 712 isolates as compared to those inoculated with *Fusarium* sp. ICIPE 717 and the control treatments (Figure 3). The number of feeding punctures by *T. tabaci* was significantly lower in all the endophyte-inoculated plants as compared to the control treatment (F = 22.71; df = 7, 21;

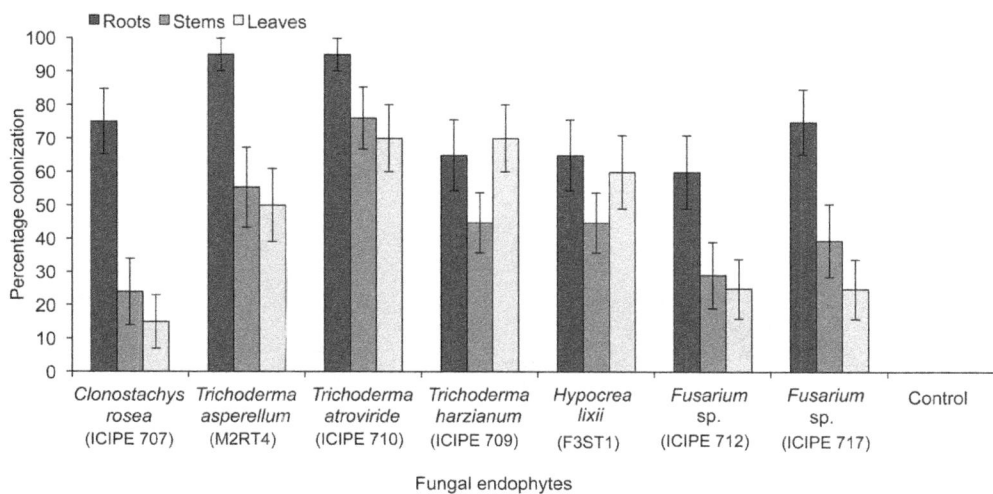

Figure 1. Endophytic colonization of onion seeds. Percentage colonization of onion plant parts (root, stem and leaves) by different fungal endophytes through seed inoculation. Data are percentage mean ± SE. (P≤0.05).

p<0.001; n = 4) (Figure 4). Plants colonized by isolates *C. rosea* ICIPE 707, *T. asperellum* M2RT4, *T. atroviride* ICIPE 710, and *H. lixii* F3ST1 had significantly lower number of feeding punctures as compared to the other treatments (Figure 4).

Highest number of eggs (18.6±2.2) was oviposited by *T. tabaci* in the control plants than in all other endophytically colonized plants (F = 16.75; df = 7, 21; p<0.001) (Figure 5). Among the isolates tested, the lowest numbers of eggs were laid by *T. tabaci* on *H. lixii* F3ST1 and *C. rosea* ICIPE 707 inoculated plants. Plants inoculated with *T. asperellum* M2RT4 and *T. atroviride* ICIPE 710 isolates were equally effective in their capacity to reduce egg laying by *T. tabaci*. *Fusarium* sp. ICIPE 717 colonized plants showed about 6 times higher number of eggs as compared to *H. lixii* F3ST1 (Figure 5).

Discussion

Plant colonization depended on inoculation methods. For instance, seed inoculation method resulted in superior colonization of onion plants as compared to the seedling inoculation. The

difference in colonization between the two may be explained in part by a reduced capacity of uninoculated seedlings to enhance endophyte proliferation due to transplantation shock [37]. Moreover, endophyte inoculation at seed stage could have the advantage of colonizing both seed radical and the plumule, which are close to one another in the seed. Tefera and Vidal [38] reported that seed inoculation of sorghum plants with *B. bassiana* resulted to good endophyte colonization in vermiculate and sterile soil substrates. Seed inoculation could be advantageous in terms of low inoculums requirement as compared to augmentative sprays [23]. Further seed treatment could provide opportunities for endophytic fungi colonization at the young seedling stage for early protection and enhanced seedling health. Backman and Sikora [39] outlined that, integrated pest management on seeds reduces costs and environmental impact, while allowing the biological agent to build up momentum for biological control. Posada et al. [40] found that direct injection of *B. bassiana* conidial suspensions had the highest post-inoculation recovery in coffee seedlings than foliar sprays, stem injections, or soil drenches. Our results show that there were differences in the level of colonization of different

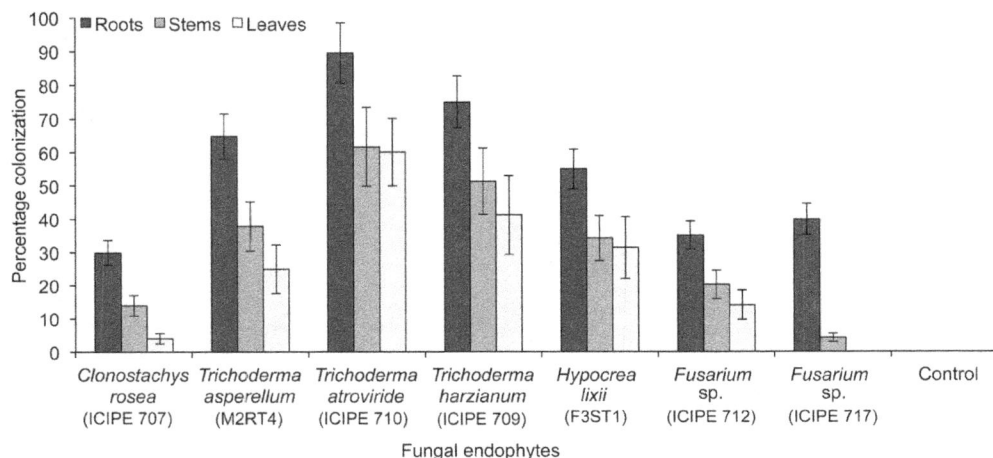

Figure 2. Endophytic colonization of onion seedlings. Percentage colonization of onion plant parts (root, stem and leaves) by different fungal endophytes through seedling inoculation. Data are mean ± SE. (P≤0.05).

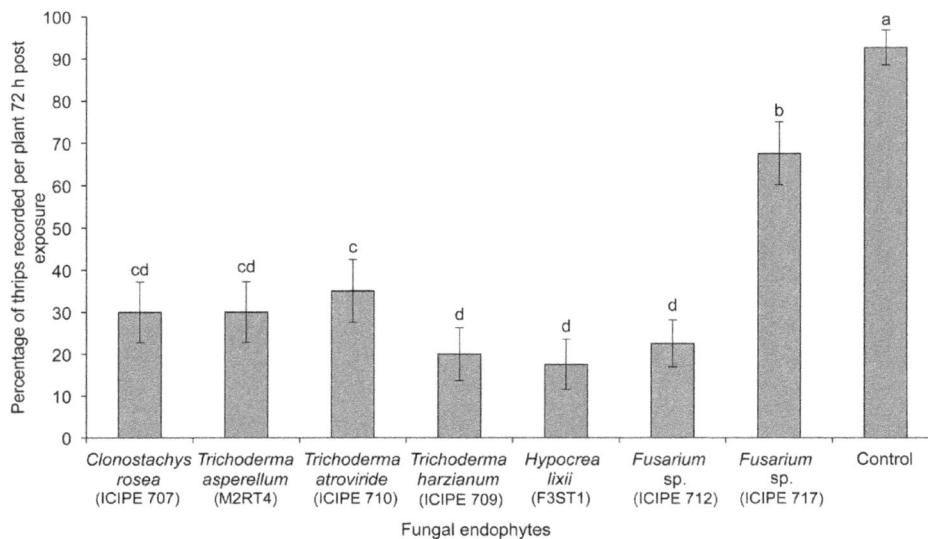

Figure 3. Effect of endophytically colonized onion plants on proportion of adult *Thrips tabaci.* An evaluation of fungal endophytes for their effect on proportion of thrips settling on inoculated onion plants after 72 h. Bars indicate means ± SE at 95% CI. Means followed by the same letter indicate no significant differences between treatments.

plant parts by fungal isolates. For instance, roots sections had higher colonization as compared to stems and leaves. These differences could be due to tissue specificity exhibited by endophytic fungi and their adaptation to particular physiological conditions of the plants [41]. Similar results were reported on French beans and Faba beans [18] and coffee [40].

Among the endophytes that colonized onion plants, *C. rosea* ICIPE 707, *H. lixii* F3ST1, *T. harzianum* ICIPE 709, *T. atroviride* ICIPE 710, and *T. asperellum* M2RT4 had significantly low proportion of thrips, number of feeding punctures and eggs. However, isolate *H. lixii* F3ST1 had the highest overall negative impact on *T. tabaci.* Lately, the impacts of fungal endophytes on suppression of different insect groups in different host plants are receiving increased attention [16], [18], [42]. The negative effect on the proportion of thrips on the endophyte-colonized plants as compared to the control could have been responsible for reduced feeding and oviposition. For instance, Akutse et al. [18] reported that Faba beans colonized endophytically by fungal endophytes of the genera *Hypocrea* and *Beauveria* had significant negative effects on leafminer, *Liriomyza huidobrensis* (Blanchard) fitness, impacting on mortality, oviposition, emergence and longevity of the pest. Cherry et al. [16] found a reduced number of *Sesamia calamistis* (Hampson) in *B. bassiana*-inoculated plants compared to non-inoculated plants. Thrips are able to distinguish among plants as suitable for feeding and/or oviposition sites to ensure fitness of their progeny [43]. *Thrips tabaci* is a key vector of *Iris yellow spot virus* (IYSV) in Kenya [44], [45] and the thrips densities are positively associated with IYSV incidence [46], [47], [48]. Hence, the reduced feeding by the thrips on endophyte-colonized plants could potentially reduce the transmission of IYSV in onions.

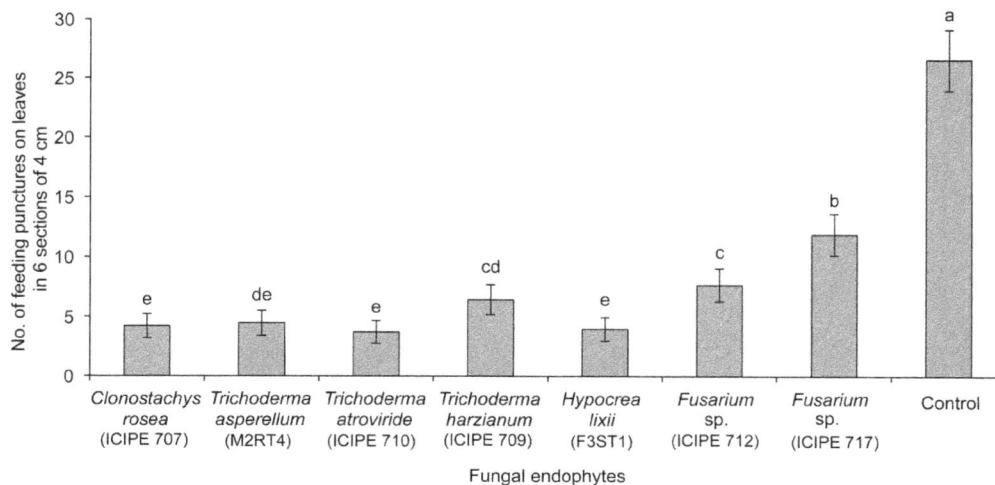

Figure 4. Effect of endophytically colonized onion plants on feeding punctures by adult *Thrips tabaci.* The figure quantifies mean feeding activity by *Thrips tabaci* exposed for 72 h on onion plants inoculated with different fungal endophytes. Bars indicate means ± SE at 95% CI. Means followed by the same letter indicate no significant differences between treatments.

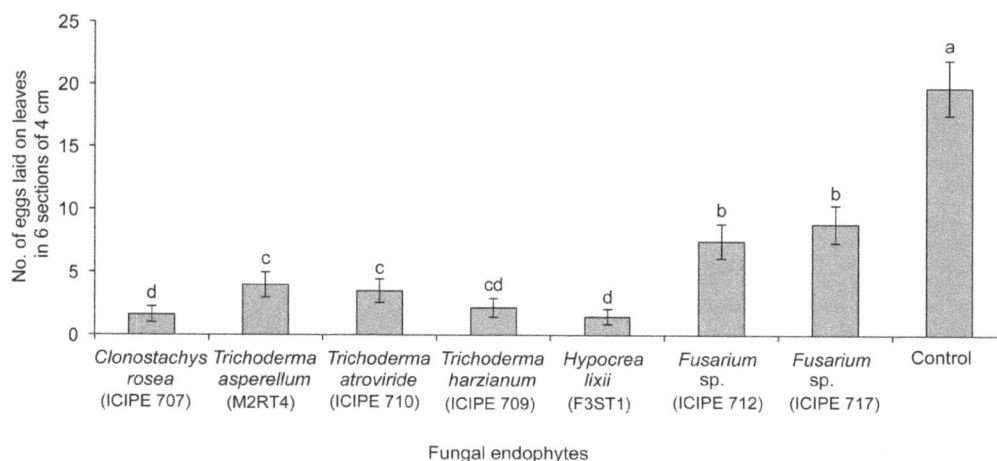

Figure 5. Effect of endophytically colonized onion plants on oviposition by adult *Thrips tabaci*. The figure shows the mean number of eggs laid by *Thrips tabaci* on onion plants endophytically colonized by different fungal isolates. Bars indicate means ± SE at 95% CI. Means followed by the same letter indicate no significant differences between treatments.

Moreover, fungal endophytes can decrease plant virus infections in plants as reported in meadow ryegrass with the *Barley yellow dwarf virus* (BYDV) [49]. The broad array of endophyte induced defence mechanisms in plants against insect pests such as production of toxic or distasteful chemicals [50] and pathogenic interaction to insects [51] could decrease insect fitness [52], a phenomenon that needs to be further investigated.

In the present study, dead insects did not present any signs of mycosis. Previous studies have also revealed that dead insects recovered from endophytically-colonized plants exhibit no signs of fungal infection [16], [18]. The influence of endophytes colonizing onions on thrips biology in terms of observable proportion, feeding and oviposition in the present study are in accordance with the findings by Cherry et al. [16] and Bittleston et al. [53] on reduced feeding and by Akutse et al. [18] on oviposition with other endophytes and pests. The reduced feeding and oviposition could have been a result of either reduced survival of thrips or antixenotic repellence of thrips, phenomena that warrant further studies to unravel the underlying mechanisms such as possible release of metabolites and/or volatiles which could have effects on thrips.

Isolates *C. rosea* ICIPE 707, *H. lixii* F3ST1, *T. harzianum* ICIPE 709, *T. atroviride* ICIPE 710, and *T. asperellum* M2RT4 effectively colonized the various plant parts of onion as compared to the *Fusarium* isolates. Consequently, isolate *H. lixii* F3ST1 had the most antagonistic impact on onion thrips and it could be used

to develop alternative and ecologically safe management strategy for onion thrips. We conclude that, onions can be successfully inoculated especially through seeds, with different fungal endophytes. However, further studies are warranted to determine the persistence of tested endophytes in the colonized plants under natural conditions and investigate potential for vertical transmission of endophytes. Additionally, being the first report of antagonistic activity of endophytes colonizing onion against *T. tabaci*, it would be crucial to determine the underlying mechanisms of such multi-trophic interactions.

Acknowledgments

We thank Drs J. Villinger, H. Fathiya, D. Salifu, J. Nyasani and Mrs. D. Osoga, respectively for fungal endophytes sequence analysis, support in molecular reagents and facilities, data analysis, useful ideas during the manuscript preparation and editorial checks. We thank Ms Elizabeth Ouna for providing assistance in morphological identification of endophytic fungus.

Author Contributions

Conceived and designed the experiments: AMM RM SS HMP SE NKM. Performed the experiments: AMM SS NKM. Analyzed the data: AMM SS RM. Contributed reagents/materials/analysis tools: NKM SS SE. Contributed to the writing of the manuscript: AMM RM SS HMP SE NKM. Obtained funding: SS RM HMP SE NKM. Co-supervised the research work: SS RM NKM HMP.

References

1. Narla RD, Muthomi JW, Gachu SM, Nderitu JH, Olubayo FM (2011) Effect of intercropping bulb onion and vegetables on purple blotch and downy mildew. J Biol Sci 11: 52–57.

2. Nawrocka B (2003) Economic importance and the control method of *Thrips tabaci* Lind. on onion. Bulletin OILB/SROP 26: 321–324.

3. Diaz-Montano J, Fuchs M, Nault BA, Fail J, Shelton AM (2011) Onion thrips (Thysanoptera: Thripidae): a global pest of increasing concern in onion. J Econ Entomol 104: 1–13.

4. Waiganjo MM, Mueke JM, Gitonga LM (2008) Susceptible onion growth stages for selective and economic protection from onion thrips infestation. Acta Hortic 767: 193–200.

5. Martin NA, Workman PJ, Butler RC (2003) Insecticide resistance in onion thrips (*Thrips tabaci*) (Thysanoptera: Thripidae). New Zeal J Crop Hortic Sci 31: 99–106.

6. Morse JG, Hoddle MS (2006) Invasion biology of thrips. Annu Rev Entomol 51: 67–89.

7. Maniania NK, Sithanantham S, Ekesi S, Ampong-Nyarko K, Baumgärtner J, et al. (2003) A field trial of the entomopathogenous fungus *Metarhizium anisopliae* for control of onion thrips, *Thrips tabaci*. Crop Prot 22: 553–559.

8. Shiberu T, Negeri M, Selvaraj T (2013) Evaluation of some botanicals and entomopathogenic fungi for the control of onion thrips (*Thrips tabaci* L.) in West Showa, Ethiopia. J Plant Pathol Microbiol 4: 161. doi: 10.4172/2157-7471.1000161.

9. Ekesi S, Maniania N (2002) *Metarhizium anisopliae*: an effective biological control agent for the management of thrips in horti- and floriculture in Africa. In: Upadhyay R, editor. Advances in microbial control of insects pests. Dordrecht, The Netherlands: Kluwer Academic Publishers. 165–180.

10. Vega FE, Goettel MS, Blackwell M, Chandler D, Jackson MA, et al. (2009) Fungal entomopathogens: new insights on their ecology. Fungal Ecol 2: 149–159.

11. Jallow MFA, Dugassa-Gobena D, Vidal S (2004) Indirect interaction between an unspecialized endophytic fungus and a polyphagous moth. Basic Appl Ecol 5: 183–191.

12. Stone JK, Bacon CW, White JF Jr (2000) An overview of endophytic microbes: endophytism defined. In: Bacon CW, White JF Jr, editors. Microbial endophytes. New York, Basel: Marcel Dekker Inc. 3–29.

13. Ownley BH, Griffin MR, Klingeman WE, Gwinn KD, Moulton JK, et al. (2008) *Beauveria bassiana*: endophytic colonization and plant disease control. J Invertebr Pathol 98: 267–270.

14. Stone JK, White JF Jr, Polishook JD, (2004) Endophytic fungi. In: Mueller G, Bills G, Foster M, editors. Measuring and monitoring biodiversity of fungi: inventory and monitoring methods. Boston, MA: Elsevier Academic Press. 241–270.

15. Strobel G, Daisy B, Castillo U, Harper J (2004) Natural products from endophytic microorganisms. J Nat Prod 67: 257–268.

16. Cherry AJ, Banito A, Djegui D, Lomer C (2004) Suppression of the stem-borer *Sesamia calamistis* (Lepidoptera: Noctuidae) in maize following seed dressing, topical application and stem injection with African isolates of *Beauveria bassiana*. Int J Pest Manage 50: 67–73.

17. Qi G, Lan N, Ma X, Yu Z, Zhao X (2011) Controlling *Myzus persicae* with recombinant endophytic fungi *Chaetomium globosum* expressing *Pinellia ternata* agglutinin: using recombinant endophytic fungi to control aphids. J Appl Microbiol 110: 1314–1322.

18. Akutse KS, Maniania NK, Fiaboe KKM, Van den Berg J, Ekesi S (2013) Endophytic colonization of *Vicia faba* and *Phaseolus vulgaris* (Fabaceae) by fungal pathogens and their effects on the life-history parameters of *Liriomyza huidobrensis* (Diptera: Agromyzidae). Fungal Ecol 6: 293–301.

19. Akello J, Dubois T, Coyne D, Kyamanywa S (2008) Effect of endophytic *Beauveria bassiana* on populations of the banana weevil, *Cosmopolites sordidus*, and their damage in tissue-cultured banana plants. Entomol Exp Appl 129: 157–165.

20. Gurulingappa P, Sword GA, Murdoch G, McGee PA (2010) Colonization of crop plants by fungal entomopathogens and their effects on two insect pests when *in planta*. Biol Cont 55: 34–41.

21. Maniania NK, Ekesi S, Löhr B, Mwangi F (2002) Prospects for biological control of the western flower thrips, *Frankliniella occidentalis*, with the entomopathogenic fungus *Metarhizium anisopliae*, on chrysanthemum. Mycopathologia 155: 229–235.

22. Akello J, Dubois T, Gold CS, Coyne D, Nakavuma J, et al. (2008) *Beauveria bassiana* (Balsamo) Vuillemin as an endophyte in tissue culture banana (*Musa* spp.). J Invertebr Pathol 96: 34–42.

23. Athman SY (2006) Host–endophyte–pest interactions of endophytic *Fusarium oxysporum* antagonistic to *Radopholus similis* in banana (*Musa* spp.), PhD thesis, The University of Pretoria. http://137.215.9.22/bitstream/handle/2263/30187/Complete.pdf?sequence=8. Accessed 2013 December 20.

24. Bing LA, Lewis LC (1991) Suppression of *Ostrinia nubilalis* (Hubner) (Lepidoptera: Pyralidae) by endophytic *Beauveria bassiana* (Balsamo) Vuillemin. Environ Entomol 20: 1207–1211.

25. Akello J, Dubois T, Coyne D, Gold CS, Kyamanywa S (2007) Colonization and persistence of the entomopathogenic fungus, *Beauveria bassiana*, in tissue culture of banana. In: Kasem ZA, Addel-Hakim MM, Shalabi SI, El-Morsi A, Hamady AMI, editors, African Crop Science Conference Proceedings. El-Minia, Egypt: Quick Color Print. Vol 8, 857–861.

26. Brownbridge M, Reay SD, Nelson TL, Glare TR (2012) Persistence of *Beauveria bassiana* (Ascomycota: Hypocreales) as an endophyte following inoculation of radiata pine seed and seedlings. Biol Cont 61: 194–200.

27. Akello J (2012) Biodiversity of fungal endophytes associated with maize, sorghum and Napier grass and their influence of biopriming on resistance to leaf mining, stem boring and sap sucking insect pests. ZEF Ecology and Development series No.86. 137.

28. Parsa S, Ortiz V, Vega FE (2013) Establishing fungal entomopathogens as endophytes: Towards endophytic biological control: J Vis Exp (74), e50360, doi:10.3791/50360.

29. Infonet-Biovision (2013) Onion. Available:. Accessed 2013 November 12.

30. Schultz B, Guske S, Dammann U, Boyle C (1998) Endophyte–host interactions. II. Defining symbiosis of the endophyte–host interaction. Symbiosis 25: 213–227.

31. Dingle J, McGee PA (2003) Some endophytic fungi reduce the density of pustules of *Puccinia recondita* f. sp. *tritici* in wheat. Mycol Res 107: 310–316.

32. Nyasani JO, Meyhöfer R, Subramanian S, Poehling H.-M (2013) Feeding and oviposition preference of *Frankliniella occidentalis* for crops and weeds in Kenyan French bean fields. J Appl Entomol 137: 204–213.

33. Bates D, Maechler M, Bolker B, Walker S (2013) lme4: Linear mixed-effects models using Eigen and S4. R package. Available: http://CRAN.R-project.org/package=lme4. Accessed 2014 May 15.

34. R Development Core Team (2011) R: A language and environment for statistical computing: 18–25. Available: http://www.r-project.org/. Accessed 2014 April 20.

35. Everitt BS, Hothorn T (2013) HSAUR: A Handbook of Statistical Analyses Using R. R package. Available: http://cran.r-project.org/package=HSAUR. Accessed 2014 May 15.

36. Venables WN, Ripley B (2002) Modern applied statistics with S, 4th edn. Available: http://cran.r-project.org/package=MASS. Accessed 2014 May 10.

37. Barrows JB, Roncadori RW (2014) Endomycorrhizal synthesis by *Gigaspora margarita* in poinsettia. Mycologia 69: 1173–1184.

38. Tefera T, Vidal S (2009) Effect of inoculation method and plant growth medium on endophytic colonization of sorghum by the entomopathogenic fungus *Beauveria bassiana*. BioControl 54: 663–669.

39. Backman PA, Sikora RA (2008) Endophytes: An emerging tool for biological control. Biol Control 46: 1–3.

40. Posada F, Aime MC, Peterson SW, Rehner SA, Vega FE (2007) Inoculation of coffee plants with the fungal entomopathogen *Beauveria bassiana* (Ascomycota: Hypocreales). Mycol Res 111: 748–757.

41. Guo LD, Huang GR, Wang Y (2008) Seasonal and tissue age influences on endophytic fungi of *Pinus tabulaeformis* (Pinaceae) in the Dongling Mountains, Beijing. J Integr Plant Biol 50: 997–1003.

42. Vega FE, Posada F, Aime MC, Pava-Ripoll M, Infante F, et al. (2008) Entomopathogenic fungal endophytes. Biol Control 46: 72–82.

43. Brown ASS, Simmonds MSJ, Blaney WM (2002) Relationship between nutritional composition of plant species and infestation levels of thrips. J Chem Ecol 28: 2399–2409.

44. Lehtonen PT, Helander M, Siddiqui SA, Lehto K, Saikkonen K (2006) Endophytic fungus decreases plant virus infections in meadow ryegrass (*Lolium pratense*). Biol Lett 2: 620–623.

45. Birithia R, Subramanian S, Pappu HR, Sseruwagi P, Muthomi JW, et al. (2011) First report of *Iris yellow spot virus* infecting onion in Kenya and Uganda. Plant Dis 95: 1195.

46. Birithia R, Subramanian S, Pappu HR, Muthomi J, Narla RD (2013) Analysis of *Iris yellow spot virus* replication in vector and non-vector thrips species. Plant Pathol 62: 1407–1414.

47. Kritzman A, Lampel M, Raccah B, Gera A (2001) Distribution and transmission of *Iris yellow spot virus*. Plant Dis 85: 838–842.

48. Schwartz HF, Gent DH, Fichtner SM, Hammon R, Cranshaw WS, et al. (2009) Straw mulch and reduced-risk pesticide impacts on thrips and *Iris yellow spot virus* on western-grown onions. Southwest Entomol 34: 13–29.

49. Hsu CL, Hoepting CA, Fuchs M, Shelton AM, Nault BA (2010) Temporal dynamics of *Iris yellow spot virus* and its vector, *Thrips tabaci* (Thysanoptera: Thripidae), in seeded and transplanted onion fields. Environ Entomol 39: 266–277.

50. Tibbets TM, Faeth S (1999) *Neotyphodium* endophytes in grasses: deterrents or promoters of herbivory by leaf-cutting ants? Oecologia 118: 297–305.

51. Marcelino J, Giordano R, Gouli S, Gouli V, Parker BL, et al. (2008) *Colletotrichum acutatum* var. *fioriniae* (teleomorph: *Glomerella acutata* var. fioriniae var. nov.) infection of a scale insect. Mycologia 100: 353–374.

52. Akello J, Dubois T, Coyne D, Kyamanywa S (2008) Endophytic *Beauveria bassiana* in banana (*Musa* spp.) reduces banana weevil (*Cosmopolites sordidus*) fitness and damage. Crop Prot 27: 1437–1441.

53. Bittleston LS, Brockmann F, Wcislo W, Van Bael SA (2011) Endophytic fungi reduce leaf-cutting ant damage to seedlings. Biol Lett 7: 30–32.

5

Transfer of Cadmium from Soil to Vegetable in the Pearl River Delta area, South China

Huihua Zhang[1]*, Junjian Chen[1]*, Li Zhu[2], Guoyi Yang[1], Dingqiang Li[1]*

1 Guangdong Institute of Eco-environmental and Soil Sciences, Guangzhou, China, **2** Management School, Jinan University, Guangzhou, China

Abstract

The purpose of this study was to investigate the regional Cadmium (Cd) concentration levels in soils and in leaf vegetables across the Pearl River Delta (PRD) area; and reveal the transfer characteristics of Cadmium (Cd) from soils to leaf vegetable species on a regional scale. 170 paired vegetables and corresponding surface soil samples in the study area were collected for calculating the transfer factors of Cadmium (Cd) from soils to vegetables. This investigation revealed that in the study area Cd concentration in soils was lower (mean value 0.158 mg kg^{-1}) compared with other countries or regions. The Cd-contaminated areas are mainly located in west areas of the Pearl River Delta. Cd concentrations in all vegetables were lower than the national standard of Safe vegetables (0.2 mg kg^{-1}). 88% of vegetable samples met the standard of No-Polluted vegetables (0.05 mg kg^{-1}). The Cd concentration in vegetables was mainly influenced by the interactions of total Cd concentration in soils, soil pH and vegetable species. The fit lines of soil-to-plant transfer factors and total Cd concentration in soils for various vegetable species were best described by the exponential equation ($y = ax^b$), and these fit lines can be divided into two parts, including the sharply decrease part with a large error range, and the slowly decrease part with a low error range, according to the gradual increasing of total Cd concentrations in soils.

Editor: Manuel Reigosa, University of Vigo, Spain

Funding: Funding was provided by Natural Science Foundation of China (NSFC No. 41171387 and 31270516), the Natural Science Foundation of Guangdong Province, China (No.S2012030006144), the Science and Technology Planning Project of Guangdong Province, China (No.2012A020100003 and 2011B030900013), and Science and Technology Planning Project of Guangzhou (No.2013J2200003). The funders had no role in study design, data collection and analysis, decision to publish, or preparation of the manuscript.

Competing Interests: The authors have declared that no competing interests exist.

* Email: hhzhang@soil.cn (HZ); jjchen@soil.gd.cn (JC); dqli@soil.gd.cn (DL)

Introduction

Cadmium is non-essential element to biota. It is known to be toxic for plants as well as animals to a much higher extent and at lower concentrations than e.g. Zn, Pb, or Cu. It especially affects humans because of their longevity and the accumulation of Cd in their organs by eating Cd-contaminated food [1–3].

Soil Cd is naturally derived from parent materials because of chemical weathering, and as a contaminant in many areas of anthropogenic activities such as mining, smelting, composts, phosphate fertilizer application, waste disposal, and vehicle exhausts [4–7]. The main forms of cadmium in soils and sediments are the exchangeable fraction, followed by the Fe-Mn oxides and residual fractions. Several studies indicated the Cd in soils contaminated by anthropogenic activates such as mining and smelting, seem to be more bioavailable than Cd from unimpacted soils [5]. To protect the safe of soil environment and prevent the soil Cd contamination, various soil quality standards were established in many countries or regions, such as 0.3 mg kg^{-1} in China [8], 0.8 mg kg^{-1} in the Netherlands and Switzerland, 0.5–1 mg kg^{-1} in Austria [9], and 5 mg kg^{-1} in Taiwan [10].

Because soil Cd is easily accumulated by plants, soil-plant-human transfer of Cd has been considered as a major pathway of human exposure to soil Cd [11,12]. Many studies have been conducted in China such as Cd in rice [10,13–15], in orchard [16,17], and in vegetable [18–21]. And the maximum permissible concentrations of Cd of 0.05 mg kg^{-1} for non-environmental pollution vegetable – so called "No-Polluted vegetable" [22] and maximium concentration of 0.2 mg kg^{-1} in leaf vegetables for food security – so called "Safe vegetable" [23] are used in China.

Although undoubtedly soil Cd is the primary source of vegetable Cd, and total soil Cd concentration was commonly used for soil environmental quality estimation, its usefulness to predict soil-to-plant transfer was often questioned since the phytoavailability of Cd in soils [12,21,24]. Yang et al [25] proposed that various vegetable species showed significantly different accumulation capacity for Cd in the same soil sample sites; Soil-to-vegetable transfer factor (TF) of Cd was much higher in a pot experiment than that in a field trial. Previous studies have showed that the TF is decided especially by the soil properties and plant species [26]. Wang et al [12] pointed out that TFs can be considered as a useful index of metals potentially transfer abilities from soil to plant, and the TFs of Cd for leafy vegetables were higher than those for non-leafy vegetables.TF has been widely used in the evaluation of potential health risk of human exposure to metals from soil [3,5,12,15,16].

In the PRD area, the vegetable planting area is about 4830 km^2. The vegetable production was 1129.2$\times10^4$ tons in 2011. The leaf vegetable is the dominant consumption for Guangdong people, the planting area is about 33% of the total planting area [27]. As a report of the Statistics Bureau of Guangdong Province, the export of fresh vegetable was up to 78.3 $\times10^4$ tons [28], mainly sold to Japan, Korea, Malaysia, Russia, and America. Hence, the quality safety of vegetable is not only

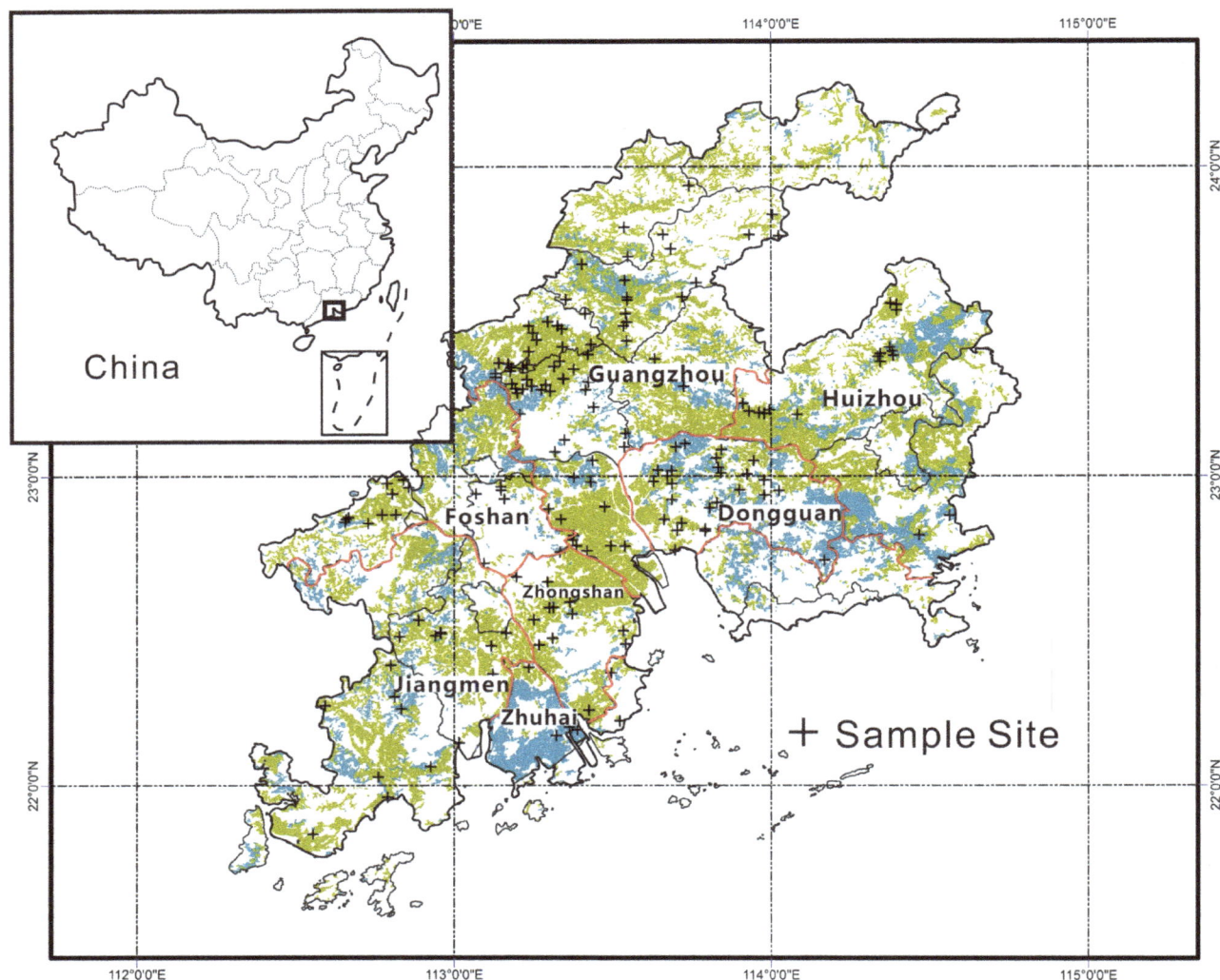

Figure 1. Locations of soil and vegetable samples in the Pearl River Delta area, Guangdong, China. (The yellow and blue areas are the dry lands and the paddy fields, respectively; the red line is the boundary of the city).

concerned by the local government, but it should receive much more attention from the international community for the increasing export and import trade.

In this study, we conducted a systemic environmental quality survey of the vegetable soil and various vegetables in the PRD area. The aim of this study was to: (1) obtain the information on concentrations and spatial patterns of Cd in soils and leaf vegetables across the PRD area; (2) investigate the relationship between soil Cd and vegetable Cd on a regional scale; (3) reveal the transfer characteristic of Cd from soil to various leaf vegetable species.

Materials and Methods

Ethics Statement

This study was carried out on collective-owned lands, and the owners of the lands gave us permission to conduct the study on these sites. The field studies did not involve endangered or protected species.

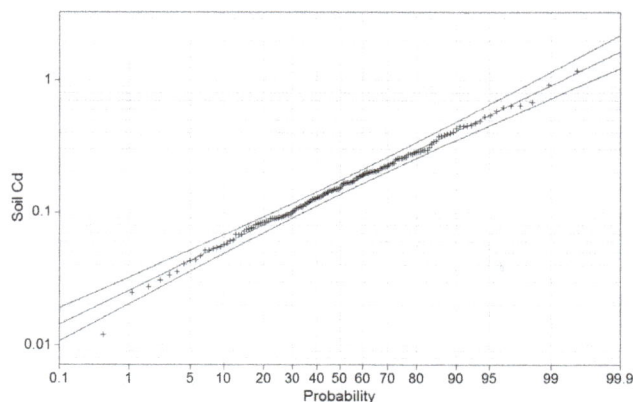

Figure 2. Lognormal probability plot for total Cd concentration in vegetable soils in the Pearl River Delta area, Guangdong, China ($mg\ kg^{-1}$, DW).

Table 1. Descriptive statistics of soil Cd concentrations (mg kg⁻¹, DW) and vegetable Cd concentration (mg kg⁻¹, FW) in the PRD area.

Sample Site	n	Soil Cd concentration			Vegetable Cd concentration			pH (H₂O)	OM[e] (%)
		Mean ± SD	Range	C.V. (%)	Mean ± SD	Range	C.V. (%)	Mean ± SD	Mean ± SD
Dongguan	27	0.138±0.079	0.041–0.433	57.2	0.024±0.012	0.004–0.057	48.4	5.76±0.87	2.58±0.70
Foshan	16	0.291±0.200	0.025–0.685	69.0	0.017±0.011	0.002–0.037	62.5	5.84±1.12	2.61±1.01
Guangzhou	73	0.226±0.190	0.027–1.180	84.3	0.027±0.024	0.002–0.082	86.8	6.34±0.96	2.44±0.66
Jiangmen	17	0.188±0.157	0.012–0.644	83.9	0.014±0.016	0.002–0.068	117.8	5.49±0.67	2.28±0.90
Zhuhai	6	0.204±0.160	0.044–0.474	78.5	0.008±0.005	0.001–0.014	58.1	6.11±0.99	2.34±0.63
Zhongshan	13	0.316±0.150	0.083–0.530	47.3	0.022±0.021	0.001–0.067	92.7	6.02±0.97	2.45±0.90
Huizhou	18	0.107±0.052	0.052–0.200	39.6	0.026±0.018	0.010–0.082	69.1	6.69±0.43	2.41±0.62
Total	170	0.208±0.169 0.158 (1.39)[d]	0.012–1.180	81.2	0.024±0.019 0.015 (1.57)[d]	0.002–0.082	84.3	6.12±0.96	2.42±0.74
Threshold		0.3[a]			0.05[b], 0.2[c]				

[a]The maximum permissible concentrations of Cadmium for agriculture soils (SEPAC, 1995).
[b]the maximum permissible concentrations of Cadmium for non-environmental pollution vegetable (AQSIQ, 2001).
[c]The maximum level of Cadmium for leaf vegetables (MOH, 2012).
[d]Geometric Mean (Geometric standard deviation).
[e]Orgnic matter.

Study area

The PRD area is located in the south of Guangdong province occupying 41698 km^2 of land area (Figure 1). The area has a subtropical – tropical monsoon climate with an average annual temperature of 21–22°C and average annual rainfall of 1600–2000 mm. The main soil types in the PRD area are Ultisol, mostly developed on the granite parent materials in the local hills, and paddy soils developed on the fluvial sediments. In the study area, 10–15 crops of vegetables are planted annually, and the rotation of vegetable and rice is applied in mostly paddy soils.

Over the last 30 yr, rapid urbanization and industrialization has taken place in this area. Heavy metal contents in soils and sediments were elevated compared with historical monitoring results [29]. As a result, heavy metal accumulation in agricultural soils has also become increasingly serious in this area because of increasing reliance on fertilizers and agrochemicals [30].

Field sampling and preparation

Both surface soil samples and vegetable samples used in this study were collected from locations shown in Figure 1. All sample sites were far away at least 100 m from the obviously polluted area such as industries, feedlot, wastewater and highway for avoiding the directly anthropogenic influence. The planting area of each site was larger than 1000 m^2 for.a certain vegetable species. In this study the vegetable species included Pakchoi (*Brassica rapa chinensis*) (n = 31), Chinese flowering cabbage (*Brassica campestris L. ssp. chinensis*) (n = 37), Leaf mustard (*Brassica juncea Coss*) (n = 16), Romaine lettuce (*Lactuca sativa L. var. longifolia*) (n = 31), Chinese lettuce (*Lactuca sativa L. var. asparagina*) (n = 15), Cauliflower (*Brassica oleracea L.var. botrytis L.*) (n = 7), Water spinach (*Ipomaea aquatic Forssk*) (n = 7), Celery (*Apium graveolens*) (n = 6), Chinese chives (*Allium tuberosum*) (n = 5), Spinach (*Spinacia oleracea*) (n = 5), Amaranth (*Amaranthus mangostanus L.*) (n = 5), watercress (*Nasturtium officinale*) (n = 5).

Vegetables and field soils at a soil depth of 0–15 cm which was the root concentrated layer in this study were collected when the vegetable were suitable for harvest. Each vegetable and soil sample consisted of five subsamples, and combined and mixed well. The fresh vegetable samples were put in clean plastic bags and immediately transported to the laboratory for sample treatment.

The vegetable samples were cleaned with tap water and Milli-Q water, and then the edible parts (including leaves and steams) were separated, weighed (Fresh weight). The washed samples were then dried in an oven at 60°C, and their dry weights (DW) were recorded. The dry vegetable samples were ground to pass through a 250 µm sieve in a steel grinder. Soil samples were air-dried at room temperature (25°C) and ground to pass through a 2 mm nylon sieve. The fine vegetable and soil sample powders were stored in polythene zip bags.

Chemical analyses

The sieved soil samples were ground further to pass through a 150 µm nylon sieve. The prepared soil samples were digested to dryness using an acid mixture of 10 ml HF, 5 ml HClO$_4$, 2.5 ml HCl, and 2.5 ml HNO$_3$. Total Cd concentrations in soils were determined by Inductively Coupled Plasma-Atomic Emission Spectrometry (ICP-AES) (Model PS 1000 AT, USA). The recovered soil Cd concentration of the National Research Center for GeoAnalysis soil standard reference materials (SRM) (ESS-4, Beijing, China; Standard value: 0.083 mg kg^{-1}) was 0.081 mg kg^{-1} (n = 32). The analytic precision was 2.4% for soil Cd.

Seventy five soil samples were analyzed for exchangeable Cd using the first step of Tessier sequential extraction[31]. The extraction procedure is 2.00 g of air-dried soil (<2 mm) were

mixed with 16 ml 1 M MgCl$_2$ solution (pH = 7.0), the mixture was shaken for one hour on 25°C. the suspension was filtered immediately after shaking. The Cd concentration in the filtrate solutions were measured by Inductively Coupled Plasma-Atomic Emission Spectrometry (ICP-AES) (Model PS 1000 AT, USA).

The ground vegetable samples were ashed in a muffle furnace for 16 h at 500°C, dissolved in 0.5 M HNO$_3$, and diluted to 25 mL with deionized water. Cd concentrations in vegetables were determined by Inductively Coupled Plasma-Atomic Emission Spectrometry (ICP-AES) (Model PS 1000 AT, USA). A plant standard reference material (GSV-4, Beijing, China; Standard value:0.057 mg kg^{-1}) was used in order to control the determination quality. The recovered Cd concentration was 0.059 mg kg^{-1} (n = 10). The analytic precision 3.5% for vegetable Cd. The Cd concentration in vegetable samples was expressed on a fresh weight basis.

The pH of soil was measured by taking 10 g of sample into 25 ml of deionized water [32]. The soil organic matter content was measured using potassium bichromate oxidation process [33].

Data analysis

The transfer factors (TFs) of Cd from soil to vegetable (edible part) were calculated using the following equation [12,15–17,20,34]:

$$TF = \frac{C_{vegetable}}{C_{soil}}$$

Where $C_{vegetable}$ is Cd concentration (FW) in the edible parts of vegetables, and C_{soil} is the total Cd concentration (DW) in the soil where the vegetable was grown.

The data statistical analysis was performed using the Minitab 16 statistical software (Minitab Inc., USA). Spatial interpolation was performed using ordinary kriging. For the low-density sampling, the ordinary kriging estimate can be thought of simply as an optimally weighted average of the data [35]. It provides a best linear unbiased prediction of spatial distribution. The spatial interpolation and contour maps displaying the spatial distribution of Cd concentrations in soils and vegetables, and transfer factors were produced based on the geostatistical analysis by using the software of ArcGIS 9.0.

Results and Discussions

Cd concentrations in vegetable soils

Total Cd concentrations in soils range from 0.012 to 1.18 mg kg^{-1} in the PRD area (Table 1). These values fit a log normal distribution (Figure 2), therefore the geometric mean (GM) value of 0.158 mg kg^{-1} and geometric standard deviations (GSD) of 1.39 were used to represent the central tendency and variations of the data. The present GM value was much higher than its background concentration (0.04 mg kg^{-1}) of soil in Guangdong province [35] and significantly lower than Cd mean concentrations of 0.858 mg kg^{-1} in the reclaimed tidal flat soil and 1.4–1.8 mg kg^{-1} in some Cd-contaminated vegetable soils in the PRD area [20,21]. The present results reflect the overall level of soil Cd concentration on the region al scale, not is responsible for local point-source contamination.

According to the maximum permissible concentrations of cadmium for agriculture soils of 0.3 mg kg^{-1} [8], thirty vegetable soils were contaminated with Cd in the study area, mainly located in the Guangzhou, Zhongshan, and Foshan cities. The soil Cd mean concentration was up to 0.316 mg kg^{-1} in the Zhongshan city, followed by 0.291 mg kg-1 in the Foshan city, 0.226 mg kg-1

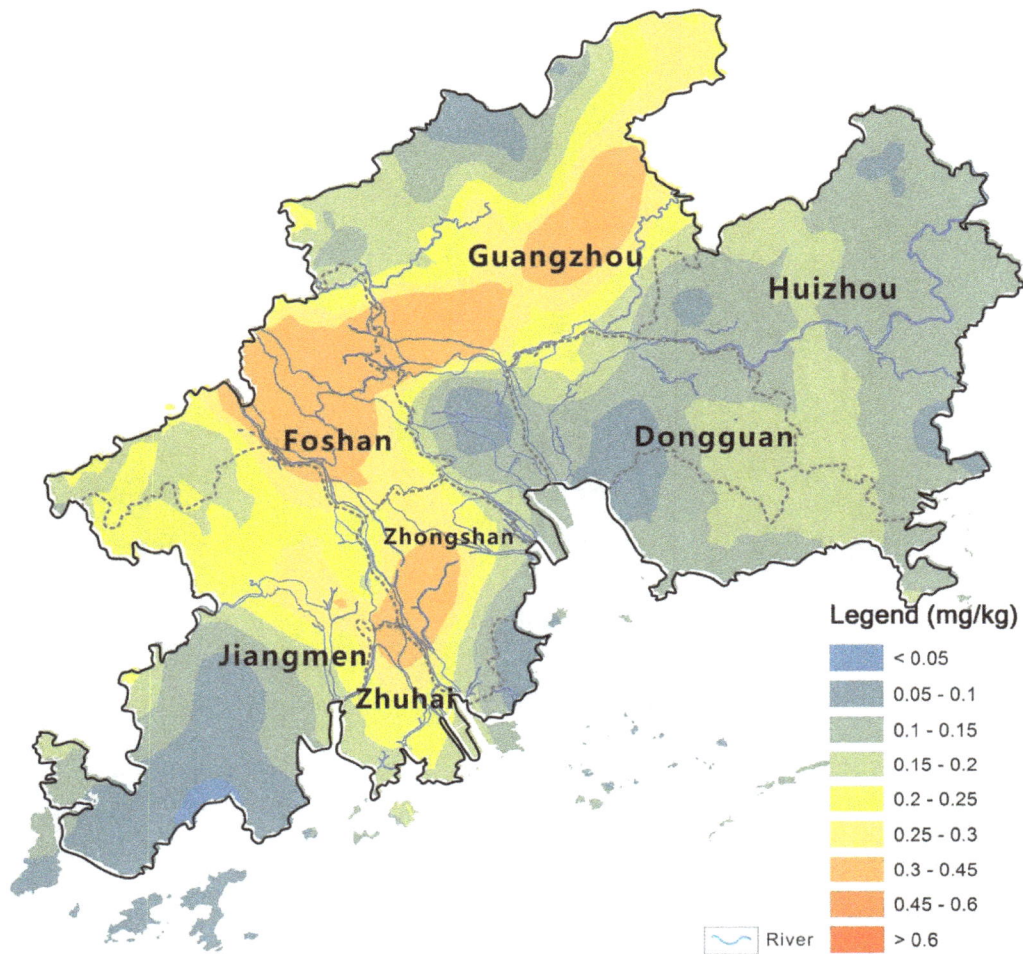

Figure 3. Spatial pattern of total Cd concentration in the soils in the PRD area (mg kg^{-1}, DW).

in the Guangzhou City, and 0.204 mg kg-1 in the Zhuhai City (Table 1). Generally, spatial characteres of soil Cd show that the western areas (including Foshan, Guangzhou, Zhongshan, Zhuhai)

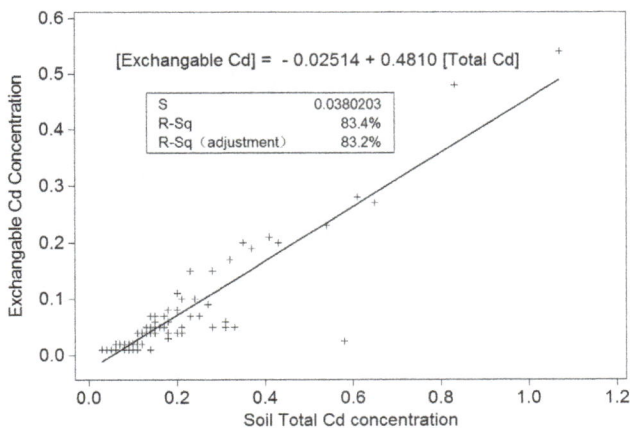

Figure 4. Relationship between exchangeable Cd concentration and total Cd concentration in vegetable soils (mg kg^{-1}, DW).

influenced by the West River and North River has the higher Cd concentration than the eastern areas (including Dongguan and Huizhou) influenced by the East River (Figure 3). The results coincided with spatial characters of other heavy metal concentrations (Pb, Zn) in sediments in the PRD area, that is, there were higher heavy metal concentrations in sediments of the West River and North River than that of the East River [36]. The spatial correlation between soil Cd cocnetrtiaons and regional rivers suggested that the cadmium produced by the mining and smelting activities in the upper reaches of the West River and North River was the main source of Cd in soils of the PRD area, and the West River and North River was the important transfer approach for Cd entering into surrounding soils through river irrigation.

Although many literatures pointed out that metal availability in soil is of main concern, because the available concentration is an indication of the amount available for plant uptake, and provided some methods to determine plant availability of heavy metals [12,21,24]. Our results reveal that the significant positive correlation between soil total Cd and exchangeable Cd was shown in Figure 4, which can be described by a linear equation:

$$y = 0.4810x - 0.02514 \ (r^2 = 0.834; p < 0.001)$$

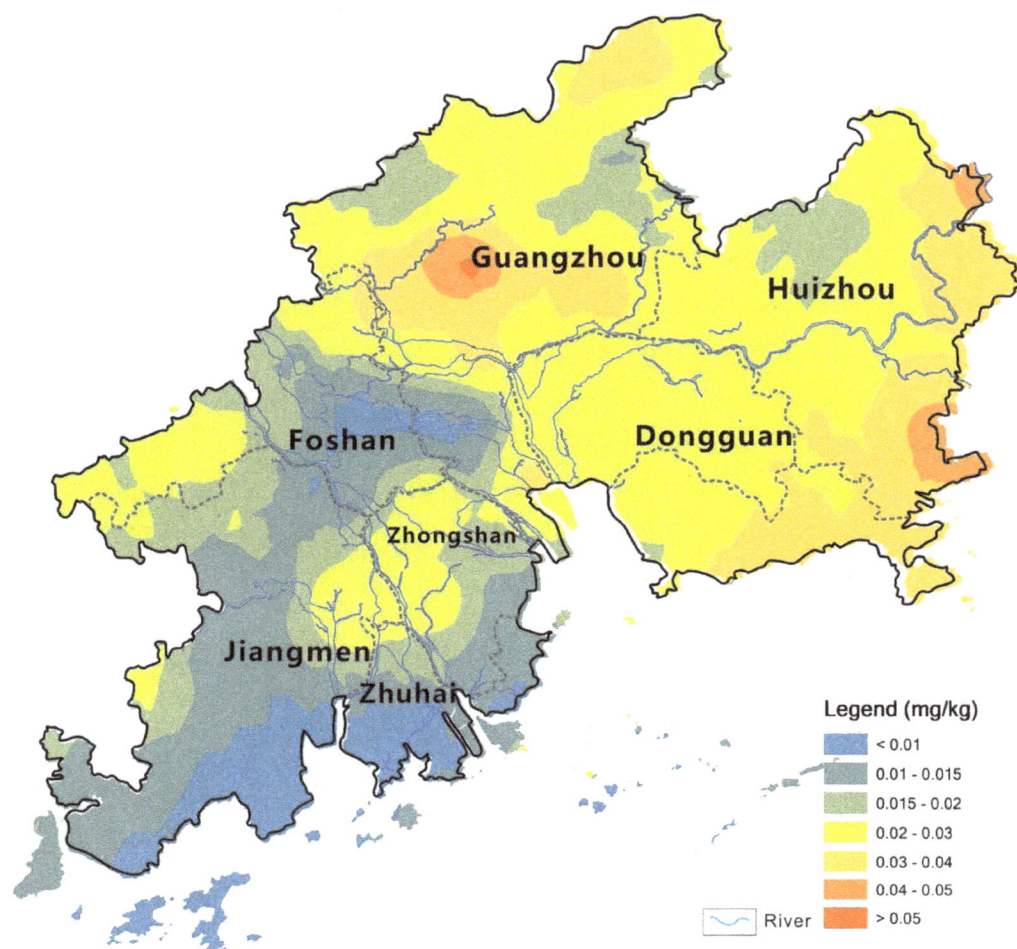

Figure 5. Spatial pattern of Cd concentrations in the vegetables in the PRD area (mg kg^{-1}, FW).

Table 2. Cd concentration (mg kg^{-1}, FW) in different vegetable species and the TF values in the PRD area.

Vegetable species	n	Cd		TF	
		Mean ± SD	Range	Mean ± SD	Range
Pakchoi (*Brassica rapa chinensis*)	31	0.023±0.015	0.002–0.074	0.222±0.220	0.010–0.920
Chinese flowering cabbage (*Brassica campestris L. ssp. chinensis*)	37	0.022±0.015	0.001–0.068	0.160±0.142	0.025–0.774
Leaf mustard (*Brassica juncea Coss*)	16	0.023±0.014	0.001–0.054	0.532±1.087	0.022–4.50
Romaine lettuce (*Lactuca sativa L. var. longifolia*)	31	0.026±0.021	0.003–0.082	0.148±0.111	0.001–0.371
Chinese lettuce (*Lactuca sativa L. var. asparagina*)	15	0.031±0.026	0.005–0.078	0.188±0.219	0.017–0.821
Cauliflower (*Brassica oleracea L.var. botrytis L.*)	7	0.010±0.013	0.002–0.032	0.048±0.042	0.014–0.103
Water spinach (*Ipomaea aquatic Forssk*)	7	0.011±0.019	0.001–0.053	0.045±0.052	0.003–0.117
Celery (*Apium graveolens*)	6	0.043±0.020	0.019–0.064	0.234±0.115	0.110–0.376
Chinese chives (*Allium tuberosum*)	5	0.027±0.032	0.001–0.074	0.229±0.241	0.001–0.563
Spinach (*Spinacia oleracea*)	5	0.037±0.027	0.006–0.067	0.213±0.034	0.182–0.262
Amaranth (*Amaranthus mangostanus L.*)	5	0.060±0.017	0.030–0.082	0.753±0.442	0.258–1.577
watercress (*Nasturtium officinale*)	5	0.021±0.020	0.002–0.039	0.087±0.046	0.003–0.129
Total	170	0.024±0.019	0.001–0.082	0.189±0.211	0.001–4.50

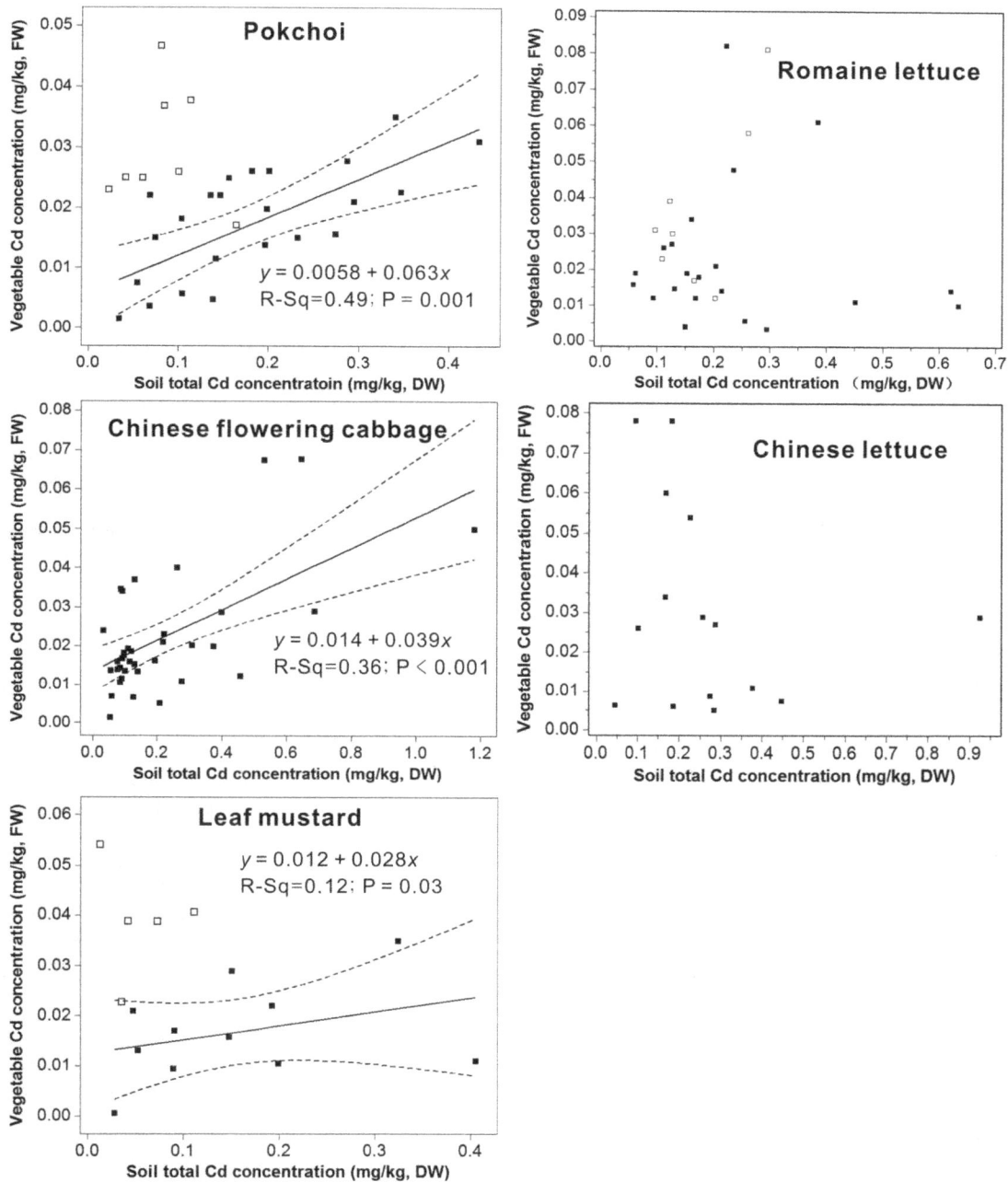

Figure 6. Relationships between soil total Cd concentrations (mg kg^{-1}, DW) and vegetable Cd concentrations (mg kg^{-1}, FW) for five species of main vegetables (soil pH: 3.79–7.72) (hollow diamonds represented samples with lower pH (3.79–5.00) removed for the regression analysis; the solid line is the sample regression line; dotted lines indicated the 95% confidence interval).

Where y is the soil exchangeable Cd concentration, x refers to soil total Cd concentration.

This indicated that exchangeable Cd concentration in soils in the PRD area was mainly controlled by the soil total Cd concentration.

Cd concentrations in vegetables

The present results show that in the PRD area the range of Cd concentration in all vegetable samples was 0.002–0.082 mg kg^{-1} with a geometric mean value of 0.015 mg kg^{-1} (Table 1). In this study Cd concentrations in all vegetable samples met the national standard of Safe vegetables (0.2 mg kg^{-1}), 150 vegetable samples met the standard of No-Polluted vegetables (0.05 mg kg^{-1}). Twenty vegetable samples were in between Safe vegetables and No-polluted vegetables, of which 14 vegetable sample sites were in the Guangzhou.

According to the different cities, the mean vegetable Cd concentration, was in the order of Guangzhou (0.027 mg kg^{-1})> Huizhou (0.026 mg kg^{-1})> Dongguan (0.024 mg kg^{-1})> Zhongshan (0.022 mg kg^{-1})> Foshan (0.017 mg kg^{-1})> Jiangmen (0.014 mg kg^{-1})> Zhuhai (0.008 mg kg^{-1}) (Table 1). This order is significantly different from Cd concentrations in soils.

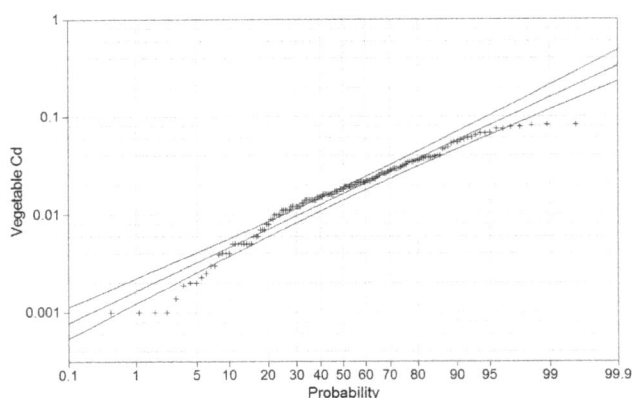

Figure 7. Lognormal probability plot for vegetable Cd concentrations in the Pearl River Delta area, Guangdong, China (mg kg^{-1}, FW).

The spatial pattern of Cd concentrations in vegetables showed that no spatial correlation with that of total Cd cocentrtaions in soils on the regional scale (Figure 3 and Figure 5).

The mean and range of Cd concentrations in various vegetable species were listed in Table 2. In the study area the dominant vegetable species are the pakchoi, Chinese flowering cabbage, leaf mustard, Romaine lettuce and Chinese lettuce for the local residents. The mean Cd concentrations in these vegetables range from 0.022 mg kg^{-1} to 0.031 mg kg^{-1} (Table 2). The amaranth vegetables had the highest Cd concentration of 0.082 mg kg^{-1} with mean value of 0.06 mg kg^{-1}. This value was closed to 0.078 mg kg^{-1} in amaranth on the reclaimed tidal flat soil reported by Li et al [20]. Compared with the previous researches on the Cd-contaminated soil [20,21], the present results revealed that there were low Cd concentrations of various vegetables in the study area.

Relationships between Cd in soils and in vegetables

Many studies have shown that the concentration of heavy metals in vegetables is influenced by many factors, such as total concentration of heavy metals in soils, vegetable species, soil pH value, soil organic matter, climate, atmospheric depositions and temperature [19,21,37–39]. McBride [40] suggested that combination of soil pH and soil total Cd concentration was reasonably predictive of Cd concentration in the above-ground plant tissue.

In this study, we discussed the correlation between Cd concentrations in five species of dominant leaf vegetables and Cd concentrations in soils. The results revealed that in the condition of soil pH>5, Cd concentrations in both pakchoi and Chinese flowering vegetables had obvious positive correlations

with Cd concentrations in soils; and Cd concentrations in leaf mustard vegetables and Cd concentrations in soils had a weak positive correlation (Figure 6). These can be described by linear equations, respectively:

$$y = 0.063x + 0.0058 \ (Pakchoi : r^2 = 0.49; p = 0.001)$$

$$y = 0.039x + 0.014$$

$$(Chinese \ flowering \ cabbage : r^2 = 0.36; p < 0.001)$$

$$y = 0.028x + 0.012 \ (Leaf \ mustard : r^2 = 0.12; p = 0.03)$$

Where y is the Cd concentration in the various vegetable, x is the soil total Cd concentration.

The results of the line regression indicated that in the condition of soil pH>5, soil total Cd concentrations explained 49%, 36% and 12% of the variability of Cd concentration in Pakchoi and Chinese flowering cabbage and leaf mustard, respectively. But Cd concentrations in the Romaine lettuce and Chinese lettuce had no correlations with total Cd concentrations in soils (Figure 6). This phenomenon should be related with various vegetable families. Pakchoi, Chinese flowering cabbage and leaf mustard belong to the brassica of cruciferous plants; Romaine lettuce and Chinese lettuce belong to the Lactuca of Compositae plants. The difference of correlations between Cd concentrations in vegetables and total Cd concentrations in soils should be the result of different absorption capacity of various vegetable families to Cd. And the result showed that the influence of soil total Cd concentrations on the Cd in various vegetable species is obviously different [39,41].

When soils pH is less than 5, total Cd concentrations in soils were low because the environmental capacity of soil Cd was small in acid soils due to the strong eluviation, but Cd concentrations in the pakchoi and leaf mustard vegetables were much higher than the reference line (Figure 6). Obviously, low soil pH can raise the available Cd content for pakchoi and leaf mustard vegetables even in the soils with low Cd concentrations.

The analysis of Pearson correlations among soil Cd, vegetable Cd, soil pH, organic matter and transfer factor showed that there was no obvious correlation among soil Cd, vegetable Cd, organic matter and soil pH; significant correlations were be found between transfer factor and soil Cd ($r^2 = -0.295$) and between transfer factor and vegetable Cd ($r^2 = 0.435$) (Table 3). Our results confirmed that soil pH played important role to Cd absorption of vegetables, but there was no linear correlation between soil pH and Cd concentration in vegetables in this study. Moreover, The lognormal probability distribution graph shows that some vegetable samples of upper tail had less high value than would be

Table 3. Pearson correlations of soil Cd, vegetable Cd, soil pH, organic matter and Transfer factor.

	soil Cd	vegetable Cd	pH	Organic matter
vegetable Cd	0.205(0.088)			
pH	0.033(0.784)	0.137(0.258)		
Organic matter	0.065(0.590)	−0.007(0.952)	0.076(0.532)	
TF	−0.295(0.013)	0.435(0.000)	−0.143(0.239)	−0.103(0.396)

Cell content: Pearson correlation (p-value).

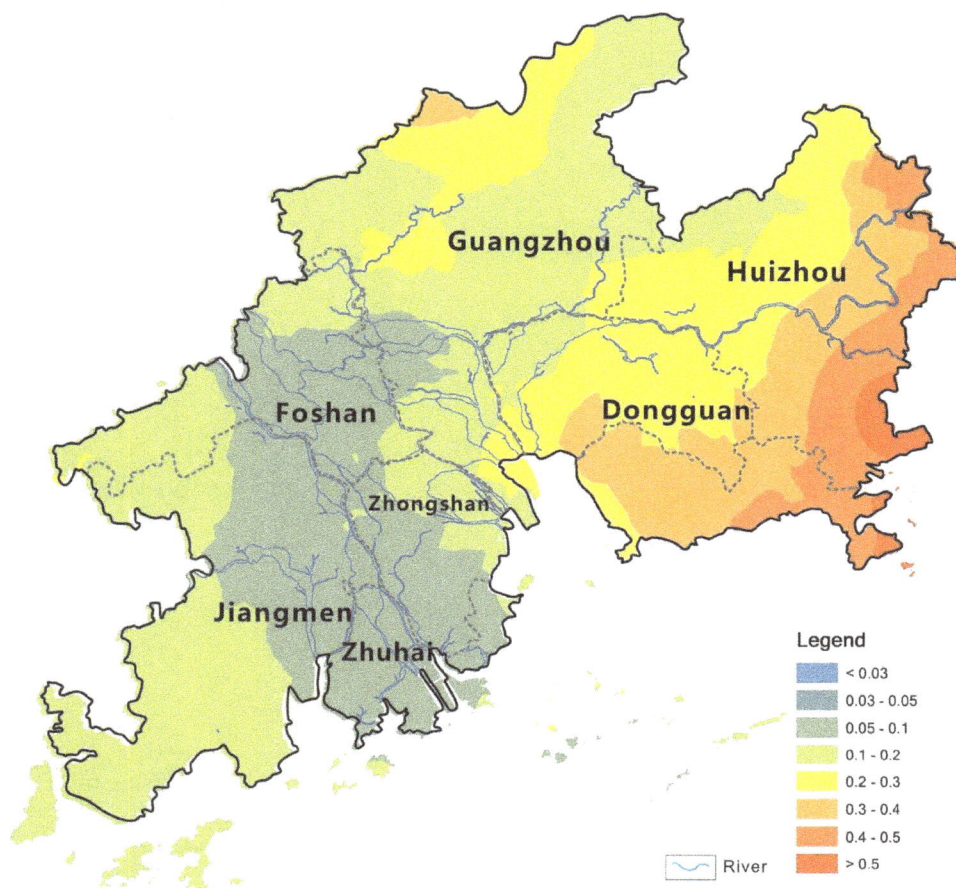

Figure 8. Spatial pattern of soil-to-vegetable transfer factor of Cd in the PRD area.

expected (Figure 7), suggesting the vegetable's ability to absorb Cd would be decreasing with the increasing the Cd content in the vegetables.

Transfer factors of Cd from soil to vegetable

The soil-to-vegetable transfer factors (TFs) reflected the ability of vegetables to take up soil metals. TFs varied significantly with plant species [12,19,21,38,39], and were commonly viewed as a "constant" for a given plant species and a given metal. TFs were used as an important character index for establishing the soil environmental quality criteria or assessing the health risk of soil contamination.

In this study, the TFs of Cd from soil to vegetable for the 12 vegetable species varied from 0.045 in Water spinach to 0.753 in Amaranth (Table 2). The TF mean values in the main five species of leaf vegetables in the study area were 0.532 for leaf mustard, 0.222 for Pakchoi, 0.188 for Chinese lettuce, 0.160 for Chinese flowering cabbage, and 0.148 for Romaine lettuce. Compared with TFs of Cd reported by Wang et al [12], and Hu et al [21], these values were much higher because of the lower Cd concentration in soils. The spatial pattern of TFs also showed that the sample sites with high TFs were always corresponding to the lower Cd concentrations in soils in the study area (Figure 3 and Figure 8).

The regression equations between the TF values and total Cd concentrations in soils can be described by power model ($y = ax^b$) for five kinds of main vegetables:

$$y = 0.0233x^{-0.883} \, (Pakchoi : r^2 = 0.433; p < 0.001)$$

$$y = 0.0392x^{-0.582}$$
$$(Chinese \, flowering \, cabbage : r^2 = 0.354; p < 0.001)$$

$$y = 0.0214x^{-0.915} \, (Leaf \, mustard : r^2 = 0.406; p = 0.008)$$

$$y = 0.0172x^{-1.047} \, (Romiane \, lettuce : r^2 = 0.358; p < 0.001)$$

$$y = 0.0173x^{-1.116} \, (Chinese \, lettuce : r^2 = 0.403; p = 0.011)$$

Where y is the soil-to-vegetable transfer factor of Cd, x is the total Cd concentration in the soils.

For main leaf vegetables, the transfer factors of Cd decreased with increasing total Cd concentrations (Figure 9), suggesting that the ability of vegetable to take up soil Cd decreased with the total soil Cd increasing. Figure 9 also illustrated that when soil Cd concentration was low (about 0.1–0.2 mg kg^{-1}), the TFs decreased sharply; with the gradual increasing of soil total Cd concentration, TFs decreased only slowly, till to a stable value.

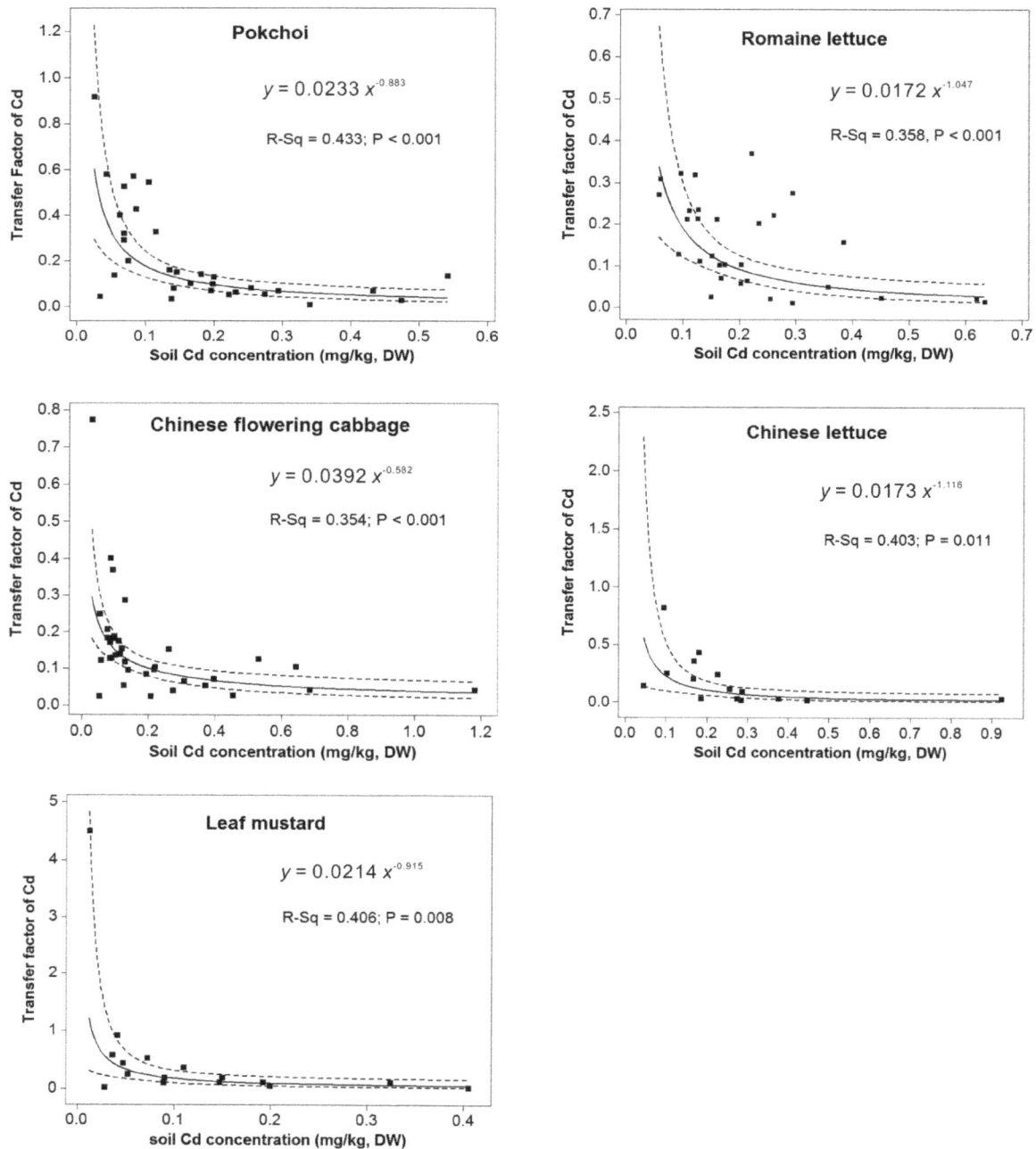

Figure 9. Relationships between soil-to-vegetable transfer factors of Cd and total Cd concentrations in soils for five species of main vegetables (the solid line is the sample regression curve; dotted lines indicated the 95% confidence interval).

Moreover, the 95% confidence interval of regression lines showed that there were a larger error range at the low Cd concentration interval of soils (0–0.1 or 0.2 mg kg^{-1}) for the various leaf vegetable species, and when soil Cd concentrations were greater than 0.1 or 0.2 mg kg^{-1}, the TFs of various vegetables would be reasonable with a lower error range for the assessment of environmental risk (Figure 9). As a result, for carrying out rational environmental health assessment in different Cd concentrations of soils, the present results indicated that the fit line of TFs for vegetable species should be discussed to divide into two parts, including the sharply decrease part and the slowly decrease part, according to the gradual increasing of soil total Cd concentration. This work is agree with the suggestion proposed by Wang et al [12] that the transfer factors of a given crop-metal system estimated from the regression model between the TF values and the corresponding soil metal concentration, are much more reasonable than using the arithmetic means or geometric means at a given soil metal concentration.

Conclusions

The present results showed that the total Cd concentration in soils is safe for planting leaf vegetables in the study area. All vegetable samples met the national standard of safe vegetables, and 88% of which met the standard of no-polluted vegetables. Total Cd concentrations in soils and soil pH played an important role to Cd absorption of vegetables, but there were no obviously linear correlations. There was no spatial correlation between total Cd concentrations in soils and Cd concentrations in vegetables. For a givenl vegetable species the regression model between the TF value and the corresponding total Cd concentrations in soils could be used to estimate the transfer abilities of Cd from soil to vegetable at a given total Cd concentration in soils.

Acknowledgments

We thank four reviewers and editors for their very useful comments that improved the quality of this manuscript.

Author Contributions

Performed the experiments: HZ JC. Analyzed the data: LZ GY. Contributed reagents/materials/analysis tools: HZ DL. Wrote the paper: HZ.

References

1. Singh BR, McLaughlin MJ (1999) Cadmium in Soils and Plants. In: Mclaughlin M., Singh BR, editors. Cadmium in soils and plants. Boston: Kluwer Academic. 357–367.
2. Tudoreanu L, Phillips CJC (2004) Modeling cadmium uptake and accumulation in plants. Adv Agron 84: 121–157. doi:10.1016/S0065-2113(04)84003-3.
3. Singh A, Sharma RK, Agrawal M, Marshall FM (2010) Risk assessment of heavy metal toxicity through contaminated vegetables from waste water irrigated area of Varanasi, India. Trop Ecol 51: 375–387.
4. WHO (1992) Cadmium: Environmental Health Criteria, 134. Geneva.
5. Chlopecka A, Bacon JR, Wilson MJ, Kay J (1996) Forms of Cadmium, Lead, and Zinc in Contaminated Soils from Southwest Poland. J Environ Qual 25: 69. doi:10.2134/jeq1996.00472425002500010009x.
6. Kirkham MB (2006) Cadmium in plants on polluted soils: Effects of soil factors, hyperaccumulation, and amendments. Geoderma 137: 19–32. doi:10.1016/j.geoderma.2006.08.024.
7. Wu L, Tan C, Liu L, Zhu P, Peng C, et al. (2012) Cadmium bioavailability in surface soils receiving long-term applications of inorganic fertilizers and pig manure. Geoderma 173-174: 224–230. doi:10.1016/j.geoderma.2011.12.003.
8. SEPAC (1995) Environmental quality standard for soil (GB 15618–1995). State Environmental Protection Administration of China, Beijing, China
9. Desaules A (2012) Critical evaluation of soil contamination assessment methods for trace metals. Sci Total Environ 426: 120–131. doi:10.1016/j.scitotenv.2012.03.035.
10. Römkens PFAM, Guo HY, Chu CL, Liu TS, Chiang CF, et al. (2009) Prediction of Cadmium uptake by brown rice and derivation of soil-plant transfer models to improve soil protection guidelines. Environ Pollut 157: 2435–2444. doi:10.1016/j.envpol.2009.03.009.
11. Sun W, Sang L, Jiang B (2012) Trace metals in sediments and aquatic plants from the Xiangjiang River, China. J Soils Sediments 12: 1649–1657. doi:10.1007/s11368-012-0596-8.
12. Wang G, Su M, Chen Y, Lin F, Luo D, et al. (2006) Transfer characteristics of cadmium and lead from soil to the edible parts of six vegetable species in southeastern China. Environ Pollut 144: 127–135. doi:10.1016/j.envpol.2005.12.023.
13. Wang Q, Dong Y, Cui Y, Liu X (2001) Instances of soil and crop heavy metal contamination in China. Soil Sediment Contam 10: 497–510.
14. Xiong X, Allinson G, Stagnitti F, Li P, Wang X, et al. (2004) Cadmium contamination of soils of the Shenyang Zhangshi Irrigation Area, China: an historical perspective. Bull Environ Contam Toxicol 73: 270–275. doi:10.1007/s00128-004-0423-z.
15. Williams PN, Lei M, Sun G, Huang Q, Lu Y, et al. (2009) Occurrence and partitioning of cadmium, arsenic and lead in mine impacted paddy rice: Hunan, China. Environ Sci Technol 43: 637–642.
16. Li JT, Qiu JW, Wang XW, Zhong Y, Lan CY, et al. (2006) Cadmium contamination in orchard soils and fruit trees and its potential health risk in Guangzhou, China. Environ Pollut 143: 159–165. doi:10.1016/j.envpol.2005.10.016.
17. Bi X, Ren L, Gong M, He Y, Wang L, et al. (2010) Transfer of cadmium and lead from soil to mangoes in an uncontaminated area, Hainan Island, China. Geoderma 155: 115–120. doi:10.1016/j.geoderma.2009.12.004.
18. Zhou Z., Fan W., Wang M (2000) Heavy metal contamination in vegetables and their control in China. Food Rev Int 16: 239–255. doi:10.1081/FRI-100100288.
19. Cui Y, Zhu Y, Zhai R, Chen D, Huang Y, et al. (2004) Transfer of metals from soil to vegetables in an area near a smelter in Nanning, China. Environ Int 30: 785–791. doi:10.1016/j.envint.2004.01.003.
20. Li Q, Chen Y, Fu H, Cui Z, Shi L, et al. (2012) Health risk of heavy metals in food crops grown on reclaimed tidal flat soil in the Pearl River Estuary, China. J Hazard Mater 227–228: 148–154. doi:10.1016/j.jhazmat.2012.05.023.
21. Hu J, Wu F, Wu S, Sun X, Lin X, et al. (2013) Phytoavailability and phytovariety codetermine the bioaccumulation risk of heavy metal from soils, focusing on Cd-contaminated vegetable farms around the Pearl River Delta, China. Ecotoxicol Environ Saf 91: 18–24. doi:10.1016/j.ecoenv.2013.01.001.
22. AQSIQ (2001) Safety qualification for agricultural product – Safety requirements for non-environmental pollution vegetable (GB 18406.1–2001). General Administration of Quality Supervision, Inspection and quarantine of China, Beijing, China.
23. MOH (2012) Maximum levels of contaminants in Foods (GB 2762–2012). Ministry of Health of the People's Republic of China.
24. Hart JJ, Welch RM, Norvell WA, Kochian LV (2002) Transport interactions between cadmium and zinc in roots of bread and durum wheat seedlings. Physiol Plant 116: 73–78. doi:10.1034/j.1399-3054.2002.1160109.x.
25. Yang Y, Zhang F-S, Li H-F, Jiang R-F (2009) Accumulation of cadmium in the edible parts of six vegetable species grown in Cd-contaminated soils. J Environ Manage 90: 1117–1122. doi:10.1016/j.jenvman.2008.05.004.
26. McLaughlin M, Parker D, Clarke J (1999) Metals and micronutrients – food safety issues. F Crop Res 60: 143–163. doi:10.1016/S0378-4290(98)00137-3.
27. Wan Z (2013) The development report of Guangdong vegetables industry in 2012. Institute of Agricultural Economics and Rural Development.
28. SBG (2012) Guangdong Statistical Yearbook –2012. Beijing: Statistics Bureau of Guangdong Province, China Statistic Press.
29. Li X, Wai OWH, Li YS, Coles BJ, Ramsey MH, et al. (2000) Heavy metal distribution in sediment profiles of the Pearl River estuary, South China. Appl Geochemistry 15: 567–581. doi:10.1016/S0883-2927(99)00072-4.
30. Wong S, Liu X, Zhang G, Qi S, Min Y (2002) Heavy metals in agricultural soils of the Pearl River Delta, South China. Environ Pollut 119: 33–44. doi:10.1016/S0269-7491(01)00325-6.
31. Tessier A, Campbell PGC, Bisson M (1979) Sequential extraction procedure for the speciation of particulate trace metals. Anal Chem 51: 844–851. doi:10.1021/ac50043a017.
32. Chinese National Standard Agency (1988) Determination of pH value in forest soil. GB 7859–87, UDC 634.0.114:631.422: 171–173.
33. Yu T, Wang Z (1988) Soil Analytical Chemistry. Beijing: Science Press.
34. Chojnacka K, Chojnacki A, Górecka H, Górecki H (2005) Bioavailability of heavy metals from polluted soils to plants. Sci Total Environ 337: 175–182. doi:10.1016/j.scitotenv.2004.06.009.
35. Zhang HH, Chen JJ, Zhu L, Li FB, Wu ZF, et al. (2011) Spatial patterns and variation of soil cadmium in Guangdong Province, China. J Geochemical Explor 109: 86–91. doi:10.1016/j.gexplo.2010.10.014.
36. Zhang C, Wang L (2001) Multi-element geochemistry of sediments from the Pearl River system, China. Appl Geochemistry 16: 1251–1259. doi:10.1016/S0883-2927(01)00007-5.
37. Voutsa D, Grimanis A, Samara C (1996) Trace elements in vegetables grown in an industrial area in relation to soil and air particulate matter. Environ Pollut 94: 325–335. doi:10.1016/S0269-7491(96)00088-7.
38. Peris M, Micó C, Recatalá L, Sánchez R, Sánchez J (2007) Heavy metal contents in horticultural crops of a representative area of the European Mediterranean region. Sci Total Environ 378: 42–48. doi:10.1016/j.scitotenv.2007.01.030.
39. Li Y, Li L, Zhang Q, Yang Y, Wang H, et al. (2013) Influence of temperature on the heavy metals accumulation of five vegetable species in semiarid area of northwest China. Chem Ecol 29: 353–365. doi:10.1080/02757540.2013.769970.
40. McBride M, Murray B (2002) Cadmium uptake by crops estimated from soil total Cd and pH. Soil Sci 167.
41. Hooda PS, McNulty D, Alloway BJ, Aitken MN (1997) Plant Availability of Heavy Metals in Soils Previously Amended with Heavy Applications of Sewage Sludge. J Sci Food Agric 73: 446–454. doi:10.1002/(SICI)1097-0010(199704)73:4<446::AID-JSFA749>3.0.CO;2-2.

6

Climate Warming May Facilitate Invasion of the Exotic Shrub *Lantana camara*

Qiaoying Zhang[1,2]*, Yunchun Zhang[2], Shaolin Peng[3]*, Kristjan Zobel[1]

1 Department of Botany, Institute of Ecology and Earth Sciences, University of Tartu, Tartu, Estonia, 2 Department of Environmental Science and Engineering, Qilu University of Technology, Jinan, Shandong, China, 3 State Key Laboratory of Biocontrol, School of Life Sciences, Sun Yat-sen University, Guangzhou, Guangdong, China

Abstract

Plant species show different responses to the elevated temperatures that are resulting from global climate change, depending on their ecological and physiological characteristics. The highly invasive shrub *Lantana camara* occurs between the latitudes of 35°N and 35°S. According to current and future climate scenarios predicted by the CLIMEX model, climatically suitable areas for *L. camara* are projected to contract globally, despite expansions in some areas. The objective of this study was to test those predictions, using a pot experiment in which branch cuttings were grown at three different temperatures (22°C, 26°C and 30°C). We hypothesized that warming would facilitate the invasiveness of *L. camara*. In response to rising temperatures, the total biomass of *L. camara* did increase. Plants allocated more biomass to stems and enlarged their leaves more at 26°C and 30°C, which promoted light capture and assimilation. They did not appear to be stressed by higher temperatures, in fact photosynthesis and assimilation were enhanced. Using lettuce (*Lactuca sativa*) as a receptor plant in a bioassay experiment, we also tested the phytotoxicity of *L. camara* leachate at different temperatures. All aqueous extracts from fresh leaves significantly inhibited the germination and seedling growth of lettuce, and the allelopathic effects became stronger with increasing temperature. Our results provide key evidence that elevated temperature led to significant increases in growth along with physiological and allelopathic effects, which together indicate that global warming facilitates the invasion of *L. camara*.

Editor: Lalit Kumar, University of New England, Australia

Funding: The study was supported by MOBILITAS Postdoctoral Research Grant (MJD305) from the European Social Fund and the Key Program of National Natural Science Foundation of China (Grant No. 2010330004103989). The funders had no role in study design, data collection and analysis, decision to publish, or preparation of the manuscript.

Competing Interests: The authors have declared that no competing interests exist.

* Email: qiaoyingzhang@163.com (QZ); lsspsl@mail.sysu.edu.cn (SP)

Introduction

Global average temperatures are increasing and are predicted to do so further in the future [1]. Changes in temperature and precipitation associated with rising concentrations of CO_2 are altering local environmental conditions, which may inhibit native species [2,3]. At the same time, this may provide some non-native species with emerging opportunities for population growth and expansions [4]. The successful invasion of new areas by non-native species can have serious ecological consequences for species interactions and ecosystem structure and functioning [5]. Therefore, it is essential to better understand the risk of invasion under climate change scenarios for effective management of invasive plants in the 21st century [6].

The abundance and distribution of plant species are tightly regulated by both climatic factors [7] and biotic interactions [8], so changes in climatic conditions are likely to cause major shifts in their population dynamics and geographic ranges [2,3,9]. Apart from changes in the potential distributions of native species, climate change may also affect the spatial distribution of invasive species [5,6,9]. Previous studies have shown that global warming has enabled alien plants to expand into regions where they previously could not survive and reproduce [9]. Any alterations of plant community structure that are caused by climate change result from underlying changes in the population dynamics of

species that make up the community [10]. Thus, understanding responses to climate change at the species level is important to the prediction of future ecosystem functioning [10].

The focal species of this study is *Lantana camara* (Verbenaceae), a small perennial shrub which can grow to around 2 m in height and forms dense thickets in a variety of environments [11]. Its native range is Central America, the northern part of South America and the Caribbean [11]. *L. camara* has been identified as one of the 100 World's Worst Invasive Alien Species [12]. Since the sixteenth century, it has been subject to intense horticultural improvement in Europe, and now it exists in many different forms and varieties around the world [13]. Its global distribution includes about 60 countries and islands between the latitudes of 35°N and 35°S [14]. *L. camara* has become a major problem in many of these areas, causing reductions in native species diversity, declines in soil fertility, allelopathic alteration of soil properties, and alteration of ecosystem processes [11,14].

The model CLIMEX has been widely used to illustrate the potential distribution of species under future climate scenarios [15]. Based on CLIMEX simulations, the potential distribution of *L. camara* will expand in some areas under current and future climate scenarios [14,16,17,18], and in China specifically its distribution could potentially expand further inland [11]. This is consistent with field investigations in southern China, where *L. camara* has recently become more prevalent [19]. Climate studies have shown that winter minimum temperatures in the region (i.e., Guangdong province) started to rise in the middle and later periods of the 1960s, and it has become warmer since the 1980s [20]. The observed increase in abundance is thus likely related to elevated temperature. We hypothesized that warming leads to positive effects on the fitness parameters of the invasive shrub *L. camara*, further facilitating its invasiveness.

We planted branch cuttings of *L. camara* in different temperature treatments (22, 26 and 30°C) in three experiments: a growth experiment, a physiological experiment and a bioassay experiment designed to assess the allelopathic potential of the species. Our goal was to describe and compare the morphological, physiological and biochemical responses of the species to future climate scenarios. Specifically, we sought to address the following questions: (1) How does the growth of *L. camara* respond to elevated temperatures? (2) How are gas exchange rates and photosynthesis affected by elevated temperatures? (3) Does the allelopathic potential of *L. camara* change with increasing temperature, as has been observed in some other plant species [21]?

Methods

Growth and morphology experiment

We collected three-year-old branches of *L. camara* on March 10, 2008 in Guangzhou, China. The sampling site ($23°02'-23°04'N$, $113°23'-113°24'E$) was neither located on farmland nor in a protected area. No specific permissions were required for these locations/activities. No endangered or protected species were involved in the sampling. The tops of the branches were cut to keep them at least 20 cm long. The cuttings were planted at the experimental field of Sun Yat-sen University, Guangzhou, Guangdong province. After three weeks, we selected uniform branch cuttings and transplanted them into plastic pots (20-cm diameter, 15-cm height, with three branches per pot). The pots were filled with equal proportions of nutrient-rich soil and vermiculite for water retention. The plants were grown in different greenhouses (14/10 h day/night cycle, $75\%\pm2\%$ relative humidity, photosynthetically active radiation (PAR) 400 mmol m^{-2} s^{-1}) at three constant temperatures (22, 26 and 30°C). There were 15 replicate pots per temperature treatment. All pots were randomly placed once a week to avoid internal effects. They were watered with diluted Hoagland solution (25% v/v) once a week, for a total of 18 weeks.

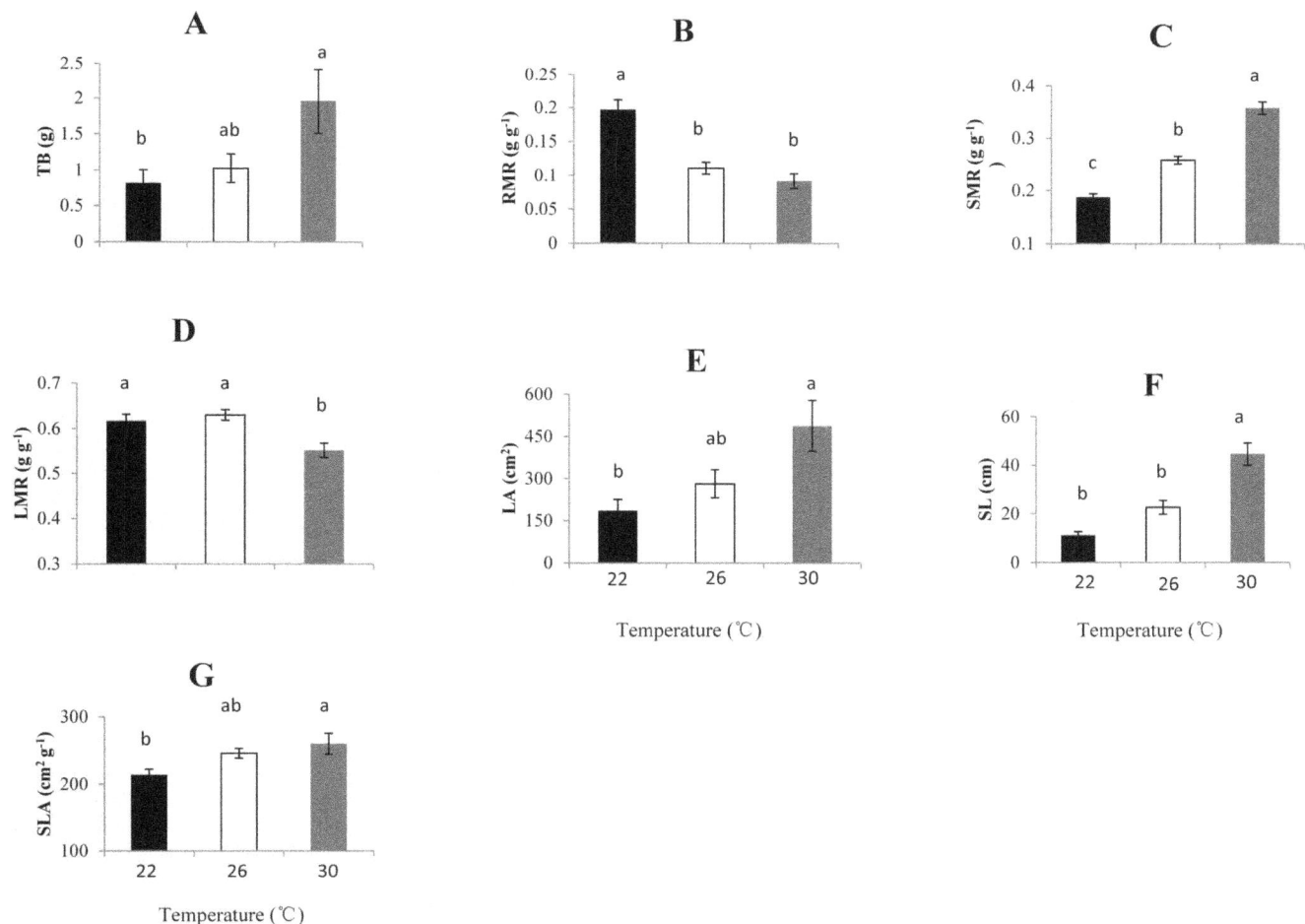

Figure 1. Responses of *Lantana camara* plants to temperature treatments (mean ± SE, *n* = 15). Different letters indicate significant differences (P<0.05) between means according to Tukey's HSD tests (The same below). (A) TB, total biomass; (B) RMR, root mass ratio; (C) SMR, support organ mass ratio; (D) LMR, leaf mass ratio; (E) LA, leaf area; (F) SL, stem length; (G) SLA, specific leaf area.

Table 1. F-values of one-way ANOVA which was used to test the effects of different temperatures on growth and morphology of *L. camara*.

	TB	RMR	SMR	LMR	LA	SL	SLA
d.f.	2, 16	2, 16	2, 16	2, 16	2, 16	2, 16	2, 16
F	4.04	22.47	91.03	8.92	5.76	27.00	4.70*
P	0.038*	<0.001***	<0.001***	0.003**	0.013*	<0.001***	0.025*

*** P<0.001,
** P<0.01,
* P<0.05.
Parameters of growth and morphology:
TB: total biomass; RMR: root mass ratio; SMR: support organ mass ratio; LMR: leaf mass ratio; LA: leaf area; SL: stem length; SLA: specific leaf area.

Then, at the end of the experiment, plants were harvested and divided into leaf blades, petioles, stems and roots, and were dried separately to a constant mass at 70°C. The total stem length was measured. Leaf area was determined using an LA meter (CI-203 Area-meter, CID, USA). The raw data were used to calculate the following growth parameters [22]: leaf mass ratio (leaves without petioles, LMR), root mass ratio (RMR), support organ mass ratio (stems and petioles, SMR), and specific leaf area (SLA). The mass ratio was calculated by dividing the dry mass by total plant dry mass.

Physiological experiment

Gas exchange. After potted plants had been allowed to grow for 18 weeks at different temperatures, we measured the gas exchange of *L. camara* on fully expanded leaves under controlled optimal conditions, using an open system with a portable photosynthesis measurement system (LI-6400, LI-COR, USA). The measurements were made on 15 plants per treatment. We found out that under greenhouse conditions, the net photosynthetic rate (Pnet) of *L. camara* was greatest at 800–1200 μmol m^{-2} s^{-1}, so we maintained PAR at 1200 μmol m^{-2} s^{-1}. We used an LI-6400 artificial light source and maintained the temperature at 22, 26 and 30°C for each treatment. In order to avoid the effect of midday photosynthetic depression [23], we completed the measurements on two sunny days from 08:00 to 11:30 and from 15:00 to 17:30. Pnet and stomatal conductance (Cond) were also measured, while intrinsic water use efficiency (WUE) was calculated by dividing Pnet by Cond [23].

Chlorophyll (Chl) fluorescence. We used the saturation pulse method [24] to measure the Chl fluorescence. Measurements were taken from the upper surface of the same leaves used in the previously described measurements, with a pulse-amplitude-modulated fluorometer (PAM 2100, Walz, Effeltrich, Germany) [25]. Before measurement, the leaves were placed in dark for at least 30 min. The intensity and duration of the saturation pulses was 4,000 μmol m^{-2} s^{-1} and 0.8 s, respectively. The "actinic light" was 600 μmol m^{-2} s^{-1}. We recorded the fluorescence parameters Fv/Fm and ΦPSII. Fv/Fm is the maximum quantum yield of photosystem II (PSII), which is assessed as (Fm - Fo)/Fm [26], where Fo and Fm are the minimal and maximal fluorescence values of a dark-adapted sample, respectively, with all of the PSII reaction centers fully open. It was measured at predawn, when plants were in the dark, to make sure that all the PSII reaction centers were open. ΦPSII is the effective quantum yield of PSII. It was calculated as ΦPSII = (Fm′−Ft)/Fm′, where Fm′ is the maximal fluorescence value reached in a pulse of saturating light with an illuminated sample, and Ft is the fluorescence value of the leaf at a given photosynthetically active radiation [27].

Bioassay experiment on allelopathic potential

After the potted *L. camara* plants had been growing for 18 weeks at three different temperatures (22, 26 and 30°C), we collected fresh leaves (10 g) randomly from plants in each greenhouse and soaked them in distilled water (100 mL) for 24 h in darkness at 22, 26 and 30°C, respectively. We then made aqueous leachates with a concentration of 0.1 g mL^{-1} from each treatment. The pH value of all leachates was adjusted to 6.8 using 1 M NaOH or HCl, and distilled water was used as a control.

Twenty uniform lettuce (*Lactuca sativa*) seeds were selected, surface-sterilised with 0.5% KMnO$_4$ for 15 min, and then washed with sterile water. The seeds were put on top of two layers of filter paper (9-cm diameter) in a glass Petri dish (9-cm diameter). Each dish contained 5 mL of aqueous leachate obtained from *L. camara*

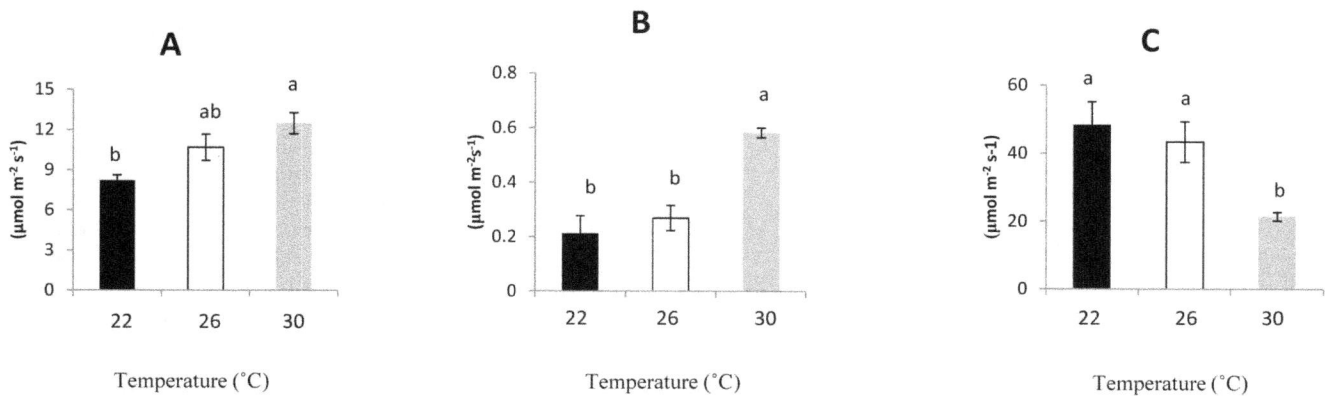

Figure 2. Gas exchange of _L. camara_ seedlings growing at different temperatures. Data are means ± SE (n = 15). (A) Pnet, Net photosynthetic rate; (B) Cond, stomatal conductance; (C) WUE, water use efficiency.

plants grown at different temperatures, while the controls contained 5 mL distilled water. The Petri dishes were kept in dark conditions at room temperature ($\sim$22°C). All of the treatments were conducted with four independent replicates. The germinated seeds (once radicle length was about 1–2 mm) were counted every 12 h for the first day, and every 24 h thereafter. Germination of _L. sativa_ was recorded up to 5 days, and seedling growth (root length and shoot length) was recorded at the end of the experiment, on the seventh day. The effect of the leachate on lettuce growth was evaluated using a response index (RI) [28] as follows:

$$RI(\%) = (T/C\text{-}1) \times 100 \qquad (1)$$

where C was the control value and T was the treatment value. RI>0 indicates a stimulatory effect, while RI<0 indicates an inhibitory effect.

Statistical analysis

The effect of different temperatures on growth, physiology and allelopathy of _L. camara_ was assessed by one-way ANOVA and means were compared by Tukey tests. All statistical analyses were performed using the software R 3.0.1 [29].

Results

Growth and morphology

The overall biomass of _L. camara_ plants, and their allocation of biomass to support organs (stem and petiole), were significantly higher in high-temperature treatments than in the control (22°C), while biomass allocation to roots (RMR) and leaves (LMR) displayed the opposite pattern (Table 1, Fig. 1). The leaf area, stem length and SLA of _L. camara_ were also significantly higher at the elevated temperatures (Table 1, Fig. 1). The stem length of plants growing at 30°C was four times that of those at 22°C, while their leaf area was double. SLA at the elevated temperature of 30°C was 22.1% higher than at 22°C.

Photosynthesis and chlorophyll fluorescence

Gas exchange. Plants growing at 30°C showed significantly higher Pnet and Cond, yet lower WUE than those growing at 22°C (Table 2, Fig. 2 A–C). No significant differences in any gas exchange parameters were found between seedlings growing at 22°C and 26°C, while seedlings growing at 26°C exhibited

significantly lower Cond and higher WUE than those growing at 30°C (Table 2, Fig. 2 B, C).

Chlorophyll fluorescence. No significant differences in Fv/Fm and ΦPSII were found among seedlings of _L. camara_ growing at the three different temperatures (Table 2, Fig. 3 A, B).

Allelopathic potential

The allelopathic effects of aqueous leachate from _L. camara_ leaves on seed germination and seedling development of lettuce at different temperatures were evaluated (Table 2, Fig. 4). The shoot and root length of lettuce significantly decreased with an increase in temperature, but there were no significant differences in germination between different temperatures. The allelopathic effects on shoot and root length were significantly greater at the higher temperatures compared to at 22°C, with the highest effect occurring at 26°C.

Discussion

Over the course of human history, people have intentionally or unintentionally moved innumerable plant species outside of their native ranges, and many of those alien plants become invasive [6]. Human activities are also partly responsible for the increase of global surface temperatures [21,30,31]. A number of recent studies on invasive plants and climate change have shown that increasing temperatures and changing precipitation might either "help" or "hinder" invasive plants, depending on the species, location and dominant forces causing changes in climate conditions [6,9,21,32]. Such variation makes it challenging to assess and understand the mechanisms that might facilitate or constrain the success of invasive species in the context of climate change [5].

As our findings illustrate, climate warming affects many aspects of the invasive species _L. camara_'s biology and ecology. Firstly, elevated temperature caused changes in the biomass allocation and morphology of plants. Plant growth is directly influenced by biomass allocation between leaves, stems, and other plant parts [33]. With rising temperature, individuals exhibited a significant increase in stem length, and biomass allocation to stems and petioles at the expense of leaves and roots. These changes may ensure greater structural support and an increased ability to capture light. Although biomass allocation to leaves decreased as in temperature increased, SLA and LA were greatest in the high-temperature treatments. SLA is a plant trait that is important for the regulation and control of functions such as carbon assimilation and carbon allocation [34,35]. Generally, the combination of

Table 2. F-values of one-way ANOVA which was used to test the effects of different temperatures on gas exchange and chlorophyll fluorescence, and allelopathic potential of *L. camara*.

	Pnet	Cond	WUE	Fv/Fm	ΦPSII	Germ	SH	RL
d.f.	2, 16	2, 16	2, 16	2, 16	2, 16	2, 16	2, 16	2,16
F	9.047	15.61	6.63	0.627	0.452	0.756	7.893	9.608
P	0.0035*	<0.001***	0.010*	0.547	0.645	0.497	0.001**	<0.001***

*** $P<0.001$,
** $P<0.01$,
* $P<0.05$.

Parameters of gas exchange and chlorophyll fluorescence:
Pnet: net photosynthetic rate; Cond: stomatal conductance; WUE: intrinsic water use efficiency; Fv/Fm; ΦPSII.
Parameters of allelopathic potential: Germ: germination; SH: shoot length; RL: root length.

increased SLA and LA results in increased light absorption, and shading of other species [36]. This may be a light utilization strategy that could enhance the competitive ability of *L. camara*, because this species cannot survive under the dense, continuous canopies of taller native forest species due to the lack of light [14]. *L. camara* usually flowers in the first growing season after its establishment and, if adequate moisture and light are available, it can flower in all seasons [37]. If the mean global temperature will rise by 1.4–5.8°C over the period of 1990–2100 as predicted [30], *L. camara* may increase in height more rapidly than its neighbours, and then suppress their growth by shading them. It may also flower more often and for longer periods of time, enabling it to produce more offspring, which could cause substantial damage to other species and their ecosystems. The responses to the increase in temperature observed in our study suggest that warming may help *L. camara* to reach further into the upper layer of the plant community, and expand its leaves as much as possible for better light capture and assimilation so as to facilitate its invasion.

Secondly, elevated temperature induced changes in the physiological parameters of *L. camara*. We found that plants exhibited a significantly higher photosynthetic capacity at higher temperatures (26 and 30°C) than at 22°C, which may be ascribed to a higher Pnet. High temperatures tended to increase stomatal conductance (Fig. 2 B, [38]), which can augment water loss. Consequently, the instantaneous water-use efficiency (WUE) of plants decreased with increasing temperature (Fig. 2 C). This response may explain why *L. camara* mostly invades wetter habitats [14]. High temperatures can influence photosynthesis in different ways, such as enhancing membrane fluidity and oxidative stress [38], or by changing the activity of the Calvin cycle and photorespiration [39]. High temperature may also inhibit the repair of PSII [40]. In this study, there were no significant differences in Fv/Fm and ΦPSII among different temperature conditions, and *L. camara* showed optimal functioning of its PSII with very low photoinhibition levels (Fv/Fm from ~0.750 to 0.870) after exposure to higher temperatures. This suggested that higher temperature did not lead to stress in *L. camara*. Higher photosynthesis can increase invasive plants' growth rates and biomass accumulation, which may enable invasive species to out-compete slower growing species and hence facilitate their colonization [41,42]. The responses we observed to elevated temperatures can be viewed as positive effects of warming on the physiological parameters of *L. camara*, i.e., increased rates of photosynthesis at higher temperatures could facilitate its invasive success. Of course, this enhancement of the plant's growth is often a "negative" effect at the ecosystem level.

Thirdly, elevated temperature induced changes in the allelopathic effects of *L. camara*. Those effects have been well-documented to cause severely reduced seedling recruitment in almost all species exposed to *L. camara*, and a reduction in the DBH growth of mature trees and shrubs [43,44]. The allelopathic effects of *L. camara* may explain why it can survive secondary succession and form monospecific thickets [14]. In our experiment, we found that its phytotoxicity increased with temperature, with respect to both seed germination and seedling growth of the receptor plants (Table 2, Fig. 4). This result is consistent with other research about how elevated temperature influences the allelopathic effects of invasive plants [21,45]. Allelopathic biochemicals produced by invasive plants function as their "novel weapons" since they can inhibit the growth of native plants in the invaded communities [46]. The increased phytotoxicity of *L. camara* in higher temperatures may be a result of the plant producing more allelochemicals, or of its allelochemicals becoming

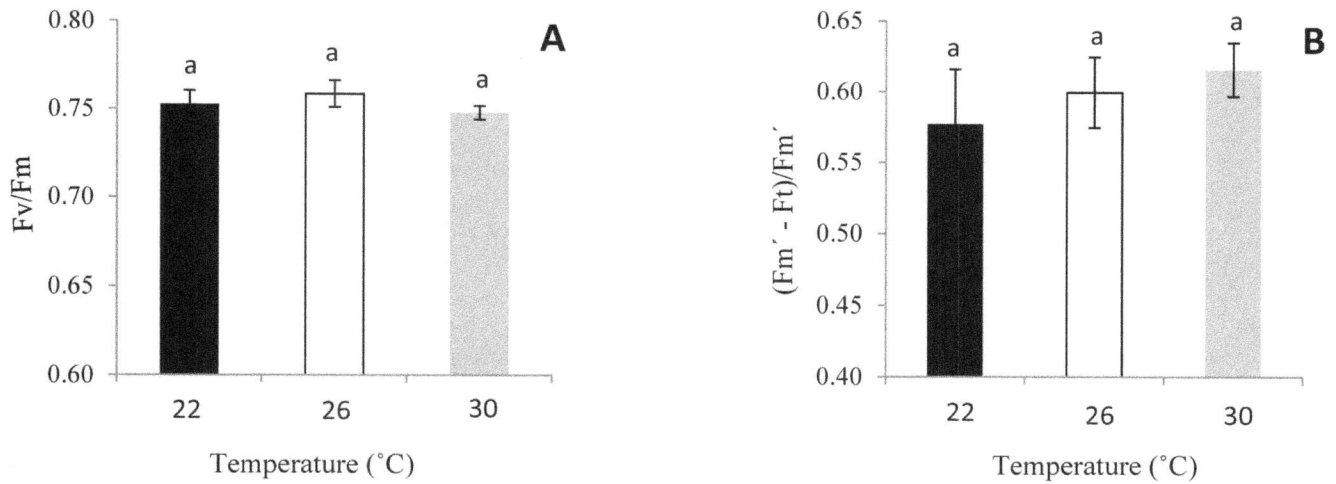

Figure 3. Chlorophyll fluorescence of *L. camara* seedlings growing at different temperatures. Data is means ±SE (n = 15). (A) Fv/Fm; (B) ΦPSII.

more phytotoxic under elevated temperatures [21]. As such, we can conclude that warming also enhanced the allelopathic potential of *L. camara*.

Biological invasions and climate change are key factors that are currently affecting global biodiversity [9] and the relationship between them is very complex [6]. In this study, we chose to study temperature, one of the most important elements of climate change, to understand its effects on plant invasion. The results

showed that elevated temperature resulted in significant increases in biomass allocation and beneficial changes in morphology, photosynthesis and allelopathic effects of *L. camara*, indicating that global warming could facilitate the invasion of this plant. Based on the predictions of climate models, a 1°C increase in mean annual temperature could result in a pole-ward shift of each of the world's vegetation zones by approximately 200 km [2]. If

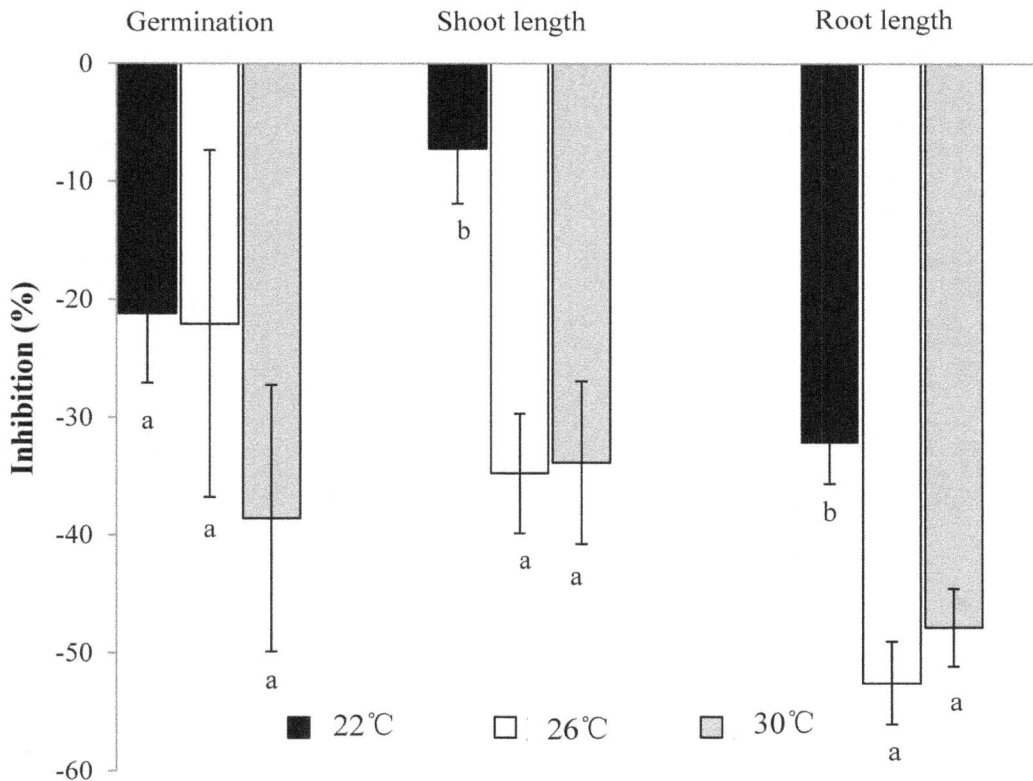

Figure 4. Allelopathic effects of aqueous leachate from fresh leaves of *L. camara* on seed germination, shoot length and root length of *Lactuca sativa*, expressed as a response index at different temperatures. Each bar represents a mean ± SE.

the global temperature increases by 1.4–5.8°C as predicted [30], *L. camara* will likely migrate toward higher latitudes.

Our experiment was conducted in the absence of competition from surrounding plants. In fact, competition with surrounding native species is one of the most important factors that influence the outcomes of invasion by alien plants. Under future climate scenarios, both native and invasive species are likely to grow more vigorously [42], which could affect competitive interactions in the invaded habitats. Future studies should also address the biotic factors that affect the invasiveness of *L. camara*.

References

1. IPCC (2007) Climate Change 2007: Synthesis Report. Summary for Policy-makers. Intergovernmental Panel on Climate Change, Cambridge University Press.
2. Walther GR, Post E, Convey P, Menzel A, Parmesan C, et al. (2002) Ecological responses to recent climate change. Nature 416: 389–395. doi:10.1038/416389a.
3. Parmesan C (2006) Ecological and evolutionary responses to recent climate change. Annual Review of Ecology, Evolution, and Systematics 37, 637–669.
4. Sorte CJB, Ibàñez I, Blumenthal DM, Molinari NA, Miller LP, et al. (2013) Poised to prosper? A cross-system comparison of climate change effects on native and non-native species performance. Ecology Letters 16(2): 261–270. doi: 10.1111/ele.12017.
5. Weitere M, Vohmann A, Schulz N, Linn C, Dietrich D, et al. (2009) Linking environmental warming to the fitness of the invasive clam *Corbicula fluminea*. Global Change Biology 15, 2838–2851. doi: 10.1111/j.1365-2486.2009.01925.x
6. Bradley B A, Blumenthal DM, Wilcove DS, Ziska LH (2010) Predicting plant invasions in an era of global change. Trends in Ecology & Evolution 25(5): 310–318.
7. Woodward FI (1987) Climate and Plant Distribution. Cambridge University Press.
8. Araújo MB, Luoto M (2007) The importance of biotic interactions for modelling species distributions under climate change. Global Ecology and Biogeography 16(6): 743–753.
9. Walther GR, Roques A, Hulme PE, Sykes MT, Pysek P, et al. (2009) Alien species in a warmer world: risks and opportunities. Trends in Ecology and Evolution 24(12): 686–693.
10. Williams AL, Wills KE, Janes JK, Vander Schoor JK, Newton PC, et al. (2007) Warming and free-air CO$_2$ enrichment alter demographics in four co-occurring grassland species. New Phytologist 176: 365–374.
11. Taylor S, Kumar L, Reid N, Kriticos DJ (2012) Climate Change and the Potential Distribution of an Invasive Shrub, *Lantana camara* L. PLoS ONE 7(4): e35565. doi:10.1371/journal.pone.0035565
12. Invasive Species Specialist Group, IUCN (2001) 100 World's Worst Invasive Alien Species. Available: http://www.issg.org/database/species/reference_files/100 English.pdf
13. Thomas SE, Ellison CA (2000) A Century of Classical Biological Control of *Lantana camara*: Can pathogens make a significant difference? pp. 97–104. In: Spencer, N.R. [ed.], Proceedings of the X International Symposium on Biological Control of Weeds, Bozeman, USA, 1999.
14. Day MD, Wiley CJ, Playford J, Zalucki MP (2003) Lantana: Current management Status and Future Prospects. ACIAR Monograph series, Canberra.
15. Shabani F, Kumar L, Taylor S (2012) Climate change impacts on the future distribution of Date Palms: A modelling exercise using CLIMEX. PLoS ONE 7(10): e48021. doi:10.1371/journal.pone.0048021.
16. Taylor S, Kumar L (2013) Potential distribution of an invasive species under climate change scenarios using CLIMEX and soil drainage: A case study of *Lantana camara* L. in Queensland, Australia. Journal of Environmental Management 114: 414–422.
17. Taylor S, Kumar L (2012) Sensitivity analysis of CLIMEX parameters in modelling potential distribution of *Lantana camara* L. PLoS ONE 7(7): e40969. doi:10.1371/journal.pone.0040969.
18. Taylor S, Kumar L, Reid N (2012) Impacts of climate change and land-use on the potential distribution of an invasive weed: a case study of *Lantana camara* in Australia. Weed Research 52: 391–401.
19. Shan JL (2003) Preliminary studies on exotic plant communities in Hainan. Chinese Journal of tropical agriculture 23(3):1–4;51.
20. Liang JY, Wu SS (2000) Climatological diagnosis of winter temperature variations in Guangdong. Journal of Tropical Meteorology 6: 37–45.
21. Wang RL, Zeng RS, Peng SL, Chen BM, Liang XT, et al. (2011) Elevated temperature may accelerate invasive expansion of the liana plant *Ipomoea cairica*. Weed research 51: 574–580.
22. Poorter L (1999) Growth responses of 15 rain-forest tree species to a light gradient: the relative importance of morphological and physiological traits. Functional Ecology 13(3): 396–410. doi:10.1046/j.1365-2435.1999.00332.x.
23. Zhang YC, Zhang QY, Luo P, Wu N (2009) Photosynthetic response of *Fragaria orientalis* in different water contrast clonal integration. Ecological Research 24(3): 617–625.
24. Schreiber U, Bilger W, Hormann H, Neubauer C (1998) Chlorophyll fluorescence as a diagnostic tool: basics and some aspects of practical relevance. In: Photosynthesis: A Comprehensive Treatise (ed. A. S. . Raghavendra) pp. 320–336. Cambridge University Press, Cambridge.
25. Brugnoli E, Björkman O (1992) Chloroplast movements in leaves - influence on chlorophyll fluorescence and measurements of light-induced absorbency changes related to Delta-Ph and Zeaxanthin formation. Photosynthesis Research 32: 23–35.
26. Bolhar-Nordenkampf HR, Long SP, Baker NR, Oquist G, Schreiber U, et al. (1989) Chlorophyll fluorescence as a probe of the photosynthetic competence of leaves in the field: A review of current instrumentation. Functional Ecology 3: 497–514.
27. Genty B, Briantais JM, Baker NR (1989) The relationship between the quantum yield of photosynthetic electron transport and quenching of chlorophyll fluorescence. Biochimica et Biophysica Acta 990: 87–92.
28. Williamson GB, Richardson D (1988) Bioassays for allelopathy: measuring treatment responses within dependent control. Journal of Chemical Ecology 14(1): 181–187.
29. R Development Core Team (2013) R: A Language and Environment for Statistical Computing. R Foundation for Statistical Computing, Vienna, Austria. ISBN 3-900051-07-0, URL http://www.R-project.org
30. Houghton JT, Ding Y, Griggs DJ, Noguer M, van der Linden PJ, et al. (2001) The projections of the earth's future climate. In: Climate Change 2001: The Scientific Basis (eds JT . Houghton, Y . Ding, DJ . Griggs, M . Noguer, PJ . Van Der Linden et al), 62–77. Cambridge University Press, Cambridge, UK.
31. Li MH, Kräuchi N, Gao SP (2006) Global warming: can existing reserves really preserve current levels of biological diversity? Journal of Integrative Plant Biology 48: 255–259.
32. Bradley BA, Oppenheimer M, Wilcove DS (2009) Climate change and plant invasion: restoration opportunities ahead? Global Change Biology 15: 1511–152.
33. Reich PB, Tjoelker MG, Walters MB, Vanderklein DW, Buschena C (1998) Close association of RGR, leaf and root morphology, seed mass and shade tolerance in seedlings of nine boreal tree species grown in high and low light. Functional Ecology 12(3): 327–338. doi:10.1046/j.1365-2435.1998.00208.x.
34. Reich PB, Walters MB, Ellsworth DS (1997) From tropics to tundra: global convergence in plant functioning. Proc Natl Acad Sci USA 94(25): 13730–13734. doi:10.1073/pnas.94.25.13730. PMID:9391094.
35. Lambers H, Poorter H (2004) Inherent variation in growth rate between higher plants: a search for physiological causes and ecological consequences. Academic Press, London, New York.
36. Qin Z, Mao DJ, Quan GM, Zhang JE, Xie JF et al (2012) Physiological and morphological responses of invasive *Ambrosia artemisiifolia* (common ragweed) to different irradiances. Botany 90: 1284–1294.
37. Sharma GP, Raghubanshi AS, Singh JS (2005) Lantana invasion: an overview. Weed Biology and Management 5: 157–167.
38. Carrion-Tacuri J, Rubio-Casal AE, de Cires A, Figueroa ME, Castillo JM (2013) Effect of low and high temperatures on the photosynthetic performance of *Lantana camara* L. leaves in darkness. Russian Journal of Plant Physiology 60(3): 322–329.
39. Yordanov I (1992) Response of Photosynthetic Apparatus to Temperature Stress and Molecular Mechanisms of Its Adaptation. Photosynthetica 26: 517–531
40. Allakhverdiev SI, Kreslavskii VD, Klimov VV, Los DA, Carpentier R, et al. (2008) Heat stress: An overview of molecular responses in photosynthesis. Photosynthesis Research 98: 541–550.
41. Lambers H, Poorter H (1992) Inherent variation in growth rate between higher plants: A search for physiological causes and ecological consequences. Adv Ecol Res 23: 187–261.
42. Anderson LJ, Cipollini D (2013) Gas exchange, growth, and defense responses of invasive *Alliaria petiolata* (Brassicaceae) and native *Geum vernum* (Rosaceae) to elevated atmospheric CO$_2$ and warm spring temperatures. American Journal of Botany 100(8): 1544–1554.

Acknowledgments

We thank Marina Semchenko and John Davison for their helpful comments and linguistic corrections. Furong Li and Na Zhao kindly helped with the experiments.

Author Contributions

Conceived and designed the experiments: QZ YZ SP. Performed the experiments: QZ YZ. Analyzed the data: YZ QZ. Wrote the paper: QZ YZ KZ.

43. Gentle CB, Duggin JD (1997) Allelopathy as a competitive strategy in persistent thickets of *Lantana camara* L. in three Australian forest communities. Plant Ecology 132: 85–95.

44. Zhang QY, Peng SL, Zhang YC (2009) Allelopathic potential of reproductive organs of exotic weed *Lantana camara*. Allelopathy Journal 23:213–220.

45. Lobón NC, Gallego JC, Daz TS, Garcia JC (2002) Allelopathic potential of Cistus ladanifer chemicals in response to variations of light and temperature. Chemoecology 12: 139–145.

46. Callaway RM, Aschehoug ET (2000). Invasive plants versus their new and old neighbors: A mechanism for exotic invasion. Science 290(5491): 521–523.

7

Appraisal of Artificial Screening Techniques of Tomato to Accurately Reflect Field Performance of the Late Blight Resistance

Marzena Nowakowska[1], Marcin Nowicki[1], Urszula Kłosińska[1], Robert Maciorowski[2], Elżbieta U. Kozik[1]*

1 Department of Genetics, Breeding, and Biotechnology of Vegetable Crops, Research Institute of Horticulture, Skierniewice, Poland, 2 Unit of Economics and Statistics, Research Institute of Horticulture, Skierniewice, Poland

Abstract

Late blight (LB) caused by the oomycete *Phytophthora infestans* continues to thwart global tomato production, while only few resistant cultivars have been introduced locally. In order to gain from the released tomato germplasm with LB resistance, we compared the 5-year field performance of LB resistance in several tomato cultigens, with the results of controlled conditions testing (i.e., detached leaflet/leaf, whole plant). In case of these artificial screening techniques, the effects of plant age and inoculum concentration were additionally considered. In the field trials, LA 1033, L 3707, L 3708 displayed the highest LB resistance, and could be used for cultivar development under Polish conditions. Of the three methods using controlled conditions, the detached leaf and the whole plant tests had the highest correlation with the field experiments. The plant age effect on LB resistance in tomato reported here, irrespective of the cultigen tested or inoculum concentration used, makes it important to standardize the test parameters when screening for resistance. Our results help show why other reports disagree on LB resistance in tomato.

Editor: Mark Gijzen, Agriculture and Agri-Food Canada, Canada

Funding: This research was supported by The Polish Ministry of Agriculture and Rural Development (Grant # HOR hn-801-15/13; "Searching for new sources of resistance to tomato late blight with regard to pathogenicity changes in Phytophthora infestans populations and attempts of markers identification linked with resistance genes"), carried out in the period 2008–2013. The funders had no role in study design, data collection and analysis, decision to publish, or preparation of the manuscript.

Competing Interests: The authors have declared that no competing interests exist.

* Email: elzbieta.kozik@inhort.pl

Introduction

Tomato (*Solanum lycopersicum* L.) is the fourth most economically important crop in the world: after rice, wheat, and soybean [1,2]. Of the 200 pathogens affecting tomato production worldwide, *Phytophthora infestans* (Mont.) de Bary, the oomycete causing late blight (LB), is the primary cause of tomato crop loss [2]. Losses are more frequent and more severe in areas where tomato is grown near potato [3].

Research on plant age-dependent expression of *P. infestans* resistance in tomato is needed, especially for the method of inoculation, plant evaluation, and inoculum concentration. In contrast, numerous studies have been done on the methods for testing *P. infestans* resistance in potato [4–14]. Tomato germplasm exist with resistance against *P. infestans* [15–20]; however, little information is available on standardized methods for evaluating LB resistance in tomato. Information on the correlation between seedlings and mature plant resistance is also limited [21–23]. Moreover, the last systematic comparison of methods for testing tomato LB resistance was reported over 40 years ago [24]. Lack of standardization already causes problems. For example, the wild tomato accession *S. pimpinellifolium* L 3708 was reported to have LB resistance conferred by one incompletely dominant gene [15,25]. That was amended a mere seven years later, when other studies reported complex inheritance of the trait [26–29]. Similarly, the accession *S. habrochaites* LA 1033 was designated *Ph - 4* and used as a LB resistance standard by several research centers [30–32], despite evidence of multiple QTLs conferring the trait [17,33–35]. Such discrepancies make it difficult for tomato breeders to use resistant germplasms.

The objective of this study was to (i) systematically compare the *P. infestans* resistance in the previously and recently reported resistant tomatoes using an array of methods, both in the field (naturally occurring infections) and with artificial inoculation (employing whole plant and detached leaf/leaflet bio-assays); (ii) determine the importance of the age-dependent expression of LB resistance; and (iii) investigate the level and stability of LB resistance of four sources of this trait in a multi-year field experiment.

Materials and Methods

Plant material

Tomato germplasm used in this study included commercial cultivars, breeding lines, landraces, and wild tomato accessions from the tomato germplasm collection at the Research Institute of Horticulture (RIH; Skierniewice, Poland), collectively referred to as 'cultigens'. Cultigens included three accessions of *S. pimpinelli-*

Table 1. Tomato cultigens used in the study, their origin, and LB resistance status (if known).

Cultigen[a]	Origin			Seed source
	country	province	collection site	
S. pimpinellifolium				
West Virginia 700 (WVa 700)[b]	-	-	-	INRA, Montfavet, France
L 3707 (PI 365951)[c]	Peru	-	-	Bar-Ilan University, Ramat-Gan, Israel
L 3708 (PI 365957)[d]	Peru	Lima	Pisiquillo	Bar-Ilan University, Ramat-Gan, Israel
S. habrochaites				
LA 1033 (6326A)[e]	Peru	Lambayeque	Hacienda Tanlis	NCSU, Raleigh, USA
LA 1353 (365934)	Peru	Cajamarca	Contumasa	TGRC, Davis, USA
LA 2552 (PE 36)	Peru	Cajamarca	Las Flores	TGRC, Davis, USA
LA 2650 (PI 503514)	Peru	Prura	Ayabaca	TGRC, Davis, USA
LA 407	Ecuador			TGRC, Davis, USA
S. huaylasense				
LA 1360 (PI 365952)	Peru	Ancash	Apricot	TGRC, Davis, USA
LA 1365 (PI 365953)	Peru	Ancash	Carnaquilloc	TGRC, Davis, USA
LA 1983	Peru	Ancash	Rio Manta	TGRC, Davis, USA
S. corneliomuelleri				
LA 1395 (PI 379014)	Peru	Amazonas	Chachapoyas	TGRC, Davis, USA
LA 1910	Peru	Huancavelica	Tambillo	TGRC, Davis, USA
S. peruvianum				
LA 1929	Peru	Ica	La Yapana	TGRC, Davis, USA
LA 2581	Chile	Arica i Parinacota	Chacarilla	TGRC, Davis, USA
LA 2744	Chile	Arica i Parinacota	Sobraya	TGRC, Davis, USA
S. lycopersicum				
LA 1673	Peru	Lima	Nana	TGRC, Davis, USA
LA 2416	-	-	-	TGRC, Davis, USA
LA 3845 (NC EBR–5)	-	-	-	NCSU, Raleigh, USA
LA 3846 (NC EBR-6)	-	-	-	NCSU, Raleigh, USA
'New Yorker' ('NY')[f]	-	-	-	INRA, Montfavet, France
LB susceptible control				
'Rumba'	-	-	-	PNOS Ożarów, Poland

[a]Plant introductions numbers (PI or PE) were added where available.
[b]Source of *Ph-2*; single incomplete-dominant gene mapped to the long arm of chromosome 10 [19,24,38].
[c]Source of race-non-specific LB resistance conferred by two independent genes: A partially-dominant gene and a dominant epistatic gene, both mapped to chromosome 9 [46,47].
[d]Source of race-specific LB resistance dubbed *Ph-3*; originally reported as conferred by single incomplete-dominant gene, corrected by subsequent research; resistance conferred by at least two QTLs [15,25–29,31–33,46,67].
[e]Source of polygenic LB resistance, conferred by several QTLs [16,17,31–35].
[f]Source of LB resistance dubbed *Ph-1*, conferred by a single completely dominant gene mapped to the distal end of chromosome 7 with morphological markers [36–39].

folium, five of *S. habrochaites*, three of *S. huaylasense*, two of *S. corneliomuelleri*, three of *S. peruvianum*, and four of *S. lycopersicum*. 'Rumba' (*S. lycopersicum*; PNOS Ożarów, Poland) served as the LB susceptible control, and it was readily infected by *P. infestans* under field conditions. The cultigens used in this study, their origin, and LB resistance background (if known) are listed in Table 1.

Plants were grown from seeds, transplanted at first true leaf stage into ⌀ 10 cm plastic pots containing Classman Potgrond substrate (Lasland, Grady, Poland), and placed in a greenhouse. Growing plants were kept at approximately 24/18°C day/night for the tests under controlled conditions, or grown for 4 to 5 weeks in the greenhouse, and then transplanted to the field in the second half of May each year.

Field evaluations

In 2007 and 2008, all cultigens were evaluated for resistance against *P. infestans* under epiphytotic conditions at the Department of Genetics, Breeding, and Biotechnology experimental field area (RIH, Skierniewice, Poland). Four- to five-week old plants were transplanted to the field in a randomized complete block design, with three replications. Each plot consisted of ten *S. lycopersicum* plants or five plants of the wild tomato cultigens, spaced 50×100 cm (within rows × between rows, respectively). Additionally, plants of the LB susceptible 'Rumba' were planted around the border of the experimental field to ensure high and uniform pathogen pressure. We also included a *S. lycopersicum* cv. New Yorker ('NY') which carries the LB resistance gene *Ph-1* [36–

39]. No fungicide control was applied throughout the growing seasons in any of the field trials.

Field plots were inspected weekly throughout the season. When the susceptible control 'Rumba' foliage developed maximal area symptomatic for LB in a given year, disease ratings for each cultigen were collected. Observations were made by the same individual throughout the experiment to avoid that source of variation. LB lesions on the leaves and stems of each plant were rated using the modified scale of Zarzycka [40]:

- 1 – infection over 97.1 to 100% of the investigated plant organ
- 2 – infection over 87.1 to 97% of the investigated plant organ
- 3 – infection over 75.1 to 87% of the investigated plant organ
- 4 – infection over 50.1 to 75% of the investigated plant organ
- 5 – infection over 30.1 to 50% of the investigated plant organ
- 6 – infection over 18.1 to 30% of the investigated plant organ
- 7 – infection over 10.1 to 18% of the investigated plant organ
- 8 – infection over 3.1 to 10% of the investigated plant organ
- 9 – no infections or small lesions.

At the beginning of this study, such arithmetically biased methods of LB assessment were employed predominantly [5,8,15,19,20,24,25,31,35], and keep on being used until this day [9,16,21,26], although more accurate methods were developed [7,11,29].

Subsequently, the disease severity index (DSI) was calculated for each cultigen, respectively, as a mean of ratings for the plants, similar to other studies of this pathosystem [26,40].

In the initial field screenings performed in 2007 and 2008, four cultigens (WVa 700, L 3708, L 3707, LA 1033) exhibited high *P. infestans* resistance. In order to assess the stability of resistance in these cultigens, tests were conducted in 2009 to 2013, under natural LB infestations. Field trials were run in two locations: RIH, Skierniewice (Central Poland), and Boguchwała (Southern Poland), which are 300 km apart from one another. In these experiments, tomato plants were evaluated for resistance against *P. infestans* as described earlier.

Pathogen cultures and inoculum preparation

P. infestans isolates used in this study, collected from tomato plants grown in different regions of Poland and tested in a pilot study (Fig. 1), were deposited at RIH (Skierniewice; n = 19) or obtained from IHAR (the Plant Breeding and Acclimatization Institute - National Research Institute, Młochów, Poland; n = 27 isolates). Isolates were transferred from rye agar onto leaflets of cv. Rumba and cultured for at least two generations, each 7 to 8 days long, with incubations in darkness at 100% RH and 16°C [40]. Inoculum consisted of a sporangial suspension that was washed off the sporulating lesions on the 'Rumba' leaflets using distilled water. Sporangia counting was done with a haemocytometer, and final inoculum concentrations (inocula loads) were adjusted according to the assay protocol described below. Prior to dilution, the suspension was chilled for 2 h at 4°C, and then incubated at RT for 30 min.

The isolate IWP 13, collected in 2008 from our experimental field (RIH Skierniewice), was used in all subsequent tests. Detailed isolate characteristics are as follows: race according to Black [3,4,7,10,11], mating type A1, mtDNA haplotype Ia, and intermediate resistance to metalaxyl [41]. This isolate was chosen based on the results of three independent disease severity tests. These tests were performed with 46 isolates of *P. infestans* on the detached leaflets of cv. Rumba, where it induced the highest disease severity (DSI = 1.7 ± 0.1). Moreover, it produced the

smallest variation in symptoms among the isolates tested in a pilot study on 'Rumba' (Fig. 1).

Seedling tests

Resistance screening experiments under controlled conditions were conducted in growth chambers and in greenhouse at the RIH, Skierniewice, Poland. These assays included detached leaflet, detached leaf, and whole plant bio-assays, and are described in detail in the paragraphs that follow. In all test methods we evaluated the influence of plant developmental stage (age in weeks) and inoculum concentration on the disease severity in four tomato cultigens with various levels of *P. infestans* resistance/susceptibility: WVa 700, L 3708, LA 1033, and 'Rumba'. All tests were conducted in the spring and summer (April to September) of 2011 to 2013.

The detached leaflet assay

We determined the effects of the plant age (4- to 8-week old plant, in one week increments) and the inoculum concentration (5×10^3, 10^4, 5×10^4 sporangia/ml) on disease symptoms development. Each treatment combination (inoculum concentration vs. plant age) was tested in a series of three independent trials with 25 leaflets per cultigen. The third to fifth fully expanded leaves, counting from the plant's top, were collected. This criterion was established by our earlier experiments as well as by previous studies [42] on the relationship between the intensity of LB symptoms and leaf position on plants. Side leaflets were detached with scissors and immediately placed on wet cellulose wadding in a plastic tray. Sporangial suspension (40 μl) was placed on the center of the abaxial side of each leaflet. Trays with leaflets were covered with glass to maintain 90 to 100% RH and incubated at 16°C. After an initial 24 h incubation in the dark, the leaflets were turned abaxial side down and incubated at the same temperature and RH for 6 days under 12 h photoperiod and a light intensity (PPFD) of 650 μmol/m²/s. Symptoms of LB were assessed 7 days after inoculation, using the DSI scale described earlier.

In our preliminary tests with 20 local isolates of the pathogen, individual leaflets of the resistant cultigens displayed high variability in LB reaction. Additionally, the size of necrotic lesions did not correspond with the intensity of sporulation. Within susceptible cultigens (i.e., compatible reactions) that was not observed, and in agreement with other studies [43] the sporulation often preceded the necrosis. These differences caused difficulties in classifying cultigens using this method. Therefore, at the time of assessment by the DSI scale described above, the inoculated leaflets were tested for sporulation intensity by shaking off the sporangia (vortexing each sample for 30 sec in 1 ml of water) and counting the sporangia with a haemocytometer.

The detached leaf assay

The severity of LB symptoms on detached tomato leaves was studied in 11 stages of development (5- to 15-week old plants, in one week increments), using three concentrations of inoculum (5×10^3, 10^4, 5×10^4 sporangia/ml). The third to fifth fully expanded leaves (counting from the plant's top) were excised from greenhouse-grown plants. The petiole of each leaf was immediately placed into distilled water (100 ml) in a plastic container with a ⌀ 1 cm hole in the lid center. The leaves in plastic containers were then placed in a plastic box and hand-sprayed with sporangial suspension, until the upper surface of each excised tomato leaf was completely covered. The inoculated leaves were incubated in darkness for 24 h at 18°C, and 95 to 100% RH. Following this initial incubation, samples were then incubated for 6 days under conditions similar to the detached leaflet method.

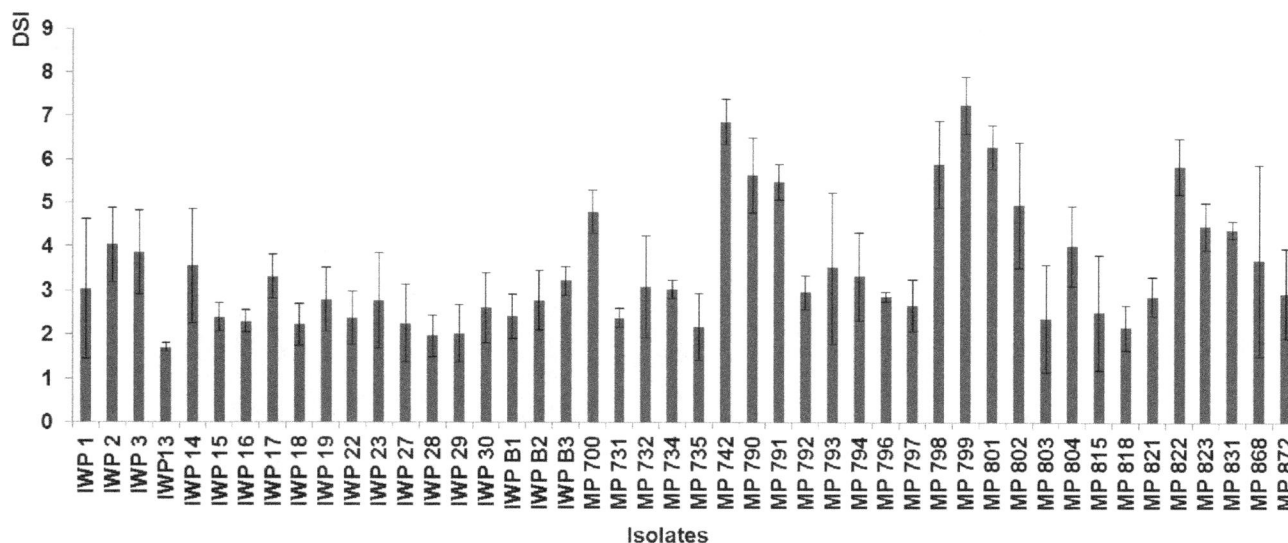

Figure 1. DSI from artificial *P. infestans* inoculations of 'Rumba' detached leaflets. A total of 46 local isolates were used. Each isolate was tested on 10 leaflets, in a series of three independent assays. Inoculum concentration of 5×10^4 sporangia/ml was used. Assay evaluations were performed on the 7th day after inoculation. Disease assessment scale was based on the % of leaflet area being infected; 1 = 100% area infected; 9 = lack of disease symptoms or few and small necrotic spots. Error bars indicate standard deviation (SD).

Leaves of each cultigen were evaluated using the scale described earlier (1 = 97.1–100% leaf area covered with lesions or dead; to 9 = no lesions). For each of the four cultigens undergoing evaluation, and for each treatment combination (inoculum concentration vs. plant age), a different number of leaves (12 to 24) per cultigen was used due to availability of plant materials. In addition, each treatment combination was examined in three independent sets of experiments.

The whole plant assay

LB severity on whole plants was tested at six developmental stages (3- to 8-weeks of age, in one week increments) of greenhouse-grown plants, using three concentrations of inoculum (5×10^3, 10^4, 5×10^4 sporangia/ml). Plants were hand-sprayed with the sporangial suspension until complete leaf coverage and excess run-off was observed. The inoculated seedlings were incubated in the dark at growth chamber for 24 h, at 16°C and 100% RH. After this initial incubation, the inoculated seedlings were grown at 16°C with 12 h of light. The plants were rated individually seven days after inoculation. The symptomatic area of both leaves and stems were evaluated using the DSI scale described earlier. Each treatment combination (developmental stage vs. inoculum concentration) was repeated three to seven times, with each cultigen represented by 12 to 72 plants depending on the experiment.

Statistical analyses

Data from all experiments were analyzed by means of the general linear model with year, location, cultigen, plant age, inoculum concentration, and their interaction as the tested variables. Means were separated with the Tukey multiple comparison procedure at significance level of 0.05. Regression and correlation analyses were used to compare results from the different testing methods. All calculations were done with the statistical software STATISTICA 8.0 (StatSoft, Inc. 2009).

Results

Field evaluations

Based on the results from the initial field experiments in 2007 and 2008 (Table 2), four tomato cultigens were chosen (LA 1033, L 3708, L 3707, WVa 700) from the original experimental group of 20 cultigens, to verify their *P. infestans* resistance in the field at two separate locations (Skierniewice, Boguchwała).

Comparative analysis of the LB resistance levels among the tested tomato cultigens indicated significant differences across the five years of study (2009 to 2013), at both locations (Table 3). The highest and most stable level of LB resistance in the field, comparable with the baseline years 2007 and 2008, was found in LA 1033 *S. habrochaites* (DSI ranging from 7.2 to 9.0, depending on the year). Plants of this cultigen were either free from any LB symptoms, or developed only slight infection symptoms (classes 6 to 8). Indeed, this cultigen demonstrated high levels of resistance in the field, even under strong *P. infestans* pressure. For example, in 2011, both locations experienced conditions particularly conducive to LB, as confirmed by disease intensities recorded for both susceptible tested tomato cultigens ('Rumba', 'NY'), while LA 1033 showed no LB symptoms (Table 3). Moreover, LA 1033 plants displayed a moderate infection intensity on leaves and stems only in the first week of September 2012 (DSI = 7.2±0.8) and 2013 (DSI = 7.8±0.7), in the Skierniewice experimental field.

Intensity of LB symptoms on the *S. pimpinellifolium* accessions L 3708 and L 3707 which ranged from DSI = 7.4 to 8.8, was generally comparable with *S. habrochaites* LA 1033, and depended on the year and the location. The WVa 700 plants displayed significant differences in LB intensity levels, depending on the location and the year (Table 3). In Skierniewice, we recorded partial infection of this cultigen in 2011 and 2013 (DSI = 4.9±1.0 and 4.5±1.1, respectively). In the remaining years of the study, the WVa 700 plants exhibited very low LB intensities (DSI = 8.0 to 8.4). In the Boguchwała study, we observed consistently high levels of disease severity in this cultigen in all years of testing (DSI ranging from 4.3 to 6.1). The 'NY' (*Ph-1*)

Table 2. Tomato cultigens' responses to *P. infestans* under natural infection in the Skierniewice experimental field.

Cultigen	DSI[a]		Mean
	2007	2008	
WV 700	8.0±0.5	8.3±0.3	8.2
L 3707	8.0±0.4	9.0±0.0	8.5
L 3708	8.1±0.6	9.0±0.0	8.5
LA 1033	8.8±0.1	9.0±0.0	8.9
LA 1353	6.3±1.1	6.9±0.7	6.6
LA 2552	7.0±0.2	5.5±0.3	6.3
LA 2650	5.0±1.1	5.8±0.8	5.4
LA 407	4.7±1.2	5.3±1.1	5.0
LA 1360	3.1±1.2	2.7±1.1	2.9
LA 1365	3.8±1.5	2.8±0.8	3.3
LA 1983	1.9±0.2	1.5±0.2	1.7
LA 1395	2.6±1.2	2.8±0.7	2.7
LA 1910	1.5±0.2	1.6±0.3	1.6
LA 1929	1.4±0.2	1.8±0.2	1.6
LA 2581	1.0±0.0	1.0±0.0	1.0
LA 2744	1.5±0.2	1.4±0.3	1.5
LA 1673	2.6±1.5	3.0±1.1	2.8
LA 2416	2.1±1.0	2.4±0.7	2.3
LA 3845	1.2±0.2	1.1±0.2	1.2
LA 3846	1.0±0.0	1.0±0.0	1.0
'NY'	2.6±1.1	1.4±0.2	2.0
'Rumba'	2.1±1.6	1.2±0.2	1.7

[a]The disease assessment scale is based on the % area of leaf and stem infection (1 = 100% percent area infected; 9 = lack of symptoms or few small necrotic spots). Disease symptoms were scored yearly, when LB susceptible control 'Rumba' turned fully symptomatic for the infection. Data presented is the mean ± SD from DSI recorded on leaves and stems. $LSD_{0.05}$ calculated according to Tukey procedure for comparing the cultigens in each year: 1.59, and for comparing the years for each cultigen: 0.99.

developed intensive LB symptoms, at levels comparable with the susceptibility control 'Rumba', irrespective of the location or year of study (DSI = 1.0 to 4.3).

The detached leaflet assay

Experimental results indicated significant effects of the plant age on the LB intensity levels across all tested cultigens. Leaflets from 4- to 5-week old plants of both *S. pimpinellifolium* cultigens tested (WVa 700, L 3708) exhibited higher degrees of infection than those from 7- to 8-week old plants, irrespective of the inoculum concentration used (Fig. 2). The LA 1033 leaflets showed the highest intensity of LB symptoms when inoculated with the 5×10^4 sporangia/ml (DSI = 4.8 to 5.8, dependent on plant age). Comparatively, at the two lower inoculum concentrations used, the disease symptoms proved significantly decreased at all developmental stages tested.

We found a broad range of intra-cultigen variation in the lesion area of individual leaflets of resistant cultigens (WVa 700, L 3708, LA 1033) inoculated with *P. infestans*. This variation was observed even in the oldest specimens (8-week old) tested with the lowest inoculum load (Fig. 2). The observed high variability of this assay makes it impossible to unequivocally distinguish genotypes exhibiting specific LB reactions. Regardless of the differences in the necrotic area, the LB resistant cultigens showed a low intensity of sporulation and were not significantly different from each other. 'Rumba' displayed sporulation intensity of

23,4±5.3 thousands of sporangia/mm^2, exceeding those of the resistant cultigens approximately 50-fold (WV 700: 0.7±0.6; L 3708: 0.5±0.5; LA 1033: 0.5±0.5 thousands of sporangia/mm^2, respectively). Also, only in case of 'Rumba', did the size of necrotic spots correspond well with the intensity of sporulation. Finally, in contrast to the resistant cultigens tested, 'Rumba' showed consistently high and uniform disease symptoms in all detached leaflet assays (Fig. 2), regardless of plant age or inoculum load.

The detached leaf assay

Significant effects of all variables tested (cultigen, plant age, and inoculum concentration) on the LB intensity were observed. The highest variation was observed for the interaction between cultigen and plant age, and the lowest was between plant age and inoculum concentration.

'Rumba' leaves displayed the highest degree of LB infection (DSI = 1.0 to 1.1), regardless of the plant age, and lacked significant differences in their respective DSI scores (Fig. 3). Relationships between plant age and degree of disease symptoms were observed for all three remaining cultigens in the study (WVa 700, L 3708, LA 1033). Additionally, we noted a different trend in each of the three resistant cultigens. The WVa 700 leaves showed low levels of LB intensity, irrespective of plant age or inoculum concentration. The lowest LB symptoms levels displayed by L 3708 were similar to those of WVa 700, but proved comparatively more dependent on plant age and inoculum concentration (Fig. 3).

Table 3. Late blight symptoms severity on chosen tomato cultigens under natural *P. infestans* field infections.

Year[a]	Cultigen	DSI[b]	
		Skierniewice	Boguchwała
2009	LA 1033	8.9±0.1	8.7±0.4
2010		8.8±0.2	8.8±0.2
2011		9.0±0.0	8.0±1.1
2012		7.2±0.8	9.0±0.0
2013		7.8±0.7	8.2±0.6
2009	WVa 700	8.0±0.4	6.1±0.9
2010		8.4±0.3	5.7±1.0
2011		4.9±1.0	5.0±0.7
2012		8.0±0.4	5.7±0.5
2013		4.5±1.1	4.3±1.3
2009	L 3708	7.9±0.4	7.9±0.3
2010		8.6±0.2	8.5±0.2
2011		8.7±0.0	8.0±0.3
2012		8.3±0.4	8.8±0.2
2013		8.2±0.4	7.9±0.7
2009	L 3707	7.8±0.4	7.4±0.6
2010		8.5±0.3	7.5±0.8
2011		8.3±0.7	8.0±0.3
2012		8.3±0.2	8.2±0.2
2013		7.4±0.5	8.3±0.7
2009	'NY'	4.3±1.4	1.0±0.0
2010		2.4±1.2	1.2±0.2
2011		1.0±0.0	1.2±0.2
2012		1.0±0.0	2.6±0.3
2013		1.7±1.3	1.0±0.0
2009	'Rumba'	3.6±1.4	1.0±0.0
2010		2.7±1.7	1.4±0.2
2011		1.0±0.0	1.1±0.2
2012		1.0±0.0	1.0±0.0
2013		1.8±0.9	1.0±0.0

[a]Data were collected from 2009 to 2013 in two locations (Skierniewice, Boguchwała).
[b]The disease assessment scale is based on the % area of leaf and stem infection (1 = 100% percent area infected; 9 = lack of symptoms or few small necrotic spots). Disease symptoms were scored yearly, when LB susceptible control 'Rumba' reached maximal intensity of LB symptoms. Data presented is the mean ± SD from DSI recorded on leaves and stems. $LSD_{0.05}$ calculated according to Tukey procedure for comparing cultigens in each year and in either location: 0.76, and for comparing cultigens in each year and between locations: 0.55.

In L 3708, at the lowest inoculum concentration tested, the impact of plant age on LB intensity was evident, as response dropped below the significance level in 10-week old and older plants. Cultigen LA 1033 also showed varied reactions to *P. infestans* inoculation, dependent on plant age; the observed differences greatly surpassed those noted for WVa 700 or L 3708 (Fig. 3). The lowest degree of disease symptoms in LA 1033 was observed in the leaves of 15-week old plants (DSI = 8.8±0.4), after application of the lowest concentration of inoculum.

The whole plant assay

As in the detached leaf experiments, also in the whole plant assays we found significant effects of all variables tested (cultigen, plant age, inoculum concentration) on the intensity of LB symptoms. This necessitated an independent analysis for each variable.

In 'Rumba', we observed a significant interaction between the disease severity and the inoculum concentration (Fig. 4). Plants inoculated with the highest inoculum concentration exhibited high disease severity levels (DSI = 1.0 to 1.6, depending on plant age) in all tested stages of development (3 to 8 weeks of age). In contrast, plants treated with the lowest inoculum concentration showed lower intensity of disease symptoms (DSI = 2.6 to 5.4, depending on plant age). This wide DSI range was dependent on plant age, and for each developmental stage tested was distributed over at least two of the nine severity rating classes (reaching max. 7 classes of spread). A low degree of disease symptoms coupled with large variation in the LB susceptible 'Rumba' under 5×10^3 sporangia/

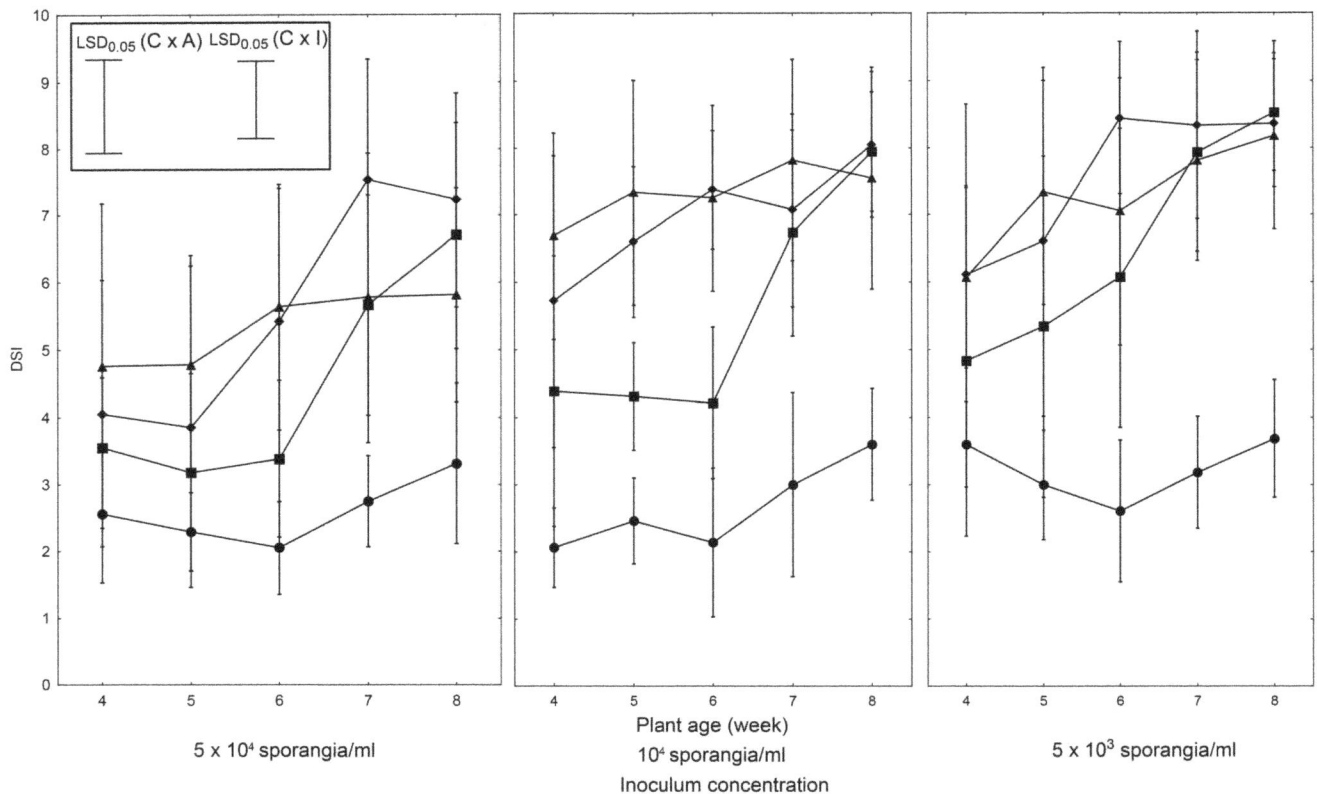

Figure 2. Severity of LB symptoms on tomato cultigens in the detached leaflet assay. Detached leaflets of tomato cultigens (● 'Rumba'; ■ WVa 700; ◆ L 3708; ▲ LA 1033) were inoculated with the *P. infestans* isolate IWP 13, at indicated age (weeks), and concentration. Assay evaluations were performed on the 7th day after inoculation. Data shown are the means of ratings from at least three independent experiment sets for each combination; 25 leaflets per cultigen were inoculated per experiment. Vertical bars at each data point signify the standard deviations (SD). $LSD_{0.05}$ calculated according to Tukey procedure (inset) for comparing cultigens at each plant age and inoculum concentration (C×I): 1.34, and for comparing inoculum concentration for each cultigen and plant age (C×A): 1.43.

ml, rendered this inoculum concentration unsuitable to correctly differentiate between the LB resistant and LB susceptible plants.

Of the resistant cultigens, WVa 700 had the lowest variation in LB symptoms at different phases of development. Indeed, WVa 700 exhibited a low degree of infection (DSI>7) at all stages of development, regardless of inoculum concentration (Fig. 4), with one exception: the 3-week old plants had increased disease severity with DSI ratings spread over all nine rating classes at the highest inoculum concentration used. We found the lowest, most uniform levels of LB intensity in 7-week old and older WVa 700 plants, regardless of inoculum concentration.

In L 3708, we observed variable levels of LB symptoms intensity, dependent on plant age and inoculum concentration (Fig. 4). Generally, younger plants (3- and 4-week old) exhibited distinctly higher levels of LB sysmptoms than older plants (7- and 8-week old). We also observed higher variation in the range of severity ratings in younger plants compared with older plants. Moreover, the range of variation in severity ratings depended on the inoculum concentration used and correspondingly increased. In addition, we observed an inverse relationship between plant age and inoculum concentration on LB intensity levels. Older plants showed a more uniform intensity of disease symptoms across inoculum loads tested, than did the younger plants. Indeed, among the L 3708 developmental stages tested, the lowest and most uniform levels of LB symptoms were observed in 7- and 8-week old plants at all inoculum concentrations tested.

LA 1033 displayed a more diverse response to *P. infestans* challenges than WVa 700 or L 3708 plants. This reaction, however, remained dependent on plant age at all tested inoculum concentrations (Fig. 4). A comparative analysis of DSI values for all inoculum concentrations at all developmental stages examined in this cultigen showed greater severity of LB symptoms, compared with WVa 700 or L 3708. The lowest level of LB intensity (DSI = 8.2±0.6) was found in 8-week old plants inoculated with the lowest concentration of inoculum. Higher inoculum concentrations caused an increase in the severity of symptoms, with only sporadic significant increases in the variation of severity ratings (Fig. 4).

In general, all control conditions assays indicated an impact of plant age on the intensity of LB symptoms at levels specific for a given tomato cultigen. Symptoms of LB tended to decrease with plant age. High variability noted for the detached leaflet assay makes this method unreliable for standard testing of the LB resistance, or requires additional assessments (e.g., sporulation intensity), but the assays may be used for efficient identification of the LB susceptible individuals, such as during the early stages of selection and breeding.

Cross-test comparison

Pooled data from the controlled condition tests (detached leaflet, detached leaf, whole plant) from all treatment combinations (cultigens, inoculum concentrations, plant ages), were individually

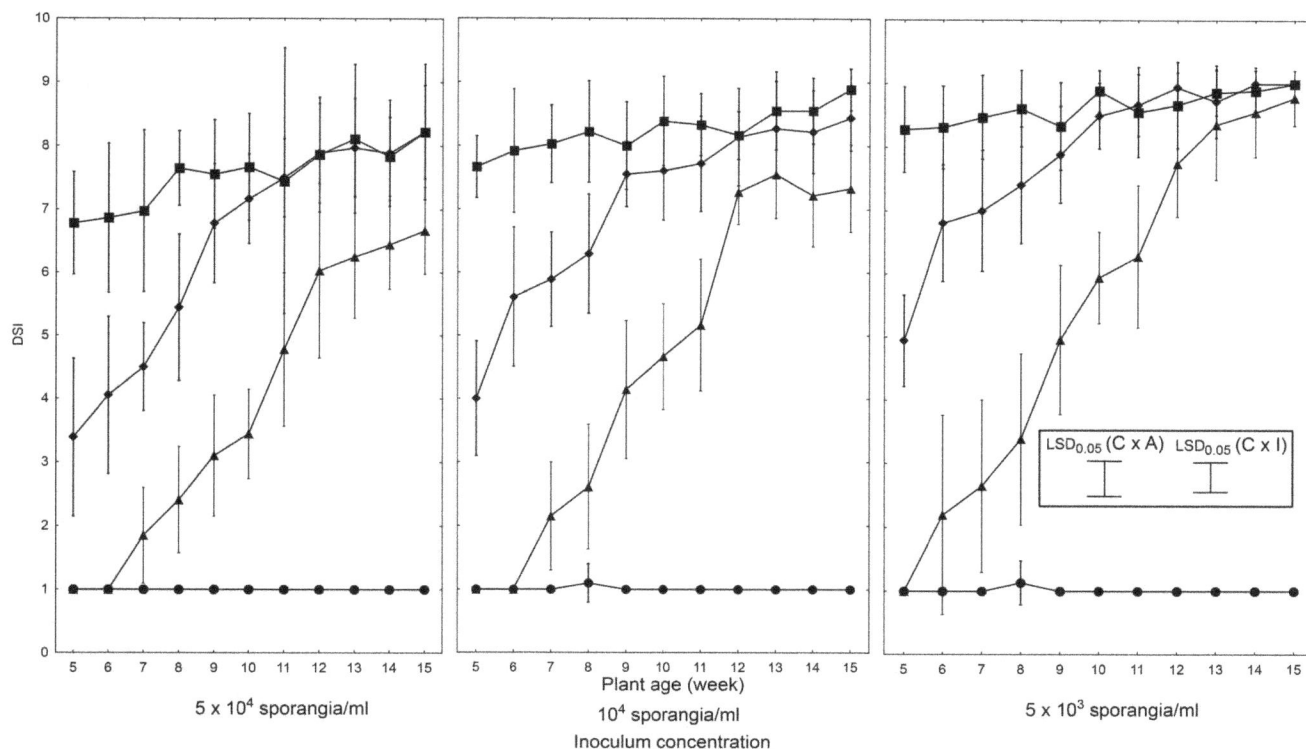

Figure 3. Severity of LB symptoms on tomato cultigens in the detached leaves assay. Third to fifth fully expanded leaves were collected for testing from the plants (● 'Rumba'; ■ WVa 700; ◆ L 3708; ▲ LA 1033) at indicated ages (weeks). Detached leaves of the tested tomato cultigens (12 to 24 leaves per cultigen and per developmental stage) were inoculated with suspension of the *P. infestans* isolate IWP 13, by spraying at indicated concentration. Presented data, for each treatment combination, are the means of ratings from three independent experiment sets. Vertical bars at each data point signify the standard deviations (SD). $LSD_{0.05}$ calculated according to Tukey procedure (inset) for comparing cultigens at each plant age and inoculum concentration (C×I): 0.58, and for comparing inoculum concentration for each cultigen and plant age (C×A): 0.74.

compared with the data of field experiments. All tested laboratory techniques showed significant linear relationships with the field results. But, the detached leaf assay and whole plant assay correlated better with the field assay than did the detached leaflet assay. The determination coefficients were: 0.94, 0.83, and 0.41, respectively (Fig. 5). The stronger relationship observed for the detached leaf assay and whole plant assay with reference to the field data resulted from lower variation within the plant materials, and the domination of extreme DSI values from the field observations.

In summary, this indicates that the detached leaf and the whole plant assays may serve as reliable tools for testing tomato LB resistance. These assays could thus replace the field trials at the early stages of testing. Additionally, it is noteworthy that at high disease intensity (DSI = 1 to 3) the detached leaflet assay showed good correlation with the field results, which further supports our suggestion to use this method for initial testing in tomato LB resistance breeding.

Discussion

The choice of methods for accurate testing of genotypes for LB resistance is an important area for plant breeding programs, including the tomato-*P. infestans* pathosystem. Lack of agreement among published reports for LB resistance in several tomato cultigens (e.g., L 3708, LA 1033) prompted the current study. Here, our aim was to standardize and compare several methods for testing the *P. infestans* resistance under controlled conditions

using four tomato cultigens, over several plant ages and inoculum concentrations. This would allow us to establish a uniform approach to assess the benefit of each cultigen for use in breeding for LB resistance. We also attempted to optimize the methods and to relate the results to those from natural field infection experiments.

Results of the multi-year (2008 to 2013) field experiments, from two locations (approx. 300 km apart), show that *S. habrochaites* LA 1033 had the lowest and most stable *P. infestans* infection levels of the cultigens tested. These findings are in agreement with other studies for LA 1033 reporting high resistance under natural field infection against a diverse set of *P. infestans* isolates, in the USA [17,33,35]. In other studies, isolates from Taiwan succeeded in infecting LA 1033 [16,31,34]. Our results of LA 1033 showing modest LB symptoms in the field could not be confirmed with detached leaf assay using *P. infestans* isolates derived from symptomatic LA 1033 plants (Nowakowska et al., unpublished data). The latter observation is in agreement with potato-LB studies, where *P. infestans* isolates derived from LB-symptomatic potato plants, carrying the *Rpi-phu1* gene conferring high levels of potato LB resistance, and grown in the field [44], failed to induce the disease symptoms in the laboratory [45]. This demonstrates the complexity of the pathosystem under field conditions.

Other tomatoes with outstanding LB resistance in the field, including the *S. pimpinellifolium* cultigens L 3707 and L 3708 [46,47], performed insignificantly lower than LA 1033. We observed, however, their stable low LB intensities in the field at all study years and in both locations, even with high *P. infestans*

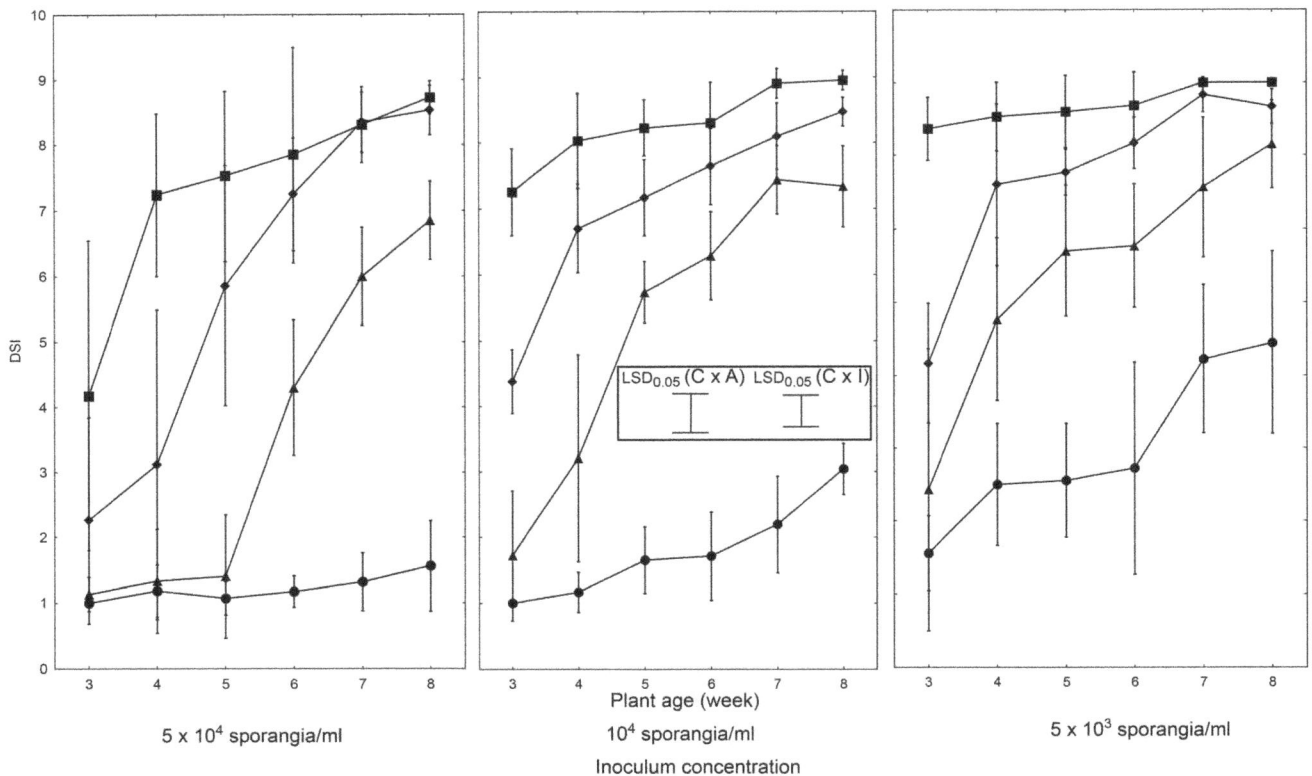

Figure 4. Severity of LB symptoms on tomato cultigens in the whole plant assay. Plants (● 'Rumba'; ■ WVa 700; ◆ L 3708; ▲ LA 1033) at indicated ages (3- to 8-weeks) were inoculated with suspension of the *P. infestans* isolate IWP 13, by spraying at indicated concentration (12 to 72 plants per cultigen and per developmental stage). Each combination was tested in three independent experiment sets. Vertical bars at each data point (means of the ratings) signify the standard deviations (SD). LSD$_{0.05}$ calculated according to Tukey procedure (inset) for comparing cultigens at each plant age and inoculum concentration (C×I): 0.55 and for comparing inoculum concentration for each cultigen and plant age (C×A): 0.61.

incidence. High LB resistance has been reported in these cultigens [15,46,47], despite a lack of clear elucidation of the trait's genetic background [3,26–29]. These cultigens have been successfully used as sources for pyramiding *Ph-3* with *Ph-2* LB resistance into new tomato cultivars [48–51]. Only few research groups reported rare instances, when high pathogen pressure, under highly conducive conditions, led to overcoming the resistance of L 3708, particularly in controlled condition assays [15,30,31,52].

In contrast to the aforementioned LB resistant cultigens, WVa 700 [19,38,53] exhibited varying levels of LB symptoms in the field, depending on the study year and location. These findings indicate the differences in the local pathogen populations. Consequently, such unstable expression of LB resistance in WVa 700 may occur due to a simpler genetic background for this trait, compared with those found in the other resistant cultigens. Similarly to our findings, this cultigen has failed to display stable LB resistance in other studies [19,30,31,54,55]. Finally, the *Ph-1* gene present in 'NY' [36–39] provides no reliable protection against *P. infestans* for field tomatoes grown in Poland, as well as in other locations [2,24,30,32,54,56,57].

Collectively, the field studies indicated that under Polish conditions, LA 1033, L 3707, and L 3708 could be considered promising sources for breeding tomato for LB resistance. This has implications for Central Europe, with field production of both tomatoes and potatoes [1].

In the controlled conditions testing methods used, LA 1033 showed the largest variability in age-dependent reaction to *P. infestans* inoculation. Furthermore, in contrast to its consistently

superior performance in the field assays, LA 1033 proved inferior to both WVa 700 and L 3708 cultigens in the detached leaf and whole plant tests. This observation underscores that the assays under controlled conditions may differ from the field tests (weather conditions, plant age, heterogenic isolate mixture, constant pathogen pressure, presence of other (a)biotic stresses). These results also suggest that full expression of LB resistance in LA 1033 occurs later than the oldest developmental stages investigated (3- to 8-week old). Our observation of lower LB intensities in the detached leaves of the 13- to 15-week old plants of this cultigen further supports this hypothesis. The abundant growth of this cultigen may pose problems with generation of appropriate plant materials in the greenhouse for large scale bio-assays. This problem, however, can be solved using the detached leaf tests.

In our study, L 3708 showed higher symptoms variability relative to plant age compared with the WVa 700 cultigen. For L 3708, the trait reached stable expression in 7- to 8-week old plants in the whole plant assays, or in 10-week old leaves. This is in contrast to other studies on L 3708 [15,25], which described high levels of LB resistance when testing only 5-week old plants. Possible reasons for these discordant results include differences in the assumed methodology (isolates used, inoculum preparation and load, conditions of the assays) or in the climatic conditions related to location (e.g., intensity of sun exposure, length of day).

In the detached leaf and whole plant assays, we recorded the lowest variability for WVa 700, with low LB symptoms levels in plants of this cultigen older than 3 weeks. Moderate LB intensities were seen in whole plant assays. Turkensteen [24] reported age-

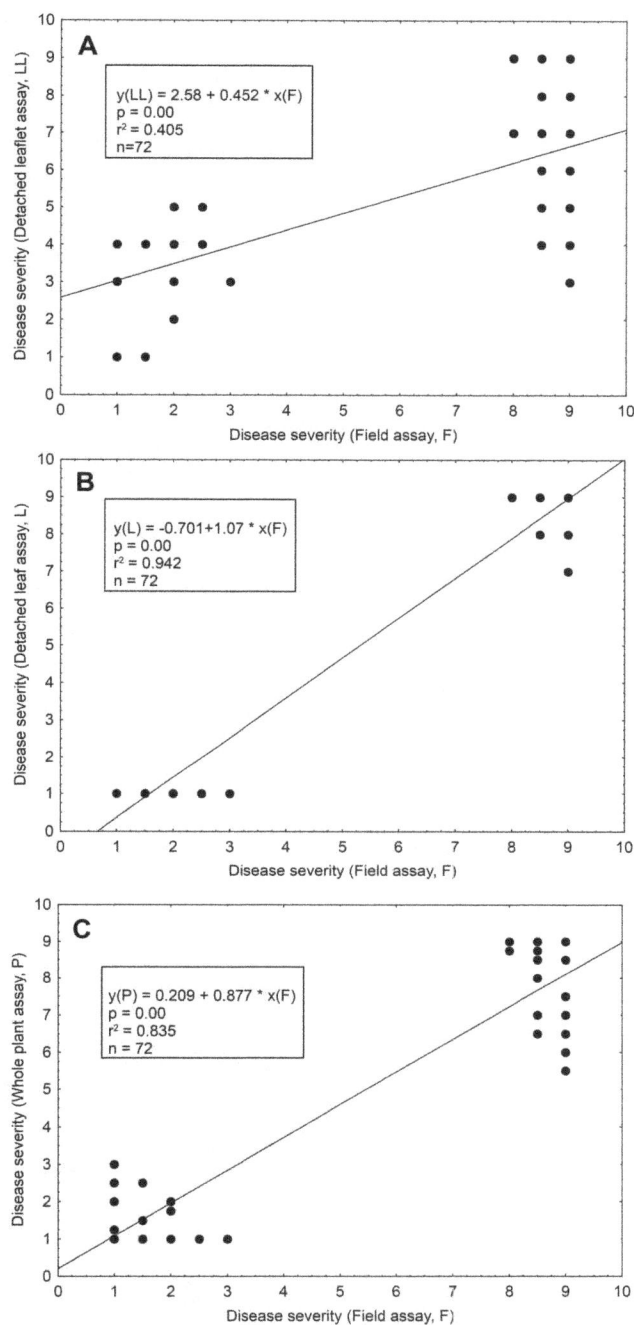

Figure. 5. Cross-comparison of the methods for testing the tomato LB resistance. Data from the field experiments (F) were pairwise compared with the results of each of the controlled-conditions method used, under the conditions optimized towards the maximal LB resistance expression (LL: Detached leaflets tested on 8-week old plants, under 5×10^4 sporangia/ml; L: Detached leaves tested on 15-week old plants, under 5×10^3 sporangia/ml; P: Whole plants tested on 8-week old plants, under 5×10^4 sporangia/ml). Calculated trend lines, with respective determination coefficients (r^2) and P-values are indicated. A: Comparison of detached leaflet assays with field assays (LL and F); B: Comparison of detached leaf assays with field assays (L and F); C: Comparison of whole plant assays with field assays (P and F).

dependent and progressively increasing LB resistance in the detached leaf assays of 6- to 8-week old WVa 700. Similarly, the 6-week old seedlings of this cultigen exhibited high resistance against both pathogen isolates tested by Moreau et al. [19]. In contrast to our results showing high LB infection in 3- to 4-week old WVa 700 plants, under 5×10^4 sporangia/ml, previous studies reported low LB intensity in this cultigen, under comparable developmental stages and inoculum concentrations [58,59]. The most likely reasons for the observed discrepancies are the pathogen isolates used or the assumed methodology, including the double isolate activation employed in our study.

Apart from the significant influence of plant developmental stage and inoculum concentration on the intensity of LB symptoms in the detached leaflet assay, we observed that LB intensity decreases with plant age. Variability of lesion size on leaflets observed using this assay, particularly in the resistant cultigens, indicated the need for improved methods for testing LB resistance (e.g., sporulation intensity). Thus, the detached leaflet assays were an inefficient testing method, although it is fast and easy. The detached leaflet assays can be used only for identification of susceptible genotypes in initial stages of breeding. Our results of tomato LB intensities are in line with the studies of LB in potato [5,9,60,61], where reported problems regarded high variability, especially for potato genotypes with moderate resistance. Although several tomato LB studies used the detached leaflet method to evaluate LB intensity [16,33,34,42,43,62], they generally included additional evaluations, such as sporulation intensity. In contrast to the detached leaflet assays, other methods used in this study (detached leaf and whole plant assays) proved more reliable for testing tomato LB resistance under controlled conditions. The observed variability among cultigens tested with these methods was significantly lower than this observed for the detached-leaflet assays. These two most effective methods, however, required different inoculum concentrations for successful separation of resistant genotypes. Both detached leaf and whole plant assays suggested an age-related LB intensity. Our results are similar to those using detached leaflet assays, except for more accurate distinction of the resistant genotypes.

Inoculation and incubation conditions influenced the reproducibility of our results. Here, infection and subsequent development of LB symptoms clearly depended on changes in temperature or RH, in accordance with the biology of the pathogen [5,8,33,57,60,63]. Standardizing the test methods (choice of isolate, preparation of inoculum, concentration of inoculum, and conditions of incubation after inoculation) resulted in greater precision in distinguishing LB resistant plants. In the light of our findings, we postulate it very useful to study the plant LB resistance in an age-dependent manner and under the controlled conditions, for each cultigen being reported, in order to better reflect field performance. This might provide an explanation for the differences in mapping of the genes or QTLs controlling the LB resistance trait [15,25,26,28,64–67], and might be of help in detailed analyses of the emerging cultigens reported as LB resistant [16,18,20,34,50].

Of the methods studied, the detached leaf and the whole plant tests resulted in the lowest discrepancies, when compared with the field experiments. This may be due to inoculation of larger plant surface area, which may generate lower assessment errors and permit a more accurate evaluation of LB resistance. Both leaf and leaflet assays may additionally exhibit reactions to *P. infestans* inoculation different from those of whole plants. These may be due to differences in environmental conditions or influences on the physiological and/or biochemical processes, as a result of detachment from the plant. Thus, we propose the whole plant assays as the most reliable method of testing the tomato LB resistance under controlled conditions. This, in the case of some cultigens (notably LA 1033), may pose other challenges, to be circumvented by using alternative methods, such as the detached leaf assays. We agree with previous reports on potato-LB [4–12,14] and tomato-LB pathosystems [21–23], that the ultimate assessment of LB resistance should be performed with field tests.

Conclusions

Our five-year study under natural field infection, in two distinct Polish locations, confirmed low and stable levels of LB symptoms in LA 1033, L 3708, and L 3707 tomato cultigens. These cultigens are useful for resistance breeding programs for Central Europe. Based on field results, we consider the cultigens carrying the *Ph-1* gene (e.g., 'New Yorker') an unsuitable source of LB resistance in Poland. The same is true for cultigens carrying *Ph-2* (e.g., WVa 700) if they are used as the only source of LB resistance. Our comparison of three methods for assessing tomato resistance against *P. infestans* under controlled conditions indicated that each method may be used for different purposes in the resistance breeding. The detached leaflet assay proved useful only to separate the LB susceptible genotypes (such as within the segregating populations) and to maintain the pathogen isolates, but was unreliable for systematic screens due to high variability. Tests on detached leaves and whole plants (greenhouse) generated lower variability than those performed on the detached leaflets and were also found to be highly correlated with field tests. Our results indicate congruent trends for age-dependent expression of LB resistance in all tested tomato cultigens, irrespective of the testing method. The plant age-related LB resistance in tomato reported here, shows the need to optimize and standardize the testing parameters when reporting new sources of resistance. As documented in this study, the reliable comprehensive evaluation of a given cultigen, by means of the optimized and well-suited assays, remains crucial to maximize the benefits from the best performing crops.

Acknowledgments

The Authors are thankful to Krystyna Szewczyk and Marzena Czajka for skillful technical assistance. Dr. Dorothy M. Tappenden PhD (Michigan State University) and Dr. Todd C. Wehner (North Carolina State University) are gratefully recognized for copy editing and critical reading of this manuscript.

Author Contributions

Conceived and designed the experiments: M. Nowakowska EUK. Performed the experiments: M. Nowakowska EUK. Analyzed the data: M. Nowakowska M. Nowicki RM UK EUK. Contributed reagents/materials/analysis tools: M. Nowakowska M. Nowicki EUK. Wrote the paper: M. Nowakowska M. Nowicki EUK.

References

1. Anonymous (2011) FAOSTAT final 2011 data. pp. http://faostat.fao.org/site/339/default.aspx.
2. Nowicki M, Kozik EU, Foolad MR (2013) Late blight of tomato. In: Varshney RK, Tuberosa R, editors. Translational genomics for crop breeding: John Wiley & Sons Ltd. pp. 241–265.
3. Nowicki M, Foolad MR, Nowakowska M, Kozik EU (2012) Potato and tomato late blight caused by *Phytophthora infestans*: An overview of pathology and resistance breeding. Plant Disease 96: 4–17.
4. Darsow U (2004) The use of four different assessment methods to establish relative potato tuber blight resistance for breeding. Potato Research 47: 163–174.
5. Dorrance AE, Inglis DA (1997) Assessment of greenhouse and laboratory screening methods for evaluating potato foliage for resistance to late blight. Plant Disease 81: 1206–1213.
6. Dorrance AE, Inglis DA (1998) Assessment of laboratory methods for evaluating potato tubers for resistance to late blight. Plant Disease 82: 442–446.
7. Douches DS, Kirk WW, Bertram MA, Coombs JJ, Niemira BA (2002) Foliar and tuber assessment of late blight (*Phytophthora infestans* (Mont.) de Bary) reaction in cultivated potato (*Solanum tuberosum* L.). Potato Research 45: 215–224.
8. Huang S, Vleeshouwers VGAA, Visser RGF, Jacobsen E (2005) An accurate in vitro assay for high-throughput disease testing of *Phytophthora infestans* in potato. Plant Disease 89: 1263–1267.
9. Michalska AM, Zimnoch-Guzowska E, Sobkowiak S, Plich J (2011) Resistance of potato to stem infection by *Phytophthora infestans* and a comparison to detached leaflet and field resistance assessments. American Journal of Potato Research 88: 367–373.
10. Park T-H, Vleeshouwers VGAA, Kim J-B, Hutten RCB, Visser RGF (2005) Dissection of foliage and tuber late blight resistance in mapping populations of potato. Euphytica 143: 75–83.
11. Sharma BP, Forbes GA, Manandhar HK, Shrestha SM, Thapa RB (2013) Determination of resistance to *Phytophthora infestans* on potato plants in field, laboratory and greenhouse conditions. Journal of Agricultural Science 5: p148.
12. Tai G (1998) Relationship between resistance to late blight in potato foliage and tubers of cultivars and breeding selections with different resistance levels. American Journal of Potato Research 75: 173–178.
13. Visker M (2005) Association between late blight resistance and foliage maturity type in potato: Physiological and genetic studies. 160 pp. p.
14. Vleeshouwers VGAA, van Dooijeweert W, Keizer LCP, Sijpkes L, Govers F, et al. (1999) A laboratory assay for *Phytophthora infestans* resistance in various *Solanum* species reflects the field situation. European Journal of Plant Pathology 105: 241–250.
15. Chunwongse J, Chunwongse C, Black L, Hanson P (2002) Molecular mapping of the *Ph-3* gene for late blight resistance in tomato. Journal of Horticultural Science & Biotechnology 77: 281–286.
16. Li J, Liu L, Bai Y, Finkers R, Wang F, et al. (2011) Identification and mapping of quantitative resistance to late blight (*Phytophthora infestans*) in *Solanum habrochaites* LA1777. Euphytica 179: 427–438.
17. Lough RC, Gardner RG (2000) Inheritance of tomato late blight resistance derived from *Lycopersicon hirsutum* LA1033 and identification of molecular markers. Hortscience 35: 490.
18. Merk HL, Ashrafi H, Foolad MR (2012) Selective genotyping to identify late blight resistance genes in an accession of the tomato wild species *Solanum pimpinellifolium*. Euphytica 187: 63–75.
19. Moreau P, Thoquet P, Olivier J, Laterrot H, Grimsley N (1998) Genetic mapping of *Ph-2*, a single locus controlling partial resistance to *Phytophthora infestans* in tomato. Molecular Plant-Microbe Interactions 11: 259–269.
20. Smart CD, Tanksley SD, Mayton H, Fry WE (2007) Resistance to *Phytophthora infestans* in *Lycopersicon pennellii*. Plant Disease 91: 1045–1049.
21. Akhtar KP, Saleem MY, Asghar M, Ali S, Sarwar N, et al. (2012) Resistance of *Solanum* species to *Phytophthora infestans* evaluated in the detached-leaf and whole-plant assays. Pak J Bot 44: 1141–1146.
22. Kim BS (2012) Evaluation of tomato genetic resources for the development of resistance breeding lines against late blight. Research in Plant Disease 18(1): 342–350.
23. Luo D, Zhang XC, Wen XH (2013) Identification and screening of introduced tomato varieties resistant to late blight. Advanced Materials Research 610: 3472–3477.
24. Turkensteen LJ (1973) Partial resistance of tomatoes against *Phytophthora infestans*, the late blight fungus. Wageningen, The Netherlands: Wageningen University.
25. Chunwongse J, Chunwongse C, Black LL, Hanson P (1998) Mapping of the *Ph-3* gene for late blight from *L. pimpinellifolium* L3708. Rep Tomato Genet Cooperative 48: 13–16.
26. Chen A-L, Liu C-Y, Chen C-H, Wang J-F, Liao Y-C, et al. (2014) Reassessment of QTLs for late blight resistance in the tomato accession L3708 using a restriction site associated DNA (RAD) linkage map and highly aggressive isolates of *Phytophthora infestans*. PLoS One 9: e96417.
27. Frary A, Graham E, Jacobs J, Chetelat RT, Tanksley SD (1998) Identification of QTL for late blight resistance from *L. pimpinellifolium* L3708. Tomato Genetics Cooperative Report 48: 19–21.
28. Kim MJ, Mutschler MA (2005) Transfer to processing tomato and characterization of late blight resistance derived from *Solanum pimpinellifolium* L. L3708. Journal of the American Society for Horticultural Science 130: 877–884.
29. Kim M-J, Mutschler MA (2006) Characterization of late blight resistance derived from *Solanum pimpinellifolium* L3708 against multiple isolates of the pathogen *Phytophthora infestans*. Journal of the American Society for Horticultural Science 131: 637–645.
30. AVDRC (2005) 2005 Progress Report. Shanhua, Tainan, Taiwan: Asian Vegetable Research and Development Center. 49–51 p.

31. Chen C-H, Sheu Z-M, Wang T-C (2008) Host specificity and tomato-related race composition of *Phytophthora infestans* isolates in Taiwan during 2004 and 2005. Plant Disease 92: 751–755.

32. Wang TC, Chen CH (2005) The variation of race composition of *Phytophthora infestans* in Taiwan during 1991–2004. 2005 APS Annual Meeting. Austin, TX: APS: Phytopathology. pp. S109.

33. Kim M, Mutschler M (2000) Differential response of resistant lines derived from the *L. pimpinellifolium* accession L3708 and *L. hirsutum* accession LA 1033 against different isolates of *Phytophthora infestans* in detached leaf lab assays. Tomato Genetics Cooperative Report 540: 23–25.

34. Li J (2010) Exploration of wild relatives of tomato for enhanced stress tolerance. Wageningen, The Netherlands: Wageningen University. 158 p.

35. Lough RC (2003) Inheritance of tomato late blight resistance in *Lycopersicum hirsutum* LA1033. Raleigh: North Carolina State university.

36. Bonde R, Murphy EF (1952) Resistance of certain tomato varieties and crosses to late blight. Bull Me Agric Exp Sta 497: 15.

37. Gallegly ME (1952) Sources of resistance to two races of the tomato late blight fungus. Phytopathology 42.

38. Gallegly ME, Marvel ME (1955) Inheritance of resistance to tomato Race-0 of *Phytophthora infestans*. Phytopathology 45: 103–109.

39. Peirce LC (1971) Linkage tests with *Ph* conditioning resistance to race 0, *Phytophthora infestans*. Rep Tomato Genet Coop 21: 30.

40. Zarzycka H (2001) Evaluation of resistance to *Phytophthora infestans* in detached leaflet assay. Preparation of the inoculum. The methods of evaluation and selection applied in potato research and breeding Monografie i rozprawy naukowe IHAR, 10a [in Polish]: 75–77.

41. Nowicki M, Lichocka M, Nowakowska M, Kłosińska U, Kozik EU (2012) A simple dual stain for detailed investigations of plant-fungal pathogen interactions. Vegetable Crops Research Bulletin 77: 61–74.

42. Nelson HE (2006) Bioassay to detect small differences in resistance of tomato to late blight according to leaf age, leaf and leaflet position, and plant age. Australasian Plant Pathology 35: 297–301.

43. Smart C, Myers K, Restrepo S, Martin G, Fry W (2003) Partial resistance of tomato to *Phytophthora infestans* is not dependent upon ethylene, jasmonic acid, or salicylic acid signaling pathways. Molecular Plant-Microbe Interactions 16: 141–148.

44. Śliwka J, Jakuczun H, Lebecka R, Marczewski W, Gebhardt C, et al. (2006) The novel, major locus *Rpi-phu1* for late blight resistance maps to potato chromosome IX and is not correlated with long vegetation period. TAG Theoretical and Applied Genetics 113: 685–695.

45. Stefańczyk E, Świątek M, Chmielarz M, Tomczyńska I, Śliwka J (2013) Influence of chosen factors on the expression of the *Rpi-phu1* gene, conferring the potato resistance against *Phytophthora infestans*. In: Śmiałowski T, Strzembicka A, Szecówka PS, editors. Ogólnopolska konferencja naukowa - Nauka dla hodowli i nasiennictwa roślin uprawnych Zakopane Poland. Plant Breeding and Acclimatization Institute (IHAR) - National Research Institute in Radzików and IHAR Cereals Department in Cracow.pp. 67–70.

46. Black LL, Wang TC, Hanson PM, Chen JT (1996) Late blight resistance in four wild tomato accessions: Effectiveness in diverse locations and inheritance of resistance. Phytopathology 86: S24.

47. Irzhansky I, Cohen Y (2006) Inheritance of resistance against *Phytophthora infestans* in *Lycopersicon pimpenellifolium* L3707. Euphytica 149: 309–316.

48. Gardner RG, Panthee DR (2010) NC 1 CELBR and NC 2 CELBR: Early blight and late blight-resistant fresh market tomato breeding lines. Hortscience 45: 975–976.

49. Gardner RG, Panthee DR (2010) 'Plum Regal' Fresh-market plum tomato hybrid and its parents, NC 25P and NC 30P. HortScience 45: 824–825.

50. Ojiewo C, Swai I, Oluoch M, Silué D, Nono-Womdim R, et al. (2010) Development and release of late blight-resistant tomato varieties 'Meru' and 'Kiboko'. International Journal of Vegetable Science 16: 134–147.

51. Panthee DR, Gardner RG (2010) 'Mountain Merit': A late blight-resistant large-fruited tomato hybrid. Hortscience 45: 1547–1548.

52. Scott J, Gardner R (2007) Breeding for resistance to fungal pathogens. In: Razdan MK, Mattoo AK, editors. Genetic improvement of Solanaceous crops: Taylor & Francis Science Publishers. pp. 421–456.

53. Gallegly ME (1960) Resistance to the late blight fungus in tomato. Proceedings of Plant Science Seminar, Camden, New Jersey 1960. Camden, New Jersey pp. 113–135.

54. Foolad MR, Merk H, Ashrafi H, Kinkade M (2006) Identification of new sources of late blight resistance in tomato and mapping of a new resistance gene. In: Moyer J, editor. 22nd Annual Tomato Disease Workshop. Mountain Horticultural Crops Research & Extension Center Fletcher, NC, USA. North Carolina State University. pp. 4–8.

55. Laterrot H (1975) Selection for the resistance to *Phytophthora infestans* in tomato. Annales de l'Amelioration des Plantes 25: 129–150.

56. Cohen Y (2002) Populations of *Phytophthora infestans* in Israel underwent three major genetic changes during 1983 to 2000. Phytopathology 92: 300–307.

57. Klarfeld S, Rubin A, Cohen Y (2009) Pathogenic fitness of oosporic progeny isolates of *Phytophthora infestans* on late-blight-resistant tomato lines. Plant Disease 93: 947–953.

58. Michalska A, Pazio M (2002) A new method for evaluating tomato leaf resistance to *Phytophthora infestans* using seedling test. Plant Breeding and Seed Science 46: 3–21.

59. Michalska AM, Pazio M (2005) Inheritance of tomato leaf resistance to *Phytophthora infestans* - new information based on laboratory tests on seedlings. Plant Breeding and Seed Science 51: 31–42.

60. Mizubuti ES, Fry WE (1998) Temperature effects on developmental stages of isolates from three clonal lineages of *Phytophthora infestans*. Phytopathology 88: 837–843.

61. Sobkowiak S, Zarzycka H, Lebecka R, Zimnoch-Guzowska E (2009) Reaction of potato standards to *Phytophthora infestans* in resistance tests. Biuletyn Instytutu Hodowli i Aklimatyzacji Roślin: 253–268.

62. Legard DE, Fry WE (1996) Evaluation of field experiments by direct allozyme analysis of late blight lesions caused by *Phytophthora infestans*. Mycologia: 608–612.

63. Hardham AR, Shan W (2009) Cellular and molecular biology of *Phytophthora*-plant interactions. In: Deising HB, editor. The Mycota. Berlin Heidelberg: Springer Verlag. pp. 4–27.

64. Huang X-M, Xu X-Y, Li J-F, Chen X-L, Xu Y-H (2009) Construction of tomato molecular genetic map and QTL analysis of resistance of gene cluster *ph-3* to tomato *Phytophthora infestans*. Scientia Agricultura Sinica 42: 3571–3580.

65. Park HP, Chae Y, Kim H-R, Chung K-H, Oh D-G, et al. (2010) Development of a SCAR marker linked to *Ph-3* in *Solanum* ssp. Korean J Breed Sci 42: 139–143.

66. Park Y, Hwang J, Kim K, Kang J, Kim B, et al. (2013) Development of the gene-based SCARs for the *Ph-3* locus, which confers late blight resistance in tomato. Scientia Horticulturae 164: 9–16.

67. Zhang C, Liu L, Wang X, Vossen J, Li G, et al. (2014) The *Ph-3* gene from *Solanum pimpinellifolium* encodes CC-NBS-LRR protein conferring resistance to *Phytophthora infestans*. TAG Theoretical and applied genetics 127: 1353.

8

Molecular Design and Synthesis of Novel Salicyl Glycoconjugates as Elicitors against Plant Diseases

Zining Cui[1,2,4]*, Jun Ito[3], Hirofumi Dohi[2], Yoshimiki Amemiya[3], Yoshihiro Nishida[2]*

1 Guangdong Province Key Laboratory of Microbial Signals and Disease Control, Department of Plant Pathology, College of Natural Resource and Environment, South China Agricultural University, Guangzhou, China, **2** Department of Nanobiology, Graduate School of Advanced Integration Science, Chiba University, Chiba, Japan, **3** Department of Environment Science for Bioproduction, Graduate School of Horticulture, Chiba University, Chiba, Japan, **4** State Key Laboratory for Biology of Plant Diseases and Insect Pests, Institute of Plant Protection, Chinese Academy of Agricultural Sciences, Beijing, China

Abstract

A new series of salicyl glycoconjugates containing hydrazide and hydrazone moieties were designed and synthesized. The bioassay indicated that the novel compounds had no *in vitro* fungicidal activity but showed significant *in vivo* antifungal activity against the tested fungal pathogens. Some compounds even had superior activity than the commercial fungicides in greenhouse trial. The results of RT-PCR analysis showed that the designed salicyl glycoconjugates could induce the expression of *LOX1* and *Cs-AOS2*, which are the specific marker genes of jasmonate signaling pathway, to trigger the plant defense resistance.

Editor: Daniel Doucet, Natural Resources Canada, Canada

Funding: Financial support was provided by the National Key Project for Basic Research (2015CB150605), the National Natural Science Foundation of China (21102173), the State Key Laboratory for Biology of Plant Diseases and Insect Pests (SKLOF201411), the Specialty and Innovation Projects of Guangdong Province University, the President Science Foundation of South China Agricultural University (4200-K13014), the Japan Society for the Promotion of Science (JSPS, ID No. P10100), Advanced Multi-Career Training Program for Postdoctoral Scholars from JST, and the VBL project funding from Chiba University. The funders had no role in study design, data collection and analysis, decision to publish, or preparation of the manuscript.

Competing Interests: The authors have declared that no competing interests exist.

* Email: ziningcui@scau.edu.cn (ZC); YNishida@faculty.chiba-u.jp (YN)

Introduction

In the past two decades, the goal of sustainable and green agriculture had been inspiring researchers to explore the feasibility of restricting toxic agrochemical usage to reduce their impact on environment and food chains. One of the alternatives, which had been studied intensively in recent years, was to make use of plant defense potentials. Induction of plant defense resistance in crops by chemical or biological elicitors had drawn increasing attentions and was considered as a prospective strategy for disease control [1,2].

During the long process of co-evolution, plants had evolved lots of defense mechanisms to deal with pests and pathogens. Following plant-pathogen interaction, a number of plant defense responses could be induced (e.g., callus deposition, PR-protein accumulation, *et al.*) at the site of infection, and also in uninfected tissues, activated by signal molecules associated with defense responses, which resulted in increased resistance to subsequent infections. The systemic acquired resistance is a "whole-plant" defense response that occurred following an earlier localized exposure to a pathogen. Activation of systemic acquired resistance required the accumulation of endogenous salicylic acid [3–5]. Besides the salicylic acid dependent defense signaling pathway, the others had also been reported. For example, endogenous jasmonic acid and methyl jasmonate were also the potent signaling molecules which could induce a large set of defense responses [6]. Systemic acquired resistance

possessed low specificity, was not easily overcome by new pathogens which emerged frequently.

Chemical elicitors are agrochemicals which do not show a direct effect on pathogens and lacked fungicidal activity themselves but induce defense mechanisms, which clearly distinguish them from conventional pesticides [7]. Some of these agrochemicals are known to have signaling functions *in planta*, such as benzothiadiazole [8–13], which is a functional analog of salicylic acid, while others may mimic the attack of a pathogen, such as harpin [14] or flagellin [15].

Saccharides are known as potent elicitors [16]. The fragments of chitin and chitosan, which act as elicitors in many plants, could induce the production of nitric oxide and hydrogen peroxide in some plant epidermal cells [17–20]. Even neutral saccharides, such as β-glucans derived from cellulose or laminarin [21,22], are capable of enhancing plant resistance. The accumulation of phytoalexins could be induced by branched hexa (β-D-glucopyranosyl)-D-glucitols in soybean [23,24]. Oligoglucans with polymerization between 8 and 17 could induce the chitinase activity in tobacco BY-2 suspension cells [25,26]. The phenolic pathway could be rapidly induced by the mannose and glucose disaccharides in *Rubus* cells [27]. It is evident that saccharides have the ability to trigger defense responses in plants, enhance resistance toward infection, and even support plant growth [28,29].

In our previous work, some 1,3,4-oxadiazole [30], benzoylureas [31–33], acylhydrazones [34,35], diacylhydrazines [36–40], semicarbazide [41], pyrazole and 1,2,4-triazole [42] derivatives

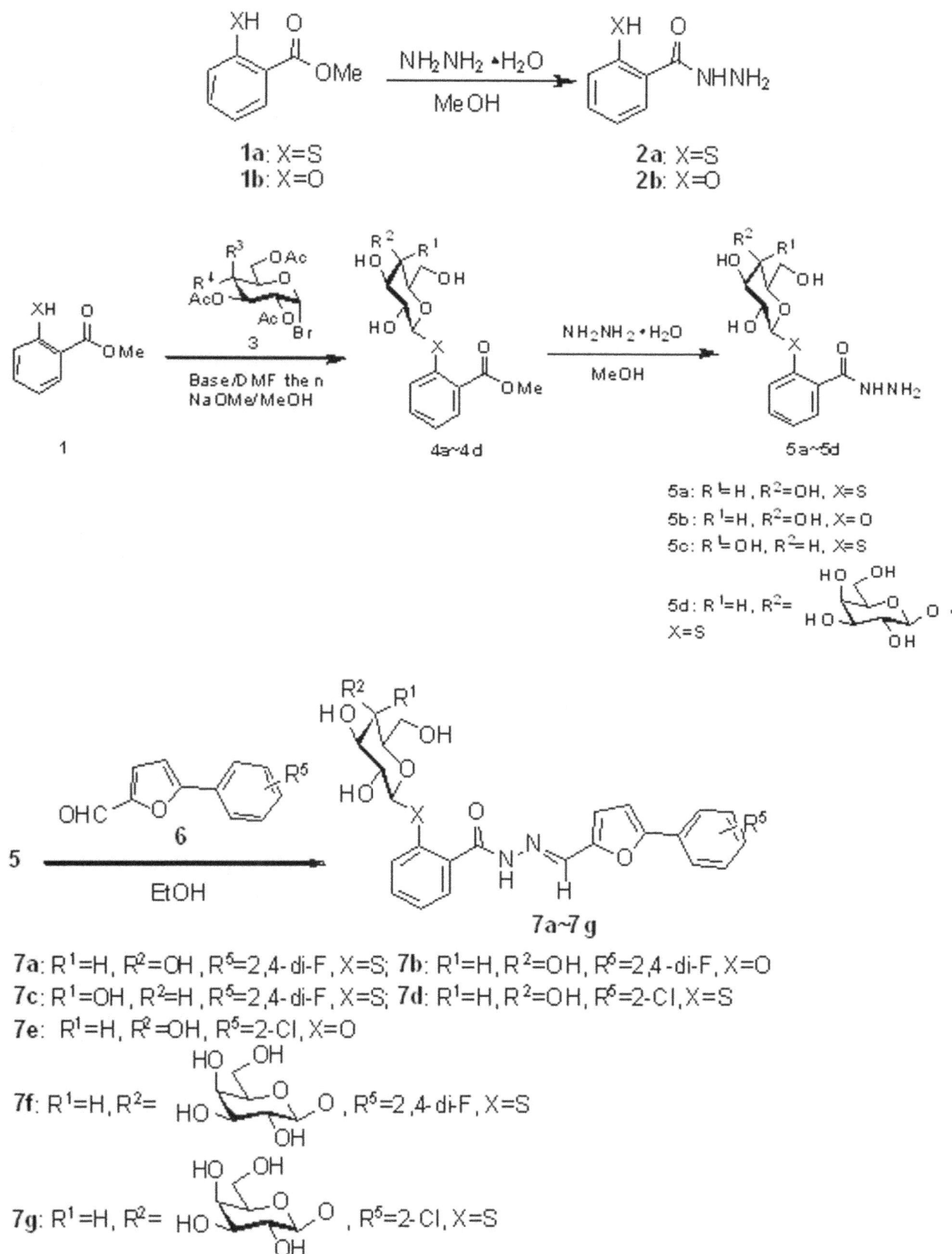

Figure 1. General synthetic procedure for salicylic glycoconjugates.

Table 1. *In vitro* fungicidal activity against five fungus species at 50 µg/mL.

| Compd. | Inhibitory rate (%) | | | | |
	C. orbiculare	F. oxysporum	S. fuliginea	R. solanii	P. capsici
2a	97.3±2.0	73.0±2.2	73.1±2.1	86.5±2.3	56.2±2.2
2b	95.4±1.3	95.5±2.3	77.7±2.3	79.2±1.7	61.7±1.8
4a	11.3±1.0	12.2±1.1	28.1±2.0	27.9±1.6	13.2±1.1
4b	28.6±1.2	28.0±1.5	10.3±1.5	28.9±2.0	28.4±1.3
4c	12.9±1.2	9.3±0.4	6.5±0.6	18.1±2.0	12.9±1.7
4d	8.1±0.4	11.6±1.0	11.5±2.1	23.0±2.0	2.0±0.4
5a	15.2±0.9	7.4±0.2	10.4±1.7	26.1±1.0	21.6±1.0
5b	19.7±1.1	9.4±0.7	25.9±2.6	12.9±1.1	26.7±1.6
5c	2.2±0.3	3.2±0.1	1.3±0.8	16.3±2.4	23.8±1.2
5d	17.6±1.3	13.3±1.2	15.2±1.4	19.5±1.3	28.4±2.3
7a	24.4±1.2	35.3±1.7	28.5±2.0	15.4±1.0	21.4±1.3
7b	15.7±1.8	24.7±1.4	37.5±2.3	47.6±1.5	31.6±1.8
7c	25.5±1.7	33.3±1.6	17.5±1.1	39.4±2.0	38.6±1.2
7d	13.3±1.9	25.3±1.3	36.5±1.6	38.4±1.3	29.0±1.4
7e	25.5±1.0	21.5±1.1	19.6±1.7	19.8±1.0	41.5±1.5
7f	6.0±1.0	12.6±1.1	27.6±1.6	37.6±2.5	24.5±1.2
7g	11.3±0.6	8.2±0.8	12.4±1.6	21.5±1.5	15.7±1.1
DMSO	1.0±0.3	1.9±0.7	1.0±0.1	1.4±0.5	1.0±0.2
Fungicides[a]	91.0±1.3 a	98.2±1.2 b	97.5±2.1 c	91.0±2.1 d	91.2±2.5 e

[a]Control fungicides: a, thiophanate-methyl; b, benomyl; c, chlorothalonil; d, validamycin; e, dimethomorph.

Table 2. *In vivo* antifungal activity against five fungus species at 500 µg/mL.

| Compd. | Inhibitory rate (%) | | | | |
	C. orbiculare	F. oxysporum	S. fuliginea	R. solanii	P. capsici
2a	51.8±2.0	55.2±2.2	48.1±3.1	51.7±2.2	33.2±1.3
2b	61.7±1.1	64.5±2.4	43.5±2.2	51.8±3.1	26.8±1.1
4a	45.6±2.6	43.6±2.3	41.9±1.1	49.7±2.3	12.9±1.2
4b	41.6±1.2	28.1±1.0	39.4±1.7	21.5±1.5	11.3±0.6
4c	51.6±1.7	59.7±2.1	49.3±2.3	17.4±1.9	3.2±0.9
4d	47.5±1.8	34.5±1.6	34.7±1.3	43.8±1.6	6.6±0.3
5a	50.3±1.3	54.5±1.2	50.8±1.7	26.0±1.3	40.4±2.2
5b	53.3±1.8	61.8±2.0	53.7±2.5	45.0±2.0	11.3±1.7
5c	34.6±1.6	55.6±0.8	54.5±1.3	34.5±1.4	24.2±1.0
5d	68.6±1.3	71.0±2.3	73.9±2.6	31.2±1.4	12.9±1.2
7a	41.3±0.5	53.8±1.5	52.3±1.1	45.5±2.1	76.0±2.2
7b	48.3±1.3	33.6±1.5	54.5±1.5	36.8±1.9	68.6±1.5
7c	54.8±1.9	62.6±1.6	34.5±0.7	28.6±1.5	83.5±1.3
7d	12.6±0.3	34.5±1.0	37.8±1.3	36.6±1.9	78.5±1.6
7e	54.3±2.5	54.6±0.9	23.3±1.2	33.2±1.8	77.5±2.0
7f	59.6±1.8	74.9±1.3	14.7±0.8	14.5±1.0	25.6±1.1
7g	54.8±1.5	40.3±1.2	42.8±1.3	33.9±1.5	34.7±1.0
DMSO	2.2±0.6	2.9±0.2	2.4±0.4	2.2±0.8	3.1±0.6
Fungicides[a]	76.8±2.3 a	74.5±2.3 b	94.6±1.7 c	81.0±2.7 d	91.2±2.4 e

[a]Control fungicides: a, 70% thiophanate-methyl WP; b, 70% benomyl WP; c, 50% chlorothalonil WP; d, 3% validamycin AS; e, 50% dimethomorph WP.

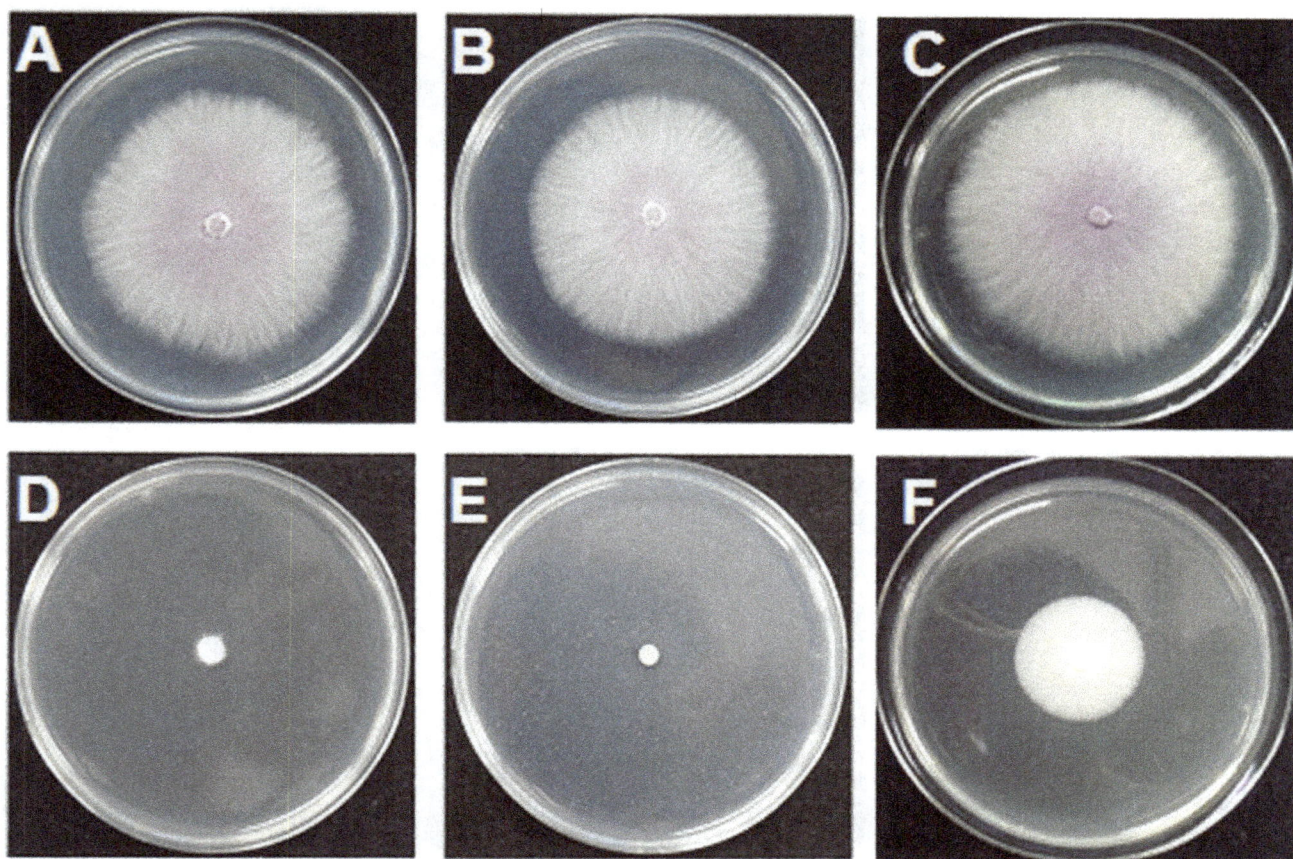

Figure 2. *In vitro* **fungicidal activity against** *Fusarium oxysporum*. A: blank control, B: **5d**, C: DMSO, D: **2b**, E: benomyl, F: **2a**.

containing 5-phenyl-2-furan were designed and synthesized. All the compounds had considerable and diverse bioactivities such as insecticidal, fungicidal, and antitumor activities. Thus, 5-phenyl-2-furan was regarded as an active scaffold in drug design. In this study, we focused on the molecular design and synthesis of novel salicyl glycoconjugates as elicitors against plant diseases. We present here the preparation and characterization of the new elicitors based on salicylic acid and 5-phenyl-2-furan moiety (Figure 1), and show that these compounds could induce the systemic acquired resistance against pathogenic infections in cucumber.

Materials and Methods

Instruments

All the melting points were determined with a Cole-Parmer melting point apparatus (Cole-Parmer, Vernon Hills, Illinois, USA) while the thermometer was uncorrected. Optical rotation data were recorded on a KRUSS P8000 instrument (KRUSS, Karlsruhe, Germany). IR spectra were recorded on a Nicolet NEXUS-470 FTIR spectrometer (International Equipment Trading Ltd., Vernon Hills, Illinois, USA) with KBr pellets. ^{1}H NMR spectra were recorded with Bruker DPX300 (Bruker, Fallanden, Switzerland) and JEOL JNM-ECS400 (JEOL Ltd., Tokyo, Japan), while tetramethylsilane was used as an internal standard. Analytical thin-layer chromatography was carried out on silica gel 60 F254 plates, and spots were visualized with ultraviolet light. Elemental analysis was carried out with a Flash EA 1112 elemental

analyzer (Thermo Finnigan, Bremen, Germany). Mass spectra were measured on a Bruker APEX IV spectrometer (Bruker, Fallanden, Switzerland).

Synthetic procedures

General synthetic procedure for hydrazides 2a and 2b. Preparation of hydrazides **2a** and **2b**: Esters **1a** and **1b** (30 mmol) was suspended in 100 mL methanol and reacted with 98% hydrazine monohydrate (60 mmol, 2.9 mL) under reflux for 12 h. The solid was filtered, washed with methanol and dried to afford hydrazides **2a** and **2b**.

2-mercaptobenzohydrazide (2a). Light yellow solid: yield 90%. m.p. 114–115°C. IR (KBr): v_{max} 3342, 3123, 1664, 1574, 1505, 1454, 1323, 1223, 1053 cm^{-1}. ^{1}H NMR (300 MHz, DMSO-d_6): 4.65 (s, 2H, NH$_2$), 5.16 (s, 1H, SH), 7.29–7.31 (m, 1H, PhH), 7.42–7.45 (m, 1H, PhH), 7.65–7.69 (m, 2H, PhH), 9.89 (s, 1H, CONH). ESI-MS: m/e 169.1 [M+H]$^+$. Anal. Calcd. (%) for C$_7$H$_8$N$_2$OS: C, 49.98; H, 4.79; N, 16.65. Found: C, 50.16; H, 4.91; N, 16.45.

2-hydroxybenzohydrazide (2b). White solid: yield 92%. m.p. 147–148°C. IR (KBr): v_{max} 3623, 3468, 1667, 1549, 1531, 1464, 1245, 1062 cm^{-1}. ^{1}H NMR (300 MHz, DMSO-d_6): δ 4.53 (s, 2H, NH$_2$), 5.23 (s, 1H, OH), 6.72–6.75 (m, 1H, PhH), 7.30–7.32 (m, 1H, PhH), 7.76–7.79 (m, 1H, PhH), 7.85–7.88 (m, 1H, PhH), 9.79 (s, 1H, CONH). ESI-MS: m/e 153.1 [M+H]$^+$. Anal. Calcd. (%) for C$_7$H$_8$N$_2$O$_2$: C, 55.26; H, 5.30; N, 18.41. Found: C, 55.52; H, 5.14; N, 18.59.

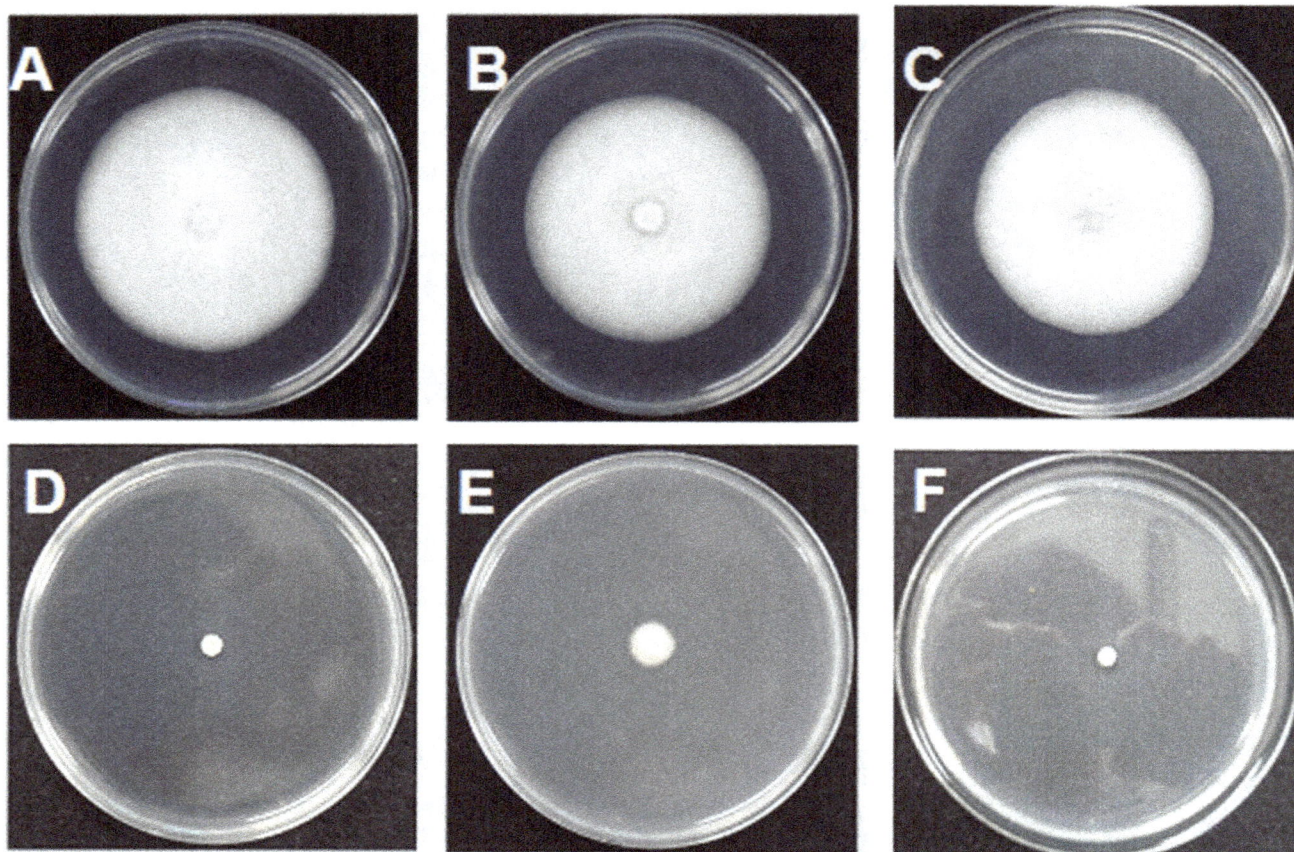

Figure 3. *In vitro* fungicidal activity against *Colletotrichum orbiculare*. A: blank control, B: **5d**, C: DMSO, D: **2b**, E: thiophanate-methyl, F: **2a**.

General synthetic procedure for hydrazides 5a–d and hydrazones 7a–g. The key intermediates *hydrazides 5a–d* were obtained almost quantitatively by the hydrazinolysis of compounds **4a~d** in alcohol. Compounds **5a~d** were condensed with 5-substituted phenyl-2-furfural to form the glycosyl hydrazones **7a–g**. All the chemical characterization was given in reference [35].

Bioassays

***In vitro* fungicidal activity.** *In vitro* fungicidal activity of the salicylic glycoconjugates against *Colletotrichum orbiculare*, *Fusarium oxysporum*, *Rhizoctonia solanii*, and *Phytophthora capsici* were evaluated using mycelium growth rate test [43–45]. The tested compounds were dissolved in DMSO (dimethyl sulfoxide) and mixed with sterile molten potato dextrose agar to a final concentration of 50 μg/mL. *In vitro* fungicidal activity of the salicylic glycoconjugates against *Sphaerotheca fuliginea* was evaluated using colonized detached leaves method [43–45]. The conidial suspensions were prepared by seeding about 2×10^5 spores mL^{-1} conidia in a 0.05% Tween 80 solution, and the DMSO solution of compounds (5000 μg/mL) was diluted with conidial suspension to a final concentration of 50 μg/mL. The solution was sprayed with a hand sprayer on the surface of the detached leaves which were inoculated with *S. fuliginea*.

P. capsici was maintained on oat medium at 17°C. *C. orbiculare*, *F. oxysporum*, and *R. solanii* were maintained on potato dextrose agar medium at 4°C. Five commercial fungicides: thiophanate-methyl, benomyl, chlorothalonil, validamycin, and dimethomorph were used as controls against the above mentioned

fungal pathogens under the same conditions. Three replicates were performed. The relative inhibition rate of the synthetic compounds compared to blank control was calculated *via* the following equation:

$$I = (C - T)/C \times 100\%$$

In which, I stands for the rate of inhibition (%), C is the diameter of mycelia in the blank control test (in mm), and T is the diameter of mycelia in the presence of tested compounds (in mm).

***In vivo* Antifungal Activity.** Using the pot culture test [42,46], the *in vivo* antifungal activities of the salicylic glycoconjugates against *C. orbiculare*, *F. oxysporum*, *S. fuliginea*, *R. solanii*, and *P. capsici* were evaluated in greenhouse along with five commercial fungicides: 70% thiophanate-methyl WP, 70% benomyl WP, 50% chlorothalonil WP, 3% validamycin AS, and 50% dimethomorph WP as controls.

The culture plates were cultivated at $24 \pm 1°C$. Germination was conducted by soaking cucumber seeds in water for 2 h at 50°C and then keeping the seeds moist for 24 h at 28°C in an incubator. When the radicles were 0.5 cm, the seeds were grown in plastic pots containing a 1:1 (v/v) mixture of vermiculite and peat. Cucumber plants used for inoculations were at the stage of two seed leaves. Ten plants were used for each treatment.

Tested compounds were confected to 2.5% EC (emulsifiable cocentration) formulations, in which pesticide emulsifier 500 (0.375%) and pesticide emulsifier 600 (2.125%) were the additives,

Leaf position

Figure 4. *In vivo* **antifungal activity against** *Colletotrichum orbiculare.*

DMSO (0.1%) was the solvent, and xylene was the co-solvent. The formulation was diluted to a concentration of 500 µg/mL with water. The solution was sprayed with a hand sprayer on the surface of seed leaves which were then inoculated with *C. orbiculare, S. fuliginea,* and *R. solanii,* respectively. Tested compounds and commercial fungicides were applied by irrigation at seedling stage, which were then inoculated with *F. oxysporum* and *P. capsici,* respectively. Three replicates for each treatment were applied.

Inoculations of *C. orbiculare* and *S. fuliginea* were carried out by spraying conidial suspension, and inoculation of *R. solanii* was carried out by spraying a mycelial suspension. *F. oxysporum* assay was carried out by embryo root inoculation, and *P. capsici* assay was carried out by irrigation inoculation.

Three replicates for each treatment were applied. After inoculation, the plants were maintained at $24\pm1°C$ and above 80% relative humidity.

The fungicidal activity was evaluated when the untreated cucumber plant (blank control) fully developed symptoms. The area of inoculated leaves covered by disease symptoms was assessed and compared to that of untreated ones to determine the average disease index. The relative control efficacy of compounds compared to the blank assay was calculated *via* the following equation:

$$I(\%) = [(CK\text{-}PT)/CK] \times 100\%$$

where I is relative control efficacy, CK is the average disease index during the blank assay and PT is the average disease index after treatment during testing.

RT-PCR for detection of pathogenesis-related gene expression. Tested compounds (500 µg/mL) were sprayed with a hand sprayer on the surface of the cucumber (*Cucumis sativus*) seed leaves, which were collected after 24 h, 48 h, and 72 h. The leaves were treated by liquid nitrogen. RNA isolation was performed with the RNAiso Plus Kit (Takara Bio). First-strand cDNA was synthesized from 100 µg/mL total RNA, which was quantified with QuantiT RNA Assay Kit (Invitrogen), by reverse transcription using the QuantiTect Reverse Transcription Kit (QIEGEN). Gene-specific primers (Table S1 in File S1) were designed and *actin* was used as the housekeeping gene [47,48]. Each reaction mixture (30 µL) contained 1 µL of the cDNA template, 100 pmol of each primer, 10 µL of Premix Ex Taq HS (Takara Bio), and 20 µL reaction buffer. The thermal cycling conditions were as follows: initial denaturation (94°C, 5 min), followed by 40 cycles of denaturation (94°C, 30 s), annealing (30 s) and extension (72°C, 30 s), and one final cycle of extension (72°C, 5 min). Finally, RT-PCR products were separated by electrophoresis and visualized in 1% agarose gel.

Leaf position

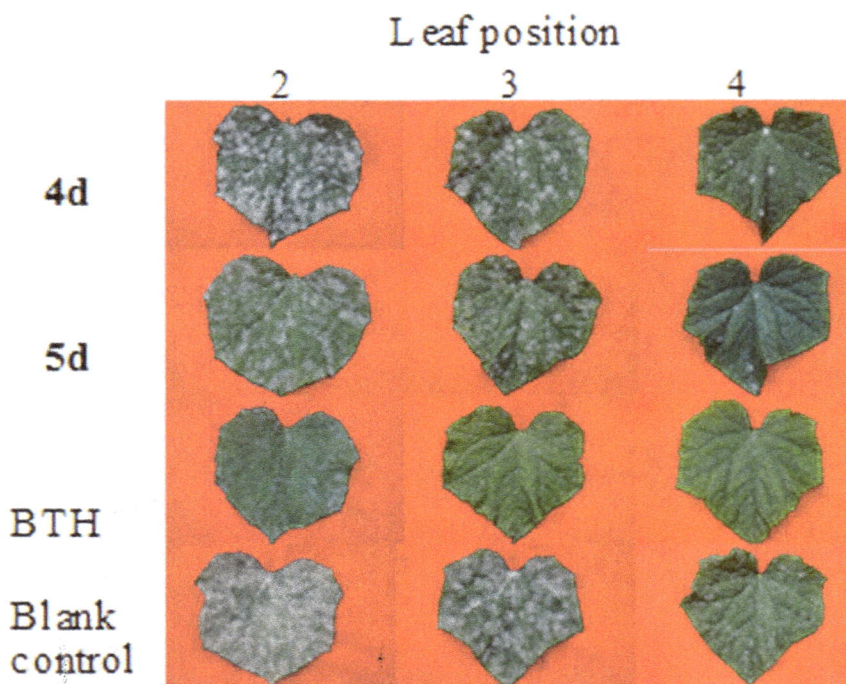

Figure 5. *In vivo* **antifungal activity against** *Sphaerotheca fuliginea.*

Ethics statement

No specific permits were required for the described field studies. No specific permissions were required for these locations. We confirm that the location is not privately-owned or protected in any way. We confirm that the field studies did not involve endangered or protected species.

Results and Discussion

Synthesis

The synthetic routes of 2-mercaptobenzohydrazide **2a**, 2-hydroxybenzohydrazide **2b** and glycosyl hydrazides **5a–d** were shown in Figure 1. The hydrazides **5a–d** were obtained almost quantitatively by hydrazinolysis of the esters **4a–d** in alcohol.

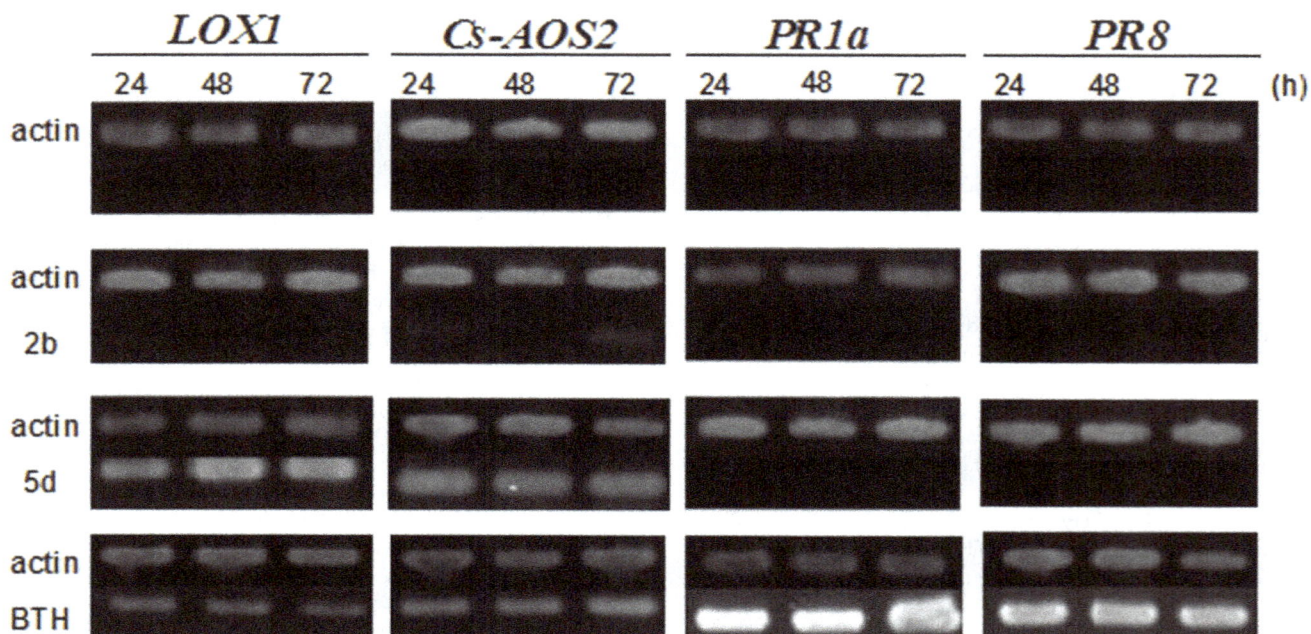

Figure 6. Effect of designed compounds on inducing the expression of pathogenesis-related genes in *Cucumis sativus.*

Finally, the hydrazides **5a–d** were reacted with aldehyde **6** by condensation to form the glycosyl hydrazones **7a–g**.

Fungicidal activity

The *in vitro* fungicidal results were shown in Table 1. The hydrazides **2a** and **2b** showed excellent activity against the tested fungi (Figures 2 and 3). For example, the inhibitory rates of the hydrazides **2a** and **2b** against *C. orbiculare* were 97.3% and 95.4%, which were better than thiophanate-methyl (91.0%). After modification of sugars, the *in vitro* activity of all the derivatives was decreased and they exhibited poor inhibitory rates. Although the *in vitro* activity of these glycosides was not encouraging, the *in vivo* tests gave promising results (Table 2), with all the carbohydrate derivatives showing considerable activity, especially against *F. oxysporum* (Table 2), *C. orbiculare* (Figure 4), and *S. fuliginea* (Figure 5). Among them, hydrazide **5d** and hydrazone **7f** had activity of 71.0% and 74.9% on *F. oxysporum*, respectively, which is similar to the control benomyl (74.5%) against the same pathogen. **5d** also showed good activity of 68.6% and 73.9% against *C. orbiculare* and *S. fuliginea*, respectively. Some hydrazones **7** exhibited promising activity against *P. capsici*. For examples, **7c** showed an inhibitory rate of 83.5%, and the inhibitory rates of **7a**, **7d** and **7e** were more than 75%.

The bioassay results showed that the tested compounds had *in vivo* antifungal activity against pathogenic fungi of Ascomycota (*C. orbiculare*, *F. oxysporum* and *S. fuliginea*), Basidiomycota (*R. solanii*), and Oomycete (*P. capsici*). The observed *in vivo* antifungal activity also had some association with the issue of pathogen biology. The tested compounds exhibited activity not only against the obligatory parasite pathogen (*S. fuliginea*), but also against the facultative parasite pathogens (*C. orbiculare*, *F. oxysporum*, *R. solanii* and *P. capsici*). The tested compounds also showed good activity against the soil-borne fungal disease (*F. oxysporum*, *R. solanii* and *P. capsici*). Also, we confirmed that all of these test compounds were safe for the host plants.

Defense activity of designed compound in plant

There are two important defense signaling pathways in plant system. One is mediated by salicylic acid and the other is mediated by jasmonic acid. In each defense pathway, there are specific marker genes which expression could be influenced by corresponding signaling molecules. In order to unveil the mode of action of our designed compounds, RT-PCR was performed to check the expression patterns of pathogenesis-related genes (*PR1a*, *PR8*, *LOX1*, *Cs-AOS2*) (Figure 6). Among them, *PR1a* and *PR8* were the specific marker genes mediated by salicylic acid, whereas *LOX1* and *Cs-AOS2* were the specific marker genes mediated by jasmonic acid. Our results showed that expressions of the *LOX1* and *Cs-AOS2* genes were significantly induced by hydrazide **5d**, and the expression level was comparable with that mediated by BTH (*S*-methyl benzo [1,2,3]thiadiazole-7-carbothioate). However, hydrazide **5d** had no obvious effect on the expressions of *PR1a* and *PR8*.

Conclusions

A new series of glycosyl hydrazines and hydrozone derivatives were designed and synthesized. Their antifungal tests indicated that most of the salicylic glycoconjugates had no *in vitro* fungicidal activity but showed considerable *in vivo* antifungal activity. The plant defense activity showed that expressions of the *LOX1* and *Cs-AOS2* genes were significantly induced by hydrazide **5d**, but the compound had no effect on the expressions of *PR1a* and *PR8*. Intriguingly, although the designed compounds were the derivatives of salicylic acid, they did not mimic the mode of action of salicylic acid, but seem to follow the jasmonic acid mediated pathway to induce the plant defense resistance.

Author Contributions

Conceived and designed the experiments: ZC YA YN. Performed the experiments: ZC JI HD. Analyzed the data: ZC HD YN. Contributed to the writing of the manuscript: ZC YN.

References

1. Franco G (2003) Systemic acquired resistance in crop protection: from nature to a chemical approach. J Agric Food Chem 51: 4487–4503.
2. Terry LA, Joyce DC (2004) Elicitors of induced resistance in postharvest horticultural crops: a brief review. Postharvest Biol Technol 32: 1–13.
3. Malamy J, Carr JP, Klessig DF, Raskin I (1990) Salicylic acid a likely endogenous signal in the resistance response of tobacco to viral infection. Science 250: 1002–1004.
4. Gaffney T, Friedrich L, Vernooij B, Negrotto D, Nye G, et al. (1993) Requirement of salicylic acid for the induction of systemic acquired resistance. Science 261: 754–756.
5. Vlot AC, Dempsey DA, Klessig DF (2009) Salicylic acid, a multifaceted hormone to combat disease. Annu Rev Phytopathol 47: 177–206.
6. Beckers G, Spoel S (2006) Fine-tuning plant defence signalling: Salicylate versus jasmonate. Plant Biol 8: 1–10.
7. Tamm L, Thürig B, Fleissbach A, Goltlieb AE, Karavani S, et al. (2011) Elicitors and soil management to induce resistance against fungal plant diseases. NJAS—Wageningen J Life Sci 58: 131–137.
8. Gorlach J, Volrath S, Knauf-Beiter G, Hengy G, Beckhove U, et al. (1996) Benzothiadiazole, a novel class of inducers of systemic acquired resistance, activates gene expression and disease resistance in wheat. Plant Cell 8: 629–643.
9. Fan Z, Shi Z, Zhang H, Liu X, Bao L, et al. (2009) Synthesis and biological activity evaluation of 1,2,3-thiadiazole derivatives as potential elicitors with highly systemic acquired resistance. J Agric Food Chem 57: 4279–4286.
10. Zuo X, Mi N, Fan Z, Zheng Q, Zhang H, et al. (2010) Synthesis of 4-methyl-1,2,3-thiadiazole derivatives *via* Ugi reaction and their biological activities. J Agric Food Chem 58: 2755–2762.
11. Fan Z, Yang Z, Zhang H, Mi N, Wang H, et al. (2010) Synthesis, crystal structure, and biological activity of 4-methyl-1,2,3-thiadiazole-containing 1,2,4-triazolo [3,4-b][1,3,4] thiadiazoles. J Agric Food Chem 58: 2630–2636.
12. Xu YF, Zhao ZJ, Qian XH, Qian ZG, Tian WH, et al. (2006) Novel, unnatural benzo-1,2,3-thiadiazole-7-carboxylate elicitors of taxoid biosynthesis. J Agric Food Chem 54: 8793–8798.
13. Du Q, Zhu W, Zhao Z, Qian X, Xu Y (2012) Novel benzo-1,2,3-thiadiazole-7-carboxylate derivatives as plant activators and the development of their agricultural applications. J Agric Food Chem 60: 346–353.
14. Wang J, Bi Y, Zhang Z, Zhang H, Ge Y (2011) Reduction of latent infection and enhancement of disease resistance in muskmelon by preharvest application of harpin. J Agric Food Chem 59: 12527–12533.
15. Felix G, Duran JD, Volko S, Boller T (1999) Plants have a sensitive perception system for the most conserved domain of bacterial flagellin. Plant J 18: 265–276.
16. Yamaguchi T, Ito Y, Shibuya N (2000) Oligosaccharide elicitors and their receptors for plant defense responses. Trends Glycosci Glyc 12: 113–120.
17. Kombrink A, Sanchez-Valle A, Thomma BPHJ (2011) The role of chitin detection in plant-pathogen interactions. Microbes Infect 13: 1168–1176.
18. Li Y, Heng Y, Zhao X, Du Y, Li F (2009) Oligochitosan induced *Brassica napus* L. production of NO and H$_2$O$_2$ and their physiological function. Carbohydr Polym 75: 612–617.
19. Yin H, Zhao X, Du Y (2010) Oligochitosan: a plant diseases vaccine–a review. Carbohydr Polym 82: 1–8.
20. Bautista-Banos S, Hernandez-Lauzardo AN, Velazquez-del Valle MG, Hernandez-Lopez M, Ait Barka E, et al. (2006) Chitosan as a potential natural compound to control pre and postharvest diseases of horticultural commodities. Crop Prot 25: 108–118.
21. Aziz A, Gauthier A, Bezler A, Poinssot B, Joubert JM, et al. (2007) Elicitor and resistance-inducing activities of β-1,4 cellodextrins in grapevine, comparison with β-1,3 glucans and α-1,4 oligogalacturonides. J Exp Bot 58: 1463–1472.

22. Klarzynski O, Plesse B, Joubert JM, Yvin JC, Kopp M, et al. (2000) Linear β-1,3 glucans are elicitors of defense response in tobacco. Plant Physiol 124: 1027–1037.

23. Sharp JK, McNeil M, Albersheim P (1984) Purification and partial characterization of a β-glucan fragment that elicits phytoalexin accumulation in soybean. J Biol Chem 259: 11312–11320.

24. Sharp JK, McNeil M, Albersheim P (1984) The primary structures of one elicitor-active and seven elicitorinactive hexa(β-D-glucopyranosyl)-D-glucitols isolated from the mycelial walls of *Phytophthora megasperma* f.sp. *glycinea*. J Biol Chem 259: 11321–11336.

25. Jamois F, Ferrieres V, Guegan JP, Yvin JC, Plusquellec D, et al. (2005) Glucan-like synthetic oligosaccharides: iterative synthesis of linear oligo-β-(1,3)-glucans and immunostimulatory effects. Glycobiology 15: 393–407.

26. Shinya T, Ménard R, Kozone I, Matsuoka H, Shibuya N, et al. (2006) Novel β-1,3, β-1,6-oligoglucan elicitor from *Alternaria alternata* 102 for defense response in tobacco. FEBS J 273: 2421–2431.

27. Nita-Lazar M, Heyraud A, Gey C, Braccini I, Lienart Y (2004) Novel oligosaccharide from *Fusarium oxysporum* L., rapidly induces PAL activity in *Rubus* cells. Acta Biochim Pol 51: 625–634.

28. Liu H, Cheng S, Liu J, Du Y, Bai Z, et al. (2008) Synthesis of pentasaccharise and heptasaccharide derivatives and their effects on plant growth. J Agric Food Chem 56: 5634–5638.

29. Kano A, Gomi K, Yamasaki-Kokudo Y, Satoh M, Fukumoto T, et al. (2010) A rare sugar, D-allose, confers resistance to rice bacterial blight with upregulation of defense-related genes in *Oryza sativa*. Phytopathology, 100: 85–90.

30. Cui ZN, Shi YX, Zhang L, Ling Y, Li BJ, et al. (2012) Synthesis and fungicidal activity of novel 2,5-disubstituted-1,3,4-oxadiazole derivatives, J Agric Food Chem 60: 11649–11656.

31. Yang XL, Wang DQ, Chen FH, Ling Y, Zhang ZN (1998) The synthesis and larvicidal activity of *N*-aroyl-*N'*-(5-aryl-2-furoyl) ureas. Pestic Sci 52: 282–286.

32. Yang XL, Ling Y, Wang DQ, Chen FH (2002) The synthesis and biological activity of *N*-phenyl-*N'*-(5-phenyl-2-furoyl) ureas. Chin J Synth Chem 10: 510–512.

33. Cui ZN, Zhang L, Huang J, Li Y, Ling Y, et al. (2008) 3D-QSAR studies on diacyl urea derivatives containing furan moiety. Acta Chim Sinica 66: 1417–1423.

34. Cui ZN, Li Y, Huang J, Ling Y, Cui JR, et al. (2010) New class of potent antitumor acylhydrazone derivatives containing furan. Eur J Med Chem 45: 5576–5584.

35. Cui ZN, Yang XL, Shi Y, Uzawa H, Cui J, et al. (2011) Molecular design, synthesis and bioactivity of glycosyl hydrazine and hydrazone derivatives: Notable effects of the sugar moiety. Bioorg Med Chem Lett 21: 7193–7196.

36. Cui ZN, Wang Z, Li Y, Zhou XY, Ling Y, et al. (2007) Synthesis of 5-(chlorophenyl)-2-furancarboxylic acid 2-(benzoyl)hydrazide derivatives and determination of their insecticidal activity. Chin J Org Chem 27: 1300–1304.

37. Cui ZN, Huang J, Li Y, Ling Y, Yang XL, et al. Synthesis and bioactivity of novel *N*,*N'*-diacylhydrazine derivatives containing furan(I). Chin J Chem 26: 916–922.

38. Li XC, Yang XL, Cui ZN, Li Y, He HW, et al. (2010) Synthesis and bioactivity of novel *N*,*N'*-diacylhydrazine derivatives containing furan(II). Chin J Chem 28: 1233–1239.

39. Cui ZN, Zhang L, Huang J, Yang XL, Ling Y (2010) Synthesis and bioactivity of novel *N*,*N'*-diacylhydrazine derivatives containing furan(III). Chin J Chem 28: 1257–1266.

40. Zhang L, Cui ZN, Yin B, Yang GF, Ling Y, et al. (2010) QSAR and 3D-QSAR studies of the diacyl-hydrazine derivatives containing furan rings based on the density functional theory. Sci China Chem 53: 1322–1331.

41. Cui ZN, Ling Y, Li BJ, Li YQ, Rui CH, et al. (2010) Synthesis and bioactivity of *N*-benzoyl-*N'*-[5-(2'-substituted phenyl)-2-furoyl] semicarbazide derivatives. Molecules 15: 4267–4282.

42. Cui ZN, Shi YX, Cui JR, Ling Y, Li BJ, et al. (2012) Synthesis and bioactivities of novel pyrazole and triazole derivatives containing 5-phenyl-2-furan. Chem Biol Drug Des 79: 121–127.

43. Li XH, Wu DC, Qi ZQ, Li XW, Gu ZM, et al. (2010) Synthesis, fungicical activity, and structure-activity relationship of 2-oxo and 2-hydroxycycloalkyl-sulfonamides. J Agric Food Chem 58: 11384–11389.

44. Li XH, Pan Q, Cui ZN, Ji MS, Qi ZQ (2013) Synthesis and fungicidal activity of *N*-(2,4,5-trichlorophenyl)-2-oxo- and 2-hydroxycycloalkyl-sulfonamides. Lett Drug Design Discov 10: 353–359.

45. Li XH, Cui ZN, Chen XY, Wu DC, Qi ZQ, et al. (2013) Synthesis of 2-acyloxycyclohexylsulfonamides and evaluation on their fungicidal activity. Int J Mol Sci 14: 22544–22557.

46. Wang BL, Shi YX, Ma Y, Liu XH, Li YH, et al. (2010) Synthesis and biological activity of some novel trifluoromethyl-substituted 1,2,4-triazole and bis(1,2,4-triazole) mannich bases containing piperazine rings. J Agric Food Chem 58: 5515–5522.

47. Bovie C, Ongena M, Thonart P, Dommes J (2004) Cloning and expression analysis of cDNAs corresponding to genes activated in cucumber showing systemic acquired resistance after BTH treatment. BMC Plant Biol 4: 15.

48. Ferreira RB, Monteiro S, Freitas R, Santos CN, Chen ZJ, et al. (2007) The role of plant defence proteins in fungal pathogenesis. Mol Plant Pathol 8: 677–700.

9

Seed Transmission of *Pseudoperonospora cubensis*

Yigal Cohen[1]*, Avia E. Rubin[1], Mariana Galperin[1], Sebastian Ploch[3], Fabian Runge[2], Marco Thines[3]

1 Faculty of Life Sciences, Bar-Ilan University, Ramat Gan, Israel, **2** Institute of Botany (210), University of Hohenheim, Stuttgart, Germany, **3** Biodiversity and Climate Research Centre, Frankfurt, Germany

Abstract

Pseudoperonospora cubensis, an obligate biotrophic oomycete causing devastating foliar disease in species of the Cucurbitaceae family, was never reported in seeds or transmitted by seeds. We now show that *P. cubensis* occurs in fruits and seeds of downy mildew-infected plants but not in fruits or seeds of healthy plants. About 6.7% of the fruits collected during 2012–2014 have developed downy mildew when homogenized and inoculated onto detached leaves and 0.9% of the seeds collected developed downy mildew when grown to the seedling stage. This is the first report showing that *P. cubensis* has become seed-transmitted in cucurbits. Species-specific PCR assays showed that *P. cubensis* occurs in ovaries, fruit seed cavity and seed embryos of cucurbits. We propose that international trade of fruits or seeds of cucurbits might be associated with the recent global change in the population structure of *P. cubensis*.

Editor: Mark Gijzen, Agriculture and Agri-Food Canada, Canada

Funding: The authors received no specific funding for this work.

Competing Interests: The authors have declared that no competing interests exist.

* Email: yigal.cohen1@gmail.com

Introduction

Downy mildew is a major disease of cucurbits [1–3]. The pathogen, *Pseudoperonospora cubensis* (Berk. & Curt.) Rost. (*Oomycota, Peronosporaceae*), attacks over 40 cucurbitaceous host species representing about 20 genera [1]. Infection occurs on cotyledons and true leaves. There are only two reports on the occurrence of this downy mildew on other plant parts. Van Haltern in the USA [4] found sporangiophores on the stem, petioles, tendrils, and peduncles of blossoms of heavily infected cantaloupe vines, but not on small fruits, and D'Ercole in Italy [5] recorded sporulation of the mildew on cucumber fruits grown under cover. There is no evidence that the pathogen spreads systemically in its hosts, nor that it is seed-borne or seed-transmitted. Here we report for the first time that *P. cubensis* may be fruit-borne, seed-borne, and seed-transmitted in cucurbits. Our preliminary report was published earlier [6].

Materials and Methods

Plants

Nine cucurbit species were grown in 4 net houses # 2, 3, 5 and 6 (6×50 m each) located on the north-east part of the campus, at Bar-Ilan University Farm, Israel in six seasons during 2012–2014. These species were also grown in a controlled glasshouse #3 located in the south-west part of the campus. The species were cucumber (*Cucumis sativum*, cvs.SMR-18 and Nadiojny), melon (*Cucumis melo* var. *reticulatus*, cvs. Ananas Yokneam and Ein-Dor), pumpkin (*Cucurbita maxima* cvs. Tripoli and Armonim), squash (*Cucurbita pepo* cvs.Beruti and Arlika), butternut gourd (*Cucurbita moschata* cv.Waltham), watermelon (*Citrullus lanatus* cv. Mallali), bottle gourd (*Lagenaria vulgaris*, local cultivar), sponge gourd (*Luffa cylindrica*, local cultivar) and bitter gourd

(*Momordica balsamina*, local cultivar). Planting took place on February (Spring season), August (Autumn season) and November (Winter season). Plants were fertilized weekly with 0.5% N:P:K and sprayed with fungicides against powdery mildew when required. Natural infection with downy mildew occurred in net-houses # 2, 3, 5 and 6 at all seasons on leaves of all cucurbits except watermelon, sponge gourd and bitter gourd. No downy mildew showed up in glasshouse #3 due to the lack of free moisture on the leaves. Seeds for all experiments describe herein were collected from greenhouse #3. Frequent PCR assays done with seeds samples taken from glasshouse #3 were proved negative for *P. cubensis*.

The procedures describe below were all similarly applied to healthy plant material derived from greenhouse #3 and infected (symptomless flowers, ovaries, fruits, seeds) plant material derived from net-houses #2, 3, 5 and 6.

Recovery of *P. cubensis* from flowers, fruits, and seeds

Strict hygiene measures were undertaken while attempting to recover *P. cubensis* from flowers, ovaries, fruits and seeds. Female flowers for ovaries were taken from cucumber and squash. Fruits were collected from cucumber, butternut gourd, pumpkin, bottle gourd and squash. Some fruits carried empty seeds because of lack of adequate pollination. The ovaries and fruits were washed with excessive soap water, washed thoroughly with tap water; surface sterilized by dipping in 4% hypochlorite solution for 10 minutes; dipped momentarily in ethanol and washed with sterile water. All further processing steps, including PCR, were carried out under strict sterile conditions. Ovaries, fruit seed cavity, empty seeds and mature seeds were homogenized in sterile cold water and used for inoculation of healthy detached leaves (taken from greenhouse #3). Mature seeds were sown (see below) in sterile soil mixture or

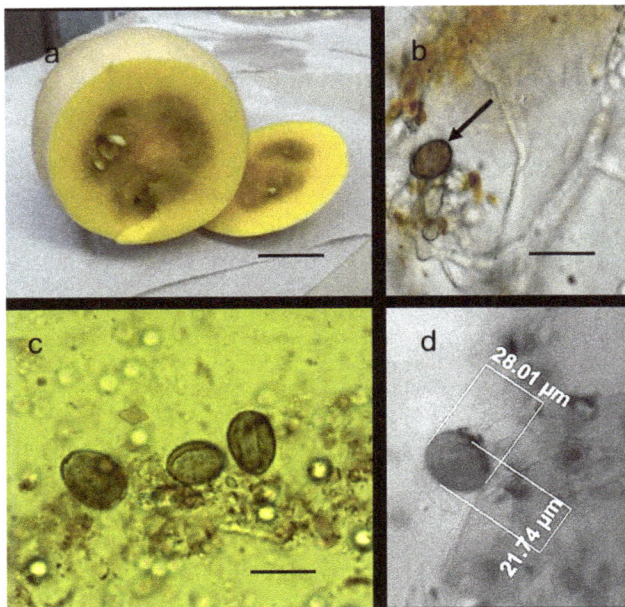

Figure 1. Infection of butternut gourd (*Cucurbita moschata*) fruit with *Pseudoperonospora cubensis*. a, transversally-cut fruit showing a dark seed cavity. **b, c, d** coenocytic mycelium with sporangia in the seed cavity. Bar in **a** = 3 cm, **b** = 30 µm, **c** = 20 µm.

used for microscopy. Three 50 µl droplets were taken from each homogenate, placed on a glass slide, covered with a cover slip, and examined with a dissecting microscope for the presence of mycelia or sporangia of *P. cubensis*. Fifty droplets, 50 µl each, of each homogenate were pippeted onto the lower surface of a detached cucumber leaf laying on a wet sterile filter paper in a 14 cm diameter Petri dish, and fifty such droplets were similarly inoculated onto a detached leaf of butternut gourd. Inoculated leaves were incubated at 20°C under 12 h photoperiod for 3 days, then washed with sterile water to remove the homogenate droplets, and thereafter kept for three weeks at 20°C under 12 h photoperiod to allow for downy mildew development. Detached leaves inoculated with droplets of sterile water served as controls. Symptoms of downy mildew with sporulation of *P. cubensis* appeared on some detached leaves inoculated with plant material (fruit seed-cavity homogenates, fruit tissue slices, ovary homogenates, and leaf pieces taken from seedlings developed from seeds) derived from net-houses #2, 3, 5 and 6 but never in similar plant material derived from greenhouse #3. Sporangia were propagated on detached leaves of cucumber laying on wet filter paper in 14 cm Petri dishes. The pathotype of the isolates was determined by inoculation of detached leaves of 9 cucurbits species as described before [7,8]. Mating type was determined by inoculation of melon and cucumber leaf discs with sporangial mixtures of the test isolate and a tester isolate of known mating type as described before [9].

Recovery of *P.cubensis* from hypocotyls

Seeds were surface sterilized and sown in sterile pots (60 ml) filled with a sterile soil mixture (peat: vermiculite, 1:1, v/v), 1 seed per pot. Plants were grown in greenhouse #3 and when reached the first true leaf stage (2 weeks after sowing) their hypocotyl was removed with a sterile scalpel, surface sterilized, placed in sterile water (1 hypocotyl/5 ml) and homogenized with a sterile blade (2 minute, 7000 rpm, 4°C). The homogenate was used for

inoculation of detached leaves of cucumber and butternut as described above.

Microscopy

Free hand sections were taken with a sterile razor blade from surface sterilized ovaries or fruits. Slices were placed on detached leaves of cucumber or butternut gourd to allow infection. Other slices were boiled in ethanol for 10 minutes, placed for 24 h in basic aniline blue solution (0.05%, pH 8.9) at 4°C, stained with 0.01% calcofluor (Sigma), and examined with Olympus A70 epifluorescent microscope for the presence of sporangia and mycelia [10]. A similar procedure was employed to embryos taken from mature seeds.

Recovery of *P.cubensis* from Seedlings

Seeds were surface sterilized, placed on sterile filter paper in sterile 14 cm petri dishes or 20×20×3 cm sterile plastic dishes (Nunk, Denmark) and incubated at 25°C under 12 h photoperiod. When cotyledons were produced (about 7 days), plants were transplanted into 0.5 L pots filled with sterile potting soil, while others were used for DNA extraction. Plants were maintained at 20°C at 12 h photoperiod to allow for downy mildew development. When symptoms appeared, plants were sealed in 1 L sterile plastic boxes (100% RH) for several days to enable sporulation of *P. cubensis*.

DNA extraction from plants or sporangia of *P. cubensis*

The method of Tinker et al [11] was employed with modifications. Samples of approximately 100–500 mg leaf, hypocotyl, root, ovary, or fruit tissue, or a sample of about 1×10^5 sporangia, were macerated in 1.5-ml micro-tubes using disposable pellet pestle grinders. Maceration was continued after adding 0.6 ml CTAB (hexadecyltrimethyl-ammonium bromide) buffer [1.4 M NaCl, 20 mM EDTA, 100 mM TRIS-Cl, 2% (W/V) CTAB pH 8.0], and the samples were incubated at 60°C for 45 min. The samples were then extracted with 0.6 ml chloroform/isoamyl alcohol (24:1) and centrifuged at 12000 g for 5 min. The aqueous phase was transferred to a 1.5-ml tube where the DNA was precipitated with an equal volume of cold (-20°C) isopropanol. DNA concentration was determined with a ND-1000 spectrophotometer (NanoDrop USA). DNA separation was done on a 1.2% agarose gel and staining with ethidium bromide.

DNA extraction from seeds

Dry seeds were placed in 2 ml tubes and rehydrated with sterile water for 15 minutes. After the water was removed, sodium hypochlorite solution (4%) containing 0.1% Tween 20 (to break surface tension) was added to the samples for 10 minutes. Seeds were rinsed with sterile water for 5 minutes. The embryo and the integument were separated and transferred to 96 well plates with 1.3 ml tubes. Care was taken not to cross-contaminate the samples and new gloves and sterile forceps were used for every seed. DNA extraction was conducted by using the BioSprint 96 DNA Plant Kit (Qiagen, Hilden, Germany) in combination with a KingFisher Flex (ThermoFisher Scientific, Waltham, USA) DNA extraction robot. The quality of the extraction was tested by conducting a PCR with the primers ITS1 and ITS4 developed by White et al. [12]. For all samples amplifiable DNA was obtained.

Primer development and molecular detection

Species-specific primers were developed based on *cox*2 sequences of *Pseudoperonospora humuli*, *Pseudoperonospora cubensis* and related species which were obtained from the database of the

Table 1. *Pseudoperonospora cubensis* was recovered from the reproductive organs of downy mildew-infected plants (A–D) but not from the reproductive organs of healthy plants (E–H).

From Downy Mildew-Infected Plants

A

Host	Ovaries Examined	Infectious	%
Cucumber	82	15	18.3
Melon	10	3	30.0
Squash	37	14	37.8
Total	**129**	**32**	**24.8**

B

Host	Fruit seed cavity Examined	Infectious	%
Cucumber	316	22	7.0
Melon	116	8	6.9
Pumpkin	29	3	10.3
Squash	49	1	2.0
Total	**510**	**34**	**6.7**

C

Host	Seeds Sown	Infected plants	%
Cucumber	400	1	0.25
Butternut gourd	400	4	1
Squash	400	6	1.5
Total	**1200**	**11**	**0.92**

D

Host	Hypocotyls Examined	Infectious	%
Cucumber	150	1	0.7
Melon	150	2	1.3
Squash	150	4	2.7
Total	**450**	**7**	**1.6**

From Control Healthy Plants

E

Host	Ovaries Examined	Infectious	%
Cucumber	40	0	0.0
Melon	10	0	0.0
Squash	30	0	0.0
Total	**80**	**0**	**0.0**

F

Host	Fruit seed cavity Examined	Infectious	%
Cucumber	100	0	0
Butternut gourd	30	0	0
Pumpkin	100	0	0
Squash	20	0	0
Total	**160**	**0**	**0.0**

G

Host	Seeds Sown	Infected plants	%
Cucumber	200	0	0
Butternut gourd	200	0	0
Squash	200	0	0
Total	**600**	**0**	**0.0**

H

Host	Hypocotyls Examined	Infectious	%
Cucumber	50	0	0
Butternut gourd	50	0	0
Squash	50	0	0
Total	**150**	**0**	**0.0**

The downy mildew-infected plants were grown in net-houses #2, 3, 5 and 6 while the healthy plants were grown in greenhouse #3.
A, E - Infectivity of crushed ovaries to detached leaves of cucumber and/or butternut gourd.
B, F - Infectivity of crushed fruit seed-cavity tissue to detached leaves of cucumber and/or butternut gourd.
C, G - Vertical transmission of *P. cubensis* from seeds to the next plant generation.
D, H - Infectivity of crushed hypocotyls to detached leaves of cucumber and/or butternut gourd.

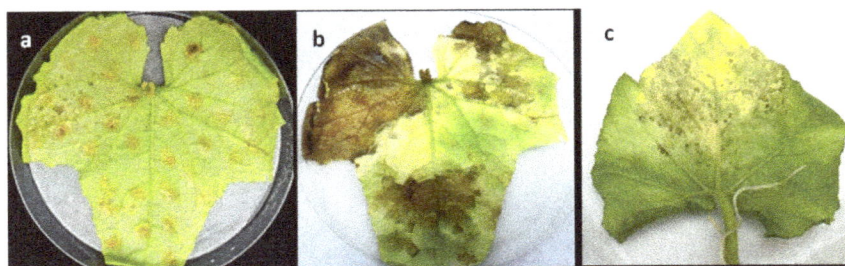

Figure 2. Tissue homogenate taken from the seed-cavity of butternut gourd or cucumber fruits are infective to detached leaves of cucumber and butternut gourd. a downy mildew starting to develop in a detached leaf of cucumber at 1 week after inoculation; **b**, **c**, downy mildew developed in a detached leaf of cucumber and butternut gourd, respectively at 3 weeks after inoculation.

National Center for Biotechnology (http://www.ncbi.nlm.nih.gov/). The primers for *P. humuli* and the two clades of *P. cubensis*, which had been reported in Runge et al. [13], as well as their specificity, are shown in Table S1. As described by Ploch et al. [14] two PCRs were carried out to detect the pathogens with high specificity and sensitivity. In the first PCR, 0.4 mM of Oomycete specific primers *cox*2-F and *cox*2-R [15] were used in a reaction mixture containing 1× Mango PCR Buffer, 0.2 mM dNTPs, 1 mM MgCl$_2$, 0.8 mg/ml BSA and 0.5 U Mango Taq DNA Polymerase (Bioline, Luckenwalde, Germany). For all three primer combinations (Table S1) a separate nested PCR was conducted with a 1 to 10 dilution of the oomycete specific PCR. Cycling temperature and times are detailed in Table S2.

Results

Downy mildew in butternut gourd fruits

Butternut gourd fruits taken from downy mildew-infected plants revealed a dark seed cavity (Fig. 1a). Microscopic examinations of free-hand sections taken from the seed cavity revealed coenocytic hyphae and a few sporangia of *P. cubensis* (Fig. 1 b–d). No sporangiophores of *P. cubensis* were seen. Between 0–10 sporangia per 50 µl droplet were detected in the seed cavity homogenate of various fruits. Butternut gourd fruits taken from healthy plants showed no discoloration of the seed cavity nor mycelia or sporangia.

Figure 3. Seed transmission of downy mildew caused by *Pseudoperonospora cubensis* in butternut gourd (*Cucurbita moschata*). a symptoms (arrows) on cotyledons at 10 days after germination. **b** symptoms with sporulation (arrows) on the first true leaf at 2 weeks after transplanting to soil.

Recovery of *P. cubensis* from ovaries, fruits, hypocotyls and seeds

Table 1 summarizes the data obtained during 2012–14. A total of 129 ovaries of cucumber, melon or squash were collected from downy mildew-infected plants, surface sterilized, crushed and inoculated onto detached healthy leaves of cucumber and butternut gourd. No ovaries of butternut gourd were used. About 25% of the ovaries were infectious, capable of producing typical downy mildew symptoms on the detached leaves (Table 1A). The isolates recovered from ovaries of cucumber or melon belonged to pathotype 3 mating type A1, while those recovered from squash belonged to pathotype 6 mating type A2.

A total of 510 fruits of cucumber, melon, pumpkin or squash were collected during 2012–2014. Fruits were surface sterilized, cut open in two halves, seeds (when available) removed, and the seed cavity tissue crushed and inoculated onto detached leaves. Thirty four fruits (6.7%) were infectious, producing typical downy mildew on the detached leaves of cucumber and/or butternut gourd (Table 1B; Fig. 2). The isolates recovered from seed cavity of cucumber and melon belonged to pathotype 3 mating type A1, while those recovered from pumpkin and squash belonged to pathotype 6 mating type A2.

Mature seeds were taken from fruits of cucumber, butternut gourd or squash, surface-sterilized, planted (20 seeds per fruit, 20 fruits from each species) in sterile soil mixture in pots and grown for one month at 25°C with 12 h light/day. As shown in Table 1C, one cucumber, four butternut gourd and six squash plants developed downy mildew symptoms with sporulation of *P. cubensis* (Fig. 3).

In a similar experiment, seeds were planted in sterile soil and grown in greenhouse #3. At 2 weeks after germination no disease was seen on the cotyledons. The hypocotyls were taken, surface-sterilized, homogenized in sterile water and inoculated onto detached leaves. Seven out of the 450 hypocotyls tested (1.6%) were infectious, producing downy mildew on detached leaves of cucumber and/or butternut gourd (Table 1D).

In parallel experiments, ovaries, fruits and seeds were collected during 2012–2014 from healthy plants growing in glasshouse #3 in which the controlled dry atmosphere prevented downy mildew development. None of the 80 ovaries, 160 fruits or 150 hypocotyls was infectious to detached leaves of cucumber or butternut gourd (Table 1 E, F, H). No seed of the 600 seeds sown developed downy mildew symptoms (Table 1 G).

Microscopy

A rare observation of *P.cubensis* sporulating on embryo of cucumber is shown in Fig. 4a. Mycelium inside the embryo of

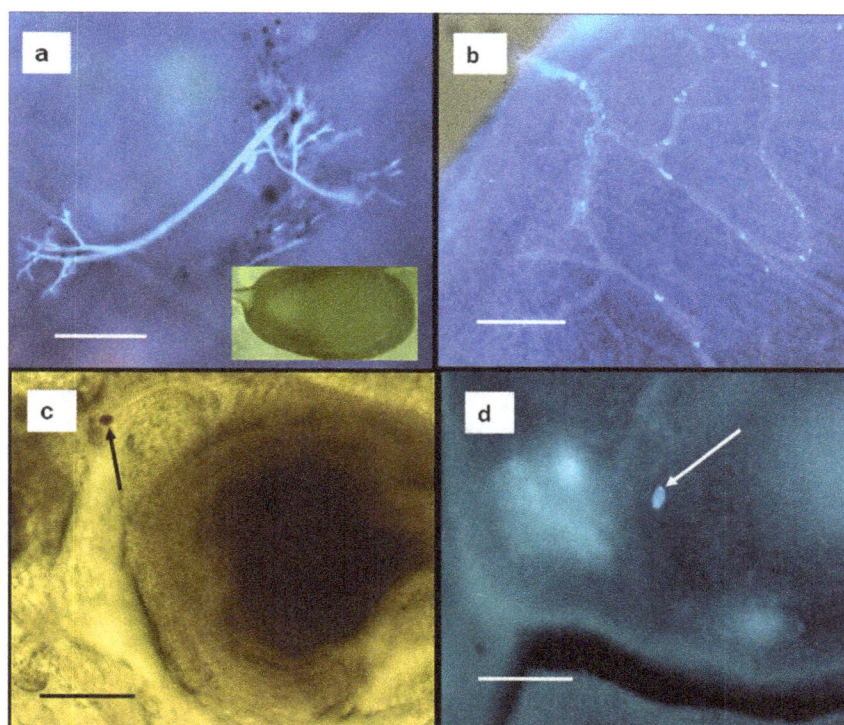

Figure 4. *P.cubensis* **in embryos and ovaries of cucumber and butternut gourd. a** sporangiophores with sporangia on a cucumber embryo (insert: a cucumber seed, actual size = 9 mm). **b** mycelium in an embryo of butternut gourd. **c** and **d** sporangia (arrow) in the ovary of cucumber near an ovule. Bar in **a** and **c** = 100 µm; in **b** and **d** = 200 µm.

butternut gourd is shown in Fig. 4b. Sporangia inside ovaries of cucumber, adjacent to the ovules, are shown in Fig. 4c and 4d.

PCR assays for *P. cubensis* in ovaries and seed cavity

About 49% of the ovaries (39 out of 80) (Table 2A) and 55% of the flower peduncles (6 out 11) (data not shown) reacted positively when tested with primers Set 1. None of the 45 healthy ovaries has tested positive (Table 2B). Of the 301 fruits examined, 35 (11.6%) tested positive with primers Set 1 (Table 2C). All 120 healthy fruits from greenhouse #3 tested negatively to *P. cubensis* (Table 2D).

Interestingly, in 12 cucumber fruits whose proximal and distal ends were tested separately, 5 tested positive, all at the proximal end (stem end) of the fruit. Similarly, in 6 squash fruits all tested positively, all at the proximal end of the fruit. PCR assays conducted with pistil tissue of squash female flowers resulted with no signal, suggesting that the pathogen moves into the fruit from the stem side and not from the pistil side. Indeed, PCR assays revealed the occurrence of *P. cubensis* in stems of cucumber, melon, squash and pumpkin. Often, petioles and stem homogenates were infectious to detached leaves of cucumber and/or butternut gourd.

PCR assays for *P.cubensis* in seed integuments and embryos

PCR assays showed that *P. cubensis* occurs in seeds collected from downy mildew-infected plants (Fig. 5, Tables 3–5). In Fig. 5, three seeds per entry were used, one slot per seed. A positive reaction to primer Set 1(detecting all 3 clades) was detected in 10 out of 15 seed samples. In 6 samples, only one seed responded positively whereas in 4 samples, 2 out of 3 seeds responded positively, indicating on the heterogeneity of infection in the seeds

in a single fruit. The same number of samples responded to primer Set 2 (detecting clades 1 and 2), but with only one seed out of three responding positively. Eight samples responded to primer set 3 (detecting clade 2), one seed per sample. Sequencing of 6 randomly chosen PCR products confirmed the identity of the organism as being *P. cubensis*.

Table 3 presents the molecular detection analyses obtained for 10 seeds of 7 fruits: B2 of squash and D2–D7 of butternut gourd. The occurrence of *P. cubensis* was tested separately in the integuments and the embryo (see Methods and Material). Two embryos (D2/7, D5/4) and three integuments (D5/8, D6/1 and D6/3) were tested positive for *P. cubensis*. One seed B2/1 tested positive for primers Set 1 only.

In another assay (Table 4), the integuments and embryo of 70 individual seeds (from 15 fruits) of cucumber were analyzed with primer Set 3 that amplifies clade 1 of *P. cubensis*. The integuments of 5 seeds and the embryo of 6 seeds tested positive. In only one seed (C4/5) both the integuments and the embryo tested positive (Table 4A), suggesting that the pathogen may colonize the integuments, the embryo, or both. No amplicons of *P. cubensis* were detected in seeds taken from healthy plants (Table 4B).

Similar results were obtained with seeds of bottle gourd (*Lagenaria vulgaris*) (Table 5). The integuments and embryo of 100 individual seeds (10 seeds/fruit) were analyzed with all three primer sets of *P. cubensis*. Eight seeds tested positive with primer set 3 in both the integument and embryo, 6 seeds tested positive with primer set 2 in the integument but not the embryo, and no seed was positive with primer set 1 (Table 5A). No amplicons of *P.cubensis* were detected in seeds taken from healthy plants (Table 5B).

Table 2. PCR assays showing the occurrence of *Pseudoperonospora cubensis* amplicons in ovaries and fruits of cucurbits.

A. Ovaries from infected plants		
Host	Ovaries	Primer Set 1
Cucumber	38	22
Squash	37	16
Melon	5	1
Total	**80**	**39**
B. Ovaries from healthy plants		
Host	Ovaries	Primer Set 1
Cucumber	15	0
Squash	15	0
Melon	15	0
Total	**45**	**0**
C. Seed cavity of fruits from infected plants		
Host	Fruits	Primer Set 1
Cucumber	236	26
Butternut gourd	28	4
Pumpkin	37	5
Total	**301**	**35**
D. Seed cavity of fruits from healthy plants		
Host	Fruits	Primer Set 1
Cucumber	80	0
Butternut gourd	20	0
Pumpkin	20	0
Total	**120**	**0**

In pumpkin, we tested 42 embryos from 5 infected fruits with primer sets 2 and 3. Two out of 5 fruits had PCR-positive embryos. In one fruit, 4 out of 10, and in the second 1 out of 10, tested positive with both primers sets 2 and 3. Thirty embryos from 4 healthy fruits were all PCR-negative.

Discussion

P. cubensis is a foliar pathogen of *Cucurbitaceae* with worldwide distribution. In the past decade major changes occurred in the population structure of this oomycete. In Israel, two major changes took place: in 2002 a new pathotype, number 6, appeared [8] and in 2010 a new mating type, A2, showed up [9]. In the

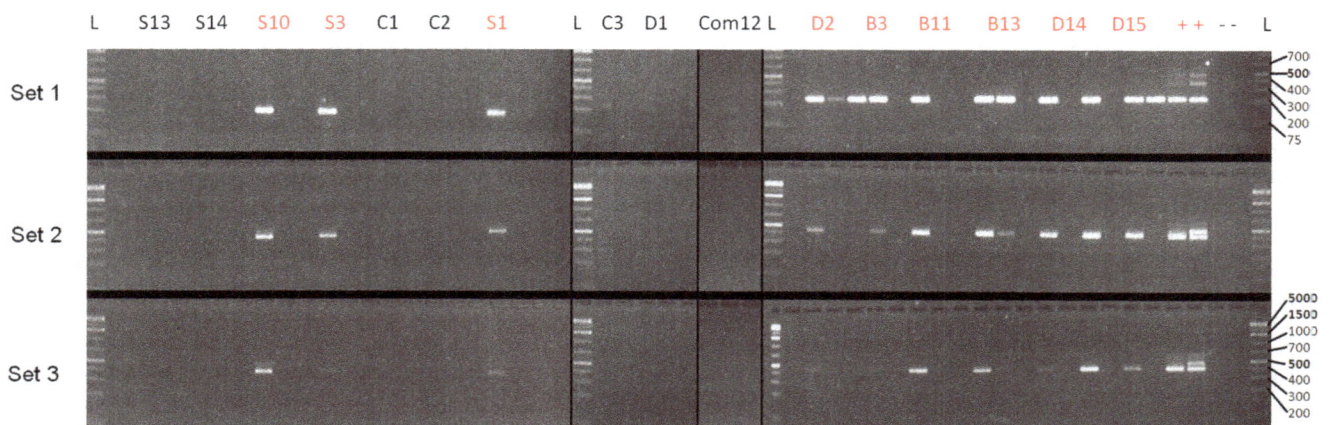

Figure 5. PCR analysis of 16 seed samples, 3 seeds per sample, one slot per seed. In the left part- embryos were tested. In the two middle parts, weak bands with only primer set 1, are seen. In the right part integuments were tested (seeds were empty). L = 1 kb plus ladder (Fermentas). S = squash Arlika. C = cucumber. Com12 = cucumber cv. Marco, Nickerson Zwaan. D = butternut gourd. B = squash Beruti. + = positive control (left: NY425 – clade 2, right: NY427 – clade 1 (Runge et al. 2011)). - = negative control (sterile water).

Table 3. PCR assays showing the occurrence of *Pseudoperonospora cubensis* amplicons in individual seeds of cucurbits.

Fruit Name	Seed Number	ITS - Test Integument	ITS - Test Embryo	cox2 Integument	cox2 Embryo	All 3 clades Integument	All 3 clades Embryo	Clades 1 and 2 Integument	Clades 1 and 2 Embryo	Clade 1 Integument	Clade 1 Embryo
B2	1	1	1	0	0	1	0	0	0	0	0
B2	2–10	1	1	0	0	0	0	0	0	0	0
D2	1–6	1	1	0	0	0	0	0	0	0	0
D2	7	1	1	0	0	0	1	0	1	0	1
D2	8–10	1	1	0	0	0	0	0	0	0	0
D3	1–10	1	1	0	0	0	0	0	0	0	0
D4	1–10	1	1	0	0	0	0	0	0	0	0
D5	1–3	1	1	0	0	0	0	0	0	0	0
D5	4	1	1	0	0	0	1	0	1	0	1
D5	5–8	1	1	0	0	0	0	0	0	0	0
D5	9	1	1	0	0	1	0	1	0	1	0
D5	10	1	1	0	0	0	0	0	0	0	0
D6	1	1	1	0	0	1	0	1	0	1	0
D6	2	1	*0	0	0	0	0	0	0	0	0
D6	3	1	*0	0	0	1	0	1	0	1	0
D6	4–10	1	*0	0	0	0	0	0	0	0	0
D7	1–10	1	1	0	0	0	0	0	0	0	0
H2O		0	0	0	0	0	0	0	0	0	0
"1953"		no positive controls used		1	1	1	1	1	1	1	1
additional positive control "21226"										1	1
Total		*empty		1	1	5	3	4	3	5	4

The seeds were taken from one fruit of squash (B) and 6 fruits of butternut gourd (D) (10 seeds per fruit). 1 = test positive; 0 = test negative.

Table 4. PCR assays showing the occurrence of *Pseudoperonospora cubensis* in individual seeds of cucumber.

A. Seeds from downy mildew-infected plants

Fruit Name	Seed Number	Clade 1	
		Integument	embryo
C1	1	0	0
	2	**1**	0
	3–5	0	0
C2	1–4	0	0
C3	1–5	0	0
C4	1	0	**1**
	2–4	0	0
	5	**1**	**1**
C5	1, 2, 4	0	0
	3	0	**1**
	5	**1**	0
C6	1, 2, 3, 5	0	0
	4	0	**1**
C7	1–5	0	0
C8	1–5	0	0
C9	1–5	0	0
C10	1–5	0	0
C11	1, 2, 3, 5	0	0
	4	0	**1**
C12	1, 3, 4, 5	0	0
	2	**1**	0
C13	1	0	**1**
	2, 3, 4	0	0
	5	**1**	0
C14	1–3	0	0
C15	1–3	0	0
Total	**70**	**5**	**6**

B. Seeds from healthy plants

Fruit Name	Seed Number	Clade 1	
		Integument	embryo
C1	1–5	0	0
C2	1–5	0	0
C3	1–5	0	0
C4	1–5	0	0
C5	1–5	0	0
C6	1–5	0	0
C7	1–5	0	0
C8	1–5	0	0
C9	1–5	0	0
C10	1–5	0	0
Total	**50**	**0**	**0**

A. Fruits taken from downy mildew-infected plants. B. Fruits taken from healthy plants.
B. 1 = test positive; 0 = test negative.

Table 5. PCR assays showing the occurrence of *Pseudoperonospora cubensis* in individual seeds of bottle gourd (*Lagenaria vulgaris*).

| | | A. Fruits from downy mildew-infected plants | | | | | |
| | | All 3 clades | | Clade 1 and 2 | | Clade 1 | |
Fruit name	Seed Number	Integument	Embryo	Integument	Embryo	Integument	Embryo
Lag1	5	0	0	1/0	0	1/0	0
	1–4, 6–10	0	0	0	0	0	0
Lag2	2	0	0	0	0	1/0	0
	7	0	0	0	0	1/0	0
	9	0	0	0	0	1/0	1/0
Lag3	1,3–5, 8, 10	0	0	0	0	0	1/0
	4	0	0	0	0	0	1/0
	5	0	0	0	0	1/0	1/0
	7	0	0	1/0	0	0	1/0
	9	0	0	1/0	0	1/0	1/0
	1–3, 6, 8, 10	0	0	0	0	0	0
Lag4	2	0	0	0	0	1/0	0
	8	0	0	1/0	0	0	0
	1, 3–7, 9, 10	0	0	0	0	0	0
Lag5	1–10	0	0	0	0	0	0
Lag6	1–10	0	0	0	0	0	0
Lag7	3	0	0	1/0	0	1/0	0
	4	0	0	0	0	1/0	1/0
	5	0	0	1/0	0	1/0	0
	1, 2, 6–10	0	0	0	0	0	0
Lag8	1–10	0	0	0	0	0	0
Lag9	1–10	0	0	0	0	0	0
Lag10	5	0	0	0	0	1/0	1/0
	6	0	0	0	0	1/0	1/0
	7	0	0	0	0	1/0	1/0
	9	0	0	0	0	0	1/0
	10	0	0	0	0	1/0	1/0
	1–4, 8	0	0	0	0	0	0

Table 5. Cont.

A. Fruits taken from downy mildew-infected plants

Fruit name	Seed Number	All 3 clades		Clade 1 and 2		Clade 1	
		Integument	Embryo	Integument	Embryo	Integument	Embryo
Total	100	0	0	6	0	14	11

B. Fruits from healthy plants

Fruit name	Seed Number	All 3 clades		Clade 1 and 2		Clade 1	
		Integument	Embryo	Integument	Embryo	Integument	Embryo
Lag1	1–10	0	0	0	0	0	0
Lag2	1–10	0	0	0	0	0	0
Lag3	1–10	0	0	0	0	0	0
Lag4	1–10	0	0	0	0	0	0
Lag5	1–10	0	0	0	0	0	0
Lag6	1–10	0	0	0	0	0	0
Lag7	1–10	0	0	0	0	0	0
Lag8	1–10	0	0	0	0	0	0
Lag9	1–10	0	0	0	0	0	0
Lag10	1–10	0	0	0	0	0	0
Total	100	0	0	0	0	0	0

A- Fruits taken from downy mildew-infected plants. B- Fruits taken from health plants.
1= test positive; 0= test negative.

USA a new race appeared in 2004 capable of destroying long-lasting resistant cucumber cultivars [16]. In Italy, a new pathotype, number 5, appeared in 2003 [17], and in the Czech Republic many new pathotype combinations appeared recently [18].

Runge et al [13] performed a molecular comparison between pre-2004 and post-2004 isolates of *P. cubensis*. They suggested that the new post-2004 genotypes have migrated (by man carrying leaf material) from South East Asia (Korea, Japan) to Europe and thereafter to the USA.

Another vehicle for such a migration could be a man carrying fruits or seeds of cucurbits, or commercial trade of fruits and seeds. We show here that fruits of cucurbits, collected from downy mildew-infected plants are symptomless but may carry mycelium and sporangia of *P. cubensis*. Fruit slices, or seed-cavity tissue homogenates made from such fruits, produced typical downy mildew symptoms with sporulation of *P. cubensis* when applied to detached healthy leaves of cucurbits. *P. cubensis* similarly occurs in symptomless stems, petioles and ovaries of infected plants. PCR assays showed that the pathogen occurs in peduncles of female flowers and at the stem end, not in the petal end, of the ovary or fruit, nor in the pistil of the flower. This suggests that penetration of the pathogen into the ovary, and thereafter into the fruit, occurs from the leaf into the stem and then through the peduncle of the female flower into the ovary.

P. cubensis was found to also occur in seeds of cucurbits. We confirmed it by microscopy, infectivity of crushed seeds to detached leaves, and species-specific PCR assays. All implicated that *P. cubensis* may be transmitted by seed.

Indeed, seeds collected from infected plants produced infected plants with typical symptoms and sporulation of *P. cubensis*. Vertical seed transfer of downy mildew occurred in cucumber (0.25%), butternut gourd (1%) and squash (1.5%). Hypocotyls produced by such seeds were infectious (1.6%) when crushed and inoculated onto detached leaves of cucumber.

These findings may now offer a new explanation to the global structural changes in the pathogen population. Cucurbit fruits (probably ornamental squash or pumpkin) were imported for commercial purposes, or transported by man, from South East Asia into Europe and/or Israel. The fruits were collected from infected plants and probably carried *P. cubensis* belonging to pathotype 6 mating type A2. Our survey [19] approved the occurrence in China of isolates that belong to the pathotype 6 mating type A2. Another survey [20] showed a clear preference of isolates belonging to pathotype 6 mating type A2 to *Cucurbita* species (pumpkin, squash, butternut gourd). It could be that the original migration took place from China and not necessarily from Korea or Japan, or that the population of *P. cubensis* in China, Korea and Japan are composed of similar genotypes. The new immigrants may have undergone sexual mating with the prevailing local isolates that belong to pathotype 3 mating type A1. Oospores were produced and the new recombinant offspring isolates were probably capable of attacking new hosts (see [9]). The new isolates may have similarly migrated to the USA [21] where they were capable of breaking down the long-lasting resistance of cultivars of cucumber.

Seed infection by downy mildew agents is common, but seed transmission of downy mildew diseases is rare and happens in a rather low proportion. Seeds collected from sunflower plants systemically infected with the homothallic oomycete *Plasmopara halstedii* contain mycelia, sporangia, oogonia and oospores of the pathogen in the inner surface of the pericarp and embryo. Only one out 276 plants grown from such seeds (0.36%) produced typical systemic infection [22]. Seed of *Camelina sativa* were reported [23] to carry sporangia of *Hyaloperonospora camelinae* on their surface and produced (no surface disinfection) 96% infected plants. Garibaldi et al [24] and Djalali et al [25] showed that basil seeds are infected with *Peronospora belbahrii*. The late blight oomycete pathogen *Phytophthora infestans* may infect seeds of tomato. Such infection occurs when fruits are inoculated on their stem scar with mixed A1+A2 sporangia [26]. The infected seeds harbor mycelia, sporangia, and oospores of the pathogen in the pericarp and embryo [26] with only one out of about 1000 seedlings showing late blight infection (Rubin and Cohen, unpublished data). Landa et al [27] demonstrated that *Peronospora arborescens*, the causal agent of downy mildew in opium poppy, can be transmitted in seeds and that commercial seed stocks collected from crops with high incidence of the disease were infected but produced a rather low incidence of infected plants. Quinoa seeds were reported to harbor oospores of the downy mildew agent *Peronospora variabilis* [28].

In conclusion, this paper demonstrates that under experimental conditions, *P. cubensis* can be transmitted from infected plants to their pedigree seeds. This fact calls for immediate action to elucidate if seed transmission may also occur in agricultural practices. Further studies are required to better understand the mechanisms by which fruits and seeds become contaminated. More research should be devoted to further confirm the vertical transmission of the pathogen to the next generation of cucurbits. The close relationships between *P. cubensis* and *P. humuli* amplicons need to be studied further. Special care should be taken by growers not to use seeds from downy mildew-infected plants.

Supporting Information

Table S1 Three sets of primers used to identify *Pseudoperonospora cubensis* in sporangia and plant material of cucurbits.

Table S2 Temperatures and cycling times of PCR reactions used to identify *Pseudoperonospora cubensis* in sporangia and plant material of cucurbits. First PCR program for primer pair *cox*2-F and *cox*2-R. Second PCR program for species-specific primers given in Table 1. Steps 2-4 were repeated 35 times.

Author Contributions

N/A. Conceived and designed the experiments: YC MT FR. Performed the experiments: AR MG SP FR. Analyzed the data: YC FR. Contributed reagents/materials/analysis tools: MT SP FR MG. Wrote the paper: YC MT FR.

References

1. Palti J, Cohen Y (1980) Downy mildew of cucurbits (*Pseudoperonospora cubensis*) - the fungus and its hosts, distribution, epidemiology and control. Phytoparasitica 8: 109–147.
2. Lebeda A, Cohen Y (2011) Cucurbit downy mildew (*Pseudoperonospora cubensis*)-biology, ecology, epidemiology, host-pathogen interaction and control. Eur J Plant Pathol 129: 157–192.
3. Savory EA, Granke LL, Quesada-Ocampo LM, Varbanova M, Hausbeck MK, et al. (2011) The cucurbit downy mildew pathogen *Pseudoperonospora cubensis*. Mol Plant Pathol 12: 217–226.
4. Van Haltern F (1933) Spraying cantaloups for the control of downy mildew and other diseases. Bulletin of Georgia Experimental Station 175.

5. D'Ercole N (1975) La peronospora del cetrioloin coltura protetta. Inftore fitopatol 25: 11–13.

6. Cohen Y, Rubin AE, Galperin M (2013) Seed transmission of *Pseudoperonospora cubensis* in dalorit (Cucurbita moschata). Phytoparasitica.

7. Thomas CE, Inaba T, Cohen Y (1987) Physiological specialization in *Pseudoperonospora cubensis*. Phytopathology 77: 1621–1624.

8. Cohen Y, Meron I, Mor N, Zuriel S (2003) A new pathotype of *Pseudoperonospora cubensis* causing downy mildew in cucurbits in Israel. Phytoparasitica 31: 458–466.

9. Cohen Y, Rubin AE (2012) Mating type and sexual reproduction of *Pseudoperonospora cubensis*, the downy mildew agent of cucurbits. Eur J Plant Pathol 132: 577–592.

10. Balass M, Cohen Y, Bar-Joseph M (1993) Temperature-dependent resistance to downy mildew in muskmelon - structural responses. Physiol Mol Plant Pathol 43: 11–20.

11. Tinker NA, Fortin MG, Mather DE (1993) Random amplified polymorphic DNA and pedigree relationships in spring barley. Theor Appl Genet 85: 976–984.

12. White TJ, Bruns T, Lee S, Taylor J (1990) Amplification and direct sequencing of fungal ribosomal RNA genes for phylo- genetics. In: Innis MA, Gelfand DH, Sninsky JJ, White TJ, editors. PCR Protocols: a Guide to Methods and Applications.New York : Academic Press. pp.315–322.

13. Runge F, Choi YJ, Thines M (2011) Phylogenetic investigations in the genus *Pseudoperonospora* reveal overlooked species and cryptic diversity in the *P. cubensis* species cluster. Eur J Plant Pathol 129: 135–146.

14. Ploch S, Thines M (2011) Obligate biotrophic pathogens of the genus Albugo are widespread as asymptomatic endophytes in natural populations of Brassicaceae. Mol Ecol 20: 3692–3699.

15. Hudspeth DSS, Nadler SA, Hudspeth MES (2000) A cox2 molecular phylogeny of the peronosporomycetes (Oomycetes). Mycologia 92: 674–684.

16. Holmes GJ, Wehner TC, Thornton A (2006) An old enemy re-emerges: downy mildew rears its ugly head on cucumber, impacting growers up and down the Eastern US American Vegetable Grower.

17. Cappelli C, Buonaurio R, Stravato VM (2003) Occurrence of *Pseudoperonospora cubensis* pathotype 5 on squash in Italy. Plant Disease 87: 449.

18. Lebeda A, Cohen Y (2012) Fungicide resistance in *Pseudoperonospora cubensis*, the causal agent of cucurbit downy mildew. In: Thind TS, editor. Fungicide Resistance in Crop Protection: Risk and Management: CAB International. pp.45–63.

19. Cohen Y, Rubin AE, Liu XL, Wang WQ, Zhang YJ, et al. (2013) First report on the occurrence of A2 mating type of the cucurbit downy mildew agent *Pseudoperonospora cubensis* in China. Plant Disease 97: 559–559.

20. Cohen Y, Rubin AE, Galperin M (2013) Host preference of mating type in *Pseudoperonospora cubensis*, the downy mildew causal agent of cucurbits. Plant Disease 97: 292–292.

21. Thomas A, Carbone I, Ojiambo P (2013) Occurrence of the A2 mating type of *Pseudoperonospora cubensis* in the United States. Phytopathology 103: 145–145.

22. Cohen Y, Sackston WE (1974) Seed infection and latent infection of sunflowers by *Plasmopara halstedii*. Can J Bot 52: 231–238.

23. Babiker EM, Hulbert SH, Paulitz TC (2012) *Hyaloperonospora camelinae* on *Camelina sativa* in Washington State: Detection, seed transmission, and chemical control. Plant Dis 96: 1670–1674.

24. Garibaldi A, Minuto G, Bertetti D, Gullino ML (2004) Seed transmission of Peronospora sp of basil. J/Plant Dis Protection 111: 465–469.

25. Farahani-Kofoet RD, Roemer P, Grosch R (2012) Systemic spread of downy mildew in basil plants and detection of the pathogen in seed and plant samples. Mycol Progress 11: 961–966.

26. Rubin E, Baider A, Cohen Y (2001) *Phytophthora infestans* produces oospores in fruits and seeds of tomato. Phytopathology 91: 1074–1080.

27. Landa BB, Montes-Borrego M, Munoz-Ledesma FJ, Jimenez-Diaz RM (2007) Phylogenetic analysis of downy mildew pathogens of opium poppy and PCR-Based in planta and seed detection of Peronospora arborescens. Phytopathology 97: 1380–1390.

28. Testen AL, Jimenez-Gasco MD, Ochoa JB, Backman PA (2014) Molecular detection of *Peronospora variabilis* in quinoa seed and phylogeny of the quinoa downy mildew pathogen in South America and the United States. Phytopathology 104: 379–386.

10

Variation in Virus Symptom Development and Root Architecture Attributes at the Onset of Storage Root Initiation in 'Beauregard' Sweetpotato Plants Grown with or without Nitrogen

Arthur Q. Villordon[1]*, **Christopher A. Clark**[2]

1 Sweet Potato Research Station, Louisiana State University Agricultural Center, Chase, Louisiana, United States of America, **2** Department of Plant Pathology and Crop Physiology, Louisiana State University Agricultural Center, Baton Rouge, Louisiana, United States of America

Abstract

It has been shown that virus infections, often symptomless, significantly limit sweetpotato productivity, especially in regions characterized by low input agricultural systems. In sweetpotatoes, the successful emergence and development of lateral roots (LRs), the main determinant of root architecture, determines the competency of adventitious roots to undergo storage root initiation. This study aimed to investigate the effect of some plant viruses on root architecture attributes during the onset of storage root initiation in 'Beauregard' sweetpotatoes that were grown with or without the presence of nitrogen. In two replicate experiments, virus-tested plants consistently failed to show visible symptoms at 20 days regardless of nitrogen treatment. In both experiments, the severity of symptom development among infected plants ranged from 25 to 118% when compared to the controls (virus tested plants grown in the presence of nitrogen). The presence of a complex of viruses (*Sweet potato feathery mottle virus*, *Sweet potato virus G*, *Sweet potato virus C*, and *Sweet potato virus 2*) was associated with 51% reduction in adventitious root number among plants grown without nitrogen. The effect of virus treatments on first order LR development depended on the presence or absence of nitrogen. In the presence of nitrogen, only plants infected with *Sweet potato chlorotic stunt virus* showed reductions in first order LR length, number, and density, which were decreased by 33%, 12%, and 11%, respectively, when compared to the controls. In the absence of nitrogen, virus tested and infected plants manifested significant reductions for all first order LR attributes. These results provide evidence that virus infection directly influences sweetpotato yield potential by reducing both the number of adventitious roots and LR development. These findings provide a framework for understanding how virus infection reduces sweetpotato yield and could lead to the development of novel strategies to mitigate virus effects on sweetpotato productivity.

Editor: Boris Alexander Vinatzer, Virginia Tech, United States of America

Funding: This research was supported by the Louisiana Sweetpotato Advertising and Development Fund. The funders had no role in study design, data collection and analysis, decision to publish, or preparation of the manuscript.

Competing Interests: The authors have declared that no competing interests exist.

* Email: avillordon@agcenter.lsu.edu

Introduction

Sweetpotato (*Ipomoea batatas*) is propagated vegetatively and in the process, propagating material gradually accumulates pathogens, especially viruses that cause decline in yield and quality. At least 30 different viruses have been isolated from sweetpotato [1], but only a few of these consistently cause problems in sweetpotato production. *Sweet potato chlorotic stunt virus* (SPCSV) is a phloem-restricted, whitefly-transmitted virus that is widespread in Africa and South America. SPCSV by itself causes significant yield reduction, but also represses resistance in sweetpotato to other viruses leading to synergistic interactions that reduce yields 80–90% [1,2]. In the U.S., sweetpotatoes are universally infected by a complex of four aphid-transmitted potyviruses: *Sweet potato feathery mottle virus* (SPFMV), *Sweet potato virus C* (SPVC), *Sweet potato virus G* (SPVG), and *Sweet potato virus 2* (SPV2) [1]. These potyviruses infect most tissues within plants and the complex can

cause yield losses of 25–40% under some circumstances [1,3]. However, it is not yet clear if this yield reduction is due to virus effects on storage root initiation, on subsequent storage root bulking, or both.

In sweetpotatoes, the most economically important physiological process is storage root initiation, defined as the appearance of cambia around the protoxylem and secondary xylem elements [4,5,6,7]. Prior work, utilizing diagnostic anatomical features, has defined key stages of sweetpotato storage root yield determination: early "tuberous-root" thickening or storage root initiation stage (up to 25 days after transplanting, DAT), "middle stage" (25–60 DAT), and the "late stage" (from 60 DAT to harvest) [5]. Recently, it has been shown that lateral root development, a key determinant of root system architecture, is fundamentally associated with the competency of adventitious roots to undergo storage root initiation [8]. Root system architecture has been referred to as an integrative result of LR initiation, morphogenesis, emergence, and growth [9]. LRs contribute to water-use efficiency and facilitate the extraction of micro- and macronutrients from the soil [10]. This relationship between root architecture and the developmental fate of adventitious roots addresses the related issues of understanding how the sweetpotato plant modulates storage root initiation and how differential root carbon sink is determined within the root system [11,12]. Knowledge of the variables that control root architecture development can be integrated with other variables that are known to influence storage root yields, enabling a more systematic approach to determining and managing yield constraints of this globally important root crop.

It has been shown in model systems that intrinsic and environmental variables influence LR development. Internal cues of LR formation include auxin, ethylene, abscisic acid, cytokinin, and strigolactones [13,14,15,16]. External variables include growth substrate water availability [17] and soil nutrients such as ammonium (NH_4^+) [18], nitrate (NO_3^-) [19], phosphate [20], and sulfate [21]. The roles of growth substrate water status and nitrogen variability in altering root architecture during the onset of storage root initiation have recently been validated in 'Beauregard' sweetpotato [8,22]. Such findings can lead directly to the development and testing of management practices for improved economic yields, water use efficiency and nutrient foraging. Recent work in model systems has demonstrated that virus infections can lead to alterations in lateral root development [22,23,24]. These findings provide evidence about the potential influence of biotic factors on root architecture development. The primary objective of this work was to investigate the effect of selected plant viruses on root architecture characteristics as measured by LR development attributes in 'Beauregard' sweetpotato plants grown with or without the presence of nitrogen.

Materials and Methods

Viruses

Virus-tested (V0) plants of sweetpotato cv. 'Beauregard', mericlone B-14 (V0 B-14), were produced originally by meristem-tip culture [3,30] and tested by three successive grafts to seedlings of the indicator plant *Ipomoea setosa*, by PCR for geminivirus detection as described by Li et al. [27], and by real-time PCR for detection of SPCSV [28] and found to be apparently free of known viruses. V0 plants were maintained by nodal propagation in tissue culture in the Louisiana State University Agricultural Center Foundation Seed Program. To produce plants for different virus treatments, V0 B-14 plants were inoculated separately with each virus treatment (potyvirus complex [V1] and SPCSV [V2]) by grafting with sweetpotato scions infected with the appropriate virus(es) and subsequently maintaining infected plants by periodically transplanting vine cuttings from infected stocks. Two sources of viruses were used in these experiments: one to provide the common U.S. complex of potyviruses and another to provide SPCSV. The potyvirus complex source plant was originally provided by Dr. G. C. Yencho (Dept. Horticultural Sciences, North Carolina State University, Raleigh). The infected 'Beauregard' plant (B-14, G-7) had been grown in the field for seven years during which it was naturally infected. Clone B-14, G-7 was tested by grafting to *I. setosa* and testing the symptomatic *I. setosa* by ELISA on nitrocellulose membranes (NCM-ELISA) using antisera provided by S. Fuentes (International Potato Center, Lima, Peru), multiplex PCR for potyviruses [26], PCR for geminiviruses [27], and real-time RT-PCR [28] and found to be infected with SPFMV, SPVC, SPVG, and SPV2 but tested

Figure 1. Virus symptoms on 'Beauregard' sweetpotato leaves the day before the first experiment was terminated. All expanded leaves originating from nodes above the substrate surface are depicted with the bottom leaf in each column being the oldest and the top being the youngest. + N = nitrogen provided as KNO_3, - N = no nitrogen provided. V0 = plants derived from non-inoculated, virus-tested plant stock; V1 = plants derived from V0 plant stock graft inoculated with the potyvirus complex (SPFMV, SPVG, SPVC, and SPV2); V2 = plants derived from plant stock infected with SPCSV. Potyvirus symptoms consist of purple ring spots (PRS, see arrow) and purple vein banding (PVB arrow). *Sweet potato chlorotic stunt virus* symptoms include deep interveinal purpling (IVP circle) that is distinguished from the natural purple cast (NPC circle) that develops on some sweetpotato leaves in that with IVP veins remain green, and the pigmentation is deeper.

negative for *Sweet potato mild mottle virus*, *Sweet potato latent virus*, *Sweet potato chlorotic fleck virus*, *Sweet potato mild speckling virus*, *Sweet potato leaf curl virus* and other related geminiviruses, *Sweet potato chlorotic stunt virus* (SPCSV), *Sweet potato collusive virus*, and *Cucumber mosaic virus* (CMV). However, the possibility that it was infected by viruses not yet recognized in sweetpotato cannot be eliminated. 'Beauregard' plants infected with the U.S. strain of SPCSV, isolate BWFT-3 which was initially obtained by single-whitefly transmission [28], was maintained by nodal propagation in tissue culture. The V1 and V2 plants were maintained and increased by propagating infected vine cuttings and their associated vegetatively propagated viruses.

Source plants

Source plants for cuttings for experiments were produced in a greenhouse at the Louisiana State University Agricultural Center, Baton Rouge (30.411380 N, 91.172807 W). A rigorous program of insecticide application, sticky card trapping, and sanitation was routinely employed to manage potential insect vectors of viruses. Plants were propagated in 1:1:1 river silt: sand: Jiffy Mix Plus (Jiffy Products of America, Inc.) amended with Osmocote 14-14-14 (Scotts-Sierra Horticultural Products Company) at 3.5 gm Kg^{-1} soil mix. Source plants were grown in 32-cell Speedling trays (Speedling, Inc.) until 25–30 cm long terminal vine cuttings could be taken.

Experimental design and treatments

Experiments were conducted in the same greenhouse. There were two replicate experiments, and the planting dates were 13 November and 19 December 2013. In all experiments, vegetative terminal vine cuttings were selected that were 25–30 cm long, with

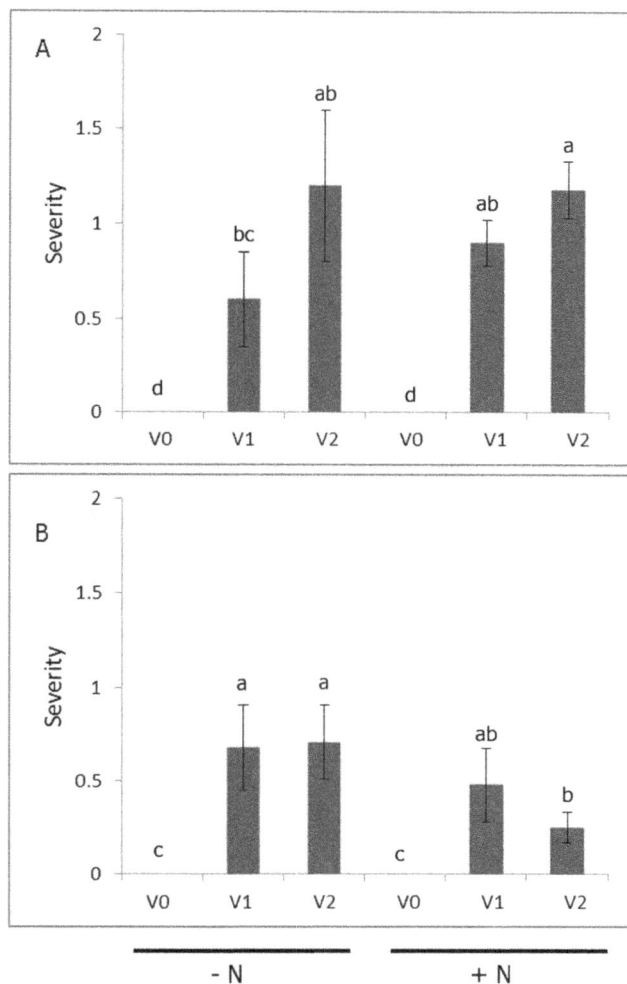

Figure 2. Virus symptom severity ratings of plants from the first (A) and second (B) experiments. Virus symptom severity was assessed the day before each experiment was terminated by rating each leaf by visual estimation of the proportion of the leaf showing symptoms using a 0 to 3 scale in which: 0 = no symptoms, 1 = <1/3 of the leaf involved, 2 = 1/3–2/3 of leaf area involved, and 3 = >2/3 of the leaf involved. + N = nitrogen provided as KNO₃, - N = no nitrogen provided. V0 = plants derived from non-inoculated, virus-tested plant stock; V1 = plants derived from V0 plant stock graft inoculated with the potyvirus complex (SPFMV, SPVG, SPVC, and SPV2); V2 = plants derived from plant stock infected with SPCSV. Severity ratings were transformed using log 10 and Fisher's LSD test at the 0.05 probability level was used to test for statistical significance. The data are expressed as means ± SE from non-transformed data.

five to six fully opened leaves, approximately 5 mm diameter at the basal cut, and with uniform distribution of nodes. Cuttings were planted (2–3 nodes under the growth substrate surface) in 10 cm-diameter polyvinyl chloride pots (height = 30 cm) with detachable plastic bottoms. The growth substrate and experimental conditions used in this study were based on previously developed methodology for measuring the effect of growth substrate moisture and nitrogen content on root architecture development at the onset of storage root initiation in 'Beauregard' [8,22]. The growth substrate for all experiments was washed river sand. The diameter of sand particles varied from 0.05 to 0.9 mm with the majority (83%) in the 0.2- to 0.9-mm range. For plants

grown in the presence of nitrogen (+N), the nutrient was provided as KNO_3 at the rate of 50 kg·ha^{-1} while each of P_2O_5 and K_2O was supplied at the rate equivalent to 134 kg·ha^{-1}. Similar P_2O_5 and K_2O rates were used for plants grown without nitrogen (-N). The greenhouse temperature regime was 29°C for 14 hr (day) and 18°C for 10 hr (night). Photosynthetic photon flux (PPF) for plant production and subsequent experiments ranged from 300 to 800 m^{-2} s^{-1}. High intensity mercury vapor lamps were used to extend daylength to 14 h per day. PPF was measured at the canopy level with a quantum sensor (Model QSO-S, Decagon Devices Inc., Pullman, WA). The relative humidity averaged 60%. Temperature and relative humidity were monitored at the canopy level using an integrated temperature and relative humidity sensor (Model RHT, Decagon Devices Inc.). The moisture of the growing substrate was maintained ≈65 to 75% of field capacity (≈12% volumetric water content). Growth substrate moisture was measured with ECH2O soil moisture sensors (Model EC-5, Decagon Devices Inc.) inserted vertically at the 2–7 cm depth.

During the experiments, plants infected with the potyvirus complex developed typical symptoms of chlorotic spotting followed by development of purple borders around the spots on leaves. Plants infected with SPCSV developed interveinal purple blotches. In both cases, symptoms developed primarily, but not exclusively on the older leaves. Virus symptom severity was assessed the day before each experiment was terminated by rating each leaf by visual estimation of the proportion of the leaf showing symptoms using a 0 to 3 scale in which: 0 = no symptoms, 1 = < 1/3 of the leaf involved, 2 = 1/3–2/3 of leaf area involved, and 3 = >2/3 of the leaf involved. The mean rating for all leaves on a plant was used for statistical analysis and comparison of treatments.

All experiments were arranged in a randomized complete block design where a pot (1 plant per pot) was considered a replicate. There were four replicates in the first experiment but one replicate was lost prior to data collection. There were five replicates in the second experiment. All experiments were terminated after 20 days by removing the detachable plastic bottoms tilting the pots and removing the growth substrate gradually using a stream of water.

Root architecture measurements

Data collection followed the procedures described in previous work [8,22]. Intact adventitious roots that were 20 cm or greater in length were floated on waterproof trays and scanned using a specialized Dual Scan optical scanner (Regent Instruments Inc., Quebec, Canada). Based on previous work, adventitious roots that were less than 20 cm in length generally failed to show anatomical features associated with lignification or storage root initiation [6,8]. The acquisition and image analysis software was WinR-HIZO Pro (v. 2009c; Regent Instruments Inc.). Root types (main root, first order LRs, second order LRs) were automatically classified based on root diameter which was in turn based on predetermined size intervals that were dynamically adjusted between samples. In the present work, three intervals were used: 0 to 0.2, 0.2 to 1.0, and 1.0 to 20 mm. LR attributes that were measured from scanned images included first and second order LR number and length. First order LR density was calculated by dividing first order LR number by the length of the main root.

Statistical analyses

Root attribute data from each experiment were pooled after verifying the lack of planting date effects. Root length and number were transformed using log 10 and square root transformation, respectively. Statistical analyses were performed using SAS Proc Mixed (SAS v. 9.1, SAS Inc., Cary, NC).). Fisher's LSD test at the

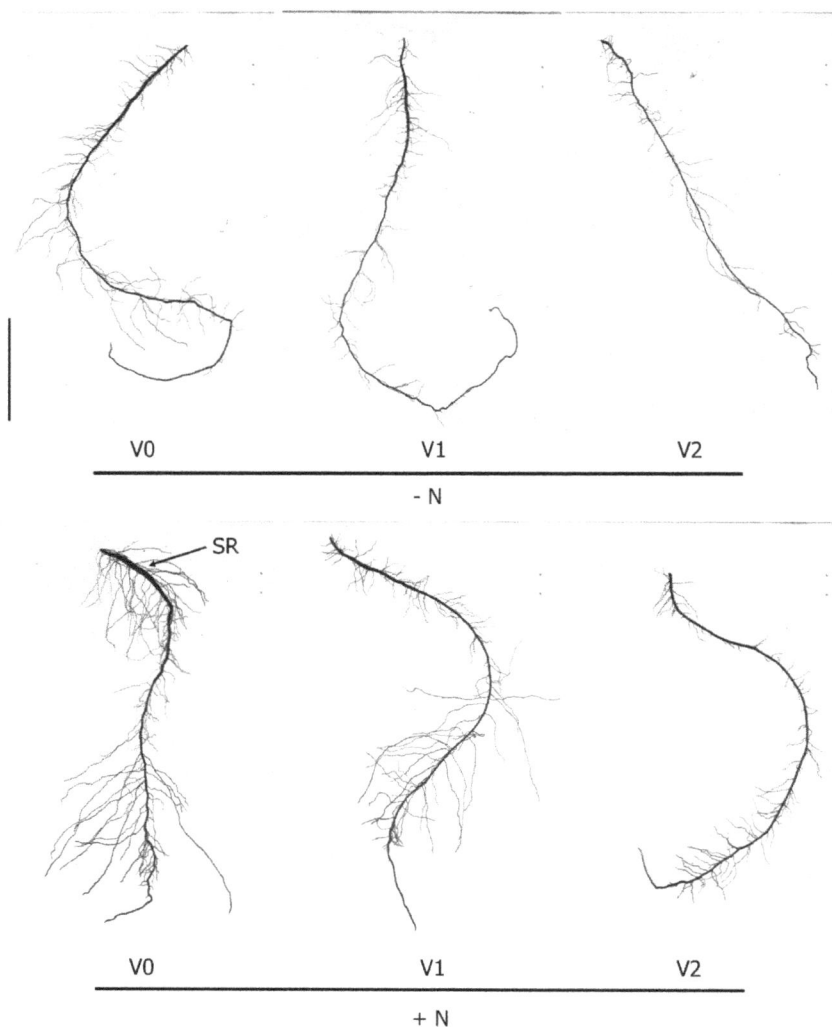

Figure 3. Representative adventitious roots from sweetpotato 'Beauregard' plants subjected to different virus treatments and grown with or without nitrogen. + N = nitrogen provided as KNO$_3$, - N = no nitrogen provided. V0 = plants derived from non-inoculated, virus-tested plant stock; V1 = plants derived from V0 plant stock graft inoculated with the potyvirus complex (SPFMV, SPVG, SPVC, and SPV2); V2 = plants derived from plant stock infected with SPCSV. SR = localized swelling indicative of successful storage root initiation. Scale bar = 5 cm.

0.05 probability level was used to test for statistical significance. The data presented were means and standard error of the means from non-transformed data.

Results

Symptom development

No virus symptoms were observed on the non-inoculated, virus-tested plants at any time during propagation of source plants or during the experiments. Virus symptoms first appeared on the infected plants at about 2 wk after transplanting and became gradually more pronounced until the experiments were terminated at 20 days after transplanting (Figs. 1,2). Virus symptoms develop earlier, are more pronounced and more consistent on plants grown in sand than in other more complex substrates (Clark, unpublished data). Although there were no significant differences in symptom severity between nitrogen treatments, symptom severity was lower in the treatments with nitrogen added in the second test (Fig. 2B).

Root architecture attributes

Representative images of LR development from each combination of nitrogen and virus treatment levels are shown in Fig. 3. The adventitious root derived from a virus-tested cutting grown in the presence of nitrogen has already manifested the initial stage of swelling in the proximal section (Fig. 3D), consistent with prior findings using this experimental system [8,22]. The presence of a complex of viruses (SPFMV, SPVG, SPVC, and SPV2) was associated with 51% reduction in adventitious root number among plants grown without nitrogen when compared to the control treatment (Fig. 4A). The effect of virus treatment on first order LR attributes depended on the presence or absence of nitrogen (Figs. 4B–D). In the presence of nitrogen, only SPCSV-infected plants showed reductions in first order LR length, number, and density, which were decreased by 33%, 12%, and 11%, respectively, when compared to the controls. In the absence of nitrogen, virus tested and infected plants manifested significant reductions for all first order LR attributes that were measured. The use of infected plant material grown with or without nitrogen

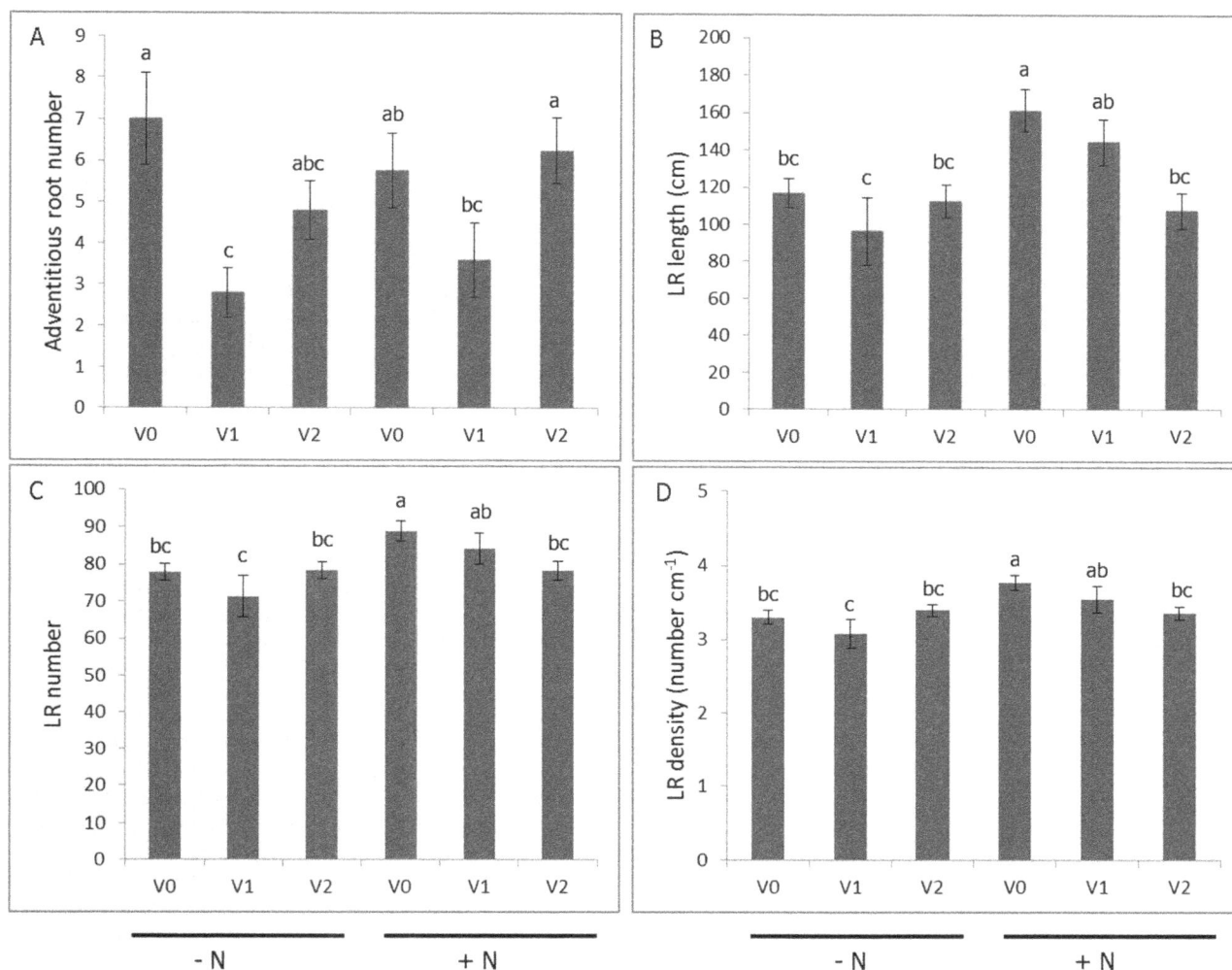

Figure 4. Variation in adventitious root number (A), first order lateral root length (B), first order lateral root number (C), and lateral root density (D) in response to virus and nitrogen treatments at the onset of storage root initiation in 'Beauregard' sweetpotato.
+ N = nitrogen provided as KNO_3, - N = no nitrogen provided. V0 = plants derived from non-inoculated, virus-tested plant stock; V1 = plants derived from V0 plant stock graft inoculated with the potyvirus complex (SPFMV, SPVG, SPVC, and SPV2); V2 = plants derived from plant stock infected with SPCSV. Root length and number were transformed using log 10 and square root transformation, respectively, and Fisher's LSD test at the 0.05 probability level was used to test for statistical significance. The data are expressed as means ± SE from non-transformed data.

was associated with 57 to 96% reduction in second order LR length compared to the control treatment (Fig. 5A). The presence of virus and the lack of nitrogen were associated with 72 to 90% reduction in second order LR number relative to the control treatment (Fig. 5B). In general, optimum conditions for LR development were associated with the use of virus tested plants grown in nitrate sufficient conditions.

Discussion

Other than demonstrating that certain strains of SPFMV can cause russet crack and determining effects of some viruses on storage root yield, little is known about how sweetpotato viruses affect the developing adventitious root system, especially during the critical storage root initiation period. Previous work has provided evidence that virus infected sweetpotato cuttings of the cultivar 'Kokei No. 14' had more adventitious roots but with lower weight when compared to virus-tested plants [29]. However, this study used a naturally infected storage root for generating

planting material and the type of virus was not specified. This work [29] also did not consider lateral root development attributes nor did it relate virus effects to storage root initiation. The findings from the present study provide a framework for understanding how virus infection reduces sweetpotato yield, using a validated phenological model for storage root yield determination [5], given a scenario where vegetative cuttings are obtained from an infected source. First, the presence of a complex of viruses (SPFMV, SPVG, SPVC, and SPV2) was associated with the reduction of adventitious root number, regardless of the presence of nitrogen. Secondly, the presence of the phloem-restricted virus, SPCSV, was associated with the reduction in LR length, number, and density, previously associated with increased lignification in the stele, thereby preventing storage root initiation [8]. Both of these situations result in the net reduction of adventitious roots that can undergo storage root initiation and subsequent bulking, thereby negatively impacting yield potential. Previously, it has been shown that nitrogen deprivation resulted in reduced lateral root development which was associated with 44% and 55% reduction

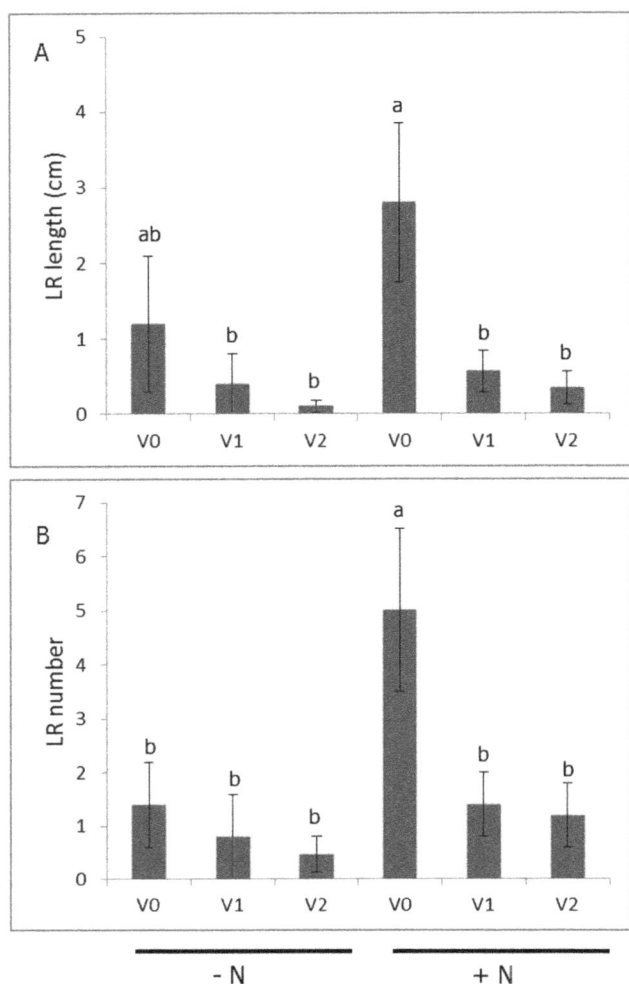

Figure 5. Variation in second order lateral root length (A) and number (B) in response to virus and nitrogen treatments at the onset of storage root initiation in 'Beauregard' sweetpotato. Treatment legend: + N = nitrogen provided as KNO_3, - N = no nitrogen provided. V0 = plants derived from non-inoculated, virus-tested plant stock; V1 = plants derived from V0 plant stock graft inoculated with the potyvirus complex (SPFMV, SPVG, SPVC, and SPV2); V2 = plants derived from plant stock infected with SPCSV. Root length and number were transformed using log 10 and square root transformation, respectively, and Fisher's LSD test at the 0.05 probability level was used to test for statistical significance. The data are expressed as means ± SE from non-transformed data.

in uptake of phosphorus and potassium, respectively [22]. Thus, there is a concomitant effect associated with diminished nutrient foraging by the root system, with potential adverse effects on shoot growth and storage root bulking. Taken together, these findings parallel previous results from field studies that have documented the reduction of storage root yield in 'Beauregard', associated with the use of virus-infected propagating materials [3,30].

It has been documented that auxin is the integrator of intrinsic and environmental signals that affect LR development [31] and that cytokinin has been shown to exert antagonistic effects [32,33]. In the past, information on morphological, hormonal, and molecular characterizations of virus infection on roots have been

limited [34,35], but new findings are beginning to reveal the integrative role of auxin in mediating virus effects on root architecture. For example, *Arabidopsis* plants that were transformed to express the *Beet necrotic yellow vein* virus p25 protein showed abnormal root branching, accompanied by significant changes in the levels of auxin, jasmonic acid and ethylene precursor, ACC [24]. In another study, *Cucumber mosaic virus*-infected *Arabidopsis* root growth patterns were accompanied by significant changes in indole-3-acetic acid (IAA), *trans*-zeatin riboside and dihydrozeatin riboside [25]. In tomato (*Solanum lycopersicon*), infection by *Tomato aspermy* virus resulted in reduced LR development, which was accompanied by increased miR164 levels [23], a microRNA family previously associated with negatively regulating LR initiation in response to auxin by limiting NAC1 expression [36]. Parallel microarray experiments on shoots and roots from tomato infected with *Tomato spotted wilt virus* showed organ-specific responses although the virus was present in similar concentrations [34]. The study showed that in tomato shoots, genes related to defense and signal transduction were induced, while there was a general repression of genes related to primary and secondary metabolism as well as amino acid metabolism. In roots, genes related to biotic stimuli were induced while those related to abiotic stress were repressed. Currently, it is not known how sweetpotato viruses affect auxin or cytokinin levels or distribution in developing adventitious roots. It has been demonstrated that gene expression in leaves was altered in sweetpotato by virus infection as soon as 5 days after inoculation [37]. Single infections with SPFMV or SPCSV altered expression of 3 or 14 genes, respectively, as opposed to 200 genes in plants infected with both viruses [38]. In sweetpotatoes, the link between virus infection and root architecture marks a new research direction toward a better understanding of the relationship between virus infection and storage root yield. Some follow-up work might include the hormonal and molecular characterization of the mechanism of lateral root suppression by viruses, for enabling the development of tools and approaches that mitigate virus effect on sweetpotato productivity.

It has been suggested that root architecture may hold the key to the next green revolution [39,40]. Root and tuber crops are second in importance to cereals as a global source of carbohydrates, with particularly high production potential in humid regions that are not suitable for cereal production [41]. Our findings in sweetpotato underscore the need to further investigate the effects of viruses on root architecture development in crops, especially those grown in low-input agricultural systems where virus diseases are a persistent threat.

Acknowledgments

We thank Mary Hoy and Theresa Arnold for help conducting this research. Approved for publication by the director of the Louisiana Agricultural Experiment Station as manuscript no. 2014-260-14045. Mention of trademark, proprietary product or method, and vendor does not imply endorsement by the Louisiana State University Agricultural Center, nor its approval to the exclusion of other suitable products or vendors.

Author Contributions

Conceived and designed the experiments: AV CC. Performed the experiments: CC. Analyzed the data: AV CC. Contributed reagents/materials/analysis tools: AV CC. Wrote the paper: AV CC.

References

1. Clark CA, Davis JA, Abad JA, Cuellar WJ, Fuentes S, et al. (2012) Sweetpotato viruses: 15 years of progress of understanding and managing complex diseases. Plant Dis 96: 168–185.

2. Gutierrez DL, Fuentes S, Salazar L (2003) Sweetpotato virus disease (SPVD): distribution, incidence, and effect on sweetpotato yield in Peru. Plant Dis 87: 297–302.

3. Clark CA, Hoy MW (2006) Effects of common viruses on yield and quality of Beauregard sweetpotato in Louisiana. Plant Dis 90: 83–88.

4. Wilson LA, Lowe SB (1973) The anatomy of the root system in West Indian sweet potato[Ipomoea batatas (L.) Lam.] cultivars. Ann Bot 37: 633–643.

5. Togari Y (1950) A study of tuberous root formation in sweet potato. Bull Nat Agr Expt Sta Tokyo 68: 1–96 [Japanese with English summary].

6. Villordon A, LaBonte DR, Firon N, Kfir Y, Pressman E, et al. (2009) Characterization of adventitious root development in sweetpotato. HortScience 44: 651–655.

7. Villordon A, LaBonte DR, Firon N (2009) Development of a simple thermal time method for describing the onset of morpho-anatomical features related to sweetpotato storage root formation. Scientia Horticulturae 121: 374–377.

8. Villordon A, Labonte DR, Solis J, Firon N (2012) Characterization of lateral root development at the onset of storage root initiation in 'Beauregard' sweetpotato adventitious roots. HortScience 47: 961–968.

9. Dubrovsky JG, Forde BG (2012) Quantitative analysis of lateral root development: pitfalls and how to avoid them. Plant Cell 24: 4–14.

10. Casimiro I, Beeckman T, Graham N, Bhalerao R, Zhang H, et al. (2003) Dissecting Arabidopsis lateral root development. Trends Plant Sci 8: 165–171.

11. Kays SJ (1985) The physiology of yield in the sweetpotato. In: Bouwkamp J, editor. Sweetpotato products: a natural resource for the tropics. CRC Press.

12. Villordon A, Ginzberg E, Firon N (2014) Root architecture and root and tuber crop productivity. Trends Plant Sci 19: 419–425.

13. De Smet I, White PJ, Glyn Bengough A, Dupuy L, Parizot B, et al. (2012) Analyzing lateral root development: how to move forward. Plant Cell 24: 15–20.

14. Ivanchenko M, Muday M, Dubrovsky J (2008) Ethylene-auxin interactions regulate lateral root initiation and emergence in Arabidopsis thaliana. Plant J 55: 335–347.

15. Lopez-Bucio J, Hernandez-Abreu E, Sanchez-Calderon L, Nieto-Jacobo MF, Simpson J, et al. (2002) Phosphate availability alters architecture and causes changes in hormone sensitivity in the Arabidopsis root system. Plant Physiol 129: 244–256.

16. Koltai H (2011) Strigolactones are regulators of root development. New Phytol 190: 545–549.

17. Deak K, Malamy J (2005) Osmotic regulation of root system architecture. Plant J 43: 17–28.

18. Lima JE, Kojima S, Takahashi H, von Wiren N (2010) Ammonium triggers lateral root branching in Arabidopsis in an ammonium transporter 1: 3-dependent manner. Plant Cell 22: 3621–3633.

19. Zhang H, Forde B (1998) An Arabidopsis MADS box gene that controls nutrient-induced changes in root architecture. Science 279: 407–409.

20. Johnson JF, Vance CP, Allan DL (1996) Phosphorus deficiency in Lupinus albus (altered lateral root development and enhanced expression of phosphoenolpyruvate carboxylase). Plant Physiol 112: 657–665.

21. Kutz A, Muller A, Henning P, Kaiser WM, Piotrowsky M, et al. (2002) A role for nitrilase 3 in the regulation of root morphology in sulphur-starving Arabidopsis thaliana. Plant J 30: 95–106.

22. Villordon A, Labonte DR, Firon N, Carey E (2013) Variation in nitrogen rate and local availability alter root architecture attributes at the onset of storage root initiation in 'Beauregard' sweetpotato. HortScience 48: 808–815.

23. Feng J, Wang Y, Lin R, Chen J (2013) Altered expression of microRNAs and target mRNAs in tomato root and stem tissues upon different viral infection. J Phytopath 161: 107–119.

24. Peltier C, Schmidlin L, Klein E, Taconnat L, Prinsen E, et al. (2011) Expression of the Beet necrotic yellow vein virus p25 protein induces hormonal changes and a root branching phenotype in Arabidopsis thaliana. Transgenic Res 20: 443–466.

25. Vitti A, Nuzzaci M, Scopa A, Tataranni G, Remans T, et al. (2013) Auxin and cytokinin metabolism and root morphological modifications in Arabidopsis thaliana seedlings infected with Cucumber mosaic virus (CMV) or exposed to cadmium. Int J Mol Sci 14: 6889–6902.

26. Li F, Zuo R, Abad J, Xu D, Bao G, et al. (2012) Simultaneous detection and differentiation of four closely related sweet potato potyviruses by a multiplex one-step RT-PCR. J Virol Methods 186: 161–166.

27. Li R, Salih S, Hurtt S (2004) Detection of geminiviruses in sweetpotato by polymerase chain reaction. Plant Dis 88: 1347–1351.

28. Kokkinos CD, Clark CA (2006) Real-time PCR assays for detection and quantification of sweetpotato viruses. Plant Dis 90: 783–788.

29. Kano Y, Nagata R (1999) Comparison of the rooting ability of virus infected and virus-free cuttings of sweet potatoes (Ipomoea batatas Poir.) and an anatomical comparison of the roots. J Hort Sci & Biotech 74: 785–790.

30. Carroll HW, Villordon A, Clark CA, LaBonte DR, Hoy M (2004) Studies on Beauregard sweetpotato clones naturally infected with viruses. Int J Pest Manage 50: 101–106.

31. Lavenus J, Goh T, Roberts I, Guyomarc'h S, Lucas M, et al. (2013) Lateral root development in Arabidopsis: fifty shades of auxin. Trends Plant Sci 18: 450–458.

32. Aloni R, Aloni E, Langhans M, Ullrich CI (2006) Role of cytokinin and auxin in shaping root architecture: regulating vascular differentiation, lateral root initiation, root apical dominance and root gravitropism. Ann Bot 97: 883–893.

33. Del Bianco M, Giustini L, Sabatini S (2013) Spatiotemporal changes in the role of cytokinin during root development. New Phytol 199: 324–338.

34. Catoni M, Miozzi L, Fiorilli V, Lanfranco L, Accotto GP (2009) Comparative analysis of expression profiles in shoots and roots of tomato systemically infected by Tomato spotted wilt virus reveals organ-specific transcriptional responses. Mol Plant-Microbe Interact 22: 1504–1513.

35. Valentine TA, Roberts IM, Oparka KJ (2002) Inhibition of Tobacco mosaic virus replication in lateral roots is dependent on an activated meristem-derived signal. Protoplasma 219: 184–196.

36. Guo HS, Xie Q, Fei JF, Chua NH (2005) MicroRNA directs mRNA cleavage of the transcription factor NAC1 to downregulate auxin signals for Arabidopsis lateral root development. Plant Cell 17: 1376–1386.

37. McGregor CE, Miano DW, LaBonte DR, Hoy M, Clark CA, et al. (2009) Differential gene expression of resistant and susceptible sweetpotato plants after infection with the causal agents of sweet potato virus disease. J Amer Soc Hort Sci 134: 658–666.

38. Kokkinos CD, Clark CA, McGregor CE, LaBonte DR (2006) The effect of sweet potato virus disease and its viral components on gene expression levels in sweetpotato. J Amer Soc Hort Sci 131: 657–666.

39. Den Herder G, Isterdael GV, Beeckman T, De Smet I (2010) The roots of a new green revolution. Trends Plant Sci 15: 600–607.

40. Lynch JP (2007) Roots of the second green revolution. Australian J Bot 55: 493–512.

41. Diop A, Calverley DJB (1998) Storage and processing of roots and tubers in the tropics. FAO, Rome.

11

Interplay between Parasitism and Host Ontogenic Resistance in the Epidemiology of the Soil-Borne Plant Pathogen *Rhizoctonia solani*

Thomas E. Simon[◊], **Ronan Le Cointe**[◊]**, Patrick Delarue, Stéphanie Morlière, Françoise Montfort, Maxime R. Hervé, Sylvain Poggi***

INRA UMR 1349 IGEPP, Le Rheu, France

Abstract

Spread of soil-borne fungal plant pathogens is mainly driven by the amount of resources the pathogen is able to capture and exploit should it behave either as a saprotroph or a parasite. Despite their importance in understanding the fungal spread in agricultural ecosystems, experimental data related to exploitation of infected host plants by the pathogen remain scarce. Using *Rhizoctonia solani / Raphanus sativus* as a model pathosystem, we have obtained evidence on the link between ontogenic resistance of a tuberizing host and (i) its susceptibility to the pathogen and (ii) after infection, the ability of the fungus to spread in soil. Based on a highly replicable experimental system, we first show that infection success strongly depends on the host phenological stage. The nature of the disease symptoms abruptly changes depending on whether infection occurred before or after host tuberization, switching from damping-off to necrosis respectively. Our investigations also demonstrate that fungal spread in soil still depends on the host phenological stage at the moment of infection. High, medium, or low spread occurred when infection was respectively before, during, or after the tuberization process. Implications for crop protection are discussed.

Editor: Ren-Sen Zeng, South China Agricultural University, China

Funding: This research was supported by the French National Research Agency (http://www.agence-nationale-recherche.fr) through funding of the SYSBIOTEL project referenced ANR-08_STRA-14. (SP TES FM RLC). Part of this research was also supported by the INRA "Plant Health & the Environment" Division (http://www.spe.inra.fr/). (SP FM RLC). The funders had no role in study design, data collection and analysis, decision to publish, or preparation of the manuscript.

Competing Interests: The authors have declared that no competing interests exist.

* Email: sylvain.poggi@rennes.inra.fr

◊ These authors contributed equally to this work.

Introduction

Predicting the spread of soil-borne pathogens in the field would prove valuable for building efficient and sustainable strategies to limit crop damage. In particular, investigating variations in pathogen spreading rates over time and space constitutes an open research area. One of the determinants of pathogen spread is the access to nutrients [1], most pathogens finding resources solely in their host. The spread of soil-borne pathogens, such as *Rhizoctonia solani*, is driven by two types of resource: 1) organic matter, when the fungus acts as a saprotroph; 2) infected tissues of the host, when the fungus acts as a pathogen [2,3].

Pathogen ability to access and use organic matter has been extensively studied in the literature. The impact of variations of resources in quality and quantity [4–6] as well as resource distribution [7–9], have been experimentally investigated. Also, modelling of mycelium growth as a function of the availability of organic matter was carried out [3,10,11], providing insights into the growth mechanisms. In each case, rich nutrient sources were shown to enhance saprotrophic growth.

In contrast, the ability of *R. solani* to access and exploit its host has received far less attention. Evidence is lacking to provide an overview of the ability of fungus to spread in soil using the resources of a host, hereby named pathogenic spread (as opposed to saprotrophic spread).

In previous work, *R. solani* pathogenic spread was quantified indirectly by measuring the number of healthy plants that were infected by mycelium spreading out of an infected host [12,13]. Though useful, these experimental data were quantified without disassociating the combined probabilities of i) infection of a susceptible plant, ii) pathogenic spread in soil and iii) infection of a neighbouring plant. It later proved difficult to analyze each effect separately [14]. Facing this shortcoming, models dealing with the spread of *R. solani* in a context of host/pathogen interaction rely on various hypotheses. For instance, Cunniffe and Gilligan [10] assumed that host substrate availability increased linearly with time. Stacey et al. [15] did not take into account the limitation of fungal growth due to depletion of host resources.

In the present study, we provide evidence linking host phenology to pathogenic spread, using *R. solani / Raphanus sativus* as a model pathosystem. Pathogenic spread depends on host/pathogen interaction, which is notably affected by the ontogenic resistance, i.e. the varying ability of the host to resist or tolerate disease as it develops [16]. In the case of radish plants, the literature suggests that older hosts are less likely to get infected [17,18]. However, we do not know the impact of host development on the pathogenic spread; older radishes are bigger and could provide more resources to an invading pathogen.

In line with previous work from Gilligan & Bailey [19] introducing components of pathozone behavior, pathogenic spread was broken down into two steps encompassing i) the ability of the pathogen to infect its host (i.e. pathogen infectivity), thus gaining access to host resources, and ii) the ability of the fungus to spread in soil as a consequence of host infection. Each step was analyzed separately by means of replicable bioassays conducted in microcosms under controlled conditions.

Our results allow us to draw some conclusions about the interplay between parasitism and ontogenic resistance, and its implication in terms of epidemic spread. We suggest that these results may be valuable for epidemiological modelling of soil-borne disease dynamics. The potential impact on disease management is also discussed.

Materials and Methods

To study the interplay between parasitism and host ontogenic resistance, three experiments in microcosms were performed under the same controlled growth conditions. The first was used to determine the successive phenological stages of a radish cultivar (Expo) according to criteria given in the extended BBCH-scale for root and stem vegetables [20]. The second examined the ontogenic resistance of radishes to *R. solani*, which can also be viewed as the impact of host phenology on pathogen infectivity. The third experiment aimed to assess the link between the host phenological stage and the spread of the pathogen in soil subsequent to host infection.

It is important to specify that, in our experiments, we did not inoculate the host directly, but rather the soil in the host vicinity (5 mm-wide area around the host). For the sake of clarity, we have shortened "soil inoculation in the host vicinity" to "host inoculation".

Host plant and inoculum used in the experiments

Radish seeds (cv. Expo F1, Vilmorin S.A.) were sown (one seed per pot) 2.5 cm deep at the center of the pots. The standard inoculum consisted of a 3 mm diameter mycelium disc (*R. solani*, AG4, strain FM1, isolated from lettuce in June 2009 at the INRA Experimental Station of Alenya, southern France), produced on malt-agar medium following the methodology described by Gilligan and Bailey [19]. Inoculum (one mycelium disc per pot) was placed 5 mm away from the host plant, 5 mm deep in the soil.

Microcosms

Soil was a 50% v/v mix of 2.25 mm sieved sand (estuary of the River Loire, Montoir-de-Bretagne, France) and 2.25 mm sieved potting soil (NFU 44551, type 992016F1, Falienor S.A., Vivy, France). Soil moisture was maintained at 30% with daily tap water sub-irrigations. Experiments were conducted in a climatic chamber. The soil was not sterilized but absence of pre-existing *R. solani* mycelium was checked during experiments using negative control pots. Environmental conditions in microcosms were set up for a 16 h light /8 h dark photoperiod at a temperature of 25°C (day) /20°C (night) and hygrometry at 50%. Blue/red neons were placed alternatively (36 W/JR, CRI830, Philips N.V.; 36 W/JR, CRI865, MazdaFluor) 17 cm above the pots. In the first experiment and during preliminary tests we used polystyrene pots 7*7*6.2 cm filled with 160 cm³ of soil. In the second experiment pots consisted of PET cable trays (section

5.8*5 cm) in two designs: 14 cm long, filled with 320 cm³ of soil to assess the probability for the fungus to spread in soil at 1 cm, 2.5 cm and 5 cm, and 28 cm filled with 640 cm³ of soil to assess the probability for the fungus to spread 10 cm. Pots were placed in trays and separated by a few centimeters to prevent the mycelium passing from one to another.

Experiments

Experiment 1: Characterization of radish phenological stages. Radishes were grown in microcosms (polystyrene pots 14.5*5.8*5 cm, filled with 320 cm³ of soil). Every three days from day 3 to day 30, destructive sampling was carried out on twenty homogeneous replicates (emerged between day 3 and day 4). Root diameter was measured with a digital caliper and the phenological stage assessed according to criteria given in the extended BBCH-scale for root and stem vegetables [20]. Tuberization was considered to begin at stage 41, i.e. when average root diameter exceeded 5 mm. It was considered complete at stage 49, when average root diameter exceeded 12 mm.

Experiment 2: Effect of host development on the pathogen infectivity. In this experiment a single host plant was challenged by an inoculum in an individual pot. It was subsequently screened for several phenological stages in order to assess the effect of the host phenology on the pathogen infectivity during a cropping period (30 days). Hosts were inoculated with R. solani every three days from sowing to day 24. There were sixteen replicates for each inoculation date. Control pots contained hosts but were not inoculated.

Thirty days after sowing, radishes were removed, carefully rinsed in water, and the presence and type of symptoms (damping-off, necrosis) were monitored on the whole plant. Damping-off symptoms were defined as a necrosis on the radish collar followed by host fall onto the soil surface, this necrosis often extending to all plant parts, including cotyledons and leaves.

Disease incidence (DI) and damping-off incidence (DO) were then calculated using the following formulas:

$$DI = (number\ of\ infected\ hosts/total\ number\ of\ emerged\ hosts) \times 100$$

$$DO = (number\ of\ hosts\ affected\ by\ damping\text{-}off/total\ number\ of\ emerged\ hosts) \times 100.$$

Experiment 3: Impact of host phenology at infection on subsequent fungal spread. The ability of R. solani to spread in soil, using infected host tissues as a resource, was quantified through the ability of the fungus to grow in soil and colonize baits located at four distances away from an infected host (1 cm, 2.5 cm, 5 cm and 10 cm). We derived colonization profiles, as defined in Bailey et al [21], after infection of (i) a young host, (ii) a tuberizing host and (iii) a tuberized host. Radishes were inoculated at four different ages: the sowing day (day zero), and four, eight and sixteen days after sowing. The number of replicates per distance and per treatment was 12, 16, 18 and 20 respectively. Pathogen spread in bare soil (with inoculation at day zero) was measured as a saprophytic spread control (16 replicates). The spread of R. solani in soil was assessed every four days during the cropping period (lasting thirty days), using millet-seeds as baits (1.6 mm sieved white millet seeds; autoclaved three times 22 min at 121°C, one day apart) placed on the surface of the soil. Two days after deposit, baits were removed and placed in Petri dishes on a semi-selective medium: KHP medium [22] without fenaminosulf and with nitrates instead of nitrites. Presence or absence of R. solani was assessed after three days in the growing chamber (darkness; 20°C) using an optical microscope (x50 magnification). The baiting design, as illustrated in Figure 1, consisted in placing four baits at a given distance (among the four distances we investigated) of the inoculum, two on each side of the inoculum. Results presented in this paper focus on the colonization of any of the two sides, referred to as zones henceforth. A zone was considered colonized when at least one of the two baits it encloses was colonized. Thirty days after sowing, the presence or absence of symptoms on the hosts was assessed. In order to disentangle the probability of the baits to be colonized from the probability of the host to be infected, we only considered data from pots with infected hosts. Pots with non-inoculated radish plants and baits were used as a double negative control to check whether i) radish plants grew without symptoms and ii) R. solani was absent from non-inoculated soil. This experiment was repeated twice.

Statistical analyses

All statistical analyses were carried out using R version 2.13.1 (R Development Core Team, 2008).

The link between root diameter (Experiment 1) and pathogen infectivity (Experiment 2) was assessed by using a generalized linear model (GLM, the "stat" and "MASS" packages in R version 2.13.1) for proportion data (distribution: quasibinomial, link: logit), separately for the disease incidence (DI) and damping-off incidence (DO). In each model, root diameter was included as a covariable. As one proportion was available for each day of sampling, the mean root diameter of the twenty measured plants was used. The impact of host phenology at infection on subsequent fungal spread (Experiment 3) was assessed by using generalized linear mixed models (GLMM; the "lme4" package version 1.0–6) for binary data (distribution: binomial, link: logit). A separate model was built for each distance (1 cm, 2.5 cm, 5 cm and 10 cm). In any of these models, age at inoculation (including control replicates as a level of the same factor) was included as a fixed factor, time as a covariable and their interaction as a fixed factor. The repetition, the pot nested into the repetition and the zone nested into the pot were included as random factors. The effect of

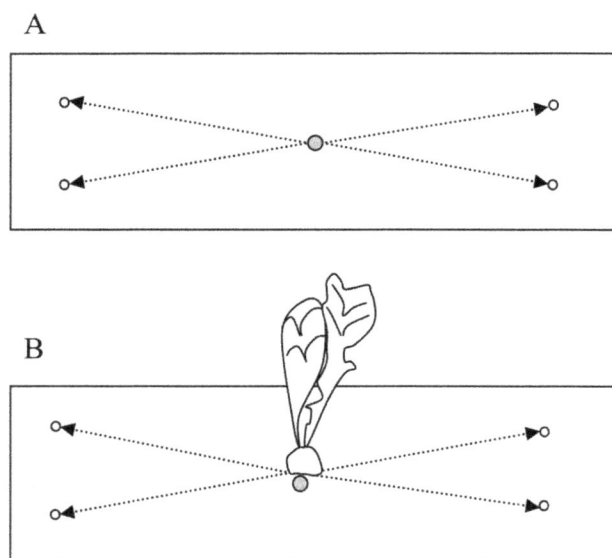

Figure 1. Experimental designs used for quantifying the spread of R. solani in soil. Placement of inoculum (grey circle), baits (empty circle) and radish. (A) Study of mycelial spread from inoculum in bare soil. (B) Study of mycelial spread from an infected host.

age at inoculation, time and their interaction was tested with a Wald test. When needed, pairwise comparisons of least square means (LSMeans; the 'lsmeans' package version 1.0–6) were performed using the False Discovery Rate for correction of P-values.

Results

Experiment 1: Characterization of radish phenological stages

Overall, the main root diameter (±SE) was 1.4 mm (±0.1 mm) at emergence, and increased from day nine to harvest, reaching 23.3 mm (±1.9 mm).

According to the BBCH-scale [20], three phenological stages were reached during the cropping period. The "Germination" stage lasted three days, from sowing to emergence of radishes (91% and 9% of radishes emerged at day 3 and day 4 respectively). It was followed by the "Leaf development" stage until day 9, at which time we observed a transition to the "Development of harvestable vegetative plant part" stage. As defined in Material & Methods, tuberization started when average root diameter exceeded 5 mm, and finished when root diameter exceeded 12 mm, i.e. at day 9 and day 15 respectively.

Experiment 2: Effect of host development on the pathogen infectivity

Disease incidence at harvest (day 30) was clearly negatively linked to host phenological stage at inoculation (Figure 2). It decreased from (incidence ±CI) 88±6% for inoculations at the seedling stage to 69±4% when tuberization occurred. It finally fell to 13% when inoculation was done twenty-four days after sowing.

The type of symptoms changed along with the phenological development (Figure 3) and appeared markedly linked to host development at inoculation. When inoculation was performed during the seedling stage, it led to a very high proportion of damping-off (84±4%). When inoculation was performed during tuberization, the proportion of damping-off symptoms decreased to 47±5%. Finally, when the host root was well developed, the proportion of damping-off symptoms fell to 3% eighteen days after sowing. The decline in necrosis observed at day 21 and day 24 might partly be attributed to the proximity of the harvest date (day 30) and corresponding data were not included in the GLM analysis which showed a significant negative effect of root diameter on damping-off incidence (t value = −4.86, P<0.01) and a significant positive effect on necrosis incidence (t value = 4.40, P<0.01).

Summarizing, we can outline epidemiological features associated with the three main phenological stages affecting the pathogen infectivity: (i) the seedling stage gives rise to high damping-off incidence, (ii) the tuberization stage is marked by a remarkable decrease in damping-off together with the appearance of necrosis symptoms, and (iii) the harvestable stage, characterized by the quasi-absence of damping-off and less necrosis that does not significantly affect host growth. Interestingly, necroses on tuberized hosts were limited and the host cuticle was penetrated by just a few millimeters.

Experiment 3: Impact of host phenology at infection on subsequent fungal spread

As explained in the Introduction, we distinguish the "saprophytic spread" (growth of the fungus in the soil, without any host to infect) from the "pathogenic spread" (growth of the fungus in soil, after infection of a host). Also, the pots used for calculating

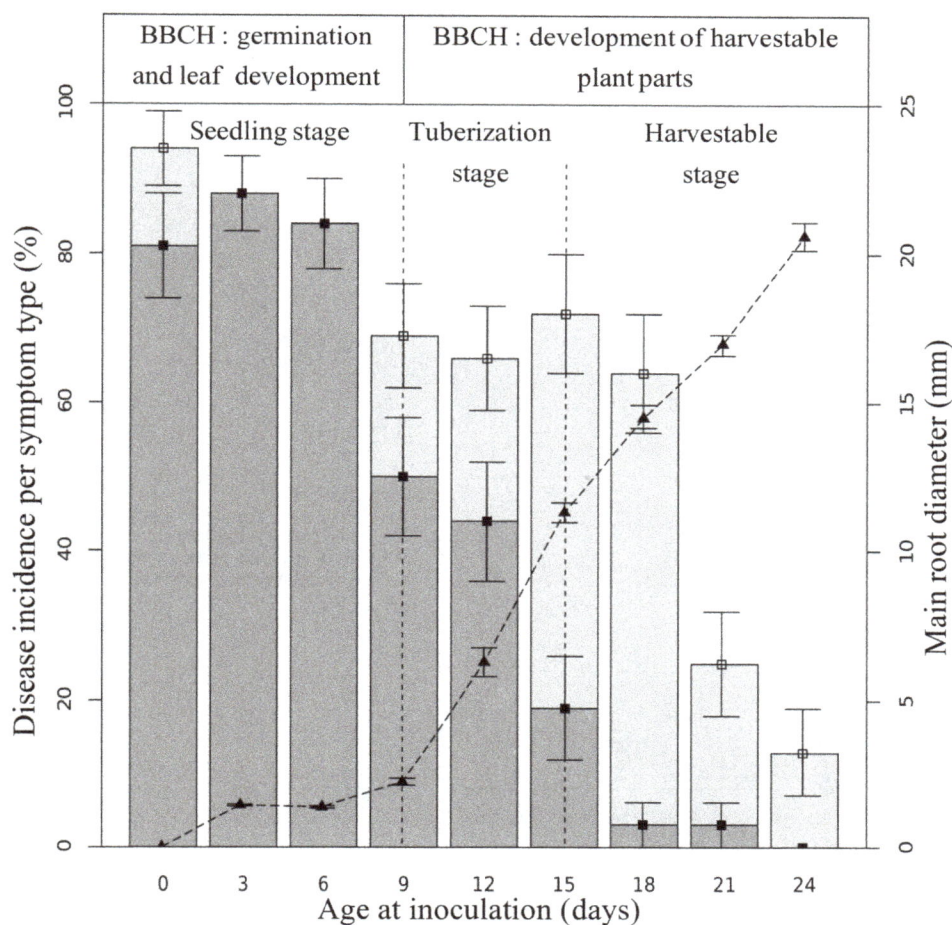

Figure 2. Effect of host development on the pathogen infectivity. Bars show the mean cumulated incidence at harvest (i.e. thirty days after sowing); dark and light grey refer to the type of symptoms, damping-off and necrosis respectively. The dotted line shows the radish root diameter.

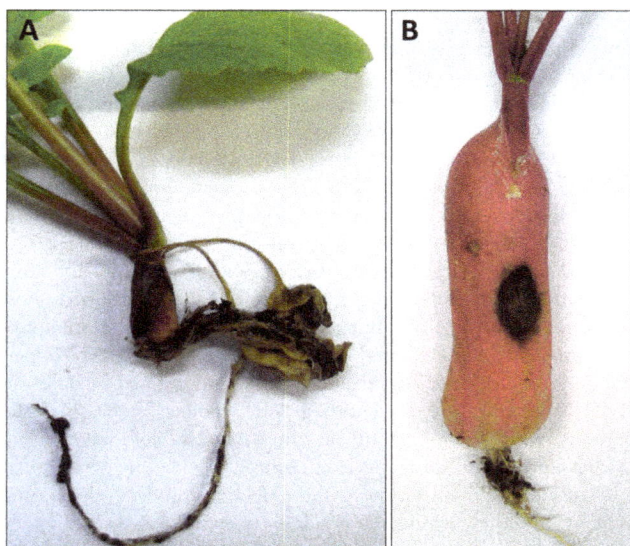

Figure 3. Typical symptoms induced by *R. solani* on radish plants. (A) Damping-off symptoms on plant inoculated 6 days after sowing and (B) necrosis symptoms on a plant inoculated 18 days after sowing.

pathogenic spread were those in which hosts were infected by the pathogen Thus, probability of host infection with host development and subsequent pathogenic spread are disentangled.

Even in the absence of any host plant, the fungus was able to spread from a mycelium disc and colonize the surrounding soil surface relatively quickly. As shown on Figure 4A (red dotted line), 74% of baits placed 1 cm away from the mycelium disc were colonized two days after inoculation, and 84% six days after inoculation. However, the saprophytic spread was limited to about 10% of baits at 2.5 cm fourteen days after soil inoculation (Figure 4B) and no colonization was recorded at 5 cm and beyond. Moreover, without any host to exploit, the fungus was unable to sustain its spread. Colonization at 1 cm started to decline after ten days and plummeted to 50% sixteen days after inoculation.

Pathogenic spread was essentially slower, yet more sustainable and more extensive than saprophytic spread. It was slower in that only 22% of baits placed at 1 cm were colonized two days after inoculation when plants were inoculated on sowing day (Figure 4A, brown line) and only 9% when plants were inoculated four days later (Figure 4A, green line). We might assume that, in the presence of a host, the pathogen allocates time to infection rather than to soil colonization.

The spread in soil is high when host infection occurs at the seedling stage. When host infection occurs during the tuberization

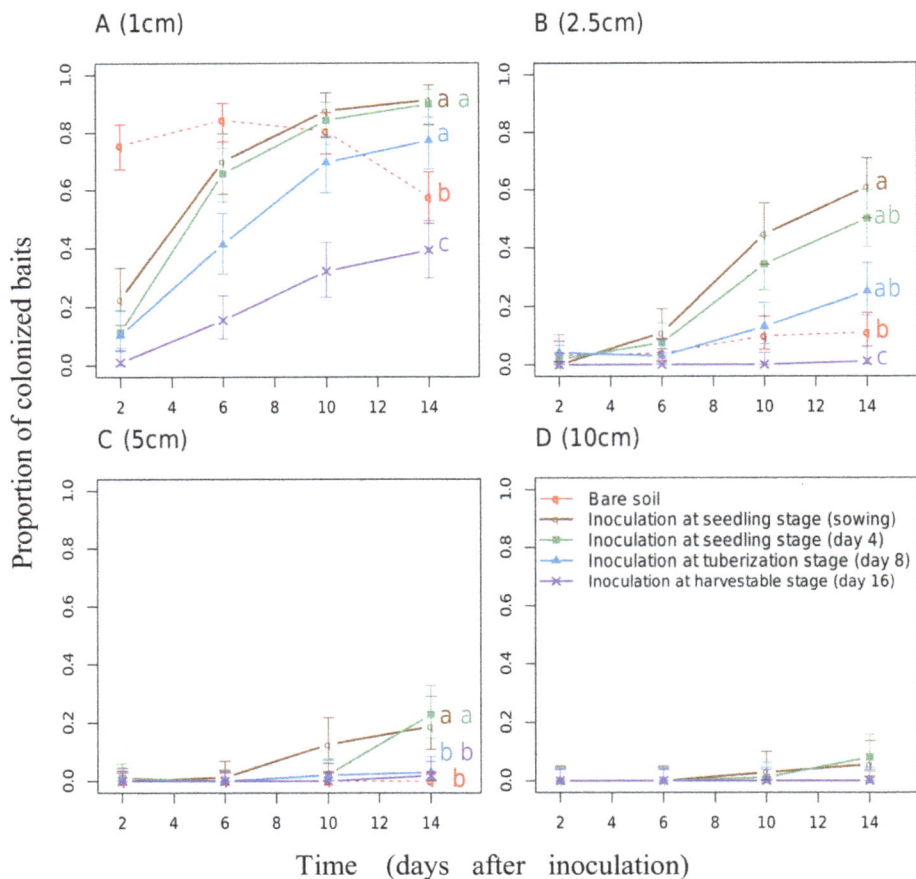

Figure 4. Proportion of baits colonized by *Rhizoctonia solani* at (A) 1 cm, (B) 2.5 cm, (C) 5 cm and (D) 10 cm. Spreading from a mycelium disc (red dotted lines), from an infected radish which were successfully inoculated during the seedling stage (at sowing (brown lines) or four days after sowing (green lines)), during the tuberization stage (eight days after sowing (blue lines)), or during the harvestable stage (sixteen days after sowing (purple lines)). Lowercase letters code the significance of differences in proportion of colonized baits according to the date of inoculation (P-value<0.05, Wald test or LSMeans with FDR correction when needed).

process (Figure 4, blue lines), colonization extends to 10 cm but with a reduced probability compared to infection occurring at the seedling stages. When host infection occurs once tuberization is completed, i.e. for soil inoculation performed at least sixteen days after sowing (Figure 4, violet lines), pathogenic spread is restrained to a short perimeter in the host vicinity, the colonization extent does not exceed 1 cm. Pairwise comparisons of least square means showed significant differences between colonization profiles after infection of a host at seedling stage and after infection of a tuberized host. There was no significant difference between pathogenic spread after infection for a host inoculated at sowing day or four days after sowing. It suggests that the fungal spread is driven by the phenological stage of the host rather than host age at infection.

These results highlight that the three phenological stages shown to affect the ability of the pathogen to infect its host (Experiment 2) also notably impact the pathogenic spread.

Discussion

In this paper, we investigated how host exploitation by a soil-borne plant pathogen sustains its spread in soil based on a two-step process: infection of the host and subsequent exploitation of the infected host. We assumed that ontogenic resistance, defined as

any change in resistance to pathogens correlated with the developmental stage of the host plant or its organs [23], would notably impact this two-step process. To our knowledge, it is one of the first such studies to examine the impact of host ontogenic resistance on the spread of a soil-borne pathogen, in a quantitative fashion. Using the radish / *R. solani* pathosystem, we characterized this impact on each step of the process. Firstly, the link between host development and its infection by the pathogen was assessed by monitoring disease incidence and symptoms for hosts inoculated at different ages. Secondly, we quantified host exploitation in terms of pathogenic spread, through the ability of *R. solani* to grow in soil and colonize baits located at determined distances. We derived colonization profiles, as defined in Bailey et al [21], after infection of a host (i) at the "seedling" stage, (ii) at the "tuberization" stage and (iii) at the "harvestable" stage, using fungal saprophytic spread as a reference.

Host development impacts the infection process, decreasing incidence, and switching the typology of symptoms caused by *R. solani* from damping-off to necrosis. Indeed, disease characteristics changed dramatically with host age at the time of inoculation. The main changes appeared during the tuberization period. Hence, we suggest that the host tuberization process leads to a remarkable change in the interaction between the host and the pathogen. A change in host metabolic profile could explain this ontogenic resistance. *R. solani* uses in priority, and probably with a greater efficiency, simple sugars rather than complex molecules [26]. Therefore, a decrease in hexose and an increase in storage sugars by tubers could deprive the pathogen from easily-usable metabolites. The combination of these mechanisms of ontogenic resistance may also increase the incubation period as evidenced by Leclerc et al [24]. In our experiments, the incubation period for damping-off was on average four days, varying between three and six days (data not shown). No clear trend was visible with host age at inoculation, but non-destructive sampling makes this type of assessment difficult.

Radish phenology impacts resource exploitation subsequent to host infection. Pathogen spread in soil from the host was shown to be high when it was infected before tuberization. Radishes challenged by inoculum from the sowing date were infected a few days later, before tuberization occurred. The colonization of their tissues by *R. solani* led to damping-off symptoms. The host biomass was largely accessible to the fungus, which could support its subsequent spread in soil. Radishes challenged by inoculum only eight or sixteen days after sowing were eventually infected during or after tuberization, and were essentially necrotic. This amount of resource was somehow insufficient or unavailable to the pathogen to support its spread in soil. We assume the fungus was able to pump resources in the diseased host and transfer them to the mycelia growing front in order to sustain its pathogenic spread, through the translocation process [7,25]. In our experiments, hyphae were clearly visible at the surface of the soil, close to the baits which turned out to be colonized by *R. solani*. Therefore, we support the idea that mycelial spread was sustained by a translocation process close to the soil surface rather than secondary root growth deeper in the soil. *R. solani* proved less efficient in exploiting tuberized hosts than non-tuberized ones, despite their relative biomass. Hence, the ability of the pathogen to spread in soil after a successful infection of a radish could not be explained by host tissue amount. Our data are in line with theoretical findings stating that it is the balance between resource investment for infection and profits through the exploitation of infected hosts that drives the survival and spread of the pathogen. Other factors such as total biomass are irrelevant if the pathogen cannot access all plant tissues [26].

Scaling up from our pathosystem to the population scale (i.e. in fields), non-tuberized (i.e. young) hosts contribute more to the disease spread than tuberized ones. They are more likely to get infected, and the pathogen may achieve an efficient exploitation of the resources to sustain its spread. This means that in the context of epidemic prevention, treatment should preferentially target young hosts. Any delay in the contact between the pathogen and its host will decrease the likelihood for the host to get infected and the ability of the pathogen to exploit the host after infection.

If our results still hold under field conditions, in spite of the above-mentioned experimental simplifications, host exploitation itself could explain the ability of the fungus to cross the inter-row gaps during a single season. The spatial structure of hosts in fields (alternates of rows/inter-rows) slows down the fungus in inter-rows, and therefore increases the importance of the age-dependent host exploitation. In French radish fields, rows are generally separated by a 10 to 12.5 cm-gap. Building upon our experimental results, and assuming near pedo-climatic conditions, it should take at least two weeks for *R. solani* to cross the inter-row between an infected host and any host belonging to an adjacent row. At that time, hosts of adjacent rows would already have tuberized, thus being less susceptible to the pathogen and less exploitable once infected.

Our results provide a better understanding of the spread of *R. solani*. However, some factors were not taken into account in the present work, and complicate the predictions of host exploitation under field conditions: Climatic conditions; different traits between *R. solani* strains; crop management, e.g. tillage, which will also modify the course of disease spread (see for example [9,27]). Also, several diseased hosts close to each other could trigger a synergy in the fungal spread of *R. solani* [28,29].

Our experimental data can account for the formation of patches in fields, i.e. the spread of *R. solani* on limited and distinct areas, rather than the entire surface. A review of patch formation [30] listed nine factors explaining their formation, one of them being an increased "host resistance". Results presented here suggest that this may be due to a combination of two factors: i) decreasing host susceptibility to the pathogen and ii) decreasing pathogen ability to exploit the infected host resources.

This study, based on highly replicated bioassays conducted under controlled conditions, improves our understanding of the epidemiology of *Rhizoctonia solani*, building on the pathozone concept [19] that we link here explicitly with the host ontogenic resistance.

To show how to link the local dynamics of a plant pathogen to the interpretation of population behavior, Kleczkowski et al. [12] compared the probability for *R. solani* to infect a host either as a primary inoculum (i.e. inoculum in soil) or a secondary inoculum (i.e. inoculum spreading from an infected host). They demonstrated that the probability of successful infection was strongly influenced by the distance of inoculum from the host (probably as the fungus spends resources building hyphae and encounters competition in soil). They also showed that the pathozone increased from 20 mm for primary inoculum to 60 mm for secondary inoculum. Our study complements this result by (i) comparing the resources provided by hosts at different phenological stages (and showing that tuberized hosts do not provide sufficient resources) and (ii) disentangling the capacity of the fungus to infect a host (depending on host age) from the capacity of the fungus to spread in soil after infection (depending on age).

Further investigations are in progress to characterize the impact of a fungicide on the pathozone the results of which will be useful for empiricists as well as theoretical epidemiologists, providing the latter with new insights to enter models designed for soil-borne epidemic simulation or invasion risk prediction (see for example [31–33]).

Supporting Information

Supporting Information S1 Dataset related to experiment 1. This text file contains raw data used in our analysis to characterize the radish phenological stages.

Supporting Information S2 Dataset related to experiment 2. This text file contains raw data used in our analysis to assess the effect of host development on the pathogen infectivity.

Supporting Information S3 Dataset related to experiment 3. This text file contains raw data used in our analysis to assess the impact of host phenology at infection on subsequent fungal spread.

Acknowledgments

We are very grateful to our missed colleague, Doug Bailey, who was initially involved in this project; his deep insight and ideas were central to the development of this work. We also thank Serge Carillo, Jean-Marie Lucas and Yannick Lucas for useful help, Vilmorin S.A. for providing the seeds, Anne Bates for providing the semi-selective medium, and Dennis Webb for text editoring.

Author Contributions

Conceived and designed the experiments: TES RLC SP FM. Performed the experiments: TES RLC PD SM. Analyzed the data: RLC MRH SP TES. Contributed reagents/materials/analysis tools: TES RLC SP. Contributed to the writing of the manuscript: TES RLC SP.

References

1. Dordas C (2008) Role of nutrients in controlling plant diseases in sustainable agriculture. A review. Agronomy for Sustainable Development 28: 33–46.
2. Garrett SD (1970) Biology of root-infecting fungi: Cambridges: Cambridge university press, London.
3. Sneh B, Jabaji-Hare S, Neate SM, Dijst G (1996) Rhizoctonia species: Taxonomy, Molecular Biology, Ecology, Pathology and Disease Control. Dordrecht, The Netherlands: Kluwer Academic Publishers.
4. Papavizas GC, Davey CB (1961) Saprophytic behavior of Rhizoctonia solani in soil. Phytopathology 51: 693: 699.
5. Pitt D (1964) Studies on sharp eyespot disease of cereals - II. Viability of sclerotia: persistence of the causal fungus, Rhizoctonia solani Kühn. Annals of applied Biology 54: 231–240.
6. Sneh B, Katan J, Henis Y, Wahl I (1966) Methods for evaluating inoculum density of Rhizoctonia in naturally infested soil. Phytopathology 56: 74–78.
7. Jacobs H, Boswell GP, Scrimgeour CM, Davidson FA, Gadd GM, et al. (2004) Translocation of carbon by Rhizoctonia solani in nutritionally-heterogeneous microcosms. Mycological Research 108: 453–462.
8. Otten W, Filipe JAN, Gilligan CA (2004) An empirical method to estimate the effect of soil on the rate for transmission of damping-off disease. New Phytologist 162: 231–238.
9. Schroeder KL, Paulitz TC (2008) Effect of inoculum density and soil tillage on the development and severity of Rhizoctonia root rot. Phytopathology 98: 304–314.
10. Cunniffe NJ, Gilligan CA (2008) Scaling from mycelial growth to infection dynamics: a reaction diffusion approach. Fungal Ecology 1: 133–142.
11. Paustian K, Schnurer J (1987) Fungal growth-response to carbon and nitrogen limitation - a theoretical-model. Soil Biology & Biochemistry 19: 613–620.
12. Kleczkowski A, Gilligan CA, Bailey DJ (1997) Scaling and spatial dynamics in plant-pathogen systems: From individuals to populations. Proceedings of the Royal Society of London Series B-Biological Sciences 264: 979–984.
13. Otten W, Filipe JAN, Bailey DJ, Gilligan CA (2003) Quantification and analysis of transmission rates for soilborne epidemics. Ecology 84: 3232–3239.
14. Otten W, Filipe JAN, Gilligan CA (2005) Damping-off epidemics, contact structure, and disease transmission in mixed-species populations. Ecology 86: 1948–1957.
15. Stacey AJ, Truscott JE, Gilligan CA (2001) Soil-borne fungal pathogens: scaling-up from hyphal to colony behaviour and the probability of disease transmission. New Phytologist 150: 169–177.
16. Ficke A, Gadoury DM, Seem RC (2002) Ontogenic resistance and plant disease management: A case study of grape powdery mildew. Phytopathology 92: 671–675.
17. Gibson GJ, Gilligan CA, Kleczkowski A (1999) Predicting variability in biological control of a plant-pathogen system using stochastic models. Proceedings of the Royal Society of London Series B-Biological Sciences 266: 1743–1753.
18. Deacon JW (1980) Introduction to modern mycology.; Oxford UBS, editor.
19. Gilligan CA, Bailey DJ (1997) Components of pathozone behaviour. New Phytologist 135: 475–490.
20. Hack H, Bleiholder H, Buhr L, Meier U, Schnock-Fricke U, et al. (1992) A uniform code for phenological growth stages of mono- and dicotyledonous plants. Extended BBCH scale, general - Einheitliche Codierung der phanologischen Entwicklungsstadien mono- und dikotyler Pflanzen. Erweiterte BBCH-Skala, Allgemein. Nachrichtenblatt des Deutschen Pflanzenschutzdienstes 44: 265–270.
21. Bailey DJ, Otten W, Gilligan CA (2000) Saprotrophic invasion by the soil-borne fungal plant pathogen Rhizoctonia solani and percolation thresholds. New Phytologist 146: 535–544.
22. Castro C, Davis JR, Wiese MV (1988) Quantitative Estimation of Rhizoctonia Solani AG-3 in soil. Phytopathology 78: 1287–1292.
23. Whalen MC (2005) Host defence in a developmental context. Molecular Plant Pathology 6: 347–360.
24. Leclerc M, Dore T, Gilligan CA, Lucas P, Filipe J (2014) Estimating the Delay between Host Infection and Disease (Incubation Period) and Assessing Its Significance to the Epidemiology of Plant Diseases Plos One 9.
25. Christias C, Lockwood JL (1973) Conservation of mycelial constituents in four sclerotium-forming fungi in nutrient deprived conditions. Phytopathology 63: 602–605.
26. Lamour A, Van den Bosch F, Termorshuizen AJ, Jeger MJ (2000) Modelling the growth of soil-borne fungi in response to carbon and nitrogen. Ima Journal of Mathematics Applied in Medicine and Biology 17: 329–346.
27. Tamm L, Thurig B, Bruns C, Fuchs JG, Kopke U, et al. (2010) Soil type, management history, and soil amendments influence the development of soil-borne (Rhizoctonia solani, Pythium ultimum) and air-borne (Phytophthora infestans, Hyaloperonospora parasitica) diseases. European Journal of Plant Pathology 127: 465–481.
28. Ludlam JJ, Gibson GJ, Otten W, Gilligan CA (2012) Applications of percolation theory to fungal spread with synergy. Journal of the Royal Society Interface 9: 949–956.
29. Zakaria AJ, Boddy L (2002) Mycelial foraging by Resinicium bicolor: interactive effects of resource quantity, quality and soil composition. FEMS Microbiol Ecol 40: 135–142.
30. Anees M, Edel-Hermann V, Steinberg C (2010) Build up of patches caused by Rhizoctonia solani. Soil Biology & Biochemistry 42: 1661–1672.
31. Ferreira IEdP, Moral RdA, Ferreira CP, Godoy WAC (2013) Modelling fungus dispersal scenarios using cellular automata. Ecological Informatics 14: 53–58.
32. Poggi S, Neri FM, Deytieux V, Otten W, Gilligan CA, et al. (2013) Percolation-based risk index for pathogen invasion: application to soil-borne disease in propagation systems. Phytopathology.
33. Poggi S, Neri FM, Deytieux V, Bates A, Otten W, et al. (2013) Percolation-based risk index for pathogen invasion: application to soilborne disease in propagation systems. Phytopathology 103: 1012–1019.

12

Field Evidence of Cadmium Phytoavailability Decreased Effectively by Rape Straw and/or Red Mud with Zinc Sulphate in a Cd-Contaminated Calcareous Soil

Bo Li[1,2], Junxing Yang[1,3], Dongpu Wei[1], Shibao Chen[1], Jumei Li[1], Yibing Ma[1]*

1 National Soil Fertility and Fertilizer Effects Long-term Monitoring Network, Institute of Agricultural Resources and Regional Planning, Chinese Academy of Agricultural Sciences, Beijing, P. R. China, **2** Institute of Plant Nutrition and Environmental Resources, Liaoning Academy of Agricultural Sciences, Shenyang, P. R. China, **3** Centre for Environmental Remediation, Institute of Geographic Sciences and Natural Resources Research, Chinese Academy of Sciences, Beijing, P. R. China

Abstract

To reduce Cd phytoavailability in calcareous soils, the effects of soil amendments of red mud, rape straw, and corn straw in combination with zinc fertilization on Cd extractability and phytoavailability to spinach, tomato, Chinese cabbage and radish were investigated in a calcareous soil with added Cd at 1.5 mg kg^{-1}. The results showed that water soluble and exchangeable Cd in soils was significantly decreased by the amendments themselves from 26% to 70%, which resulted in marked decrease by approximately from 34% to 77% in Cd concentration in vegetables. The amendments plus Zn fertilization further decreased the Cd concentration in vegetables. Also cruciferous rape straw was more effective than gramineous corn straw. In all treatments, rape straw plus red mud combined with Zn fertilization was most effective in decreasing Cd phytoavailability in soils, and it is potential to be an efficient and cost-effective measure to ensure food safety for vegetable production in mildly Cd-contaminated calcareous soils.

Editor: Wenju Liang, Chinese Academy of Sciences, China

Funding: The authors thank the Special Fund for Public Industry in China (Agriculture, 200903015) and the Natural Science Foundation of China (41201312 and 41401361) for financial supports. The funders had no role in study design, data collection and analysis, decision to publish, or preparation of the manuscript.

Competing Interests: The authors have declared that no competing interests exist.

* Email: ybma@caas.ac.cn

Introduction

Recently, increasing cadmium (Cd) accumulation in vegetables is a growing concern globally because of increased fertilizer- and biosolids-borne Cd in soils [1–3]. As a consequence, international trade organizations have sought to limit the concentration of Cd in some crops sold in international markets. The National Food Hygiene Standard of China (NFHSC, GB 15201-94) proposed maximum levels of 0.05 mg kg^{-1} of Cd for spinach and other vegetables. However, there are some areas where Cd concentrations in vegetables are over the limit [4,5]. Therefore, reduction of Cd uptake by vegetables and translocation to edible parts is one of the important strategies for proper use of mildly Cd-contaminated soils and safeguarding the safety of farm produce [1].

Nowadays, "in situ" remediation techniques of mildly Cd-contaminated soils are regarded as possible effective approaches to address the issues of excessive vegetable Cd concentrations. During the last decade, the possibility of Cd immobilization in soils through the addition of different amendments or sorbent, has been extensively investigated in order to reduce the risk of groundwater contamination, plant uptake, and exposure to living organisms [6–9]. Among these amendments or sorbents, red mud (RM), a by-product of aluminium (Al) manufacturing, can be very effective in increasing Cd sorption and decreasing soluble Cd concentrations in Cd-contaminated and acidic soils under pot trials [10–13] and field studies [14,15], and lead to a reduction in

Cd uptake by plants. Lombi et al. [16] indicated that the specific sorption of Cd by Fe and Al oxides in RM was the main mechanisms of fixation. It is therefore important to select cost-effective and feasible amendments to immobilize Cd by specific sorption. It has been documented that thiol (-SH) can reduce Cd bioavailabilty by mechanisms of chelation [17,18]. The cruciferous rape (Brassica napus L.) exhibited higher concentration of thiol (-SH) in straw than other crops [19,20]. However, the effect of the incorporation of rape straw (RS) into Cd-contaminated soil has never been investigated under field condition. In addition, use of zinc (Zn) fertilizers in soils, such as $ZnSO_4$, has been reported to decrease the accumulation of Cd in crops [21–24]. Abdel-Sabour et al. [21] reported that the Cd/Zn ratio in plant tops was significantly affected by both Cd and Zn concentrations in soil. Yang et al. [23] also found that the application of foliar Zn or seed Zn fertilizer could significantly decrease the Cd concentration in cucumber shoots by about 12–36% in Cd-contaminated soils. Köleli et al. [24] also expressed that Cd toxicity in the shoots of bread and durum wheat was alleviated by Zn treatment.

Therefore, we hypothesized that the Cd immobilizing amendments of RS plus RM in combination with Zn fertilizer might be more effective in reducing Cd accumulation in vegetables grown in Cd-contaminated and calcareous (high pH) soils under field conditions. The present study was conducted to investigate the efficiency of these amendments with Zn fertilization on Cd accumulation in the edible parts of four common vegetables (spinach, tomato, Chinese cabbage and radish) grown in mildly Cd-contaminated and calcareous soil under field condition. Furthermore, the effect of Cd immobilizing amendments (RS, RM, RS+RM) with Zn fertilizer on the Cd fractions associated with different soil components was also studied. The results will be helpful to find practical and cost-effective measures to reduce Cd accumulation in crops.

Materials and Methods

Soil characteristics and amendments used

The field experiment was conducted at long-term experiment station of the Chinese Academy of Agricultural Sciences, Dezhou (DZ) city, Shandong Province, China ($37°20'N$, $116°29'E$). The soil used in the field experiment had a pH (1:5 soil/water suspension) of 8.9 and contained 1.2% organic matter, 6.17% $CaCO_3$, 0.08% total N and 0.1% total P as measured by the standard methods given in Jackson [25]. The soil contained 64% sand, 18% clay and 18% silt. Total Zn and Cd concentrations of the soil were 54 mg Zn kg^{-1} and 0.11 mg Cd kg^{-1}, measured as described by Jackson [25], while DTPA-extractable [26] concentrations of Zn and Cd were 0.11 mg Zn kg^{-1} and <0.005 mg Cd kg^{-1}, respectively.

Red mud, RS, and corn straw (CS) were used as immobilizing amendments in the field experiment. Red mud (pH = 11.1) was from Zibo City, Shandong Province, China. The mineralogical composition of RM sample (XRD analysis) is a mixture of SiO_2 (20%), Fe_2O_3 (28%), Al_2O_3 (21%), CaO (6.2%), MgO (1.3%), TiO_2 (3.3%), K_2O (0.26%) and Na_2O (11%). The specific surface area, determined by the BET/N_2-adsorption method (Sorptomatic CarloErba), was 12.2 $m^2\,g^{-1}$ for RM. Zinc and Cd concentrations in RM were 94 mg kg^{-1} and <0.01 mg Cd kg^{-1}, respectively. The RM sample was dried overnight at 105°C, finely ground and sieved to <1 mm. The rape straw sample was obtained from rape (Allium cepa L. cv. Zheshuang No. 6) grown at long-term experiment station of the Chinese Academy of Agricultural Sciences, Jiaxing city, Zhejiang Province, China ($30°15'N$, $120°20'E$), which was oven-dried at 70°C to constant weight and then finely ground in a Retsch-grinder (Type: 1 mm, made in Germany) using a 1 mm mesh screen to ensure uniform plant tissue disruption and distribution in soil during the field experiment. The rape straw with pH of 6.41 and electrical conductivity (EC) of 398 µS cm^{-1} (straw:solution ratio 1:10), contained 19 mg Cu kg^{-1}, 23 mg Zn kg^{-1}, 0.86 mg Pb kg^{-1} and 0.67 mg Cd kg^{-1} dry weight. The corn straw sample was obtained from corn (Zea mays L. cv. Jingdan No. 28) grown at long-term experiment station of the Chinese Academy of Agricultural Sciences, Changping, Beijing, China ($40°13'N$ $116°15'E$). The sample was also oven-dried at 70°C to constant weight and then finely ground using a 1 mm mesh. The corn straw with pH of 6.10 and EC of 187 µS cm^{-1}, contained 36 mg Cu kg^{-1}, 59 mg Zn kg^{-1}, 0.76 mg Pb kg^{-1} and 0.84 mg Cd kg^{-1} dry weight.

Field experiment and plant analysis

The field experiment was a randomized complete block split-spot design with 3 replications for the control and amendment treatments (main treatments) and 2 replications for the control and amendment treatments (sub-treatments) in combination with Zn fertilization (12 g $ZnSO_4$ per plot, based on the mass of top 20 cm soil). The size of each plot was 4 m^2 (2 m×2 m). Before amendment addition, soils in the plots were added with 1.5 mg Cd kg^{-1} in the form of $CdSO_4$ on 1 February, 2009. The concentration of Cd added to soil was chosen based on preliminary experiments and represented mildly Cd contamina-

tion. To decrease the variability, the salts of $CdSO_4$ were mixed with topsoil samples (0–20 cm) separately in a container, after which the spiked soils were returned to the plots and equilibrating for 2 months. Fertilizers were then applied to all plots according to local farming practices. The equivalent nitrogen (0.2 g N kg^{-1} soil as urea), phosphorus (0.06 g P_2O_5 kg^{-1} soil as superphosphate), potassium (0.06 g K_2O kg^{-1} soil as potassium sulfate) were applied as basal fertilizers to each plot before the spinach and tomato seeding. After spinach and tomato harvest, the same equivalent phosphorus, potassium and nitrogen were applied as basal fertilizers to each plot before Chinese cabbage and radish seeding. All nutrients were mixed homogenously with soil before sowing.

The main treatments were (1) control, (2) 0.5% RM (W/W), (3) 0.1% RS (W/W), (4) 0.5% RM+0.1% RS, (5) 0.1% CS (W/W), (6) 0.5% RM+0.1% CS. The sub-treatments were applied with Zn fertilization (12 g $ZnSO_4$ per plot) before the Cd-contaminated soils in the plots had been added with the amendments mentioned above except CS treatment. The amendments were applied to the surface of each plot before being ploughed into the soil to a depth of 20 cm.

Four commonly cultivated vegetable varieties (Spinach (Spinacia oleracea L. cv. Huabo No. 1), Tomato (Lycopersicum esculentum Mill. cv. Lufen No. 3), Chinese cabbage (Brassica campestris L. cv. Degao No. 16) and Radish (Raphanus sativus Linn. cv. Qianxi No. 2) in DZ were selected in this experiment and were sown directly into the soil according to different growth periods. Spinach was sown on 3 April and harvested on 10 May, 2009. Tomato was sown on 5 May, and harvested on 15 September, 2009. Radish and Chinese cabbage were sown simultaneously on 16 October, 2009 after harvest of spinach and tomato. Sufficient seed was sown to guarantee healthy germination, then seedlings were thinned after germination). Only the edible portions were sampled after all vegetables were grown to maturity as the study was focused on the food safety. At each plot, 10 subsamples of the edible parts of vegetables were collected and combined for chemical analysis. The fresh vegetable samples were put in clean plastic bags and transported to the laboratory for sample treatment. The samples were washed with 0.2% HCl solution followed by tap water and de-ionized water, then oven-dried (not peeled) at 70°C for 6 h to constant weight and dry weights (DW) were recorded. The plant samples were ground using a Retsch-grinder (Type: 0.5 mm, made in Germany), then weighted 0.5 g to 200 mL digestion tubes with 10 ml of concentrated nitric acid (HNO_3) and digested for 9 h at 110°C after standing overnight [27]. Cadmium concentrations were determined using inductively coupled plasma mass spectrometry (ICP-MS). Blank and bush leaf material (BGW-07603) (China Standard Materials Research Center, Beijing, PR China) were used for quality control. The Cd recovery rates were 90±10%.

Soil analysis by sequential extraction procedure

After harvest, 10 subsamples of soils (0–20 cm) were evenly collected from each plot, bulked together, air-dried, and ground to pass a 0.26-mm sieve. Soil pH was measured using de-ionized water (1:5 soil/water suspension) with an ORION combined electrode. The fractions of Cd bound to the soil were determined by a sequential extraction procedure according to Basta and Gradwohl [28], in order to study the effects of the different amendments on Cd fractions. To extract the water soluble fraction (WS-Cd), each sample collected from plots (2 g) was treated with 25 mL of de-ionized water (pH 6.5) and shaken for 2 h at room temperature. It was then treated with 25 mL of 0.1 N $Ca(NO_3)_2$ solution to extract the exchangeable fraction (Exch-Cd), and with 25 mL 0.02 M EDTA solution to extract the complexed fraction

Table 1. The concentration of Cd (C_{Cd}, mg Cd kg^{-1} in dry weight) and yield (g plant^{-1} in dry weight) for edible part of spinach, tomato, Chinese cabbage and radish in Cd-contaminated soils (added Cd at 1.5 mg kg^{-1}) with different amendments with (+Zn) and without Zn fertilization.

Treatment	Spinach		Tomato		Chinese cabbage		Radish	
	C_{Cd} (mg kg^{-1})	Yield (g plant^{-1})	C_{Cd} (mg kg^{-1})	Yield (g plant^{-1})	C_{Cd} (mg kg^{-1})	Yield (g plant^{-1})	C_{Cd} (mg kg^{-1})	Yield (g plant^{-1})
CK	0.75±0.05 a	0.96±0.03 c	0.35±0.03 a	4.31±0.03 d	0.49±0.03 a	161±1 h	0.56±0.02 a	94±1 d
CK+Zn	0.53±0.03 b	1.11±0.02 b	0.31±0.03 b	4.58±0.06 bc	0.35±0.02 b	164±2 h	0.45±0.02 b	102±1 c
RM	0.40±0.04 c	1.14±0.03 b	0.18±0.02 d	4.61±0.03 bc	0.19±0.02 c	175±1 f	0.19±0.01 d	115±1 b
RM+Zn	0.29±0.03 d	1.21±0.04 ab	0.16±0.02 de	4.69±0.06 b	0.16±0.01 d	191±2 c	0.16±0.01 e	117±1 b
RS	0.41±0.05 c	0.91±0.03 c	0.14±0.02 e	4.51±0.04 c	0.16±0.01 d	179±2 e	0.17±0.01 de	103±1 c
RS+Zn	0.36±0.03 c	0.93±0.04 c	0.13±0.01 e	4.77±0.07 ab	0.12±0.01 e	185±2 d	0.16±0.02 e	108±3 c
RM+RS	0.18±0.02 e	1.25±0.02 a	0.12±0.02 e	4.83±0.05 a	0.12±0.01 e	192±3 c	0.13±0.01 f	117±2 b
RM+RS+Zn	0.12±0.02 f	1.31±0.03 a	0.09±0.02 f	4.89±0.07 a	0.09±0.01 f	201±3 a	0.10±0.01 g	123±1 a
CS	0.47±0.05 bc	0.99±0.03 c	0.23±0.02 c	4.48±0.04 c	0.20±0.02 c	171±2 g	0.22±0.01 c	95±1 d
RM+CS	0.37±0.04 c	1.25±0.04 a	0.16±0.01 de	4.69±0.05 b	0.18±0.01 cd	189±1 c	0.18±0.01 de	113±1 b
RM+CS+Zn	0.26±0.04 d	1.27±0.03 a	0.13±0.02 e	4.62±0.09 bc	0.15±0.01 d	196±1 b	0.15±0.02 e	120±1 a

The application rates in soils were at 0.5% (W/W) for red mud (RM), 0.1% (W/W) for rape straw (RS) and corn straw (CS), and 3 g ZnSO$_4$ per square meter, respectively.

Note: Within each column in the same vegetable, mean values ± standard errors with the same letter do not differ significantly at 5% level (P<0.05) according to the Fisher's least significant test.

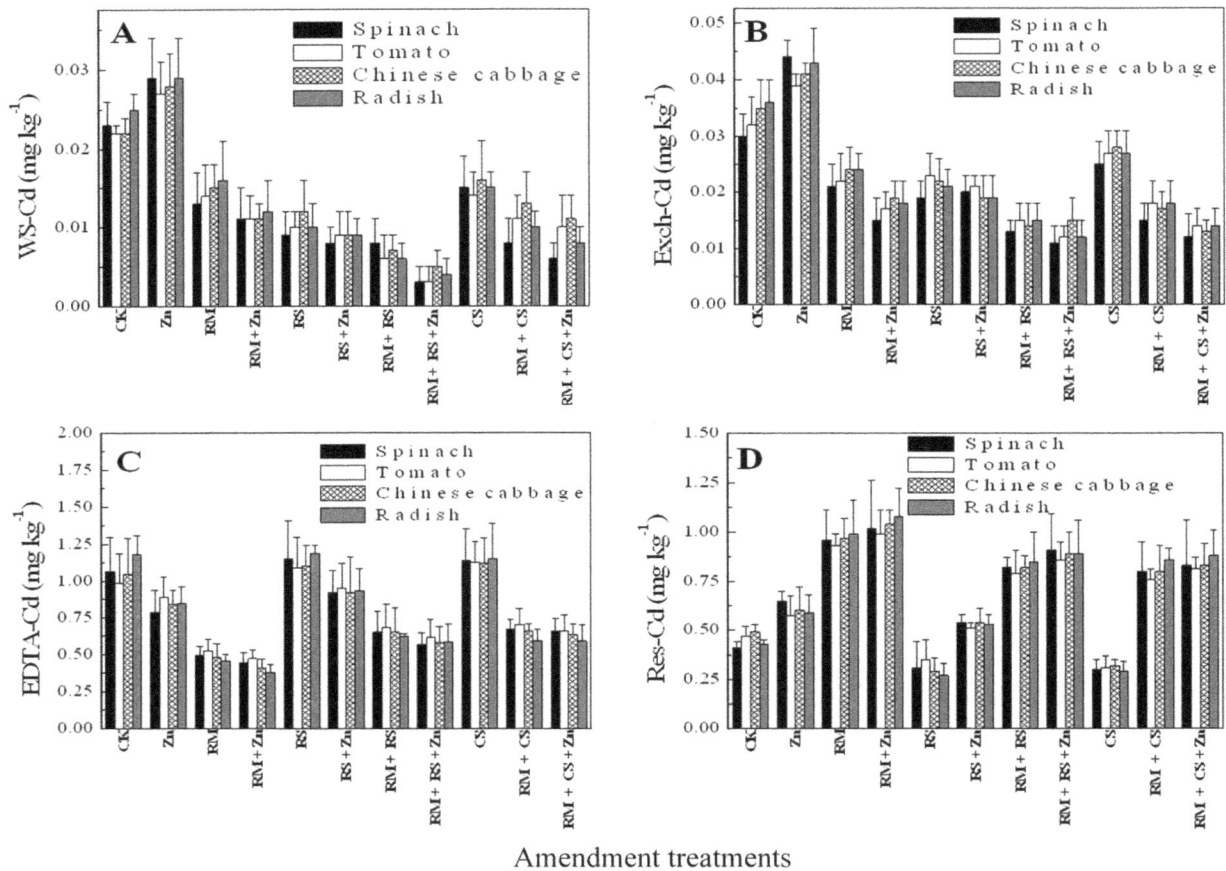

Figure 1. Concentrations (mg Cd kg^{-1} soil) of water soluble (WS-Cd, A), exchangeable (Exch-Cd, B), EDTA extractable (EDTA-Cd, C) and residual Cd (Res-Cd, D) in the soils with and without different amendments after spinach, tomato, Chinese cabbage and radish cultivation.

(EDTA-Cd). After each step of the extraction process the samples were centrifuged at 10,000 rpm for 0.5 h and filtered to separate the liquid and solid phases. After the third extraction, the residual form of Cd (Res-Cd) was determined by drying the solid phase overnight at 105°C and digesting it with HNO_3 and HCl (ratio 1:3) in a Microwave Milestone MLS 1200. The Cd concentrations in each extract or digest were determined using inductively coupled plasma mass spectrometry (ICP-MS).

Data analysis

All results were presented as arithmetic means with standard errors and analyzed by SPSS 11.0 statistical package. Statistical comparisons of means of plant data were analyzed with one way ANOVA followed by the Fisher's least significant test. Correlation coefficient analyses were conducted using program of Origin 7.0.

Results

Effects on Cd in vegetables

The concentrations of Cd and yield for edible parts of spinach, tomato, Chinese cabbage and radish with both the unamended and amended soil combined with Zn fertilization are presented in Table 1. Compared with the vegetable grown in unamended soil, the concentrations of Cd in the edible parts of the four vegetables were reduced with amendment treatments, and the reduction (% of control) was significantly different (P<0.05) among the different

treatments. The reduction of Cd in vegetables ranged from 37% to 76% for spinach, and from 34% to 63% for tomato, and from 59% to 76% for Chinese cabbage, and from 61% to 77% for radish, with the lowest for CS treatment and the highest for RS+ RM treatment. Although the yield of edible parts of spinach, tomato, Chinese cabbage and radish treated with amendments were increased by 3–36%, 4–13%, 6–25% and 1–31%, respectively (Table 1), the total Cd uptake in edible parts of the four vegetables was still significantly decreased with amendment treatments (data can be calculated using the Cd concentration multiplying by yield of vegetables in Table 1) as there was "dilution effect" of Cd in edible parts of plant. Combined with amendment treatments, Zn application further decreased the Cd concentration in vegetables up to 74–84% of those in unamended treatment.

Changes in Cd fractions in soils

The concentrations of Cd in different fractions of soils with different treatments after plant harvest were shown in Figure 1. The proportion of Cd in the control (total Cd 1.5 mg kg^{-1}) soil were 1.46% (0.023 mg kg^{-1}) in the WS-Cd, 2.11% in the Exch-Cd, 67.9% in the EDTA-Cd, and 28.5% in the Res-Cd. The concentrations of WS-Cd and Exch-Cd were significantly lower (P<0.05) in the amended soil than in the unamended soil except Zn treatment (Fig. 1). In general, addition of amendments combined with Zn fertilization to the soil can significantly decrease

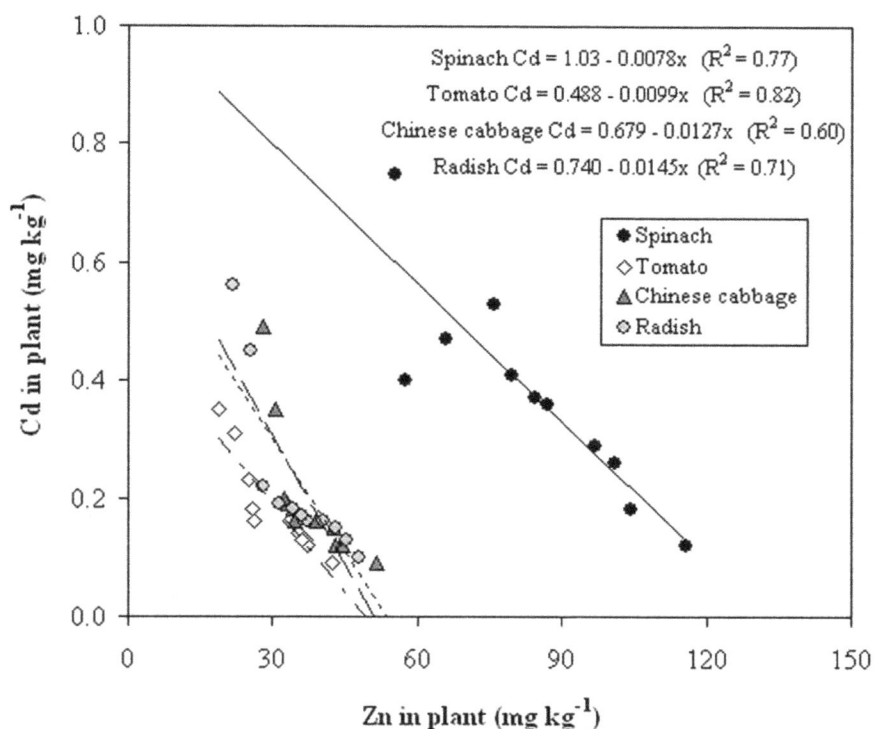

Figure 2. Concentration of Cd in edible parts of spinach, tomato, Chinese cabbage and radish as a function of the Zn concentration in these plants in Cd-contaminated soils (1.5 mg added Cd kg^{-1} soil) with different amendments with and without Zn fertilization.

WS-Cd from 34% (CS) to 84% (RS+RM+Zn) among the four plants. The addition of the RM, RM+CS and RM+RS combined with Zn fertilization also remarkably increased the Res-Cd. As for RS and CS treatments, the EDTA-Cd increased about 5%. Among the fractions, WS-Cd and Exch-Cd were more pronouncedly affected by the treatment of amendments than the other fractions, which suggested that WS-Cd and Exch-Cd were transformed to the non-extractable form in the amended soil.

Discussion

Accumulative evidence from pot trials [27,29] and field samples [4,14,15,30] clearly showed that vegetables grown on Cd-contaminated soils results in elevated Cd levels in edible parts of the vegetables, exceeding the maximum allowable limit (0.05 mg kg^{-1} fresh weight) of NFHSC. Our results showed that Cd concentrations of the edible of parts of the spinach, tomato, Chinese cabbage and radish grown in the unamended soil (1.5 mg kg^{-1} Cd exposure) was 0.75 mg kg^{-1} (0.06 mg kg^{-1} fresh weight), 0.35 mg kg^{-1} (0.06 mg kg^{-1} fresh weight), 0.49 mg kg^{-1} (0.053 mg kg^{-1} fresh weight) and 0.56 mg kg^{-1} (0.08 mg kg^{-1} fresh weight), respectively, being 1.06–1.60 fold as high as the NFHSC value (Table 1). When different amendments were added to soil, the Cd concentrations in edible parts of the four vegetables were almost decreased to the values of <0.05 mg kg^{-1} fresh weight. The present results also indicated that distinctive differences in Cd accumulation when comparing one vegetable to another, following the order: spinach (leafy vegetables, Chenopodiaceae) > radish (root vegetables, Cruciferae) > Chinese cabbage (kale vegetables, Cruciferae) > tomato (fruit vegetables, Solanaceae), which is similar with the result in soil with Cd added less than 2 mg kg^{-1} from the study of Yang et al. [31]. The differences in Cd accumulation are probably because the soil-

to-plant transfer factor of Cd (TF = M(edible part)/M(soil), M is the Cd concentration in edible part or soil) for the leafy vegetables were higher than those for the non-leafy vegetables [27,30]. When grown in the unamended soil (1.5 mg kg^{-1} Cd exposure), the TF of edible parts of spinach, radish, Chinese cabbage and tomato was 0.50, 0.37, 0.33 and 0.23, respectively. When applied with the amendments and/or Zn fertilization, the decline in TF of the four vegetables treated with the amendments was significantly different, ranging from 0.08 (RS+RM+Zn) to 0.35 (CK+Zn) for spinach, from 0.07 (RS+RM+Zn) to 0.30 (CK+Zn) for radish, from 0.06 (RS+RM+Zn) to 0.23 (CK+Zn) for Chinese cabbage, and from 0.08 (RS+RM+Zn) to 0.21 (CK+Zn) for tomato. Although all amendments played an important role of decreasing Cd transfer from soil to plant, however, in comparison with other three vegetables, the transfer ability of spinach for Cd were always stronger than others (Table 1). Not only for Cd, spinach could also accumulate other heavy metals (e.g., Ni, Pb and Cu) intensely [31–33], for instance, nickel level in spinach shoot was found to be 1.5–4.9 fold as high as those in other six plant species (including tomato and cabbage) in high-pH soil [33]. These results might be related to the characteristic of leafy vegetables, easy uptake/translocation of heavy metals from soil to shoots by passive uptake - transpiration based on bigger surface area of plant leaves and stomatal aperture [34]. Generally, accumulation for heavy metals of different parts of plant is in the order of root > shoot > fruit, which might be the reason that radish (root vegetables) could accumulate more Cd than Chinese cabbage (kale vegetables) and tomato (fruit vegetables).

Previous studies showed that addition of red mud [10–12,14,15,35] and plant materials [36,37] to Cd-contaminated and acidic soils could effectively reduce Cd bioavailability in soils. However, little information was available about the addition of red mud and/or plant materials to mildly Cd-contaminated and calcareous soils (high pH) under field conditions. Results from the present study showed that Cd concentration and uptake in edible parts of the four vegetables treated with RM, RS and CS under mildly Cd-contaminated and calcareous soils was significantly (P< 0.05) reduced. The reason might be because the RM, RS and CS application markedly reduced Cd mobility in the Cd-contaminated soil. As shown in Fig. 1, the WS-Cd and Exch-Cd in the amended soils averagely decreased from 26% (CS) to 70% (RS+ RM) except Zn treatment, while the Res-Cd averagely increased from 35% (RM+CS) to 108% (RM) among RM, RM+CS and RM+RS treatment, and the EDTA-Cd averagely increased at 5% for RS and CS treatments, suggesting that the addition of RM could lead to Cd transformation from WS-Cd and Exch-Cd to Res-Cd, while the addition of CS and RS could lead to Cd transformation from WS-Cd and Exch-Cd to EDTA-Cd. Furthermore, among the amendments RS+RM was the most effective, with the greatest reductions in WS-Cd (70%) and Exch-Cd (57%). Yang et al. [38] also showed that RS and nano-treated RM were the two best amendments in decreasing Exch-Cd in alkaline soil and total Cd in cucumber plants. The reason might be ascribed mainly to different mechanisms of bindings of Cd between these amendments. In the present study, th e application of RM (0.5%, W/W), RS (0.1%, W/W) and CS (0.1%, W/W) to the calcareous DZ soil (pH 8.9) had no obvious increase on soil pH (data not shown). The transformation of WS-Cd and Exch-Cd to Res-Cd was probably due to the specific sorption of Cd by Fe and Al oxides in RM, while the transformation of WS-Cd and Exch-Cd to EDTA-Cd was probably due to the complexation with RS or CS. Luo et al. [39] further investigated the sorption mechanism of cadmium on red mud as same as used in the study using batch sorption experiments, sequential extraction analysis and X-ray absorption near edge structure (XANES) spectroscopy and supplied evidence of the formation of inner-sphere complexes of Cd similar to XCdOH (X represents surface groups on red mud) on the red mud surfaces although outer-sphere complexes of Cd were the primary species.

Some studies revealed that plant materials could be sorbent materials for Cd due to the tendency of Cd to form stable complexes with organic ligands [36,40]. Wu et al. [41] found that the Cd concentrations in grains of rice by rotation with rape were decreased approximately by 46–80% of those for rice cultivation only, the decreasing of WS-Cd plus Exch-Cd and increasing of Org-Cd was might be related to the abundance of organic material secreted from rape roots in soil. Harada et al. [17] also found that Cd stress could result in a 3-fold increase in total thiols mainly contributing to synthesis of cysteine, glutathione and phytochelatins in *Arabidopsis*. In the present study, an obvious decrease of WS-Cd in soil was displayed with the addition of RS (Fig. 1), which might be ascribed to the high affinity for Cd induced by sulfur compounds (thiol) in rape straw. However, for CS, a relative lower affinity for Cd could be resulted from (semi)cellulose as main components in straw [42].

Results from the present study indicated that Cd concentrations in the edible parts of vegetables were significantly lower for amendment treatments with Zn fertilization than those with amendments only (Table 1). Furthermore, the results (Fig. 2) also showed that there were significant negative correlations between concentrations of Cd and Zn in edible parts of vegetables with $R^2 \geq 0.60$ (P<0.01, n = 11), and supplied the evidence of Zn antagonistic effect on Cd uptake by plants in the calcareous soils with amendments. Oliver et al. [22] also found Zn fertilization markedly reduced Cd concentration in wheat grain in areas where it was marginal to severe Zn deficiency in South Australia. These results suggested that Zn fertilization could be a practical measure combined with soil amendments to decrease Cd concentration in plants in soils where the Zn availability is low.

Recently, there is evidence that RM combined with other amendment, such as gravel sludge (GS), a waste product of the gravel industry, had higher long-term efficiency in immobilizing Cd than only RM or GS treatment under field condition [15]. Similarly, our results provided clearly evidence for a synergistic interaction between RM and RS leading to highly significant (P< 0.01) reductions in the Cd accumulation of edible parts of the four vegetables. Considering all of the above-mentioned facts, the RS+ RM is suggested to act as an efficient, economic and practical measure for mildly Cd-contaminated and calcareous soil, best in combination with Zn fertilization if soils with low Zn availability.

Conclusions

This study clearly demonstrated that water soluble and exchangeable Cd in soils was significantly decreased by red mud, rape and corn straw, which resulted in significant decrease by about 34% to 77% in Cd concentration in vegetables in Cd-contaminated and calcareous soils. Combined with the amendments, Zn fertilization further decreased Cd concentration in the edible part of vegetables up to 74% to 84%. The effect of rape and corn straw could be ascribed to formation of stable complexes with organic ligands while red mud to specific sorption of Cd. Also cruciferous rape straw was more effective than gramineous corn straw. In all treatments, rape straw plus red mud combined with Zn fertilization was most effective in decreasing Cd phytoavailability in soils, and it is potential to be an efficient and cost-effective measure to ensure food safety for vegetable production in mildly Cd-contaminated and calcareous soils.

Acknowledgments

We thank Yubao Zuo for field management of the experiment.

Author Contributions

Conceived and designed the experiments: YM BL. Performed the experiments: BL JY DW. Analyzed the data: BL JY SC JL. Contributed reagents/materials/analysis tools: DW SC. Contributed to the writing of the manuscript: BL YM JY.

References

1. McLaughlin MJ, Parker DR, Clarke JM (1999) Metals and micronutrients-food safe issues. Field Crop Res 60: 143–163.
2. Satarug S, Baker JR, Urbenjapol S, Haswell-Elkin M, Reilly PEB, et al. (2003) A global perspective on cadmium pollution and toxicity in non-occupationally exposed population. Toxicol Lett 137: 65–83.
3. Hooda PS (2010) Trace Elements in Soils. New York: Wiley-Blackwell.
4. Demirezen D, Aksoy A (2006) Heavy metal levels in vegetables in Turkey are within safe limits for Cu, Zn, Ni and exceeded for Cd and Pb. J Food Quality 29: 252–265.
5. Li J, Xie ZM, Xu JM, Sun YF (2006) Risk assessment for safety of soils and vegetables around a lead/zinc mine. Environ Geochem Hlth 28: 37–44.
6. Castaldi P, Santona L, Melis P (2005) Heavy metals immobilization by chemical amendments in a polluted soil and influence on white lupin growth. Chemosphere 60: 365–371.
7. Castaldi P, Melis P, Silvetti M, Deiana P, Garau G (2009) Influence of pea and wheat growth on Pb, Cd, and Zn mobility and soil biological status in a polluted amended soil. Geoderma 151: 241–248.
8. Tandy S, Healey JR, Nason MA, Williamson JC, Jones DL (2009) Remediation of metal polluted mine soil with compost: co-composting versus incorporation. Environ Pollut 157: 690–697.
9. Liu YJ, Naidu R, Ming H (2011) Red mud as an amendment for pollutants in solid and liquid phases. Geoderma 163: 1–12.

10. Lombi E, Zhao FJ, Wieshammer G, Zhang G, McGrath SP (2002) In situ fixation of metals in soils using bauxite residue: biological effects. Environ Pollut 118: 445–452.

11. Lombi E, Zhao FJ, Zhang GY, Sun B, Fitz W, et al. (2002) In situ fixation of metals in soils using bauxite residue: chemical assessment. Environ Pollut 118: 135–443.

12. Friesl W, Lombi E, Horak O, Wenzel W (2003) Immobilization of heavy metals in soils using inorganic amendments in a greenhouse study. J Plant Nutr Soil Sc 166: 191–196.

13. Lee SH, Kim EY, Park H, Yun J, Kim JG (2011) In situ stabilization of arsenic and metal-contaminated agricultural soil using industrial by products. Geoderma 161: 1–7.

14. Friesl W, Friedl J, Platzer K, Horak O, Gerzabek MH (2006) Remediation of contaminated agricultural soils near a former Pb/Zn smelter in Austria: Batch, pot and field experiments. Environ Pollut 144: 40–50.

15. Friesl W, Platzer K, Horak O, Gerzabek MH (2009) Immobilising of Cd, Pb, and Zn contaminated arable soils close to a former Pb/Zn smelter: a field study in Austria over 5 years. Environ Geochem Hlth 31: 581–594.

16. Lombi E, Hamon RE, McGrath SP, McLaughlin MJ (2003) Lability of Cd, Cu, and Zn in polluted soils treated with lime, beringite, and red mud and identification of a non-labile colloidal fraction of metals using isotopic techniques. Environ Sci Technol 37: 979–984.

17. Harada E, Yamaguchi Y, Koizumi N, Hiroshi S (2002) Cadmium stress induces production of thiol compounds and transcripts for enzymes involved in sulfur assimilation pathways in Arabidopsis. J Plant Physiol 159: 445–448.

18. Herbette S, Taconnat L, Hugouvieux V, Piette L, Magniette MLM (2006) Genome-wide transcriptome profiling of the early cadmium response of Arabidopsis roots and shoots. Biochimie 88: 1751–1765.

19. Jones MG, Hughes J, Tregova A, Milne J, Tomsett AB, et al. (2004) Biosynthesis of the flavour precursors of onion and garlic. J Exp Bot 55: 1903–1918.

20. Wang LQ, Remediation of Cd-contaminated soils by in situ immobilization techniques. Dissertation for the Doctoral Degree. Beijing: Capital Normal University, 2009 (in Chinese).

21. Abdel-Sabour MF, Mortvedt JJ, Kelsoe JJ (1988) Cadmium-zinc interactions in plants and extractable cadmium and zinc fractions in soil. Soil Sci 145: 424–431.

22. Oliver DP, Hannam R, Tiller KG, Wilhelm NS, Merry RH, et al. (1994) The effects of zinc fertilization on cadmium concentration in wheat grain. J Environ Qual 23: 705–711.

23. Yang JX, Wang LQ, Wei DP, Chen SB, Ma YB (2011) Foliar spraying and seed soaking of zinc fertilizers decreased cadmium accumulation in cucumbers grown in Cd-contaminated soils. Soil Sediment Contam 20: 400–410.

24. Köleli N, Eker S, Cakmak I (2004) Effect of zinc fertilization on cadmium toxicity in durum and bread wheat grown in zinc-deficient soil. Environ Pollut 131: 453–459.

25. Jackson ML (1958) Soil Chemical Analysis (2nd ed.). Baton Rouge: CRC Press.

26. Lindsay WL, Norvell WA (1978) Development of a DTPA soil test for zinc, iron, manganese and copper. Soil Sci Soc Am J 42: 421–428.

27. Alexander PD, Alloway BJ, Dourado AM (2006) Genotypic variations in the accumulation of Cd, Cu, Pb and Zn exhibited by six common grown vegetables. Environ Pollut 144: 736–745.

28. Basta NT, Gradwohl M (2000) Estimation of Cd, Pb, and Zn bioavailability in smelter-contaminated soils by a sequential extraction procedure. J Soil Contam 9: 149–164.

29. Kuboi T, Noguchi A, Yazaki J (1986) Family-dependent cadmium accumulation in higher plants. Plant Soil 92: 405–415.

30. Wang G, Su MY, Chen YH, Lin FF, Luo D, et al. (2006) Transfer characteristics of cadmium and lead from soil to the edible parts of six vegetable species in southeastern China. Environ Pollut 144: 127–135.

31. Yang JX, Guo HT, Ma YB, Wang LQ, Wei DP, et al. (2010) Genotypic variations in the accumulation of Cd exhibited by different vegetables. J Environ Sci 22: 1246–1252.

32. Li B (2010) The phytotoxicity of added copper and nickel to soils and predictive models. Ph D dissertation. Chinese Academy of Agricultural Sciences, Beijing, China.

33. Giordani C, Cecchi S, Zanchi C (2005) Phytoremediation of soil polluted by nickel using agricultural crops. Environ Manage 36: 675–681.

34. Marchiol L, Sacco P, Assolari S, Zerbi G (2004) Reclamation of polluted soil: phytoremediation potential of crop-related Brassica species. Water Air Soil Pollut 158: 345–356.

35. Garau G, Silvetti M, Deiana S, Deiana P, Castaldi P (2011) Long-term influence of red mud on As mobility and soil physico-chemical and microbial parameters in a polluted sub-acidic soil. J Hazard Mater 185: 1241–1248.

36. Cui YS, Du X, Weng LP, Zhu YG (2008) Effect of rice straw on the speciation of cadmium (Cd) and copper (Cu) in soils. Geoderma 146: 370–377.

37. Tlustoš P, Vostál J, Száková J, Balík J (1995) Direct and subsequent efficiency of selected measures on the Cd and Zn content in the biomass of spinach. Rostlinná Výroba 41: 31–37 (in Czech).

38. Yang JX, Wang LQ, Li JM, Wei DP, Chen SB, et al. (2014) Effects of rape straw and red mud on extractability and bioavailability of cadmium in a calcareous soil. Front Environ Sci Eng (in press).

39. Luo L, Ma CY, Ma YB, Zhang SZ, Lv JT, et al. (2011) New insights into the sorption mechanism of cadmium on red mud. Environ Pollut 159: 1108–1113.

40. Almas A, Singh BR, Salbu B (1999) Mobility of cadmium-109 and zinc-65 in soil influenced by equilibration time, temperature, and organic matter. J Environ Qual 28: 1742–1750.

41. Wu FL, Lin DY, Su DC (2011) The effect of planting oilseed rape and compost application on heavy metal forms in soil and Cd and Pb uptake in rice. Agr Sci China 10: 267–274.

42. Chen SB, Sun C, Wei W, Lin L, Wang M (2012) Difference in cell wall components of roots and its effect on the transfer factor of Zn by plant species. China Environ Sci 32: 1309–1313 (in Chinese).

13

Variation in Capsidiol Sensitivity between *Phytophthora infestans* and *Phytophthora capsici* Is Consistent with Their Host Range

Artemis Giannakopoulou[1], Sebastian Schornack[1,2], Tolga O. Bozkurt[1,3], Dave Haart[4], Dae-Kyun Ro[5], Juan A. Faraldos[6], Sophien Kamoun[1]*, Paul E. O'Maille[4,7]*

1 The Sainsbury Laboratory, Norwich, United Kingdom, 2 Sainsbury Laboratory, Cambridge University, Cambridge, United Kingdom, 3 Imperial College, Faculty of Natural Sciences, Department of Life Sciences, London, United Kingdom, 4 Institute of Food Research, Food & Health Programme, Norwich, United Kingdom, 5 Department of Biological Sciences, University of Calgary, Calgary, Canada, 6 School of Chemistry, Cardiff University, Cardiff, United Kingdom, 7 John Innes Centre, Department of Metabolic Biology, Norwich, United Kingdom

Abstract

Plants protect themselves against a variety of invading pathogenic organisms via sophisticated defence mechanisms. These responses include deployment of specialized antimicrobial compounds, such as phytoalexins, that rapidly accumulate at pathogen infection sites. However, the extent to which these compounds contribute to species-level resistance and their spectrum of action remain poorly understood. Capsidiol, a defense related phytoalexin, is produced by several solanaceous plants including pepper and tobacco during microbial attack. Interestingly, capsidiol differentially affects growth and germination of the oomycete pathogens *Phytophthora infestans* and *Phytophthora capsici*, although the underlying molecular mechanisms remain unknown. In this study we revisited the differential effect of capsidiol on *P. infestans* and *P. capsici*, using highly pure capsidiol preparations obtained from yeast engineered to express the capsidiol biosynthetic pathway. Taking advantage of transgenic *Phytophthora* strains expressing fluorescent markers, we developed a fluorescence-based method to determine the differential effect of capsidiol on *Phytophtora* growth. Using these assays, we confirm major differences in capsidiol sensitivity between *P. infestans* and *P. capsici* and demonstrate that capsidiol alters the growth behaviour of both *Phytophthora* species. Finally, we report intraspecific variation within *P. infestans* isolates towards capsidiol tolerance pointing to an arms race between the plant and the pathogens in deployment of defence related phytoalexins.

Editor: Mark Gijzen, Agriculture and Agri-Food Canada, Canada

Funding: The authors acknowledge support from the Biotechnology and Biological Sciences Research Council Institute Strategic Programme (ISP) Grant BB/J004561/1 (Understanding and Exploiting Plant and Microbial Secondary Metabolism) at the John Innes Centre and ISP grant BB/I015345/1 (Food and Health) at the Institute of Food Research. Work at The Sainsbury laboratory was supported by the Gatsby Charitable Foundation (http://www.gatsby.org.uk). AG received support from the Norwich Research Park Ph.D. student fellowship and Onassis Foundation (http://www.onassis.gr/en/). The funders had no role in study design, data collection and analysis, decision to publish, or preparation of the manuscript.

Competing Interests: The authors have declared that no competing interests exist.

* Email: sophien.kamoun@tsl.ac.uk (SK); paul.o'maille@jic.ac.uk (PEO)

Introduction

Plants are exposed to a variety of disease causing organisms, including viruses, bacteria, fungi, oomycetes, nematodes, insects, and parasitic plants [1,2]. Yet, one concept in plant pathology is that in general plants are resistant to most pathogens. Plants have evolved a defense system that enables them to produce compounds that affect microbes in various ways. Some of these compounds are broad spectrum, whereas others are not. Among such defence compounds are phytoalexins, which are antimicrobial specialized metabolites that are induced under stress conditions or upon infection by a pathogen [3–5]. The spectrum of action of phytoalexins remains poorly understood and, surprisingly, their contribution to species-level (also known as nonhost) resistance is not always fully appreciated.

One well-studied phytoalexin is capsidiol, which is produced by the solanaceous plants *Capsicum annuum* (pepper) or *Nicotiana tabacum* (tobacco) after infection by pathogens such as the oomycete *Phytophthora capsici* [6,7]. Remarkably, capsidiol affects diverse pathogens such as fungi and oomycetes. [3–6,8–11]. Capsidiol is a bicyclic sesquiterpenoid compound and member of the isoprenoid class of phytoalexins. Like all sesquiterpenes, capsidiol derives from a common substrate farnesyl diphosphate (FPP) [12]. Two key enzymes are responsible for the biosynthesis of capsidiol. 5-*epi*-aristolochene synthase (EAS) catalyzes the cyclization of FPP to the intermediate 5-*epi*-aristolochene, then 5-*epi*-aristolochene dihydroxylase (EAH) mediates the two hydroxylation steps at positions C-1 and C-3 of 5-*epi*-aristolochene to yield capsidiol [13]. The dihydroxylase works in parallel with a cytochrome P450 reductase (CPR; NADPH-ferrihemoprotein reductase), which transfers electron equivalents for EAH reactions.

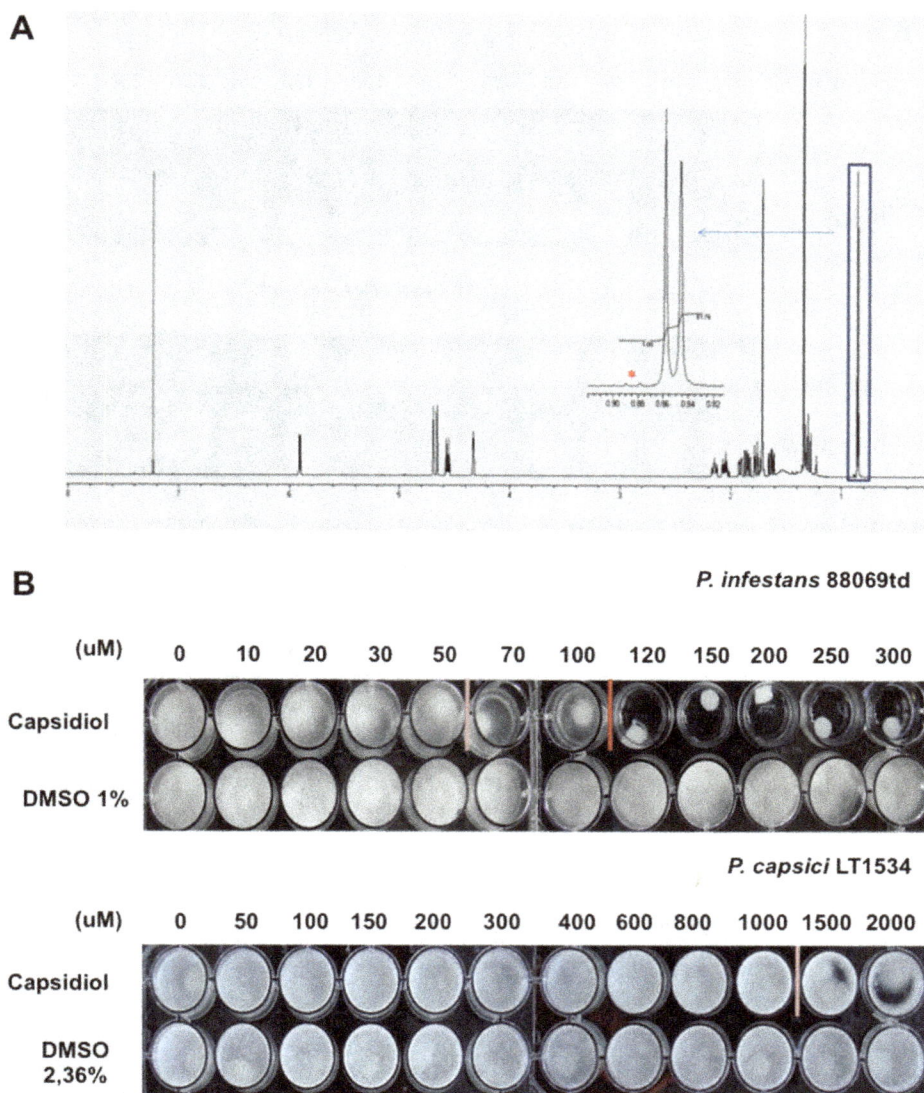

Figure 1. *P. infestans* **is more sensitive to capsidiol than** *P. capsici.* (A) Verification of capsidiol purity as tested by NMR spectroscopy (Nuclear Magnetic Resonance Spectroscopy). ^{1}H NMR (CDCl$_3$, 600 MHz) spectrum of capsidiol. NMR integrations of the diagnostic methyl doublet at δ_H 0.88 ppm (expansion) reveal a purity of greater than 98.8%. (*) Represents the impurity. (B) Growth inhibition assay of *P. infestans* and *P. capsici* after 10 days of exposure of mycelial plugs to capsidiol. Pink bar delineates the lowest concentration with an inhibitory effect and the red bar the concentration after which there is no longer growth. This experiment was performed 4 times.

The oomycete genus *Phytophthora* includes some of the most destructive plant pathogens [14]. Several *Phytophthora* spp. infect solanaceous plants, including important crops like potato, tomato and pepper. Two of the most notorious species are the potato and tomato late blight pathogen *Phytophthora infestans* and the vegetable pathogen *P. capsici*. Both *P. infestans* and *P. capsici* have emerged as model systems to study oomycete pathogens and they have been extensively studied at the genomic level [15–19]. These species follow a hemibiotrophic life style and adopt two separate phases during infection. In the early stage of infection, both pathogens need living host cells. This biotrophic phase is followed by extensive necrosis of host tissue (necrotrophic phase) [20]. The host range of *P. infestans* is limited to solanaceous plants, particularly potato and tomato, whereas *P. capsici* affects a wide range of hosts in the Cucurbitaceae, Fabaceae, and Solanaceae families [14]. Although these two *Phytophthora* species share a common host in tomato, *P. infestans* cannot infect several host plants of *P. capsici*, notably pepper. Nonhost resistance to *P. infestans* is associated with a plant localized cell death response also known as the hypersensitive response (HR) [21].

The molecular basis of host-specificity of *Phytophthora* species, such as *P. infestans* and *P. capsici* is unknown. Although disease resistance genes that operate at the nonhost level are likely to be implicated [22], early work has also suggested a role for phytoalexins. For example, in the 1970s, several studies have shown that capsidiol has differential activity against *P. infestans* and *P. capsici* [4,5]. Jones et al. showed that *P. infestans* is more sensitive (~10 fold) to capsidiol than *P. capsici*, both in spore germination and growth assays [4]. Jones et al. also showed that, below a certain threshold, capsidiol has a reversible effect on both *Phytophthora* species [4]. This level of capsidiol is only reached in vivo in pepper varieties that are resistant to *P. capsici*, which led

A

P. infestans 88069td

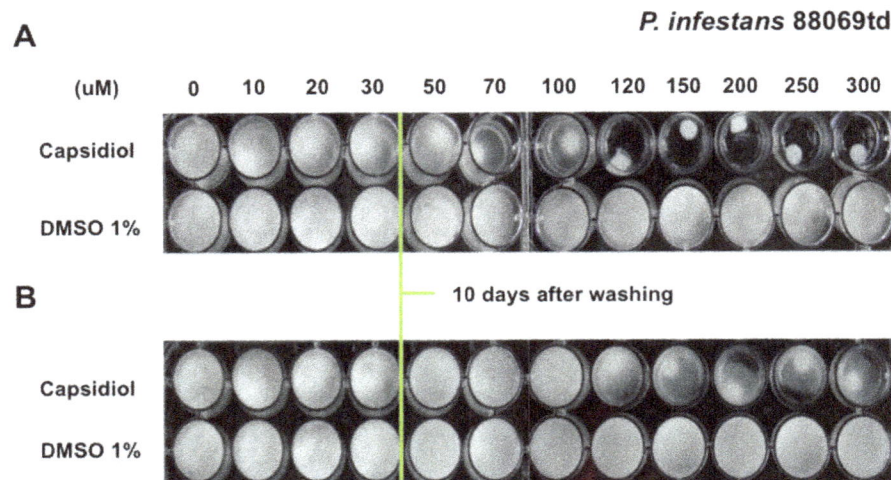

B

— 10 days after washing

Figure 2. Capsidiol inhibits *P. infestans* growth reversibly. (A) Growth inhibition assay of *P. infestans* after 10 days of exposure of mycelial plugs to capsidiol. (B) Restoration of growth after washing treatment. Green line indicates the point after which the washing treatment was applied. The experiment was performed 3 times. Picture was taken 10 days after the washing and 20 days after initial exposure to capsidiol.

the authors to suggest sensitivity to capsidiol and differential accumulation of this phytoalexin might determine host susceptibility [23]. Apart from these pioneering studies that date back to the 1970s, only few publications have examined the role of capsidiol in *Phytophthora* pathosystems except to use it as a marker for defense [6,8,10,24]. Nonetheless, Shibata et al. showed that silencing of *NbEAS* and *NbEAH*, two ethylene-regulated genes for capsidiol biosynthesis, negatively impact the resistance of *Nicotiana benthamiana* against *P. infestans* suggesting a positive role of capsidiol in this interaction [25,26].

In this study, we revisited the effect of capsidiol on *P. infestans* and *P. capsici*, and the variation in sensitivity to this phytoalexin. Compared to the earlier studies [4,11,23,27] we used highly pure preparations obtained from yeast engineered to express the capsidiol biosynthetic pathway [28]. We also assayed the effect of capsidiol on both mycelial growth and zoospores, using a novel fluorescence-based assay taking advantage of transgenic *Phytophthora* strains expressing fluorescent markers for biomass quantification. We confirmed major differences in capsidiol sensitivity between *P. infestans* and *P. capsici*. We also showed that capsidiol alters the growth behaviour of both *Phytophthora* species. Finally, we monitored the intraspecific variation within *P. infestans* isolates to capsidiol.

Results

P. infestans is more sensitive to capsidiol than *P. capsici*

To examine the effect of capsidiol on *Phytophthora* spp., we conducted inhibition assays using mycelial plugs of 2 to 3 week-old plates of *P. infestans* and *P. capsici*, which were placed in sterilised 26-well plates (Greiner Bio-one) in a rich medium, supplemented with varying concentrations of capsidiol. In our experiments, we used a metabolically engineered yeast system [28] to produce high purity capsidiol as shown by Nuclear Magnetic Resonance (NMR) Spectroscopy (Fig. 1A). Mycelial growth was assessed by visual inspection after 10 days of incubation of agar-grown mycelial plugs in capsidiol- or control-containing liquid medium at 20°C in the dark for *P. infestans* and 25°C and illumination for *P. capsici*. We observed reduced *P. infestans* growth at capsidiol concentrations of 50 μM or above and no growth was observed at concentrations of 120 μM and higher. *P. capsici* growth was affected at capsidiol

concentrations of 1.5 mM or higher, but was not fully inhibited in any of the tested capsidiol concentrations (Fig. 1B). Since capsidiol stock solution was dissolved in DMSO, we also tested whether DMSO affects mycelial growth of *Phytophthora*. We found, that DMSO did not affect *Phytophthora* growth at concentrations below 2.5% (v/v), which is equivalent to the highest relative DMSO concentration that was used during the experiment. In summary, our results confirm earlier indications that *P. capsici* displays a higher degree of resistance to capsidiol than *P. infestans*. However, in our hands complete growth inhibition of *P. infestans* was achieved with 120 μM capsidiol, a value 2 times less than previously reported [4].

Capsidiol arrest of *P. infestans* growth is reversible

It has been reported [4,23] that growth inhibiting effects of capsidiol are reversible at concentrations below 5 mM, while higher capsidiol concentrations are considered to be fungitoxic [23]. Following our plug inhibition assays, we studied the reversibility of capsidiol growth inhibition using the previously established *P. infestans* microtitre plate assay. The capsidiol-containing medium was removed from the wells and the mycelia were washed 3 times with deionised water, after which fresh liquid culture medium (Plich) was added. Growth restoration was observed 24 hours after washing and 10 days later the extent of mycelial growth was similar to the control that was grown without any capsidiol (Fig. 2B). This finding is consistent with reports that low capsidiol concentrations reversibly inhibit *Phytophthora* growth [4].

Quantitative evaluation of differential growth inhibition of *P. infestans* and *P. capsici* by capsidiol

In order to quantify the effect of capsidiol on the growth of *Phytophthora* strains, we developed and applied an inhibition assay with zoospore suspension solutions and measured the amount of growing mycelia using either optical density or emitted fluorescence of transgenic *Phytophthora* strains. For this experiment we used *P. infestans* 88069td, *P. infestans* 88069 [29], *P. capsici* LT1534 tdtom and *P. capsici* LT1534 [16] strains (td and tdtom strains are transgenic strains expressing the red fluorescent marker tandem dimer RFP, known as tdTomato). Zoospores were

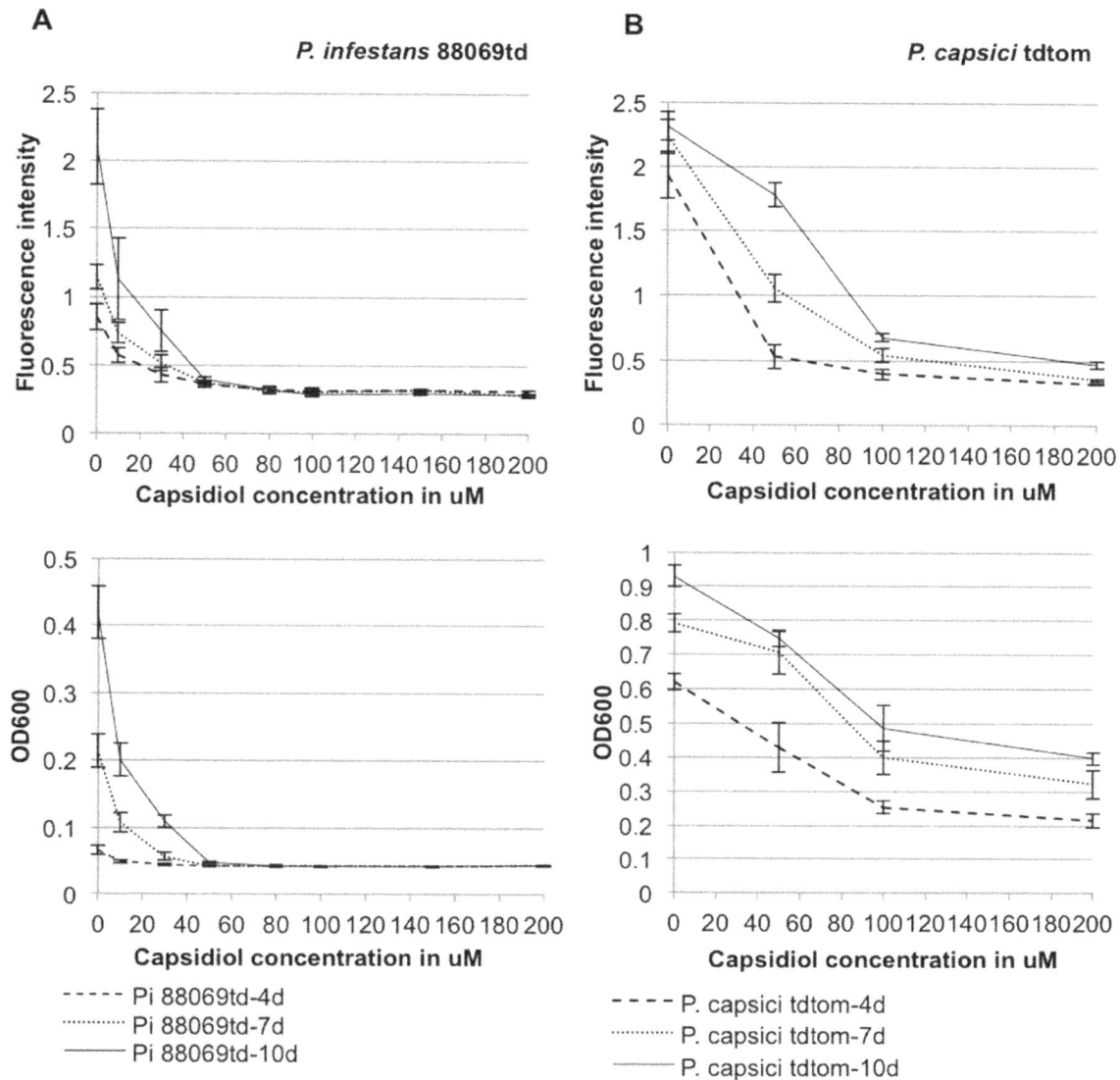

Figure 3. Capsidiol is not affecting *P. capsici* growth as severely as it does *P. infestans*. (A) Dose response curves of *P. infestans* 88069td calculated at 4, 7 and 10 days for both Fluorescence intensity and OD600. (B) Dose response curves of *P. capsici* tdtom calculated at 4, 7 and 10 days for both Fluorescence intensity and OD600.

harvested from *Phytophthora* plates and incubated with various concentrations of capsidiol in Plich medium in microtitre plates. The plates were scanned at 1 to 3 day intervals for OD600 (Optical Density at 600 nm) and red fluorescence intensity. Dose response curves were obtained by measuring both red fluorescence intensity and OD600 at increasing capsidiol concentrations to directly compare the difference in sensitivity between *P. infestans* and *P. capsici* (Fig. 3).

Results from measurements of red fluorescence intensity under capsidiol treatment, revealed statistically significant difference between *P. infestans* and *P. capsici* (Fig. 4). All concentrations of capsidiol above 50 μM dramatically affected the ability of *P. infestans* 88069td to emit red fluorescence. The given values were at a range of 0.3 red fluorescent units (RFU) after 10 days, close to the value obtained with the non-fluorescent 88069 strain (Fig. 4A and S1). On the contrary, *P. capsici* tdtom retained its ability to emit red fluorescence up to a concentration of 650 μM of

capsidiol, after which RFU levels dropped down to the non-fluorescent *P. capsici* strain values (Fig. 4B and S1).

The observed growth differences could also be reported using OD600 measurements (Fig. 5). Capsidiol concentrations of 50 μM or greater were detrimental for *P. infestans* growth, after which it was suppressed at OD600 levels lower than the control strain (Fig. 5A and S1). *P. capsici* growth was severely affected by capsidiol concentrations of 650 μM and higher, where the OD600 values were close to the ones of the control strain (Fig. 5B).

These results corroborate the findings that *P. capsici* is more resistant to capsidiol than *P. infestans* and revealed that the difference in sensitivity is almost 13 fold. DMSO did not affect the red fluorescence intensity or OD600 of any of the *Phytophthora* strains at concentrations below 2.36% (v/v), a value equivalent with the maximum capsidiol solution that was used during the experiment.

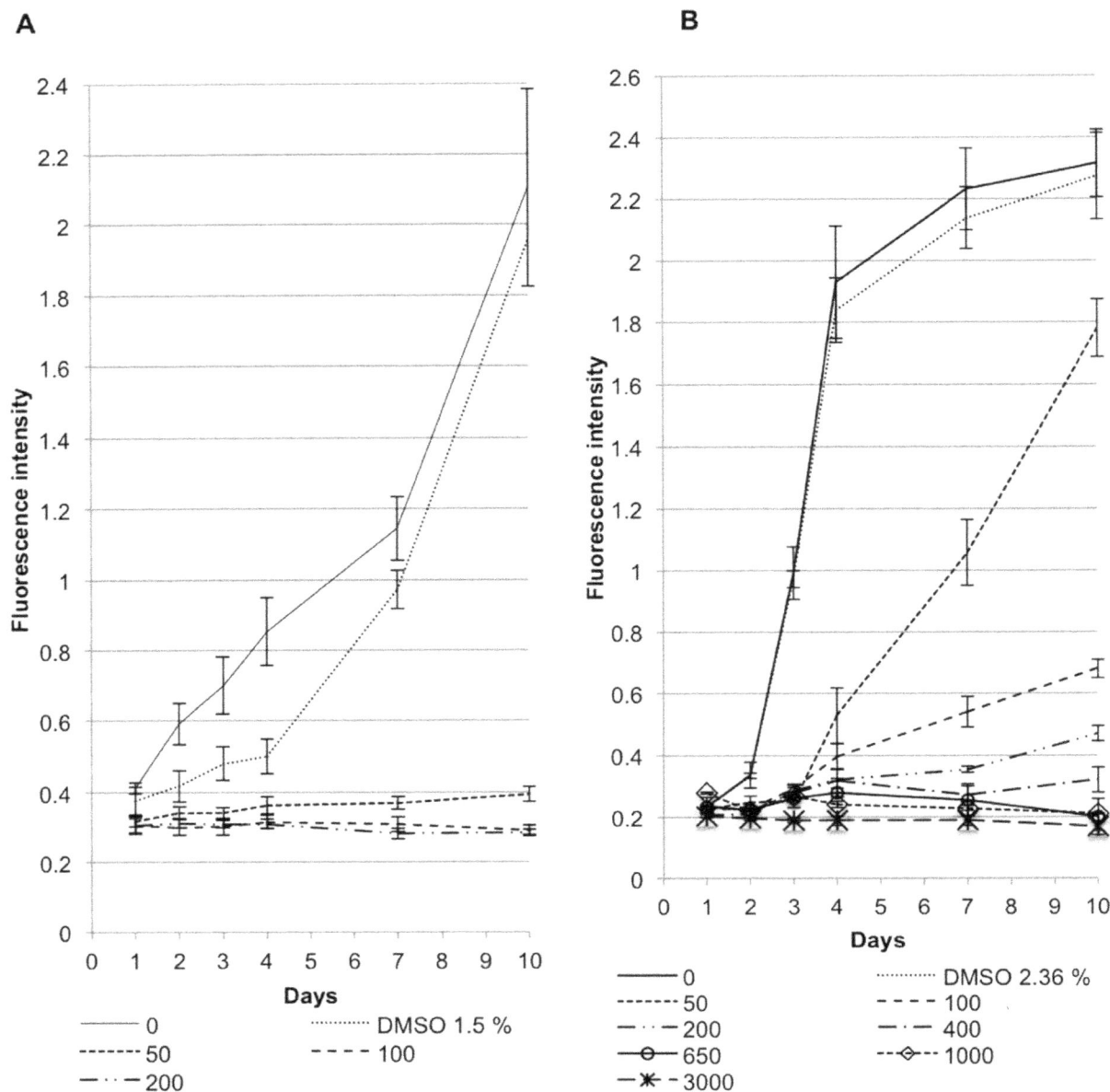

Figure 4. Scatter plots correlating fluorescence intensity and capsidiol concentration. The plots illustrate fluorescence intensity of *P. infestans* 88069td (A) and *P. capsici* tdtom (B) strains over time for a maximum of 10 days. The experiment was performed 3 times.

Capsidiol alters *P. infestans* and *P. capsici* mycelial growth

In order to study the effects of capsidiol-mediated inhibition of mycelial growth of *Phytophthora* we microscopically monitored the hyphal morphology during a capsidiol time course treatment at 2–4 day intervals in microtitre plates. When monitoring hyphal growth of *P. infestans* 88069td (Fig. 6) and *P. capsici* tdtom (Fig. 7) we observed that capsidiol alters *P. infestans* growth more severely and is effective at concentrations of 10 μM, whereas *P. capsici* remains unaffected. Stunted branching of *P. capsici* tdtom mycelia was observed at capsidiol concentrations of 400 μM. DMSO did not have any effect on growth for either *P. infestans* 88069td or *P. capsici* tdtom (Fig. S2). These results are in agreement with the limiting capsidiol concentrations obtained in zoospore inhibition assays for both species.

Variation in sensitivity to capsidiol among *P. infestans* isolates

To identify differences in sensitivity towards capsidiol between various *P. infestans* isolates, we conducted an experiment exposing mycelial plugs to various concentrations of capsidiol, as described above. For this experiment we used the following *P. infestans* isolates: 88069 [30], 88069td [31,32], T30-4 [15], 06_3928A [18], VK98014 [33], EC1-3527, EC1-3626, 2004_7804B [18], 2011_8410B [18] and NL08645 [15] (Table 1). We found that only one isolate, 06_3928A, displayed a similar level of resistance to capsidiol as our reference isolate, 88069, whereas the other isolates were more sensitive with isolate NL08645 being the most sensitive to capsidiol (Fig. 8). DMSO did not have any effect on *Phytophthora* growth in the concentrations used to dilute capsidiol. Our results support

A

B

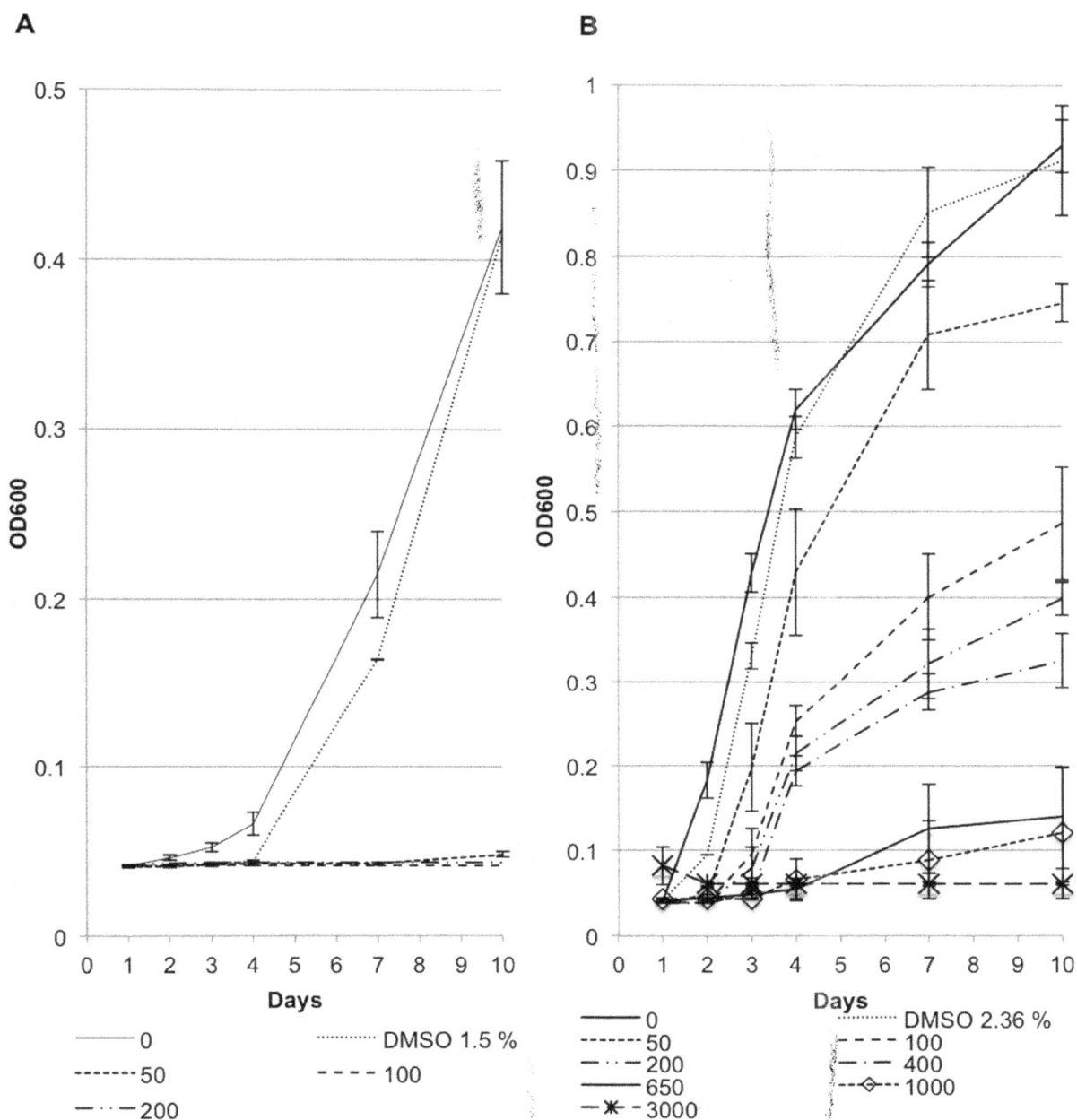

Figure 5. Scatter plots correlating 0D600 and capsidiol concentration. The plots illustrate growth of *P. infestans* 88069td (A) and *P. capsici* tdtom (B) strains over time for a maximum of 10 days. The experiment was performed 3 times.

strain-specific variation of *P. infestans* isolates to capsidiol growth inhibition, the genetic basis of which remains to be studied.

Discussion

In this work, we developed new assays to examine the effect of the phytoalexin capsidiol on two *Phytophthora* species with differing host ranges. Our results are overall consistent with a 1975 report [4] that *P. infestans* is more sensitive to capsidiol than *P. capsici* using highly pure preparations of capsidiol. A major (>10-fold) differential effect of capsidiol between species was noted using both mycelial and zoospore assays. Considering that this phytoalexin is produced by pepper but not potato, our findings are

consistent with the hypothesis that capsidiol contributes to nonhost resistance of pepper to *P. infestans*.

Previous studies used capsidiol purified from pepper fruits or tobacco cell cultures [4,11,23,27]. We used a recently developed method to produce highly pure capsidiol synthesized in yeast [28]. This reduced the likelihood that contaminating phytochemicals may have affected the experiments. It allowed us to directly test the effect of capsidiol on *Phytophthora* species and helped us to more accurately estimate the inhibitory doses of capsidiol on *Phytophthora* growth. Furthermore, we took advantage of fluorescently labelled *Phytophthora* strains to measure biomass and growth. Although our findings are consistent with the earlier studies, we could more accurately estimate the difference in

Capsidiol concentration in uM   *P. infestans* **88069td**

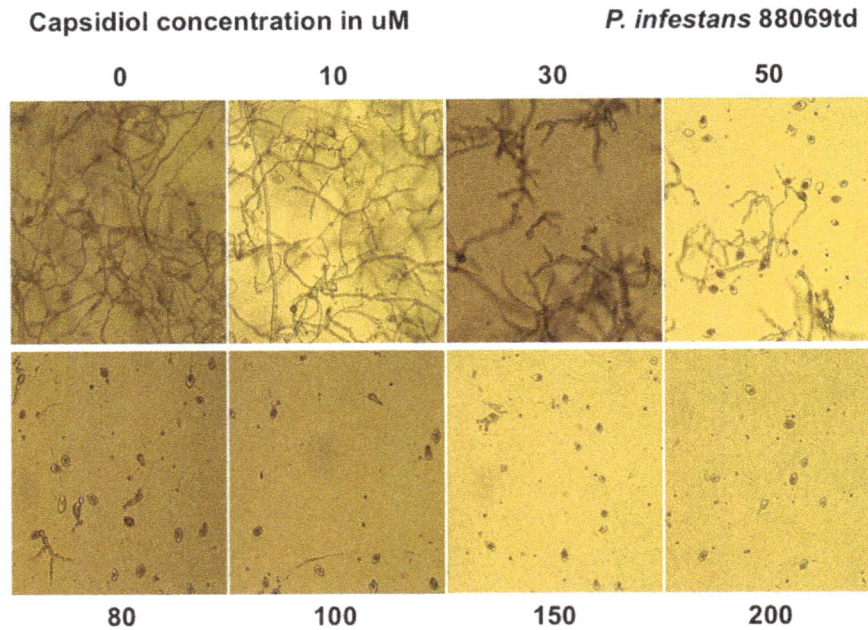

Figure 6. Growth behaviour of *P. infestans* **88069td, after 10 days of exposure to different capsidiol concentrations.** The experiment was performed 3 times.

sensitivity. We found that 120 µM of capsidiol completely inhibited growth of *P. infestans* both in mycelial and zoospore assays, whereas Jones et al. concluded that this effect started at 200 µM of capsidiol [4]. Furthermore, in our mycelial plug assays *P. capsici* was not completely inhibited even at a concentration of 2 mM capsidiol, whereas according to Jones et al. 1.5 mM is a completely inhibitory concentration [4]. However, these differences are probably due to the assays used. Our zoospore assays

were more consistent with the results of Jones et al. who concluded that capsidiol has a fungistatic effect at 3.75 mM and is fungitoxic at concentrations that exceed 5 mM [4,23]. We also found that the difference in sensitivity to capsidiol between the two *Phytophthora* species is approximately 13 fold, which is in agreement with earlier studies that showed *P. capsici* to be at least 10 times more resistant to capsidiol than *P. infestans* [4]. Finally we extended our studies and showed that the level of sensitivity between different *P.*

Capsidiol concentration in uM   *P. capsici* **tdtom**

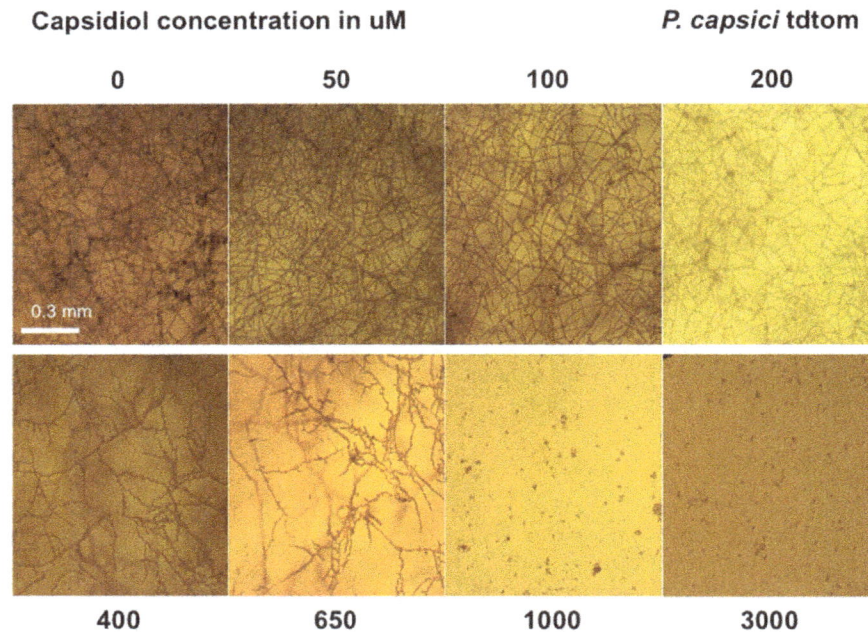

Figure 7. Growth behaviour of *P. capsici* **tdtom, after 10 days of exposure to different capsidiol concentrations.** The experiment was performed 3 times.

Table 1. Provenance of *Phytophthora* samples.

Isolate ID	Country of origin	Collection year	Host species	Reference
88069	The Netherlands	1988	*Solanum lycopersicum*	Van West et al. (1998)
88069td				Whisson SC et al. (2007)
T30–4				Haas et al. (2009)
06_3928A	United Kingdom	2006	*Solanum tuberosum*	Cooke, Cano et al. (2012)
VK98014	The Netherlands	1998	*Solanum tuberosum*	G. J. T. Kessel et al. (2012)
EC13527	Ecuador	2002	*Solanum andreanum*	World Oomycete Genetic Resource Collection at UC Riverside, CA
EC13626	Ecuador	2003	*Solanum tuberosum*	World Oomycete Genetic Resource Collection at UC Riverside, CA
2004_7804B	Scotland	2004	*Solanum tuberosum*	Cooke, Cano et. al, (2012)
2011_8410B	United Kingdom	2011	*Solanum tuberosum*	Cooke, Cano et. al, (2012)
NL08645	The Netherlands	2008	*Solanum venturii*	G. J. T. Kessel et al. (2012)

infestans isolates varies, providing a basis for studying the underlying genetic variation.

Earlier studies have showed that similar to capsidiol, other phytoalexins show a differential toxicity to phytopathogenic fungi. Hargreaves et al. [34] showed that the major phytoalexins from *Vicia faba* including isoflavoinoid medicarpin and wyerone acid had a greater impact on germ tubes produced by the necrotrophic fungus *Botrytis cinerea*, than *B. fabae*. They further highlighted a differential toxicity in wyerone derivatives than medicarpin [34]. Also, another study on the effect of the phytoalexins pisatin and maackiain from garden pea and red clover, respectively, against 19 fungal species revealed that nonhost phytoalexins have a greater effect inhibiting growth of the pathogens tested than the phytoalexins naturally occurring in the host [35]. These studies point out that differential activity of phytoalexins is a common phenomenon, and highlight the importance of understanding how different pathogens have evolved to cope with them.

What could be the nature of the differential effect of capsidiol on the two *Phytophthora* species? Ward and Stoessl [36] argued that *P. capsici* detoxification of capsidiol is unlikely and instead proposed that *P. capsici* does not induce high enough levels of capsidiol during infection of its host plant pepper [4,36,37]. Detoxification would probably involve oxidation of capsidiol to a less fungitoxic ketone, capsenone, as noted in *in vitro* assays with the fungi *Botryris cinerea* and *Fusarium spp.* [3,36]. Importantly, capsenone was not detected in pepper tissue infected with *P. capsici* indicating that the pathogen may evade the phytoalexin by limiting its induction [36]. Alternatively, ATP-binding cassette (ABC) transporters may be involved as an efflux pump. Coleman et al. showed that the rot causing ascomycete *Nectria haematococca* can overcome the effect of the pea phytoalexin pisatin using a specific ABC transporter, NhABC1, that enhances the fungus tolerance to the phytoalexin [38]. Since there is no evidence that *P. capsici* can detoxify capsidiol [36,37], *P. capsici* may rely on ABC transporters to cope with capsidiol. A more recent study on the role of ABC transporters in fungicide sensitivity in *P. infestans* failed to show correlation between up-regulation of ABC transporter genes in strains that are less sensitive to fungicides [39]. Whether inter- or intra-specific variation in expression of ABC transporter genes explains differences in capsidiol sensitivity in *Phytophthora* remains to be determined.

A genetic difference in the target of capsidiol could underpin the difference in sensitivity between the *P. capsici* and *P. infestans*. De Marino et al. [40] showed that capsidiol has a bacteriostatic effect against the human gastritis pathogen *Helicobacter pylori in vitro*

but the mode of action remains unknown. It would be interesting to identify the molecules that are targeted by capsidiol in *Phytophthora*. Given that the genome sequences of *P. capsici* and *P. infestans* are available [15,41,42], a promising approach would be to determine transcriptome dynamics in response to capsidiol. From an evolutionary perspective, it would be of great interest to examine the response of *Phytophthora* to other sesquiterpenes that emerged during the functional divergence of terpene synthases in solanaceous plants [43].

The differences in capsidiol sensitivity observed among various *P. infestans* isolates reflect the remarkable level of diversity noted in this highly adaptable plant pathogen species [18,21]. This variation is similar to what has been noted for sensitivity to fungicides in *P. infestans* and other oomycetes [44–46]. In some cases the genetic basis of chemical sensitivity has been identified. Randall et al. determined that sequence polymorphisms in the large subunit of RNA polymerase I (RPA190) contributes to *P. infestans* insensitivity to the oomycete-specific control chemical Mefenoxam [44]. Also, Blum et al. demonstrated that for two oomycetes, the causal agent of downy mildew in grape, *Plasmopara viticola* and *P. infestans*, an amino acid change in a protein known to be involved in cellulose biosynthesis (PiCESA3 and PvCESA3 in the two pathogens respectively) confers insensitivity to Mandipropamide [45,46].

We observed that sensitivity to capsidiol ranged ~5 fold in the *P. infestans* isolates tested. Is there a biological significance for these differences? Although potato does not produce capsidiol, it is possible that *P. infestans* has evolved mechanisms to tolerate other terpenoids produced by potato, which might contribute to host immunity. Indeed, potato is known to accumulate rishitin, another bicyclic sesquiterpene phytoalexin that is related to capsidiol [22]. In the future, it would be interesting to examine whether there is any correlation between aggressiveness and tolerance to capsidiol among various *P. infestans* isolates.

Finally, our work points to a biotechnological approach to engineer resistance to *P. infestans*. Genetic manipulation of capsidiol production in *Nicotiana benthamiana*, a *P. infestans* host plant that produces capsidiol, has already indicated that this phytoalexin contributes to disease resistance [25,26]. Interestingly, *P. capsici* is markedly more aggressive pathogen of *N. benthamiana* than *P. infestans* [29], possibly because it can tolerate the capsidiol produced by this plant. Ultimately, capsidiol biosynthetic genes could be transferred from pepper or tobacco to potato and tomato as a potential strategy for disease resistance against *P. infestans*.

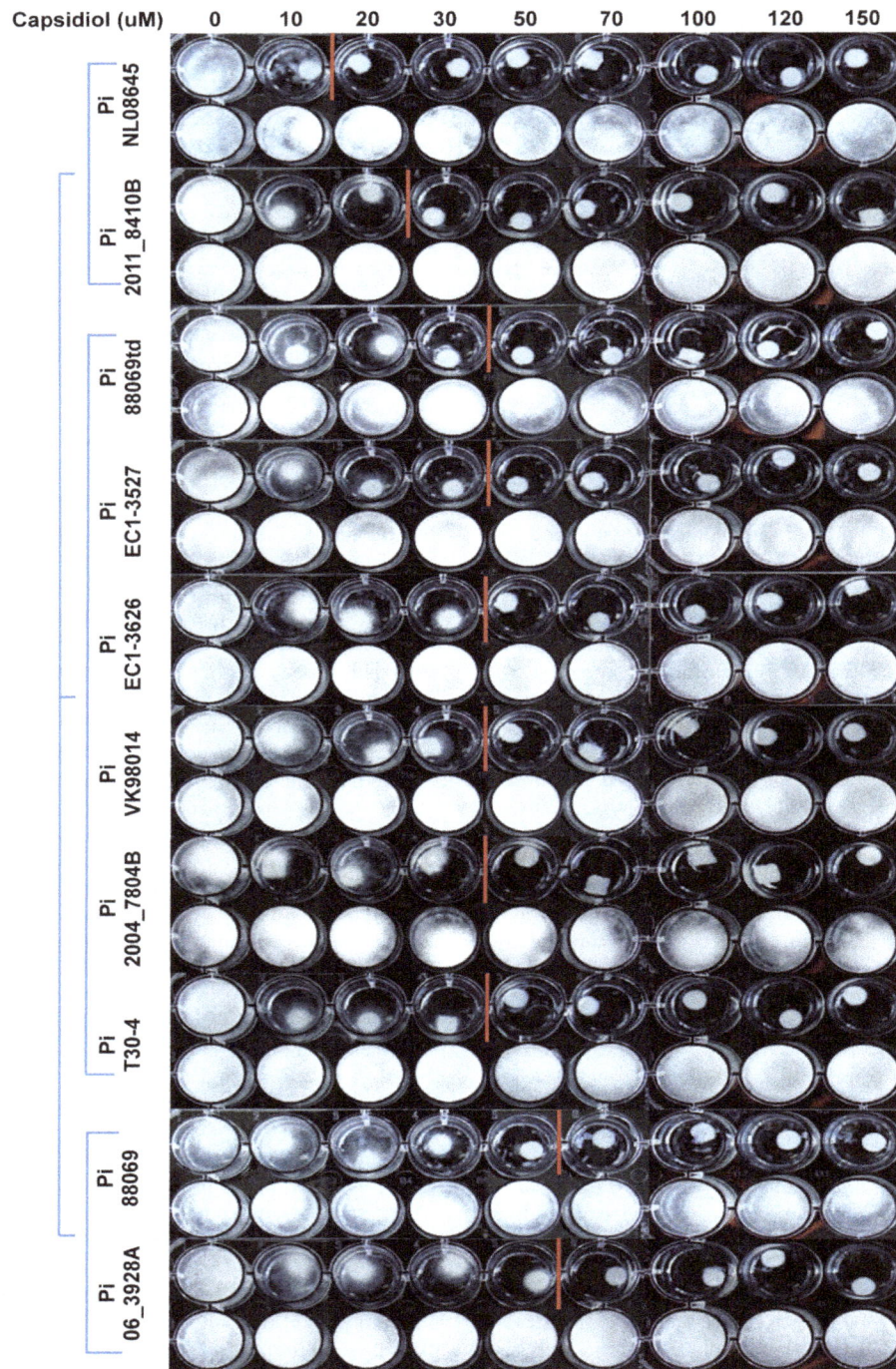

Figure 8. Different *P. infestans* isolates have different sensitivity to capsidiol. Isolates are clustered according to their sensitivity, starting from the most sensitive to the least. Top row of wells of each isolate represents capsidiol treatment (capsidiol was dissolved in DMSO/Plich media) in µM and the lower row represents treatment with 1% (v/v) DMSO/Plich (negative control). The experiment was performed 3 times.

Materials and Methods

Yeast growth

The yeast strain EPY300 was engineered to express the capsidiol biosynthetic pathway [28] and was used to produce capsidiol by fermentation. In brief a starter culture (ca. 20 ml) was prepared and inoculated into a 5 L-bioreactor containing rich media to full capacity. The media consisted in 1% Bacto yeast extract, 2% Bacto peptone (BD Biosciences, Oxford, UK), 1.8% galactose, 0.2% glucose, 150 mg/L methionine and 80 mg/L adenine hemisulphate (Sigma Aldrich Co Lt, Dorset, UK). The bioreactor was set to 30°C, with constant stirring (180 rmp) and aeration at 4 L/min. After 96 hours both stirring and aeration were stopped, and the temperature was reduced to 5°C. Once yeast cells had settled (24–48 h), the media containing the yeast-produced capsidiol was decanted for extraction.

Capsidiol extraction and purification

Capsidiol was isolated by dichloromethane extractions of the media. The combined extracts (eg 5 L total volume) were dried, filtered and evaporated to dryness using a rotary evaporator. The crude extract (ca. 1,500 mg) was re-dissolved in a minimum volume of 1:1 hexane/ethyl acetate (EA) and applied to a glass sinter funnel 40 mm×40 mm containing silica gel (previously equilibrated with hexane), and connected to an on-house pump. Purification of capsidiol was accomplished by vacuum filtration using a gradient (0–66%) of ethyl acetate/hexane. Each fraction (25 mL) was assessed for capsidiol ($R_f = 0.363$) content by analytical TLC (Merck silica gel 60 (F_{254}) 7×7 cm aluminium-coated plates), using 66% ethyl acetate/hexane as the developing solvent and visualization with CAM solution (cerium ammonium molybdate). Fractions judged to contain exclusively capsidiol were combined together and evaporated under reduced pressure. Further (and final) purification of this material was effected by preparative TLC (Merck silica gel 60 (F_{254}) 20×20 cm glass-coated plates) previously sprayed with a 0.5% berberine chloride ethanolic solution (non-destructive visualization of capsidiol by UV at 365 nm). Briefly, the silica gel TLC plates were divided horizontally in two halves by removing a thing line of silica coating and the sample (containing around 25 mg of product) was placed in a continuous line 1 cm above the bottom of the plate. After loading, the band was 'focused/concentrated' twice by standing the plate in a tank containing pure EA until the solvent front reached 2 cm. After air-drying, the plate was finally developed using 66% ethyl acetate/hexane. The band corresponding to capsidiol ($R_f = 0.363$) after visualization by UV light (365 nm) was marked and scraped off avoiding the very bottom of the band, which was shown to contain an as yet unidentified more polar terpene compound. The silica gel scrapings were loaded into a pipette-column and washed using ethyl acetate.

Typically we found that 5 L fermentation yielded around 1,500 mg of crude extract, which is reduced to about 700 mg after silica column, to produce 300–400 mg of essentially pure capsidiol (>97% by 1H-NMR) after preparative TLC.

Identification of capsidiol

The unambiguous identification and purity estimation of the yeast-produced capsidiol was carried out by NMR spectroscopy and combined liquid chromatography-mass spectroscopy (LC/MS) following the method of Literakova et al. [27] with slight modifications. In brief, we used an isocratic 75% methanol: water solvent mixture, in a C8 reverse column (Agilent 1100 MSD) with negative mode TIC and SIM at m/z 201, 219, 259.

Phytophthora cultivation

Phytophthora strains were grown on rye sucrose agar as previously described [47] at 20°C in the dark (*P. infestans*) or on V8 vegetable juice agar [7] plates (*P. capsici*) at 25°C and illumination. For the plug inhibition assays, 5 mm diameter plugs were taken from 2–3 week-old *Phytophthora* plates and placed in the wells of a 24-well plate, previously filled with 1 ml of Plich medium [2.4 gr sucrose, 0.27 gr asparagine, 0.15 g KH_2PO_4, 0.10 gr $MgSO_4$ $7H_2O$, 10 mg cholesterol, 10 mg ascorbic acid, 2 mg

thiamine HCl, 4.4 mg $ZnSO_4$ $7H_2O$, 1 mg $FeSO_4$ $7H_2O$, 0.07 mg $MnCl_2$ $4H_2O$ and 20 g agar (Difco) dissolved in 1 L deionized water [48]. For the zoospores inhibition assays, spores were harvested as previously described [47,49] and diluted to 50,000 spores/ml. Droplets of 10 μl were added to each well of a 96-well plate, previously filled with 250 μl of Plich medium. Plates were kept at 20°C in the dark and 25°C and illumination for *P. infestans* and *P. capsici* respectively. Washes were applied to the plates containing the *Phytophthora* plugs by carefully removing the Plich media from the wells, adding distilled water, expose for 1 to 2 minutes and remove. This step was repeated at least 2 times. Finally 1 ml of fresh Plich media was added and plates were kept at 20°C in the dark (*P. infestans*) and 25°C and illumination for *P. capsici*.

Spectroscopic growth assays

For the zoospore inhibition assays, 10 μl zoospore solution per well was distributed into 96-well microtitre plates (Greiner bio-one), covered with a plastic lid (Greiner bio-one), sealed with Parafilm (Pechiney Plastic Packaging Company) and incubated at 25°C in the dark for *P. infestans* and at 25°C and illumination for *P. capsici*, over 10 days. At regular intervals, mycelial growth was monitored using a Varioscan Flash Multimode Reader (Thermo Scientific) by measuring light absorption at OD600 as well as emission of red fluorescence (excitation at 360 nm, emission at 465 nm).

Light microscopy

Mycelia grown in 96-well microtitre plates were imaged using a Zeiss Axiovert 25 microscope in transmission light mode with 10x magnification. Pictures were taken using a Cannon E0S-D30 camera.

Supporting Information

Figure S1 Fluorescence intensity of the non-fluorescent stains *P. infestans* 88069 and *P. capsici* LT1534. These strains were used as controls to verify that the signal in the fluorescent strains corresponds to fluorescence.

Figure S2 Growth behaviour of *P. infestans* 88069 and *P. capsici* LT1534 after exposure to DMSO. Both Phytophthora strains were exposed to 1.5% and 2.36% (v/v) DMSO/Plich for 10 days. DMSO levels correspond to the maximum capsidiol solution that was used in each experiment. The experiment was performed 3 times.

Acknowledgments

We thank Angela Chaparro-Garcia, Marina Pais, Liliana Cano, Melissa Dokarry, Caroline Laurendon for useful discussions and Steve Whisson for providing the 88069td strain.

Author Contributions

Conceived and designed the experiments: AG SS SK PEO. Performed the experiments: AG DH JAF SS. Analyzed the data: AG DH JAF SS. Contributed reagents/materials/analysis tools: DH JAF DKR. Contributed to the writing of the manuscript: AG SS TOB DH JAF DKR SK PEO.

References

1. Westwood JH, Yoder JI, Timko MP, dePamphilis CW (2010) The evolution of parasitism in plants. Trends in Plant Science 15: 227–235.
2. Agrios GN (2005) Plant pathology: Elsevier Academic Press. 922 p.

3. Stoessl A, Unwin CH, Ward EWB (1973) Postinfectional fungus inhibitors from plants - Fungal Oxidation of Capsidiol in Pepper Fruit. Phytopathology 63: 1225–1231.

4. Jones DR, Unwin CH, Ward EWB (1975) Significance of Capsidiol Induction in Pepper Fruit during an Incompatible interaction with Phytophthora-Infestans. Phytopathology 65: 1286–1288.

5. Ward EWB (1976) Capsidiol production in pepper leaves in incompatible interactions with fungi. Phytopathology 66: 175–176.

6. Maldonado-Bonilla LD, Betancourt-Jimenez M, Lozoya-Gloria E (2008) Local and systemic gene expression of sesquiterpene phytoalexin biosynthetic enzymes in plant leaves. European Journal of Plant Pathology 121: 439–449.

7. Huitema E, Smoker M, Kamoun S (2011) A straightforward protocol for electro-transformation of Phytophthora capsici zoospores. Methods in molecular biology (Clifton, NJ) 712: 129–135.

8. Milat ML, Ducruet JM, Ricci P, Marty F, Blein JP (1991) physiological and structural-changes in tobacco-leaves treated with cryptogein, a proteinaceous elicitor from phytophthora-cryptogea. Phytopathology 81: 1364–1368.

9. Grosskinsky DK, Naseem M, Abdelmohsen UR, Plickert N, Engelke T, et al. (2011) Cytokinins Mediate Resistance against Pseudomonas syringae in Tobacco through Increased Antimicrobial Phytoalexin Synthesis Independent of Salicylic Acid Signaling. Plant Physiology 157: 815–830.

10. Keller H, Czernic P, Ponchet M, Ducrot PH, Back K, et al. (1998) Sesquiterpene cyclase is not a determining factor for elicitor- and pathogen-induced capsidiol accumulation in tobacco. Planta 205: 467–476.

11. Ma CJ (2008) Cellulase elicitor induced accumulation of capsidiol in Capsicum annumm L. suspension cultures. Biotechnology Letters 30: 961–965.

12. Cane DE (1990) Enzymatic formation of sesquiterpenes. Chemical Reviews 90: 1089–1103.

13. Ralston L, Kwon ST, Schoenbeck M, Ralston J, Schenk DJ, et al. (2001) Cloning, heterologous expression, and functional characterization of 5-epi-aristolochene-1,3-dihydroxylase from tobacco (Nicotiana tabacum). Archives of Biochemistry and Biophysics 393: 222–235.

14. Kamoun KLaS (2009) Oomycete Genetics and Genomics: Diversity, INteractions and Research tools: WILLEY-BLACKWELL.

15. Haas BJ, Kamoun S, Zody MC, Jiang RHY, Handsaker RE, et al. (2009) Genome sequence and analysis of the Irish potato famine pathogen Phytophthora infestans. Nature 461: 393–398.

16. Jupe J, Stam R, Howden AJM, Morris JA, Zhang R, et al. (2013) Phytophthora capsici-tomato interaction features dramatic shifts in gene expression associated with a hemi-biotrophic lifestyle. Genome Biology 14.

17. Raffaele S, Farrer RA, Cano LM, Studholme DJ, MacLean D, et al. (2010) Genome Evolution Following Host Jumps in the Irish Potato Famine Pathogen Lineage. Science 330: 1540–1543.

18. Cooke DEL, Cano LM, Raffaele S, Bain RA, Cooke LR, et al. (2012) Genome analyses of an aggressive and invasive lineage of the irish potato famine pathogen. PLoS pathogens 8: e1002940–e1002940.

19. Pais M, Win J, Yoshida K, Etherington GJ, Cano LM, et al. (2013) From pathogen genomes to host plant processes: the power of plant parasitic oomycetes. Genome Biology 14.

20. Kamoun S, Smart CD (2005) Late blight of potato and tomato in the genomics era. Plant Disease 89: 692–699.

21. Vleeshouwers V, van Dooijeweert W, Govers F, Kamoun S, Colon LT (2000) The hypersensitive response is associated with host and nonhost resistance to Phytophthora infestans. Planta 210: 853–864.

22. Kamoun S, Huitema E, Vleeshouwers V (1999) Resistance to oomycetes: a general role for the hypersensitive response? Trends in Plant Science 4: 196–200.

23. Egea C, Alcazar MD, Candela ME (1996) Capsidiol: Its role in the resistance of Capsicum annuum to Phytophthora capsici. Physiologia Plantarum 98: 737–742.

24. Ahmed Sid A, Perez Sanchez C, Emilia Candela M (2000) Evaluation of induction of systemic resistance in pepper plants (Capsicum annuum) to Phytophthora capsici using Trichoderma harzianum and its relation with capsidiol accumulation. European Journal of Plant Pathology 106: 817–824.

25. Shibata Y, Kawakita K, Takemoto D (2010) Age-Related Resistance of Nicotiana benthamiana Against Hemibiotrophic Pathogen Phytophthora infestans Requires Both Ethylene- and Salicylic Acid-Mediated Signaling Pathways. Molecular Plant-Microbe Interactions 23: 1130–1142.

26. Matsukawa M, Shibata Y, Ohtsu M, Mizutani A, Mori H, et al. (2013) Nicotiana benthamiana Calreticulin 3a Is Required for the Ethylene-Mediated Production of Phytoalexins and Disease Resistance Against Oomycete Pathogen Phytophthora infestans. Molecular plant-microbe interactions : MPMI 26: 880–892.

27. Literakova P, Lochman J, Zdrahal Z, Prokop Z, Mikes V, et al. (2010) Determination of Capsidiol in Tobacco Cells Culture by HPLC. Journal of Chromatographic Science 48: 436–440.

28. Trinh-Don N, MacNevin G, Ro D-K (2012) De Novo Synthesis of High-Value Plant Sesquiterpenoids in Yeast. In: Hopwood DA, editor. Natural Product Biosynthesis by Microorganisms and Plants, Pt C. 261–278.

29. Chaparro-Garcia A, Wilkinson RC, Gimenez-Ibanez S, Findlay K, Coffey MD, et al. (2011) The Receptor-Like Kinase SERK3/BAK1 Is Required for Basal Resistance against the Late Blight Pathogen Phytophthora infestans in Nicotiana benthamiana. Plos One 6.

30. van West P, de Jong AJ, Judelson HS, Emons AMC, Govers F (1998) The ipiO gene of Phytophthora infestans is highly expressed in invading hyphae during infection. Fungal Genetics and Biology 23: 126–138.

31. Bozkurt TO, Schornack S, Win J, Shindo T, Ilyas M, et al. (2011) Phytophthora infestans effector AVRblb2 prevents secretion of a plant immune protease at the haustorial interface. Proceedings of the National Academy of Sciences 108: 20832–20837.

32. Whisson SC, Boevink PC, Moleleki L, Avrova AO, Morales JG, et al. (2007) A translocation signal for delivery of oomycete effector proteins into host plant cells. Nature 450: 115–+.

33. Li Y, van der Lee TAJ, Evenhuis A, van den Bosch GBM, van Bekkum PJ, et al. (2012) Population Dynamics of Phytophthora infestans in the Netherlands Reveals Expansion and Spread of Dominant Clonal Lineages and Virulence in Sexual Offspring. G3-Genes Genomes Genetics 2: 1529–1540.

34. Hargreaves JA, Mansfield JW, Rossall S (1977) Changes in phytoalexin concentrations in tissues of broad bean plant (Vicia-Faba L) following inoculation with species of botrytis. Physiological Plant Pathology 11: 227–&.

35. Delserone LM, Matthews DE, Vanetten HD (1992) Differential toxicity of enantiomers of maackiain and pisatin to phytopathogenic fungi. Phytochemistry 31: 3813–3819.

36. Ward EWB, Stoessl A (1972) Postinfectional inhibitors from plants.3. detoxification of capsidiol, an antifungal compound from peppers. Phytopathology 62: 1186–1187.

37. Jones DR, Unwin CH, Ward EWB (1975) Postinfectional inhibitors from plants. 21. capsidiol induction in pepper fruit during interactions with phytophthora-capsici and monilinia-fructicola. Phytopathology 65: 1417–1419.

38. Coleman JJ, White GJ, Rodriguez-Carres M, VanEtten HD (2011) An ABC Transporter and a Cytochrome P450 of Nectria haematococca MPVI Are Virulence Factors on Pea and Are the Major Tolerance Mechanisms to the Phytoalexin Pisatin. Molecular Plant-Microbe Interactions 24: 368–376.

39. Judelson HS, Senthil G (2006) Investigating the role of ABC transporters in multifungicide insensitivity in Phytophthora infestans. Molecular Plant Pathology 7: 17–29.

40. De Marino S, Borbone N, Gala F, Zollo F, Fico G, et al. (2006) New constituents of sweet Capsicum annuum L. fruits and evaluation of their biological activity. Journal of Agricultural and Food Chemistry 54: 7508–7516.

41. Lamour KH, Mudge J, Gobena D, Hurtado-Gonzales OP, Schmutz J, et al. (2012) Genome Sequencing and Mapping Reveal Loss of Heterozygosity as a Mechanism for Rapid Adaptation in the Vegetable Pathogen Phytophthora capsici. Molecular Plant-Microbe Interactions 25: 1350–1360.

42. Cooke DEL, Cano LM, Raffaele S, Bain RA, Cooke LR, et al. (2012) Genome Analyses of an Aggressive and Invasive Lineage of the Irish Potato Famine Pathogen. Plos Pathogens 8.

43. O'Maille PE, Malone A, Dellas N, Hess BA Jr, Smentek L, et al. (2008) Quantitative exploration of the catalytic landscape separating divergent plant sesquiterpene synthases. Nature Chemical Biology 4: 617–623.

44. Randall E, Young V, Sierotzki H, Scalliet G, Birch PRJ, et al. (2014) Sequence diversity in the large subunit of RNA polymerase I contributes to Mefenoxam insensitivity to Phytophthora infestans. Molecular Plant Pathology.

45. Blum M, Boehler M, Randall E, Young V, Csukai M, et al. (2010) Mandipropamid targets the cellulose synthase-like PiCesA3 to inhibit cell wall biosynthesis in the oomycete plant pathogen, Phytophthora infestans. Molecular Plant Pathology 11: 227–243.

46. Blum M, Waldner M, Gisi U (2010) A single point mutation in the novel PvCesA3 gene confers resistance to the carboxylic acid amide fungicide mandipropamid in Plasmopara viticola. Fungal Genetics and Biology 47: 499–510.

47. Kamoun S, van West P, Vleeshouwers V, de Groot KE, Govers F (1998) Resistance of Nicotiana benthamiana to Phytophthora infestans is mediated by the recognition of the elicitor protein INF1. Plant Cell 10: 1413–1425.

48. Hoitink HAJ, Schmitthenner AF (1969) Rhododendron wilt caused by phytophthora citricola. Phytopathology 59: 708–&.

49. Schornack S, van Damme M, Bozkurt TO, Cano LM, Smoker M, et al. (2010) Ancient class of translocated oomycete effectors targets the host nucleus. Proceedings of the National Academy of Sciences of the United States of America 107: 17421–17426.

14

Strong Discrepancies between Local Temperature Mapping and Interpolated Climatic Grids in Tropical Mountainous Agricultural Landscapes

Emile Faye[1,2,3]*, **Mario Herrera**[3], **Lucio Bellomo**[4], **Jean-François Silvain**[1], **Olivier Dangles**[1,3,5]*

1 Institut de Recherche pour le Développement (IRD), UR 072, Laboratoire Evolution, Génomes et Spéciation, UPR 9034, Centre National de la Recherche Scientifique (CNRS), Gif sur Yvette, France et Université Paris-Sud 11, Orsay, France, **2** UPMC Univ Paris06, Sorbonne Universités, Paris, France, **3** Facultad de Ciencias Exactas y Naturales, Pontificia Universidad Católica del Ecuador, Quito, Ecuador, **4** Mediterranean Institute of Oceanography (MIO) CNRS/INSU, IRD, UM 110, Université de Toulon, La Garde, France, **5** Instituto de Ecología, Universidad Mayor San Andrés, Cotacota, La Paz, Bolivia

Abstract

Bridging the gap between the predictions of coarse-scale climate models and the fine-scale climatic reality of species is a key issue of climate change biology research. While it is now well known that most organisms do not experience the climatic conditions recorded at weather stations, there is little information on the discrepancies between microclimates and global interpolated temperatures used in species distribution models, and their consequences for organisms' performance. To address this issue, we examined the fine-scale spatiotemporal heterogeneity in air, crop canopy and soil temperatures of agricultural landscapes in the Ecuadorian Andes and compared them to predictions of global interpolated climatic grids. Temperature time-series were measured in air, canopy and soil for 108 localities at three altitudes and analysed using Fourier transform. Discrepancies between local temperatures vs. global interpolated grids and their implications for pest performance were then mapped and analysed using GIS statistical toolbox. Our results showed that global interpolated predictions over-estimate by 77.5±10% and under-estimate by 82.1±12% local minimum and maximum air temperatures recorded in the studied grid. Additional modifications of local air temperatures were due to the thermal buffering of plant canopies (from −2.7°K during daytime to 1.3°K during night-time) and soils (from −4.9°K during daytime to 6.7°K during night-time) with a significant effect of crop phenology on the buffer effect. This discrepancies between interpolated and local temperatures strongly affected predictions of the performance of an ectothermic crop pest as interpolated temperatures predicted pest growth rates 2.3–4.3 times lower than those predicted by local temperatures. This study provides quantitative information on the limitation of coarse-scale climate data to capture the reality of the climatic environment experienced by living organisms. In highly heterogeneous region such as tropical mountains, caution should therefore be taken when using global models to infer local-scale biological processes.

Editor: Michael Sears, Clemson University, United States of America

Funding: This work was partly conducted within the project "Adaptive management in insect pest control in thermally heterogeneous agricultural landscapes" (ANR-12-JSV7-0013-01) funded by the Agence Nationale pour la Recherche (ANR, http://www.agence-nationale-recherche.fr/). A financial support of the McKnight Foundation (http://www.mcknight.org/) to EF during the fieldwork of this study is greatly acknowledged. The funders had no role in study design, data collection and analysis, decision to publish, or preparation of the manuscript.

Competing Interests: The authors have declared that no competing interests exist.

* Email: ehfaye@gmail.com (EF); olivier.dangles@ird.fr (OD)

Introduction

Bridging the gap between the predictions of coarse-scale climate models and the fine-scale climatic reality of species is increasingly recognized as a key issue of climate change biology research [1,2,3,4]. Despite decades of study on microclimates [5,6,7,8] and evidence for habitat-related and topographical variations in local temperatures and their relevance for species ecology [2,9,10,11,12,13], most attempts to understand and model species distributions still do not integrate spatially-explicit fine-scale climatic data (e.g. [14,15,16]). Many work use global model of temperature interpolation to examine species vulnerability to climate change and, doing so, ignore the critical issue of habitat complexity in climate buffering [4,5,17]. Indeed, climate surfaces used in species distribution models (SDMs) are rarely generated or interpolated to a resolution finer than 1 km^2 (e.g. WorldClim database), a resolution that is still very coarse relative to the home ranges or body size of most species [13,18]. For instance, [8] showed that climate grid lengths used in SDMs are, on average, ~10,000-fold larger than studied animals, and ~1,000-fold larger than studied plants. Their meta-analysis showed that the WorldClim was the most widely used climatic dataset in global SDMs. As this commonly used coarse scale climatic data in SDMs overlook the spatiotemporal thermal heterogeneity experienced by organisms, there is an urgent need for a more sophisticated use of these datasets for making inferences about biological processes that are driven by hour to hour operative temperatures of organisms.

An important yet poorly studied issue in climate change biology is to quantify to what extent climatic conditions differ between widely used 1 km^2 interpolated grid cells of global climatic database and real-world landscapes of similar areas. While it is now well-known that most organisms, especially tiny ectotherms such as insects and other arthropods, do not experience the climatic conditions recorded at weather stations [9,12,18], there is little quantitative information on the spatial and temporal heterogeneity at the landscape scale of local climatic conditions (i.e. conditions at biologically relevant scales, e.g., from cm to km for insects) and their consequences for organisms' performance. A better quantification of the climatic conditions of ecologically-relevant habitats over relatively large landscape scales (e.g., 1 km^2) is therefore a necessary first step to better incorporate dynamical microclimate into global distribution models.

Here, we investigate the sources of variance between global interpolated and local temperatures by examining 1) how well WorldClim predicts local air temperatures in our study region (the tropical Andes), 2) to what extent temperatures in crop canopies and soils differ from local air temperatures, and 3) how relevant is to use WorldClim to infer the potential performance of an insect crop pest. Addressing these questions is not an easy task as the mosaic of climatic habitats relevant for small ectothermic species at a 1-km^2 scale in real-world landscapes may be outstandingly complex. In this study, we focused on highland agricultural landscapes of the tropical Andes as most prior similar data came from low elevation and temperate agroecosystems. In such systems, most crop pests experience, over their entire life cycle, climatic conditions in three well-defined environmental layers (air, air inside-canopy and soil) and these conditions are remarkably stable over the year [19]. In this context, we firstly decided to map over replicated 1-km^2 climatic grid cells the ecologically relevant local temperatures for ectothermic crop pests in agricultural landscapes, and to compare these maps to interpolated temperature grid cells of the widely used WorldClim database. We used Fourier analysis applied to local temperature time-series as a tool to fit daily variations of temperature and to feature microclimate discrepancies in space and in time (both in terms of amplitude and phase). We then explored the implication of our thermal landscape mapping for pest performance by comparing temperature frequencies in our grid cells with the temperature-dependent growth curve of the potato tuber moth (*Phthorimaea operculella*) a major crop pest species in the region and worldwide.

Materials and Methods

1. Study area

The Ecuadorian Andes are characterized by a low seasonality, with mean temperatures varying more within days (up to 30°K variation) than within months and years (less than 0.6°K and 0.2°K variations, respectively, see [19]). This region exhibits a marked altitudinal gradient in temperatures (between 2000 and 4000 m) with mean monthly air temperature roughly decreasing by 0.6°K every 100 m of elevation [20]. Agricultural landscapes dominate the altitudinal belt between 2600 and 3800 m, and are typically composed by small field crops (mainly potato *Solanum tuberosum* L., broad bean *Vicia faba* L., corn *Zea mays* L., alfalfa *Medicago sativa* L., and pasture), natural grasslands (páramos) and a few forest patches [21]. Under the climatic conditions of the region, crops can be planted and harvested all year round, thereby creating a landscape mosaic of a wide variety of crops at different phenological stages.

Our study area was located 115 km south from the equatorial line (01°01′36″S, 78°32′16″W) in the Cotopaxi province of Ecuador. It spread out on a 20-km^2 elevation transect (2.35×8.5 km), ranging from 2,600 to 3,800 m a.s.l. The gradient had a Southwest exposure and an average slope of 9.5° (±5.2) (based on a 30 m resolution digital elevation model). To investigate the elevation effect on local vs. global interpolated temperature variations, we divided our study area into three 400 m altitudinal belts which correspond to natural floors in the hillside (2,600–3,000 m, 3,000–3,400 m, and 3,400–3,800 m) with a mean monthly temperature of 13.2±0.4°C, 10.8±0.6°C, and 9.3±0.4°C, respectively. Beyond temperature, these belts also differed in terms of landscape composition (Appendix S1 in Supporting Information), with lower elevations dominated by small fields (0.3±0.1 Ha) of potato, corn, broad bean, and pasture while the higher band had larger fields (0.7±0.3 Ha) of mainly potato and pasture. Working in these agricultural landscapes no requires specific permissions expect the kind agreement of the field owner. The presented study did not involve endangered or protected species.

2. Temperature data collection

In each of the three-altitudinal belts, we measured temperature regimes in six habitats (five crops and natural grasslands) where insect pests can be found. In each habitat, we defined three layers: air, air inside-canopy (referred as "air canopy" in the text) and soil. These layers are all used by most insect pests over their life cycle: air layer by adults, air canopy layer by adults and leaf-eating larvae and pupae, soil layer by tuber feeding larvae and pupae. In each layer of each habitat, temperature was recorded with a 1 min time step using data loggers (Hobo U23-001-Pro-V2 internal temperature loggers, Onset Computer Corporation, Bourne, USA) with an accuracy of ±0.21°K over the 0–50°C range and a resolution of 0.02°K at 25°C. According to [4], 1) air loggers were fixed on a wooden stake at 1 m high to overstep most crop canopies and sheltered by a 20 cm^2 white plastic roof to minimize solar radiation heating; the roof was itself placed 5 cm above the logger to avoid warming by greenhouse effect, 2) air canopy loggers were placed 0.3 m high inside vegetation 5 cm bellow large leaves to minimize the effect of direct solar radiation and 3) soil loggers were buried 0.1 m into the ground where roots and tubers grow (see Appendix S2 for photographs). In each field, only one logger per layer measured the temperatures. Those triplets of loggers were located at the centre of the field to avoid edge effect (see Appendix S3 for an analysis of the spatial variability of temperatures within a field and [22]). As vegetation land cover influences microclimate beneath and around plants, see [5,6], we repeated these 54 measurements (3 elevations ×6 habitats ×3 layers) for three classes of leaf area index (LAI) [23] defined as follows: 0 (bare soil), 0.01–0.5 for and >0.5 of LAI. Minimum LAI was fixed to 0.01 to avoid confusion with bare soil and allowed enough leaf area to place the loggers underneath. At each measurement site, LAI values were visually estimated (twice) measuring the ratio of leaf area within a 1-m^2 quadrant sub-divided into 0.1 m^2 cells delimited by strings. This indirect method did not account for leaves that lie on each other however it relates to shaded areas that influence inside-canopy and soil microclimates [23].

Each of the 162 measurement combinations (3 altitudinal belts ×6 habitats ×3 layers × 3 LAI classes) was replicated 1–3 times depending on availability of habitats at a given elevation and phenology stage. In total 324 independent temperature time series were acquired over 15 days between September and December 2011 (data available in Appendices S9, S10 and S11). Importantly, under the climatic conditions of the study area, 15-days time series characteristics did not differ from those obtained over one year (see Appendix S4 for details). At each measurement site, we

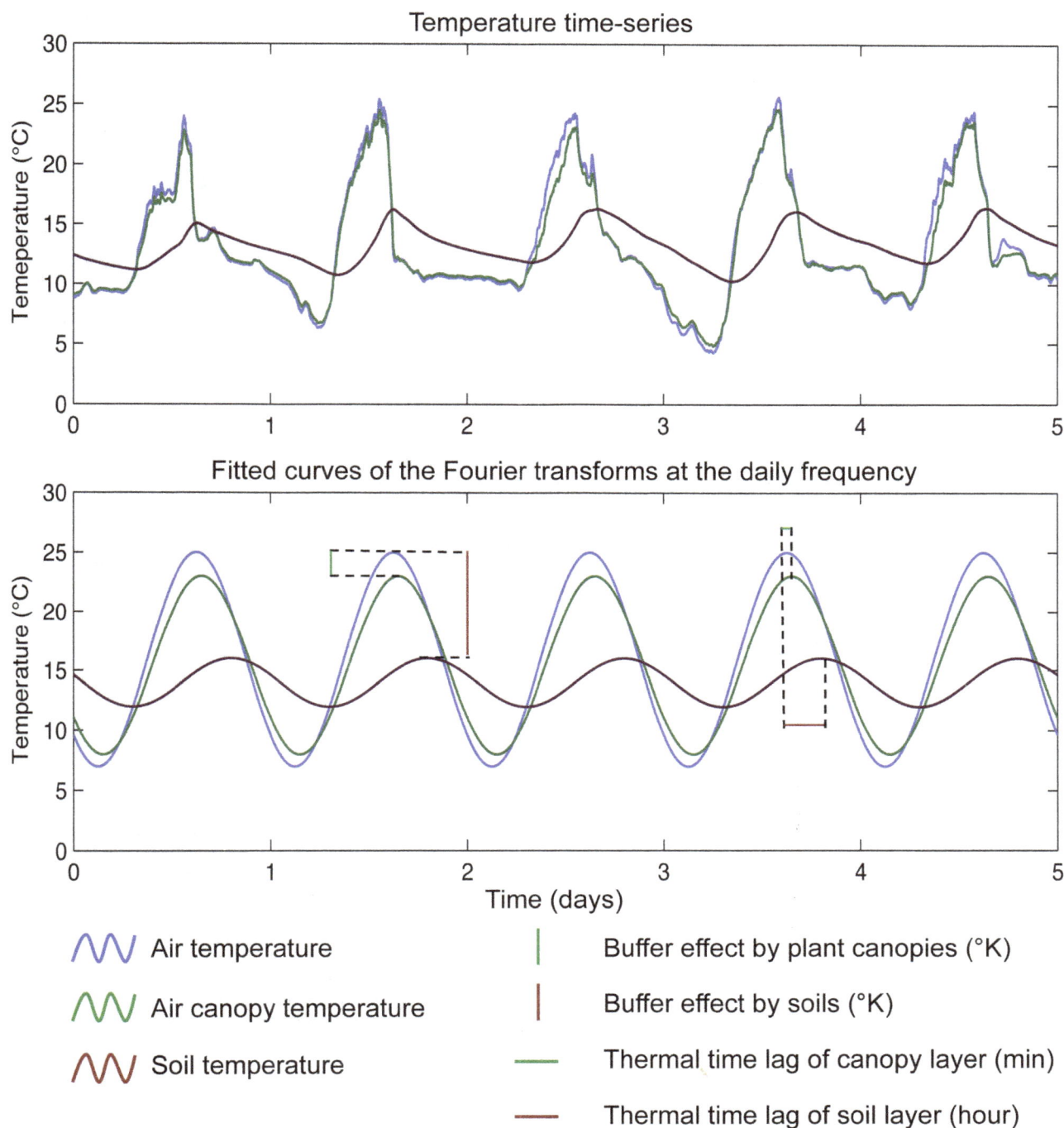

Figure 1. Fit of temperature time series with discrete Fourier transforms at the daily frequency K_d**.** Air temperatures are in blue, crop canopy temperatures are in green and soil temperatures are in brown.

recorded the UTM-WGS84 geographic coordinates with a handheld GPS Garmin Oregon 550 (Garmin, Olathe, USA).

3. Global solar radiations

Infrared and visible radiations (expressed in Watt/m^2) were monitored in each altitudinal belts using a LI-1400 LI-COR datalogger equipped with a LI-200 pyranometer sensor (LI-COR, Lincoln, USA) placed perpendicular to gravity. Between 9:00 AM and 4:00 PM, mean global solar radiations ranged from 500 to

1000 watts/m^2, with temporal variability mainly induced by short-term changes in cloud cover.

4. Data analyses

4.1. Times series analyses using Fourier transforms. Air and air canopy temperature time series showed extreme events during a few minutes that were certainly due to strong radiations experienced at the study sites – these affected loggers recording despite their plastic roofs. Therefore, we found

A. For minimum temperatures

B. For maximum temperatures

2800 m   3200 m   3600 m

☐ Crop   ▨ Storage $\Delta\, Air_L - Air_{WC}$

■ Forest   — Road   Cooler   Warmer
                                  -10°C   -1 +1   +14°C

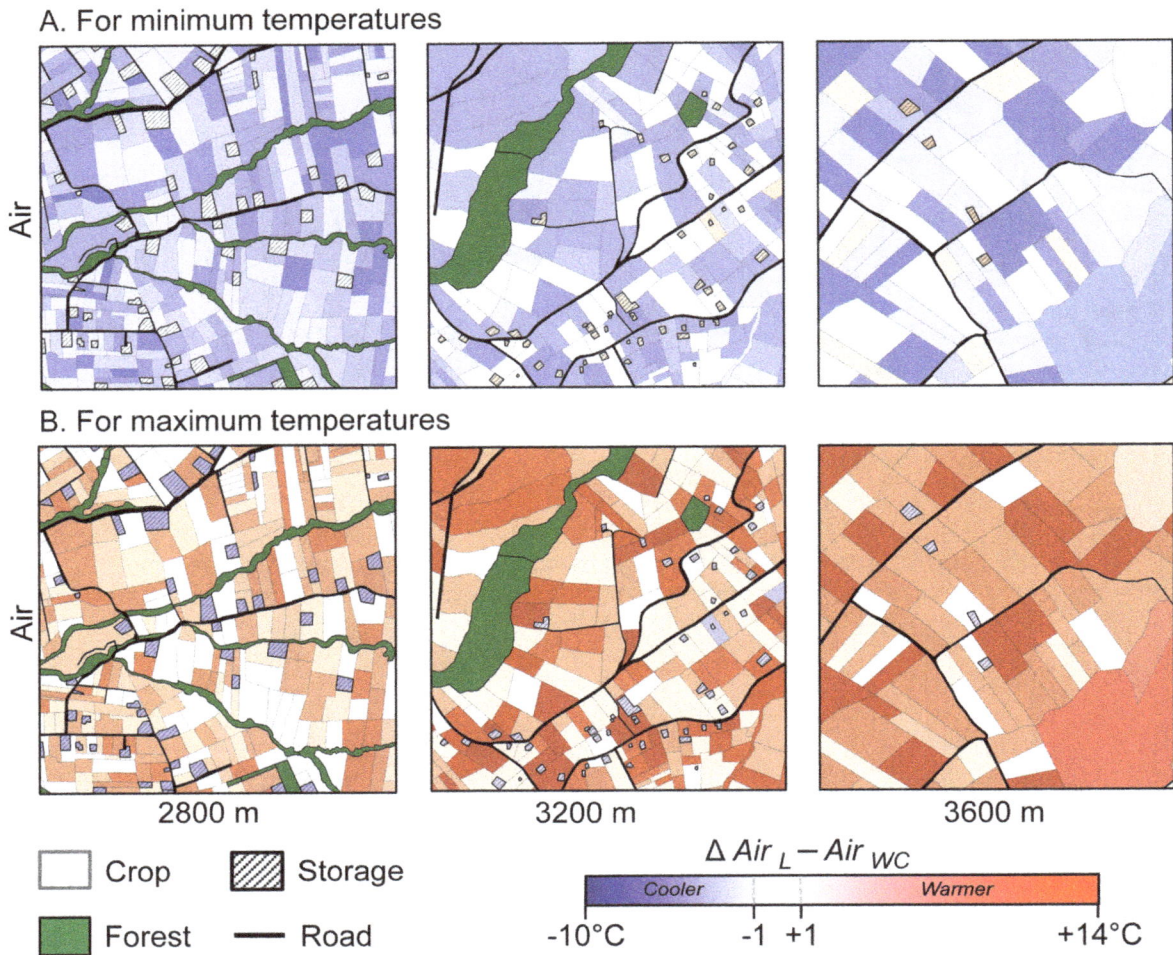

Figure 2. Maps showing the differences between local air temperatures and the WorldClim interpolated minimum (A) and maximum (B) *($\Delta\, Air_L - Air_{WC}$)*. Blue colours indicate $\Delta\, Air_L - Air_{WC} < 0$, i.e. area where local air temperatures are cooler than those gave by WorldClim. Red colours indicate $\Delta\, Air_L - Air_{WC} > 0$, i.e. area where air local temperatures are warmer than the ones gave by the WorldClim. White colours $\Delta\, Air_L - Air_{WC} = 0$ indicate areas where air WorldClim temperatures equate air local temperatures (± 1°C). The extent and position of each square is equal to the spatial resolution of the WorldClim database: 30-arc sec that is the equivalent of 0.86 km² for the study area. Temperatures in storages were obtained from [26].

relevant to fit our time series data with a discrete Fourier transform (DFT) at the daily frequency k_d (Fig. 1) as this allowed averaging daily minimum and maximum temperatures while limiting the effect of short extremes (mainly for maximum). Moreover fitting temperature time series with the DFT allowed us to circumvent (or partially resolve) the issue of comparing time series with different temporal resolution: a sinusoid built from a daily time step time series will be accurate enough to compare with another sinusoid built from a one minute time step time series (our operative temperatures vs. global climatic models).

DFT analyses allowed us estimating two important descriptors of the time series at the daily frequency k_d: the amplitude A_d and the phase ϕ_d of the DFT (see Appendix S5 for details). The thermal amplitude allowed us to measure the thermal buffer effect in Kelvin between air and canopy layers and air and soil layers (Fig. 1 and Appendix S5). The phase allowed us to measure the thermal time lag expressed in minute in inside-canopy and soil layers with respect to the air layer (Fig. 1 and Appendix S5). Thermal time lag therefore quantifies the time delay in time series to reach their maximum between air vs. canopy and air vs. soil

layers. This is an important climatic parameter to test whether microclimate conditions below canopy (canopy and soil layers) would track air conditions with some time lag depending on habitat characteristics.

We also ran DFT analyses on a four-year monitoring (2008–2012) of air temperatures (recorded at one meter high with half an hour time step with the same shelter process described above) to measure the seasonality. Analyses were performed for the three-altitudinal belts of the study area (2800, 3200, 3600 m) by reading the amplitude at the seasonal frequencies (91, 182 and 364 days, see Appendix S6). On average the Fourier transform amplitudes at 91, 182 and 364 days were 0.14 ($+/-0.01$), 0.44 ($+/-0.04$), 0.97 ($+/-0.03$)°K indicating that the seasonality was negligible in the study area [24].

All Fourier analyses were performed in MATLAB R2011a (Mathworks, Natick, USA). The effects of habitat, elevation, LAI classes and the interaction "elevation × LAI classes" on daytime and nigh-time DFT amplitudes and on DFT thermal time lag were assessed using a two-way ANOVA with Bonferroni corrections. When habitat was found significant, we ran post-hoc

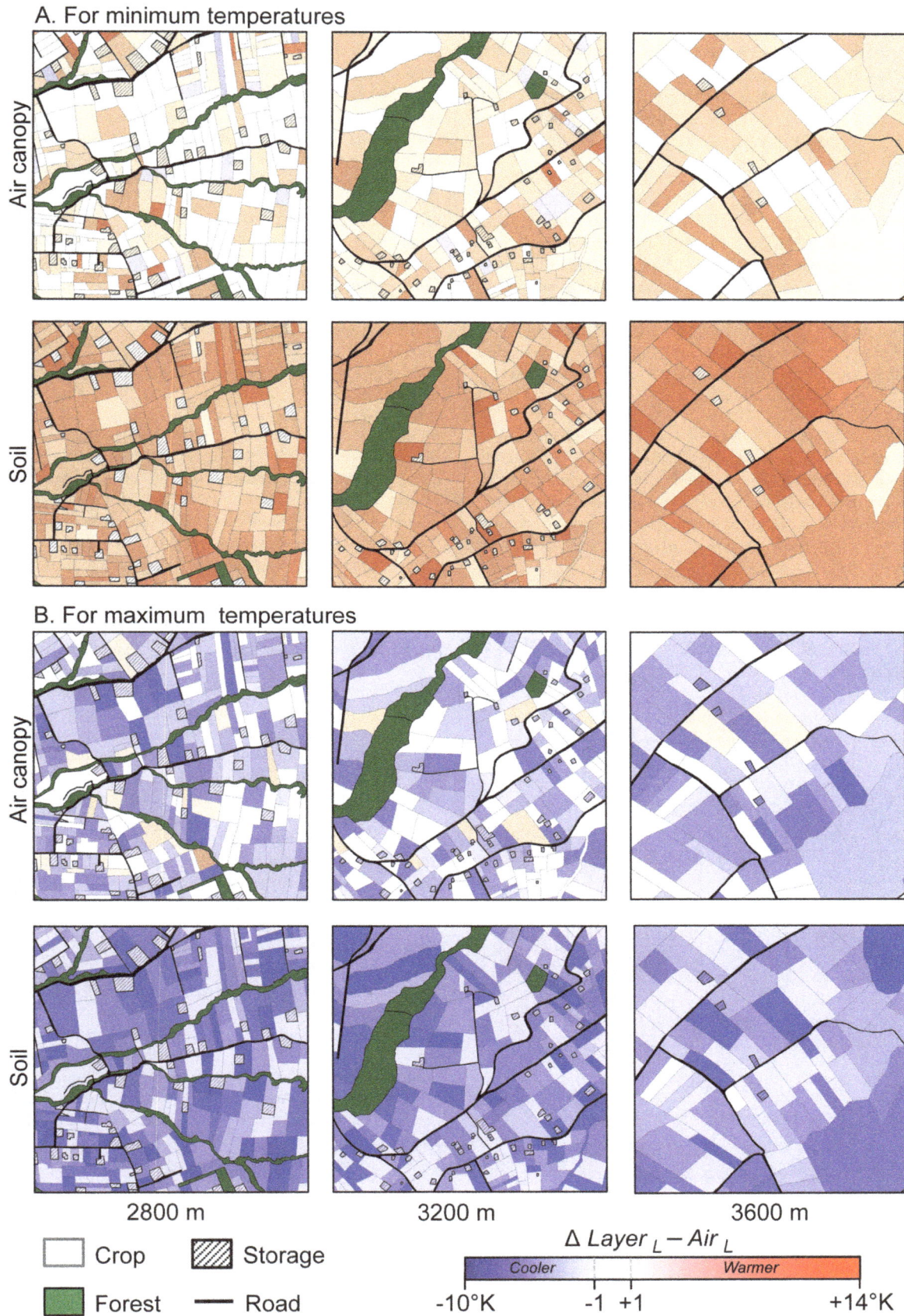

Figure 3. Maps showing the differences between local air canopy and soil temperatures with the air local for minimum (A) and maximum (B) *(Δ Layer L − Air L)*. Colour code is given in Figure 2.

multiple comparisons using a Tukey HSD test to identify differences among habitats. All statistical analyses were performed in R version 3.0.0 (R Development Core Team 2012).

4.2. Thermal landscape analyses. To compare local temperatures with global interpolated climate data employed in species distribution models, we considered one of the most widely used and readily available climate database, WorldClim [25]. The WorldClim database is a set of global climate layers (interpolated averages of monthly minimum, maximum and mean 1.5 m high air temperatures from weather stations spread out worldwide) with a spatial resolution of 30 arc seconds. Close to equator, this resolution is equivalent to squares of 0.86 km. In each altitudinal belt, we selected one WorldClim grid cell with homogenous slope (between 5.4° and 7.9°), micro-topography and exposition (south-west). Based on a digitized municipal cadastre (from the town council of Salcedo, Cotopaxi province) and a 5-m resolution digital orthophoto (Ecuadorian Military Geographical Institute, www.igm.gob.ec/site/index.php), we built the digital landscape of each grid cell in ArcGIS 10.01 (ESRI, Redlands, USA). In addition to the six studied habitats, crop storage infrastructures were also included into the digital maps as they significantly modify air temperature patterns, offering optimal conditions for crop pest development [26]. Outside air vs. inside air storage-temperature relationships for different elevations were derived from measurements made by [26] within the same area with

similar temperature data design (see Fig. 1 in Appendix A2 of their paper). Roads and woodlots were also indicated on the maps even if they were not included in the temperature comparison analysis, as they do not constitute relevant habitats for crop pests.

In order to simulate landscape thermal heterogeneity, crop habitats were attributed with one crop type (potato, broad bean, corn, alfalfa or pasture) and one LAI classes (0, 0.01–0.5, >0.5) based on a survey of 85 sites in the region, in which we quantified landscape composition (% of each crop and LAI classes) in 100-m radius sampling circles (see Appendix S7). For each habitat, we assigned the corresponding air, air canopy and soil temperature values at each elevation. Finally, since we were particularly interested in minimal and maximal values, as they are the most biologically relevant for ectothermic crop pests [4], we focused on minimum and maximum temperatures obtained from the DFT analyses and the WorldClim database.

Afterwards, we decomposed the variance of temperatures between global interpolated grids and local temperatures measured in agricultural landscapes by mapping the differences in minimum and maximum temperatures between the air local temperatures (Air $_L$) and the WorldClim interpolated temperatures (Air $_{WC}$) for the three studied grid cells. Then, to illustrate the part of the variance due to microclimate effects, we mapped the differences in minimum and maximum temperatures between measured local air canopies, soil temperatures (Layer $_L$) and the air local temperatures (Air $_L$) for the three studied grid cells.

4.3. Pest performance in thermal landscape. As a final step of our analysis, we explored the implication of our thermal landscape mapping for pest performance by comparing temperature frequencies in our grid cells with the temperature-dependent growth curve of a major crop pest species in the region: *Phthorimaea operculella* (Lepidoptera: Gelechiidae). This pest is considered one of the most important potato pests worldwide, but also attacks a wide variety of other crops such as tomato (*Solanum lycopersicum* L.), eggplant (*Solanum melongena* L.) or tobacco (*Nicotiana tabacum* L.) (see [27] for a review). *P. operculella* feeds on different part of the plant (leaves, stems, and tubers) and also tubers in storage structures [26,28]. In agricultural landscapes, *P. operculella* is abundant in virtually all types of habitats (even far from its host plant) because 1) this pest is able to fly over large distances (100–250 m) to infest suitable host plants [29] and 2) a significant quantity of tubers are left in the field after harvest, and are rapidly colonized by the moth before the following crop is planted. It is therefore common to observe infested potato plants in corn or broad bean fields. These left-over potatoes are well know by farmers and agronomists as significant obstacle to the control of these pests [28].

The temperature-dependent growth rate curve of *P. operculella* larvae (in day-1) over a 0–40°C range was obtained using published temperature-response data of laboratory experiments performed in the Andean region (see [30] for details). PTM development rate data were then modeled with the [31] equation as modified by [32]:

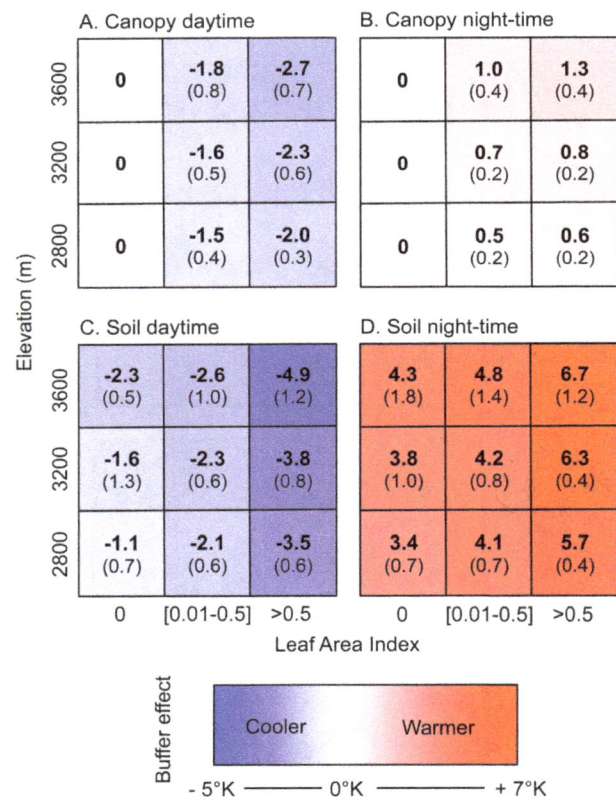

Figure 4. Mean thermal buffering from Fourier transforms at the daily frequency for canopy (A, B) and soil temperatures (C, D) as a function of elevation and leaf area index. (A, C) show the daytime temperature excursion with respect to air, whereas (B, D) are the equivalent results for night-time temperatures. The 95% interval of confidence is given between brackets. Blue colours show colder temperatures than air. Red colours show warmer temperatures than air.

$$D(T) = \frac{\dfrac{dT}{298.16}\exp\left[\dfrac{e}{R}\left(\dfrac{1}{298.16}-\dfrac{1}{T}\right)\right]}{1+\exp\left[\dfrac{f}{R}\left(\dfrac{1}{g}-\dfrac{1}{T}\right)\right]+\exp\left[\dfrac{h}{R}\left(\dfrac{1}{i}-\dfrac{1}{T}\right)\right]} \quad (1)$$

where T is temperature in Kelvin (°C+273.15), $R = 1.987$, and d, e, f, g, h, and i estimated parameters. This model has been widely used to describe the kinetics of insect development based on several assumptions about the underlying developmental control

Table 1. Results of the two-way ANOVA with a Bonferroni correction on the effects of habitat, elevation, LAI and elevation × LAI terms on daytime and nigh-time DFT amplitudes and thermal time lag on inside-canopy and soil temperature time series.

Effect	Canopy				Soil			
	Df	Mean sq	F value	P value	Df	Mean sq	F value	P value
Daytime amplitude								
Habitat	5	6.282	3.370	**0.007***	5	5.745	2.466	0.036
Elevation	2	12.491	6.701	**0.002***	2	5.722	2.456	0.089
LAI	1	40.171	21.551	**<0.001***	1	136.78	58.705	**<0.001***
Elevation × LAI	2	2.513	1.348	0.263	2	0.292	0.125	0.882
Residuals	132	1.864			127	2.330		
Night-time amplitude								
Habitat	5	0.936	3.895	**0.002***	5	1.390	0.839	0.525
Elevation	2	4.539	18.896	**<0.001***	2	2.143	1.293	0.278
LAI	1	1.083	4.509	0.035	1	157.52	95.041	**<0.001***
Elevation × LAI	2	1.754	7.302	0.010	2	0.097	0.059	0.943
Residuals	132	0.240			127	1.657		
Thermal Time Lag								
Habitat	5	0.001	1.297	0.269	5	0.009	3.881	**0.003***
Elevation	2	0.001	5.777	**0.004***	2	0.024	10.139	**<0.001***
LAI	1	0.001	29.322	**<0.001***	1	0.022	9.165	**0.003***
Elevation × LAI	2	0.001	2.374	0.097	2	0.005	2.334	0.101
Residuals	132	0.001			127	0.002		

Bold* indicates significant results ($P < 0.05$).

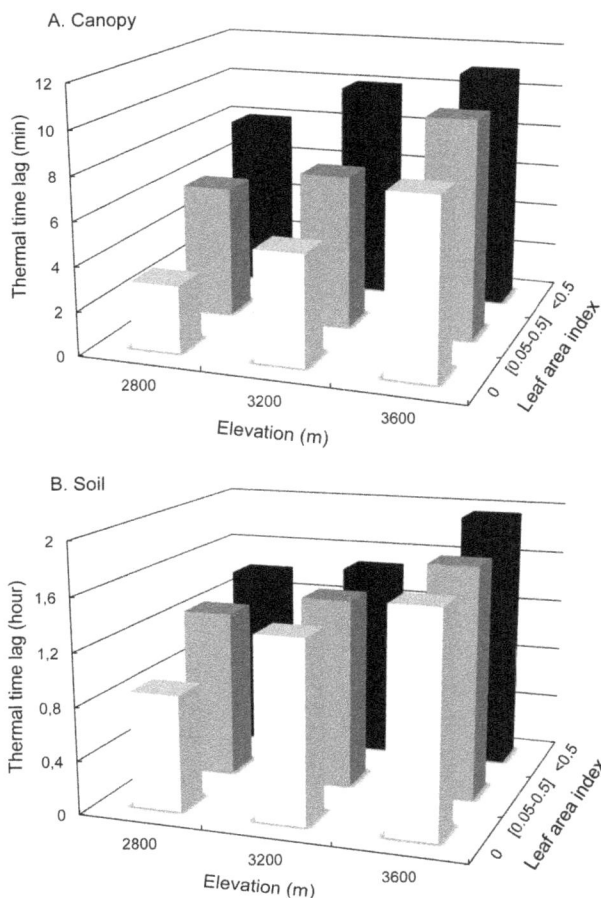

Figure 5. Thermal time lag from Fourier transforms at the daily frequency for canopy (A) and soil temperatures (B) as a function of elevation and leaf area index. The z-axis (log+1 transformed) is expressed in minutes (A) and in hours (B).

enzymes. For instance, it has been used to describe poikilotherms' temperature-dependent development [33].

We then compared the growth rate performance curve of *P. operculella* for local temperature distribution (canopy and soil layer temperatures) and for global interpolated ones (e.g., Fig. 3 in [3]). Distributions of canopy and soil minimum, maximum and mean temperatures were extracted from the three digitized landscapes using the geostatistical analyst extension of ArcGIS. Canopy and soil temperature frequencies were expressed as the percent of total grid cell area. The growth performance model of *P. operculella* given by Eqn. 1 was implemented with WorldClim minimum and maximum temperatures and the local minimum, maximum and mean temperature distribution. This allowed estimating insect growth rate within the range of WorldClim and measured field data.

Results

1. Local vs. global air temperature discrepancies in thermal landscapes

Differences in average minimum and maximum temperatures between local air temperatures and the global coarse grain interpolated air temperatures from the WorldClim (Δ Air $_L$ – Air $_{WC}$) were mapped for the three studied grid cells (Fig. 2). While

minimum local air temperatures were cooler than those predicted by WorldClim in $77.5 \pm 10\%$ of the studied areas (blue areas, average min Δ Air $_L$ – Air $_{WC} = -2.9°$K) maximum local air temperatures were warmer than extrapolated temperatures in $82.1 \pm 12\%$ of the studied areas (red areas, average max Δ Air $_L$ – Air $_{WC} = +5.6°$K). This pattern was not influenced by elevation. Notably, for all elevations, local mean air temperatures were quite well predicted by the WorldClim ($+/-1°$K) as in average $55.3 \pm 3.4\%$ of the studied areas felt in the range of Air $_L$ – Air $_{WC} \leq 1°$K (Appendix S8).

2. Temperature discrepancies due to microclimate in agricultural landscapes

Differences in average minimum and maximum temperatures between local canopy and soil temperatures and local air temperatures (Δ Layer $_L$ – Air $_L$) were mapped for the three studied grid cells (Fig. 3). Overall, canopy and soil areas were always cooler than maximum air temperature and were always warmer than air minimum temperatures resulting in a general buffer effect of minimum and maximum air temperatures by canopy and soil layers. The buffer effect on air temperatures was significantly stronger for soil than for canopy layer (see Fig. 4, Student's t-test, $t = -27.10$ and $t = 4.52$, $P < 0.001$ for night-time and daytime, respectively). Interestingly, the buffer effect on air temperatures by soil was higher during night-time than daytime (Fig. 4D) while the opposite pattern was found in crop canopy (Fig. 4A).

Elevation had a significant effect on air temperature buffering in the canopy layer but not in the soil layer (Table 1). Contrastingly, LAI had a highly significant thermal buffering effect in both soils (night and daytime) and canopies (daytime, see Table 1). Buffer effect on air temperatures by bare soil (e.g. without plant cover, LAI = 0) ranged from $-1.1°$K to $-2.3°$K for daytime and from $3.4°$K to $4.3°$K for night-time. Crop type had no significant effect on buffering patterns except for potato in which higher buffer effects were recorded (Post-Hoc HSD test, $P < 0.05$).

Overall, thermal time lag was much shorter in canopies (7.5 ± 2.6 min) than in soils (1.5 ± 0.3 hours, Fig. 5). LAI classes had a significant positive effect on thermal time lag for both canopy and soil layers (Table 1). On average, thermal time lag increased by 2 min. in canopies and 30 min. in soils between two LAI classes. Similarly, elevation had a significant positive effect on thermal time lag for both canopy and soil layers (Table 1) with an average increase of 2 ± 0.3 min. in canopies and of 60 ± 31 min. in soil between two altitudinal belts (Fig. 5).

3. Thermal performance curve using local vs. interpolated temperatures

To assess the implication of local vs. global interpolated temperature discrepancies for crop pest performances, we plotted the frequency distribution of the minimum (blue bars), maximum (red bars) and mean local (stripped bars) temperatures and those given by WorldClim (from minimum to maximum temperature, shaded region in the background) with the temperature-dependent growth rate curve of the potato moth *P. operculella* (Fig. 6). As a general pattern, global interpolated temperature ranges predicted lower growth rates of *P. operculella* than those predicted by local temperatures at all elevations, in both inside-canopy and soil layers (where the pest lives most of their time). While mean temperature distribution generally fell within the WorldClim min-max range, extreme temperatures (and especially maximum ones) largely exceeded this range.

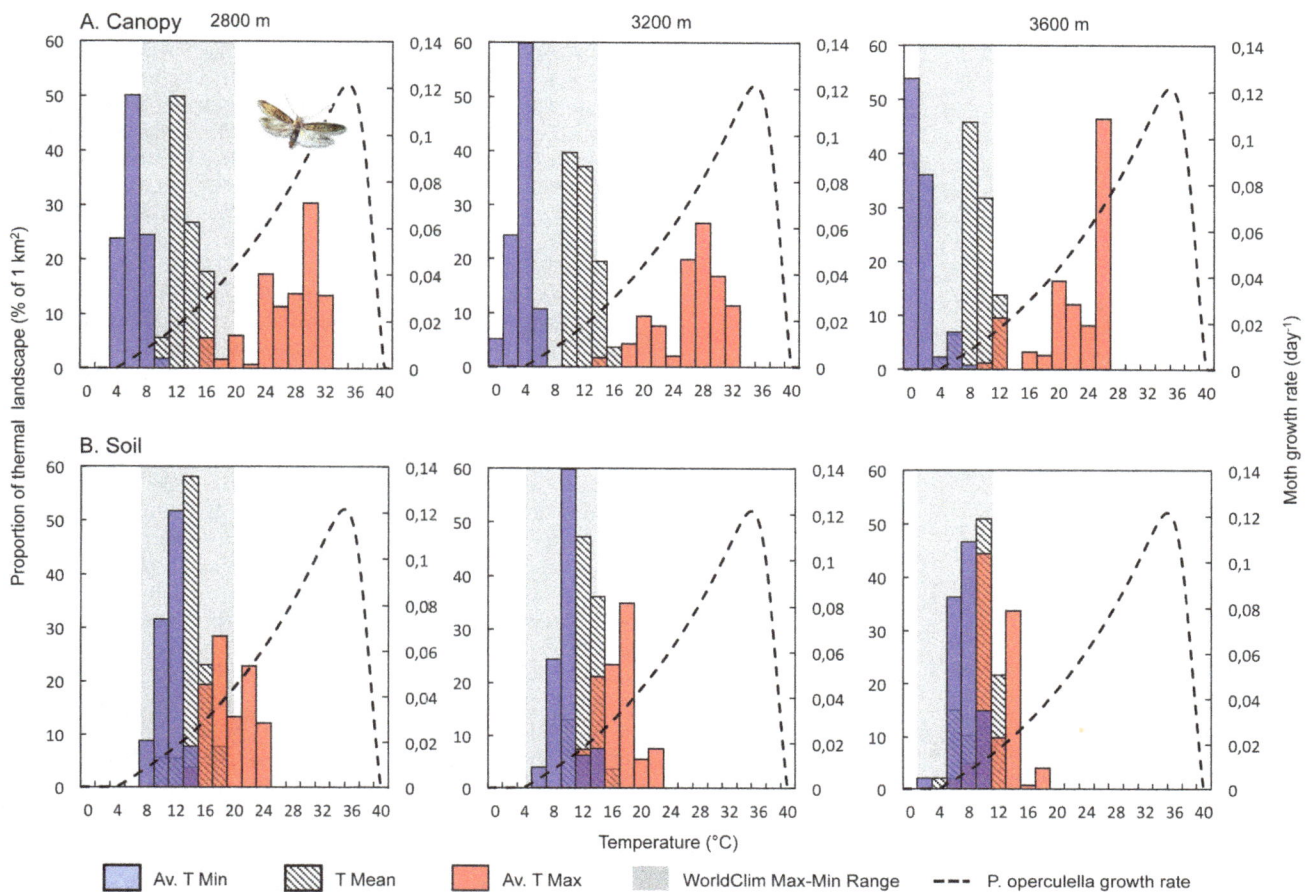

Figure 6. Superimposed plot of the temperature-dependent growth rate curve of the potato moth Phthorimaea operculella (dashed line) and the frequency distribution (% of area) of average minimum (blue), maximum (red) and mean (striped) temperatures for canopy and soil layers at the three studied elevations. Grey (shaded) bands in the background represent the WorldClim minimum and maximum temperature range.

The WorldClim estimations predicted *P. operculella* growth rates ranging between 0.007 and 0.045 day^{-1} at 2800 m, and between 0 and 0.018 day^{-1} at 3600 m, the maximum rates being slightly lower than those predicted by soil temperatures (0.068 day^{-1} at 2800 m and 0.037 day^{-1} at 3600 m). These differences were exacerbated in canopy layers where estimated maximum growth rates were 2.6–4.3 times higher than those predicted by WorldClim (0.118 day^{-1} at 2800 m and 0.079 day^{-1} at 3600 m). Discrepancies between WorldClim and local temperature-based growth rate estimations were not significantly affected by elevation (One-way ANOVA, $F = 7.79$, $P = 0.219$ and $F = 1.67$, $P = 0.419$ for canopies and soils, respectively).

Discussion

Accurate predictions of the responses of organisms to climate change using SDMs require knowledge of microclimates at spatial and temporal scales relevant for studied organisms [13,34,35]. To our knowledge, our study is the first to quantify the thermal heterogeneity among a set of agricultural habitats at fine spatial and temporal scales and to compare those thermal microhabitats to the most widely used global climatic dataset in SDMs. By documenting the mosaic of thermal habitats found in tropical agricultural landscapes, our study confirms previous evidence that microclimates strongly differ from nearby macroclimates due to

the variability of air motion and solar radiation patterns created by complex topographies with heterogeneous elevation, slope angle, exposure or roughness [1,7,18,36]. Our study therefore supports the view that results from the long tradition of agrometeorological studies on microclimates (e.g. [6,17,22]) have to be revived in the new context of microhabitat modelling for predicting the response of organisms to climate change.

1. LAI-based and elevation-based climate heterogeneity

In contrast to many previous studies (see [7] for a review), our objective was not to examine the well-documented effect of topography on local temperatures but rather to examine the less-known effects of habitat types and vegetation land cover on thermal landscape features. We found significant thermal time lag and buffer effects on air temperatures by plant and soil layers below crop canopies during night-time and daytime. The top of canopies reflects and absorbs part of the solar radiation during the day, allowing less energy to reach the layers (plants and soils) below canopies. During the night, infrared heat released from both the ground and plants is partly held back by the canopy above [5]. As a consequence plants and soils limit night-time cooling and daytime warming [6], leading to a significant buffer effect of minimum and maximum temperatures [1,4,17]. That is also why we found a buffer effect on air temperatures by soil higher during night-time than daytime and the opposite pattern for crop canopies.

Our results indicate a strong effect of elevation on thermal buffering and thermal time lag by canopy and soil layers. This could result from the combination of a negative relationship between elevation and air temperature and a positive relationship between elevation and solar radiation exposure, part of which is absorbed by plants and soils [6]. As a result, the difference between air temperature and canopy and soil temperature increased with elevation. Interestingly, the modifications of local temperatures by habitats and LAI were of the same magnitude (from -2.70 to $4.82°C$ in average) than that generated by topography-related factors [7,36], supporting the need to better consider habitat effects on microclimates.

2. Fine scale variations in temperature vs. climatic units

Our findings show that the complex agricultural mosaic resulting from habitat types and LAI classes at the landscape scale was a major modifier of the thermal patterns in the studied tropical highlands. More importantly, our findings revealed that, at best, 55% of landscape habitats had real mean air temperatures that were well estimated by WorldClim predictions while in average less than 20% of these areas had minimum and maximum air temperatures well estimated. Additional thermal discrepancies between large and fine-scale temperatures resulted from heterogeneity in crop types and phenologies. This strongly supports the view that the common use of the WorldClim database arrayed into 1-km^2 grids may not adequately capture the reality of the climatic environment experienced by living organisms, in particular tiny ectothermic species [2,3,13,18]. It is important to note that to obtain the highest level of thermal heterogeneity we chose a complex mountainous agricultural study area that provided boundary conditions for climate modelling. Indeed, these mountainous areas provide strong climatic gradients and extreme habitat fragmentation which combined with un-seasonal agrosystem make up a mosaic of thermal patches that expanded the difficulties to faithfully assess climatic parameters for modelling [25]. In view of the urgent need of fine scale climate data with large extent [2,8,35] more research is necessary to develop accurate up- or down-scaling methods, in mountainous locations where thermal heterogeneity is large, and may be needed to properly describe the ecologically significant microclimates [7,37].

3. Microclimates and species distribution models

From tiny insects to mega-herbivores, it is well recognized that species ecology is strongly influenced by micro-climatic features of the landscape [2,10,11,12,13] yet quantitative information on how thermal landscape heterogeneity may affect species performance is scarce. Short-scale differences in temperatures may provide opportunities for individual organisms, even with limited dispersal capabilities, to escape unfavourable microclimates or to maximize physiological performances by selecting preferred microclimates [38,39]. Our analysis showed that predictions on *P. operculella* growth rates strongly differed between Wordclim-based and locally-measured temperatures, suggesting that global species distribution models using global coarse-scale climatic datasets without further microclimate modelling could be strongly limited to accurately predict species occurrence and performance, in particular that of ectotherms living in habitats such as mountain slopes. Such a spatial heterogeneity in thermal patches, where climatic conditions are strongly modified, provides a mosaic of favourable, sub-optimal or lethal thermal habitats that directly influences the performance of natural populations of ectotherms.

Coarse-extent modeling of microclimate is currently one of the major obstacles to predicting how organism will react to their experienced environments and forecast their distribution under climate change [8]. To date, two main types of models have been shown to provide relatively accurate, continent-wide calculations of microclimate: statistical model and mechanistic model [13]. The first one is statistical as the variables are not deterministically but stochastically related. These models perform statistical correlation of species occurrences with climatic data and have proven to be powerful interpolative tools for defining and projecting climatic envelopes [40,41]. A disadvantage of these statistical models is that they can only be applied to the conditions under which they are fitted. On the other hand, mechanistic models of the climatic responses of organisms [13,34] use fundamental knowledge of the interactions between process variables to define the model structure. Therefore they do not require much data for model development and validation. One of them is the Microclim model recently developed by [35,42] for all terrestrial landmasses — except Antarctica— which quantify key microclimatic parameters at macro-scales, with a relatively fine spatial (15 km^2) and temporal resolution (hours). The microclimatic parameters such as wind velocity, humidity, and solar radiation allow building energy and mass budgets of organisms, and therefore serve as key inputs for biophysical models of species distributions.

It is important to highlight that a better spatiotemporal resolution in temperature patterns should go in pair with the development of more accurate temperature-based population dynamics models to integrate it [2,13,34,43]. Existing predictions of models based on insect response measured in constant temperatures may yield different and less realistic results than those from predictions of models that include the effect of real temperature fluctuation on insect biology [33]. For example, to date, we still do not know the impact of a few hours of warm temperature for the performance of ectotherm species at longer time scales [33]. In this context, fine-scale spatiotemporal temperature mapping has revealed a key step for any studies aiming at understanding, predicting and managing the responses of species distributions to climate change.

Acknowledgments

We are grateful to the city council of Salcedo (Cotopaxi, Ecuador) for providing the digital shape files of the study area cadastre. We also thank all farmers who collaborated with us during fieldwork. And finally, we gratefully acknowledge Dr. Pincebourde S. and Dr. Duyck F. for their constructive scientific comments and suggestions.

Author Contributions

Conceived and designed the experiments: EF OD. Performed the experiments: EF MH. Analyzed the data: EF LB OD. Contributed reagents/materials/analysis tools: EF LB OD. Contributed to the writing of the manuscript: EF LB JFS OD.

References

1. Scherrer D, Korner C (2011) Topographically controlled thermal-habitat differentiation buffers alpine plant diversity against climate warming. J Biogeogr 38: 406–416.
2. Bennie J, Hodgson JA, Lawson CR, Holloway CTR, Roy DB, et al. (2013) Range expansion through fragmented landscapes under a variable climate. Ecol Lett 117: 285–229.
3. Logan ML, Huynh RK, Precious RA, Calsbeek RG (2013) The impact of climate change measured at relevant spatial scales: new hope for tropical lizards. Global Change Biol 19: 3093–3102.
4. Scheffers BR, Edwards DP, Diesmos A, Williams SE, Evans TA (2013) Microhabitats reduce animal's exposure to climate extremes. Global Change Biol 20: 495–503.
5. Geiger R (1965) The climate near the ground. Cambridge, USA.
6. Jones HG (1992) Plants and microclimate: a quantitative approach to environmental plant physiology. Cambridge University Press, Cambridge.
7. Dobrowski SZ (2011) A climatic basis for microrefugia: the influence of terrain on climate. Global Change Biol 17: 1022–1035.
8. Potter KA, Woods HA, Pincebourde S (2013) Microclimatic challenges in global change biology. Global Change Biol 19: 2932–2939.
9. Cloudsley-Thompson JL (1962) Microclimates and the distribution of terrestrial arthropods. Annu Rev Entomol 7: 199–222.
10. Tracy CR (1977) Minimum size of mammalian homeotherms: role of the thermal environment. Science 198: 1034–1035.
11. Willmer PG (1982) Microclimate and the environmental physiology of insects. Adv Insect Physiol 16: 1–57.
12. Unwin DM, Corbet SA (1991) Insects, plants and microclimate. Richmond Publishing Company Ltd.
13. Kearney M, Porter W (2009) Mechanistic niche modelling: combining physiological and spatial data to predict species' ranges. Ecol Lett 12: 334–350.
14. Beaumont LJ, Pitman AJ, Poulsen M, Hughes L (2007) Where will species go? Incorporating new advances in climate modelling into projections of species distributions. Global Change Biol 13: 1368–1385.
15. Deutsch CA, Tewksbury JJ, Huey RB, Sheldon KS, Ghalambor CK, et al. (2008) Impacts of climate warming on terrestrial ectotherms across latitude. P Natl Acad Sci USA 105: 6668–6672.
16. Warren RJ, Chick L (2013) Upward ant distribution shift corresponds with minimum, not maximum, temperature tolerance. Global Change Biol 19: 2082–2088.
17. Suggitt AJ, Gillingham PK, Hill JK, Huntley B, Kunin WE, et al. (2011) Habitat microclimates drive fine-scale variation in extreme temperatures. Oikos 120: 1–8.
18. Sears MW, Raskin E, Angilletta MJ (2011) The world is not flat: defining relevant thermal landscapes in the context of climate change. Integr Comp Biol 51: 666–675.
19. Dangles O, Carpio C, Barragan AR, Zeddam JL, Silvain JF (2008) Temperature as a key driver of ecological sorting among invasive pest species in the tropical Andes. Ecol Appl 18: 1795–1809.
20. McCain CM (2007) Could temperature and water availability drive elevational species richness patterns? A global case study for bats. Global Ecol Biogeogr 16: 1–13.
21. Dangles O, Carpio FC, Villares M, Yumisaca F, Liger B, et al. (2010) Community-based participatory research helps farmers and scientists to manage invasive pests in the Ecuadorian andes. Ambio 39: 325–335.
22. Baldocchi DD, Verma SB, Rosenberg NJ (1983) Microclimate in the soybean canopy. Agr Meteorol 28: 321–337.
23. Wilhelm WW, Ruwe K, Schlemmer MR (2000) Comparisons of three-leaf area index meters in a corn canopy. Crop Sci 40: 1179–1183.
24. Fitzpatrick EA (1964) Seasonal distribution of rainfall in Australia analysed by Fourier methods. Archiv Meteorologie, Geophysik und Bioklimatologie 13: 270–286.
25. Hijmans RJ, Cameron SE, Parra JL, Jones PG, Jarvis A (2005) Very high resolution interpolated climate surfaces for global land areas. Int J Climatol 25: 1965–1978.
26. Crespo-Perez V, Rebaudo F, Silvain JF, Dangles O (2011) Modeling invasive species spread in complex landscapes: the case of potato moth in Ecuador. Landscape Ecol 26: 1447–1461.
27. Rondon S (2010) The potato tuberworm: a literature review of its biology, ecology, and control. Am J Potato Res 87: 149–166.
28. Hanafi A (1999) Integrated pest management of potato tuber moth in field and storage. Potato Res 42: 373–380.
29. Cameron PJ, Walker GP, Penny GM, Wigley PJ (2002) Movement of potato tuberworm within and between crops, and some comparisons with diamondback moth. Environ Entomo 31: 65–75.
30. Crespo-Pérez V, Dangles O, Régnière J, Chuine I (2013) Modeling temperature-dependent survival with small datasets: insights from tropical mountain agricultural pests. Bul Entomol Res 103(03): 336–343.
31. Sharpe PJH, DeMichele DW (1977) Reaction-kinetics of poikilotherm development. J Theor Biol 64: 649–670.
32. Schoolfield RM, Sharpe PJH, Magnuson CE (1981) Non-linear regression of biological temperature-dependent rate models based on absolute reaction-rate theory. J Theor Biol 88: 719–731.
33. Gilbert E, Powell JA, Logan JA, Bentz BJ (2004) Comparison of three models predicting developmental milestones given environmental and individual variation. B Math Biol 66(6): 1821–1850.
34. Buckley LB, Urban MC, Angilletta MJ, Crozier LG, Rissler LJ, et al. (2010) Can mechanism inform species' distribution models? Ecol Lett 13: 1041–1054.
35. Kearney M, Shamakhy A, Tingley R, Karoly DJ, Hoffmann AA, et al. (2013) Microclimate modelling at macro scales: a test of a general microclimate model integrated with gridded continental-scale soil and weather data. Global Change Biol Doi : 10.1111/2041–210X.12148.
36. Scherrer D, Korner C (2010) Infra-red thermometry of alpine landscapes challenges climatic warming projections. Global Change Biol 16: 2602–2613.
37. Fridley JD (2009) Downscaling climate over complex terrain: high finescale (< 1000 m) spatial variation of near-ground temperatures in a montane forested landscape. J Appl Meteorol Clim 48: 1033–1049.
38. Kinahan AA, Pimm SL, van Aarde RJ (2007) Ambient temperature as a determinant of landscape use in the savanna elephant, Loxodonta africana. J Therm Biol 32: 47–58.
39. Dillon ME, Liu R, Wang G, Huey RB (2012) Disentangling thermal preference and the thermal dependence of movement in ectotherms. J Therm Biol 37: 631–639.
40. Guisan A, Thuiller W (2005) Predicting species distribution: offering more than simple habitat models. Ecol Lett 8: 993–1009.
41. Elith J, Leathwick JR (2009) Species distribution models: ecological explanation and prediction across space and time. Annu Rev Ecol Evol S 40: 677–697.
42. Kearney MR, Isaac AP, Porter WP (2014) microclim: Global estimates of hourly microclimate based on long-term monthly climate averages. Scientific Data 1.
43. Bakken GS, Angilletta MJ (2014) How to avoid errors when quantifying thermal environments. Func Ecol 8: 96–107.
44. Bloomfield P (2004) Fourier analysis of time series: an introduction. Wiley-Interscience, New York, USA.

15

Multiple Complexes of Nitrogen Assimilatory Enzymes in Spinach Chloroplasts: Possible Mechanisms for the Regulation of Enzyme Function

Yoko Kimata-Ariga*, Toshiharu Hase

Institute for Protein Research, Osaka University, Suita, Osaka, Japan

Abstract

Assimilation of nitrogen is an essential biological process for plant growth and productivity. Here we show that three chloroplast enzymes involved in nitrogen assimilation, glutamate synthase (GOGAT), nitrite reductase (NiR) and glutamine synthetase (GS), separately assemble into distinct protein complexes in spinach chloroplasts, as analyzed by western blots under blue native electrophoresis (BN-PAGE). GOGAT and NiR were present not only as monomers, but also as novel complexes with a discrete size (730 kDa) and multiple sizes ($>$120 kDa), respectively, in the stromal fraction of chloroplasts. These complexes showed the same mobility as each monomer on two-dimensional (2D) SDS-PAGE after BN-PAGE. The 730 kDa complex containing GOGAT dissociated into monomers, and multiple complexes of NiR reversibly converted into monomers, in response to the changes in the pH of the stromal solvent. On the other hand, the bands detected by anti-GS antibody were present not only in stroma as a conventional decameric holoenzyme complex of 420 kDa, but also in thylakoids as a novel complex of 560 kDa. The polypeptide in the 560 kDa complex showed slower mobility than that of the 420 kDa complex on the 2D SDS-PAGE, implying the assembly of distinct GS isoforms or a post-translational modification of the same GS protein. The function of these multiple complexes was evaluated by in-gel GS activity under native conditions and by the binding ability of NiR and GOGAT with their physiological electron donor, ferredoxin. The results indicate that these multiplicities in size and localization of the three nitrogen assimilatory enzymes may be involved in the physiological regulation of their enzyme function, in a similar way as recently described cases of carbon assimilatory enzymes.

Editor: Douglas Andrew Campbell, Mount Allison University, Canada

Funding: This work was supported by grants-in-aid for Scientific Research on Priority Areas (23570165) from the Japan Society for the Promotion of Science (to YKA), www.jsps.go.jp. The funders had no role in study design, data collection and analysis, decision to publish, or preparation of the manuscript.

Competing Interests: The authors have declared that no competing interests exist.

* Email: a-yoko@protein.osaka-u.ac.jp

Introduction

Intracellular enzymes are pertinently distributed and/or co-localized with functionally related proteins rather than evenly dispersed within cells or organelles, and dynamically change their states in response to environmental changes, for the biological reactions to proceed efficiently in a highly controlled fashion [1,2]. Various chloroplast enzymes are subjected to light/dark modulation of their activity through redox modulation [3], whose rate is adjusted by other factors such as specific metabolites [4]. In some cases, the reversible changes in the redox and activation states are accompanied by oligomerization and re-dissociation of transient complexes. For the enzymes of photosynthetic carbon assimilation, several lines of evidence have suggested that enzymes of the Calvin cycle associate to form multiprotein complexes. A multiprotein complex including two major Calvin cycle enzymes and a small protein (CP12) has been identified [5], and its reversible dissociation allowing for rapid regulation of enzyme activity was shown to be mediated by thioredoxin in response to changes in light availability [6]. Another recent finding for the key photosynthetic enzyme, ferredoxin-NADP$^+$ oxidoreductase (FNR), is its reversible association with thylakoid binding proteins

(Tic 62 and TROL) in response to light signal and stromal pH, which regulates the stability and dynamic light-dependent membrane tethering of FNR [7,8]. Assimilation of nitrate is another major biological process in photosynthetic organisms, and has a large effect on plant growth and development [9]. Nitrate transported into cells is reduced to nitrite by nitrate reductase present in the cytosol and is further reduced to ammonium by nitrite reductase (NiR) in the chloroplast. The resulting ammonium is fixed as the amine group of Gln by glutamine synthetase (GS), and then two molecules of Glu are synthesized from Gln and 2-oxoglutarate by glutamate synthase (also called as glutamine oxoglutarate aminotransferase; GOGAT). The holoenzyme of NiR contains a siroheme and a 4Fe-4S cluster, and the GOGAT holoenzyme contains an FMN and a 3Fe-4S cluster. NiR, GS and GOGAT are nuclear-encoded and known to be located in the stromal fraction of chloroplasts in higher plants. NiR and GOGAT require reducing powers for their reactions, and their physiological electron donor is ferredoxin (Fd) (or NADH in the case of GOGAT), which is reduced by light-dependent reactions of the photosynthetic electron-transfer chain. These enzymes are known to be highly regulated during development and by external conditions such as light availability and nitrogen availability at the level of protein expression (reviewed in [9,10]). Findings of the multiprotein complexes of the Calvin cycle enzymes and FNR suggest that similar post-translational mechanisms may also exist for the regulation of nitrogen assimilatory enzymes. A recent study of protein profiles of GS gene products in the legume *Medicago truncatula* showed that the GS polypeptides assembled into organ-specific protein complexes with different molecular mass, implying organ-specific post-translational modifications under defined physiological conditions [11]. In this study, the possibility for the formation of protein complexes involving three nitrogen assimilatory enzymes, NiR, GS and GOGAT, was investigated using spinach leaves, by the method of blue native PAGE (BN-PAGE). BN-PAGE is a powerful tool for separating protein complexes from biological membranes and the soluble fraction under native conditions [12], which was also used for the above analyses of the Calvin cycle enzymes [6] and FNR [7,8]. In spinach, NiR is encoded in a single gene per haploid genome, and expressed as a single polypeptide of 65 kDa protein, whose X-ray crystal structure was recently solved [13]. GOGAT in plant plastids is present as two distinct classes, with different physiological electron donors, Fd and NADH. Fd-dependent GOGAT (Fd-GOGAT) is encoded in a single gene in spinach, expressed as a large single polypeptide of 165 kDa, and accounts for more than 95% of the total GOGAT activity in photosynthetic plant tissue [14]. NADH-GOGAT in plants is present as a 240 kDa polypeptide containing an additional NADH-binding region at the C-terminus. GS in higher plants also exists as two distinct types, cytosolic GS (GS1) and plastidic GS (GS2), with different polypeptide sizes of 38–40 kDa and 42–45 kDa, respectively [15]. Plant GS1 and GS2 polypeptides assemble into decameric complexes to be active holoenzymes [16,17]. In this study, we found each of NiR, GS and GOGAT assembled into discrete protein complexes in the chloroplasts of spinach leaves, and the dissociation profile of these complexes varied in response to the changes in the stromal conditions. The function of these complexes was evaluated, thus suggesting the involvement of the multiplicities in size and localization of these nitrogen assimilatory enzymes in the physiological regulation of enzyme function, in a similar way as the cases of the carbon assimilatory enzymes.

Materials and Methods

Isolation of stroma and thylakoid proteins

Chloroplasts were prepared essentially as described by Mach [18] from mature leaves of spinach (*Spinacia oleracea*) purchased at a local market, and then ruptured in 20 mM Hepes-KOH, pH 7.5 by repeated freeze-thawing at $-20°C$, unless otherwise specified. Following centrifugation at $10,000 \times g$ for 5 min at 4°C, the supernatant was filtered through a 0.2-μm filter membrane, and reserved as a stromal fraction. The pellet was re-suspended in the same buffer and centrifuged as above; this washing step was repeated twice in order to remove residual stromal components. The resulting pellet of the membrane fraction was re-suspended for solubilization in a Native PAGE sample buffer (50 mM Bis-Tris-HCl, pH 7.2, 50 mM NaCl and 10% glycerol) (life technologies) containing 1% n-dodecyl-ß-D-maltoside (DDM), at a chlorophyll concentration of 0.5 mg/ml, and then incubated for 30 min at 4°C with constant rotation, followed by centrifugation at $20,000 \times g$ for 15 min at 4°C. The supernatant was reserved as a fraction of thylakoid proteins (possibly containing proteins from other plastid membranes such as envelop), adjusted to 0.25% Coomassie brilliant blue (CBB) G-250, and used for BN-PAGE analysis. For the preparation of proteins from whole chloroplasts,

isolated chloroplasts were directly suspended in the Native PAGE sample buffer containing 1% DDM, and the solubilized proteins were recovered by the same procedure as the membrane fraction described above.

BN-PAGE, Native PAGE, two-dimensional SDS-PAGE and western blot analysis

BN-PAGE was performed at 4°C, by using a 4–16% linear polyacrylamide gradient of Native PAGE Novex Bis-Tris Gel system (life technologies) according to the manufacture's instruction. Native PAGE was performed according to the same protocol as BN-PAGE except that CBB was omitted from the running buffer and the sample loading buffer. Unless otherwise specified, samples loaded on the gel were derived from 10 μg on a chlorophyll basis of the chloroplasts whose proteins were extracted. For two-dimensional SDS-PAGE after BN-PAGE (2D BN/SDS-PAGE), the lanes were cut out after the run, and incubated in a denaturing solution composed of 1% SDS and 1% 2-mercaptoethanol for 15 min before applying on top of a 12% SDS-PAGE gel. For western blot analysis, proteins on the gels were transferred to an immobilon PVDF membrane (Millipore) and probed with polyclonal antibodies raised against maize recombinant proteins (Fd-GOGAT, GS1 and FNR) and spinach NiR; proteins were detected by using horseradish peroxidase conjugated to Protein A (life technologies) with enhanced chemiluminescence (Western ECL substrate, Bio-Rad).

pH treatment of chloroplasts or stromal fraction

For the preparation of stroma and thylakoid proteins under different pH conditions, chloroplasts were suspended in 20 mM potassium phosphate buffer at pH 6.0 or pH 8.0, followed by repeated freeze-thawing, and incubated on ice for 30 min before the isolation of the stroma and thylakoid proteins. For the pH shift assay of stromal proteins, the isolated stromal fraction prepared by using 20 mM Hepes-KOH, pH 7.5 buffer was diluted with three-times volume of the same buffer, or 20 mM potassium phosphate buffer at pH 6.0 or pH 8.0, and incubated for 30 min at room temperature before BN-PAGE analysis. For the pH shift assay of whole chloroplasts, chloroplasts disrupted in 20 mM Hepes-KOH, pH 7.5 by repeated freeze-thawing were diluted with three-times volume of the same buffer, or 20 mM potassium phosphate buffer at pH 6.0 or pH 8.0, and incubated for 30 min at room temperature before extraction of stromal proteins for the analysis.

DTT treatment of thylakoid membranes

Thylakoid membranes prepared by using 20 mM potassium phosphate buffer at pH 6 was solubilized in the Native PAGE sample buffer containing 1% DDM for 30 min, and then incubated for 15 min in the absence or presence of DTT at 2, 10 and 50 mM at 4°C with constant rotation. After further incubation for 15 min at room temperature, the supernatant was recovered by centrifugation at $20,000 \times g$ for 15 min at 4°C, and used for BN-PAGE and 2D BN/SDS-PAGE analyses.

GS activity assays in gel and in solution

GS enzyme activity assays in gel and in solution were performed essentially as described by Seabra et al. [11]. For an in-gel GS activity assay, Native PAGE was performed as described above in this section, and the gel was incubated in GS activity solution and then in a stop solution containing $FeCl_3$ for chelating the product [11], allowing the detection of reddish brown color corresponding to the activity. GS activity in solution was determined by quantification of γ-glutamyl hydroxamate produced by the transferase reaction of glutamine and hydroxylamine, per minute per mg of total protein [11].

Fd-binding assay

Fd-affinity resin was prepared by using recombinant Fd from *Leptolyngbya boryana* [19] and CNBr-activated Sepharose 4B (GE Healthcare), following the manufacture's directions. For a Fd-binding assay, stromal proteins prepared by using 20 mM potassium phosphate buffer at pH 6.0 or pH 8.0 were mixed with the Fd resin, and the unbound fraction was recovered by centrifugation. The resulting resin was washed with each potassium phosphate buffer (at pH 6.0 or pH 8.0), and the bound proteins were eluted with the same buffer containing 0.2 M NaCl and analyzed by western blots after BN-PAGE.

Results and Discussion

Multiple complexes of nitrogen assimilatory enzymes in spinach chloroplasts

Using stroma and thylakoid proteins isolated from the chloroplasts of mature spinach leaves, native protein profiles of GOGAT, NiR and GS were analyzed by western blots under BN-PAGE (Fig. 1B-D). Proteins from the two chloroplast fractions were well-separated as seen in their CBB-stained patterns (Fig. 1A). Multiple bands with various patterns among the three enzymes were detected as follows. In the analysis of GOGAT (Fig. 1B) using antibody which specifically recognizes Fd-GOGAT, two bands with different mobilities were detected exclusively in the stromal fraction (lane S). The faster-migrating band corresponds to the size of the holoform of Fd-GOGAT (approx. 190 kDa), and the size of the other band was estimated to be 730 kDa, which is about four times of the holoform. NiR is also present almost exclusively in stroma (Fig. 1C), and in addition to the major band at around 70 kDa of the holoenzyme, smear and faint bands at around 120, 170 and 340 kDa were observed. On the other hand, in GS (Fig. 1D), besides the band at 420 kDa of the decameric holoenzyme in stroma, a slower-migrating band at 560 kDa was detected in the thylakoid fraction. This 560 kDa protein was relatively resistant against the elution from thylakoid membranes under the condition of 0.1% DDM by which complexes of FNR with Tic62/TROL were mostly eluted (our preliminary experiment), indicating that its binding to thylakoid membranes is relatively tight under the current experimental conditions. In the analysis of FNR as a control (Fig. 1E), monomer size of the holoenzyme at 40 kDa was detected in both fractions, and additional bands around 400 kDa in thylakoids and 220 kDa in stroma are thought to be the complexes of FNR with Tic 62, Trol or some other proteins [7,8]. Next, the identities of the multiple bands observed for each enzyme were analyzed by western blots under 2D BN/SDS-PAGE (Fig. 1F–I) as follows, and their physiological significances were further analyzed.

Analysis of GOGAT complex

As shown in Figure 1F of GOGAT analysis, the two bands detected by BN-PAGE analysis exhibited the same mobility on the 2D SDS-PAGE, showing a polypeptide size of 170 kDa. Together with the previous reports that Fd-GOGAT is a single copy gene in spinach, and that NADH-GOGAT shows a very different molecular weight of 240 kDa, the 730 kDa protein was shown to be either a homomultimer (most likely a tetramer) of 190 kDa holoenzyme or its complex with other protein(s) in chloroplasts. In order to further investigate the identity and physiological significance of the 730 kDa complex, its response to the changes in the conditions of stromal solvent was analyzed (Fig. 2A).

Stromal proteins were extracted from isolated chloroplasts using either a slightly acidified solvent (at pH 6) to mimic the nightly stromal environment or a slightly alkalized solvent (at pH 8) simulating the changes in the stromal pH following illumination. As shown in Figure 2A, distribution of the two bands is clearly different between the two solvent conditions (lanes 1 and 5); the 730 kDa is the major band at pH 6, on the other hand at pH 8, it mostly disappeared, and the intensity of the 190 kDa band was somewhat increased. No GOGAT signal was detected in the thylakoid fraction. This suggests that the 730 kDa complex is converted into the 190 kDa holoenzyme, and the latter form may be relatively unstable at pH 8 under the current experimental conditions. By addition of NaCl at 0.1 M (lane 2) or DTT at 10 mM (lane 3) into the stromal solvent at pH 6, the 730 kDa complex was mostly converted into the 190 kDa protein, suggesting the involvement of electrostatic interactions and disulfide bonds for the assembly of the complex. Addition of ascorbic acid at 10 mM (lane 4), which was shown to affect the redox regulation of FNR-Tic62 complex [20], had no significant effect under the current conditions. Next, whether the pH-dependent conversion between the two forms (730 kDa and 190 kDa) of GOGAT is reversible or not was investigated (Fig. 3). Stromal proteins extracted by using a standard extraction buffer at pH 7.5 (lane 1) were diluted by a buffer at pH 6 or 8 (lanes 3 and 4). Treatment with pH 8 buffer increased the 190 kDa monomer and decreased the 730 kDa complex (lane 4), while no significant change was seen by the treatment with pH 6 buffer (lane 3), in other words, re-association was not observed. Basically the same results were obtained by the same treatments of whole chloroplasts (lanes 5–7) instead of the stromal fraction, indicating that the effect of thylakoid components for the conversion is negligible. Thus, Fd-GOGAT forms a dissociable 730 kDa complex (probably a tetramer). In this regard, there is a report that bacterial NADPH-GOGAT, which consists of two subunits (α and β) and shares considerable homology with Fd-GOGAT throughout its sequence of α subunit, appears to be a $(\alpha\beta)_4$ tetramer in an analytical gel filtration experiment [21] and by X-ray small angle scattering measurements [22]. Although Fd-GOGAT is generally considered to be monomeric in solution, both NADPH-GOGAT α subunit and Fd-GOGAT form similar dimers in crystals [23,24]. Therefore, a tetramer of Fd-GOGAT may be present under physiological conditions. This 730 kDa complex dissociates into the 190 kDa monomer enzyme in response to DTT and alkalized pH (Figs. 2&3), suggesting the involvement in the redox regulation of the function of GOGAT, which will be addressed in the latter section.

Analysis of NiR complex

The situation of NiR turned out to be partly similar to that of GOGAT. In the western analysis of 2D BN/SDS-PAGE (Fig. 1G), the multiple bands observed by BN-PAGE analysis showed the same mobility as the monomer (65 kDa), indicating that they are derived from the same NiR gene product, known to be a single copy in spinach. Therefore, the multiple bands other than the monomer NiR detected on BN-PAGE analysis were either homo-multimeric forms of NiR enzyme or its complexes with other components in the chloroplasts. Responses to the different conditions in the stromal solvent such as salt and pH (Fig. 2B) have some common features with those of GOGAT, but there also are a few intriguing differences. As in the case of GOGAT in Figure 2A, protein distributions are largely different between the treatments of pH 6 and pH 8 (lanes 1 and 5). Intensive shifts of the bands toward higher molecular weight up to around 800 kDa were observed at pH 6 (lane 1); monomer band

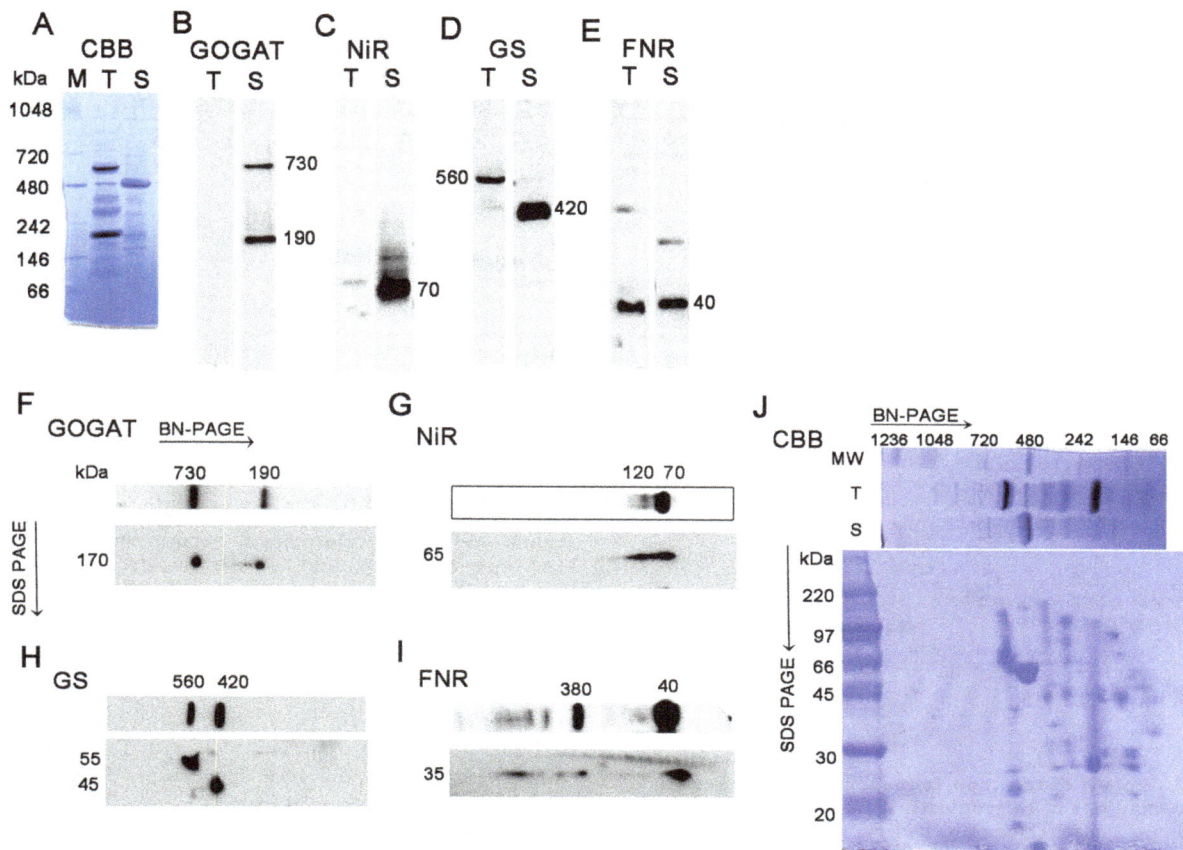

Figure 1. Protein complexes in spinach chloroplasts analyzed by blue-native PAGE (BN-PAGE) and SDS-PAGE. Upper panel; CBB-staining (A) and western blots (B–E) of BN-PAGE analysis of thylakoid (T) and stroma (S) proteins extracted from spinach chloroplasts. Lower panel; western blots (F–I) and CBB-staining (J) of 2D SDS-PAGE after separation by BN-PAGE of stromal proteins (F and G) and whole chloroplast proteins (H, I and 2D SDS-PAGE of J). Western blots were probed with polyclonal antibodies against each protein indicated. The numbers beside each band stand for the estimated molecular weights. All the samples loaded were derived from 10 μg on a chlorophyll basis of chloroplasts whose proteins were extracted.

at around 70 kDa became very faint, and instead, most abundant band of 170 kDa and less abundant, smear bands at around 340 kDa and 800 kDa are detected. The identity of the 50 kDa band is not clear but may be either a partial degradation product of NiR or a non-specifically detected band. Addition of DTT at 10 mM reduced the above shift and increased the 70 kDa monomer at both pH 6 and 8 (lanes 3 and 7), which is similar to the case of GOGAT. Unlike GOGAT, addition of NaCl at 0.1 M largely reduced the overall signals (lanes 2 and 6), and addition of ascorbic acid at 10 mM increased the monomer protein (lanes 4 and 8), implying that these agents affect the stability and/or dissociation of NiR. Another difference from GOGAT is the apparent reversible feature between NiR monomer and multiple complexes in response to the pH shifts (Fig. 3B); dilution of the stromal extract with pH 6 buffer increased the higher molecular-weight complexes, especially the 170 kDa band (lane 3 as compared to lane 1), while dilution with pH 8 buffer almost diminished the signals for those complexes (lane 4). Dilution of whole chloroplast suspension instead of the stromal extract conferred basically the same results (lanes 5–7) except that more intensive shifts of the bands were seen at lower pH, as observed in Figure 2B. Thus the conversion between monomer NiR and the higher molecular-weight complexes upon the pH shifts appears to be reversible, and their distribution varies

in response to the relatively moderate changes in the solvent conditions, which may be related to the regulation of physiological function of NiR, addressed in the latter section.

Analysis of GS complex

Another intriguing feature implying a regulatory mechanism of enzyme function was found by the analysis of the GS complex. Unlike the cases of GOGAT and NiR, the 2D SDS-PAGE analysis of the 420 kDa and 560 kDa proteins (Fig. 1H) exhibited polypeptides of different molecular weights (around 45 and 55 kDa, respectively), suggesting either that they are derived from different GS isoproteins or that a post-translational modification of the GS polypeptide occurred. The 45 kDa polypeptide corresponds to the size of plastidic GS2 protein, but the 55 kDa is rather large for the conventional plant GS proteins. In this connection, Seabra et al. [11] reported that some of the organ-specific complexes of *M. truncatula* GS2 protein, detected by western blots after Native PAGE, appeared to contain a polypeptide of higher molecular mass (50 kDa) in addition to a 42 kDa conventional GS2 polypeptide as analyzed by western blots after SDS-PAGE. Because basically only one GS2 isoprotein was shown to be expressed in this plant except in seeds, this additional polypeptide (50 kDa) was indicated to represent a post-translational modification of 42 kDa *M. truncatula* GS2 protein.

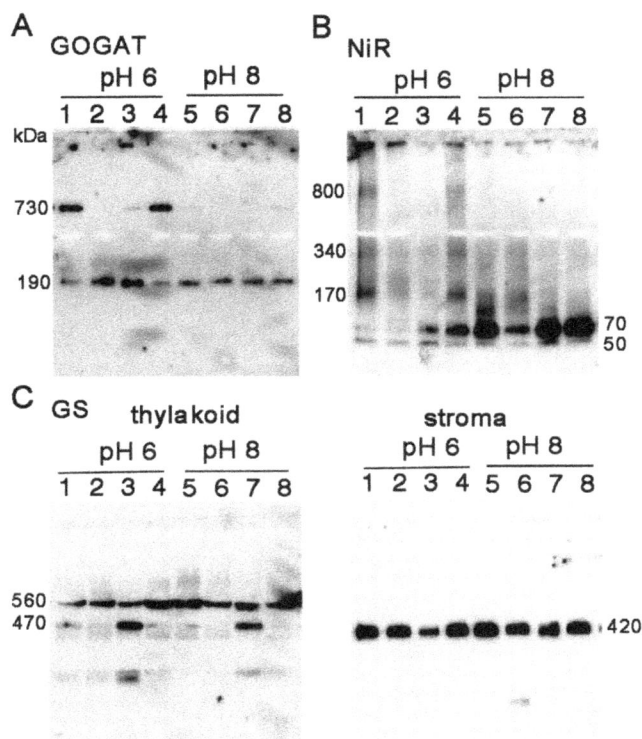

Figure 2. Protein distribution in response to the changes in solvent conditions of chloroplasts. Stroma and thylakoid proteins extracted from spinach chloroplasts under different solvent conditions of potassium phosphate buffer at pH 6 (lane 1) or pH 8 (lane 5) containing 0.1 M NaCl (lanes 2 and 6), 10 mM DTT (lanes 3 and 7) or 10 mM ascorbic acid (lanes 4 and 8) were analyzed by western blots after BN-PAGE. Results of stromal proteins for GOGAT (A) and NiR (B), and stroma and thylakoid proteins for GS (C) were presented. The numbers beside each band stand for the estimated molecular weights.

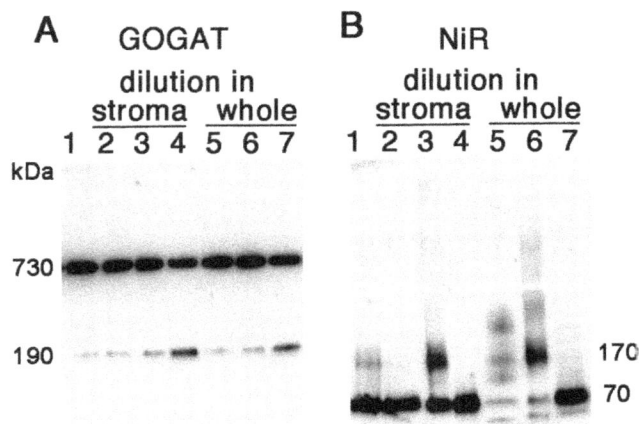

Figure 3. Protein distribution in response to the pH shift of stroma and whole chloroplasts. Stromal proteins extracted by using Hepes-KOH buffer at pH 7.5 (lane 1) were four-fold diluted with the same buffer (lane 2), potassium phosphate buffer at pH 6 (lane 3) or pH 8 (lane 4), incubated for 30 minutes at room temperature and analyzed by western blots for GOGAT (A) and NiR (B) after BN-PAGE. Stromal proteins extracted from whole chloroplasts treated with above three conditions were also analyzed (whole; lanes 5–7).

Since only one GS2 isoprotein has been reported in spinach (*S. oleracea*), it is not clear whether multiple GS2 isoproteins are expressed in spinach leaves or not. Concerning membrane-bound GS proteins, a slower migrating form of the GS2 protein was shown to accumulate in the fraction of plastid membranes in root nodules of *M. truncatula* [25], and the interaction of cytosolic GS1 protein with a symbiosome membrane protein (nodulin 26) in soybean root nodules was reported recently [26].

In contrast to the cases of GOGAT and NiR, no significant differences in the GS profile were observed between the two pH conditions in both the stroma and thylakoid fractions (Fig. 2C). Instead, addition of DTT produced an additional 470 kDa band below the 560 kDa band in thylakoids (lanes 3 and 7 in the left panel), together with the decrease in the amount of the stromal 420 kDa holoenzyme especially at pH 6 (lane 3 in the right panel). These results suggest either that the stromal decameric GS (420 kDa) may be recruited into thylakoid membranes under the conditions at 10 mM DTT via anchoring protein(s), or that the 470 kDa protein may be a cleaved or dissociated product of the 560 kDa protein complex. In order to seek out the identity of this 470 kDa band, 2D BN/SDS-PAGE analysis of thylakoid proteins treated with DTT was performed (left panel in Fig. 4). The 470 kDa protein on BN-PAGE exhibited the same mobility as that of the 560 kDa protein, suggesting that the 470 kDa band is probably derived from the 560 kDa complex although the subunit composition and integrity of the two complexes is not clear. Our

preliminary experiment shows that the 470 kDa protein is more easily released from thylakoid membranes compared to the 560 kDa protein by the mild extraction with 0.1% DDM. This 470 kDa protein may be an intermediate complex to be released into stroma as a decameric holoenzyme by unknown mechanism. The identities of these higher molecular-weight forms are not clear, but in the root nodules of *M. truncatula*, the slower migrating form of GS2 protein as described above was shown to localize in the fraction of the plastid membrane as a catalytically inactive form [25]. Also, among the multiple GS complexes (420~620 kDa) observed in the leaves of the same plant [11], only the lowest 420 kDa band was detected by an in-gel GS activity assay, implying the involvement of this multiplicity in the regulatory mechanism for GS enzyme activity. Thus, the activity of the GS complex in spinach chloroplasts was addressed in the following section.

Native PAGE analysis and GS activity assays

Native PAGE without the CBB dye was performed for an in-gel GS activity assay because the dye interferes the activity staining in BN-PAGE. Native PAGE without detergents is limited to the separation of acidic proteins with a pI below the pH of the gel and is often characterized by protein aggregation and broadening of protein bands as compared with BN-PAGE. However, it is even milder than BN-PAGE and thought to offer advantages for retaining physiological protein assemblies and oligomerization. As shown in Figure 5A, the migration pattern of molecular weight markers (lane M) was very similar to that of BN-PAGE (lane M in Fig. 1A). Minor differences were seen in the pattern of stromal proteins with slower mobility of a major band (ribulose 1,5-bisphosphate carboxylase/oxygenase complex) compared to that in BN-PAGE (lane S in Figs. 1A&5A). In contrast, the pattern of thylakoid proteins was rather different between the two PAGEs especially for the slower mobility of chlorophyll binding protein complexes (around 200, 300 and 600 kDa in BN-PAGE; lane T in Figs. 1A&5A). For the patterns of the three nitrogen assimilatory enzymes, no large difference was seen except for the apparent lack

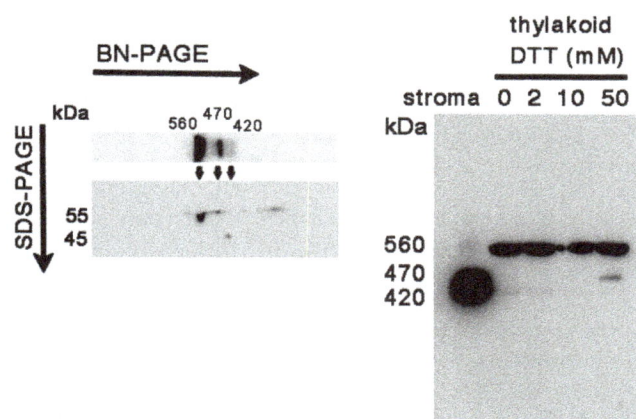

Figure 4. Western analysis of GS distribution in thylakoid proteins treated with DTT. Thylakoid proteins treated with DTT were analyzed by western blots of BN-PAGE (right) and 2D SDS-PAGE after BN-PAGE (left) using the GS polyclonal antibody. In the pattern of 2D SDS-PAGE (left), a faint spot at about 45 kDa is probably derived from the contamination of the stromal GS holoenzyme which is fairly abundant in the stromal fraction.

of the 190 kDa GOGAT monomer, and instead, smear bands migrating around the 480 kDa marker were observed in addition to the major band around 720 kDa marker on Native PAGE (Fig. 5B). Thus, the behavior of the monomer form of GOGAT appears to be different between the two native PAGEs. For NiR (Fig. 5C), more intensive shifts, as compared to that on BN-PAGE (Fig. 1C), were observed even at neutral pH. These results may represent more physiological situation than those in BN-PAGE. As for GS (Fig. 5D), the distribution pattern was basically the same as those on BN-PAGE (Fig. 1D) except that the mobility of the band in the thylakoid fraction was slightly slower. The identity of a very faint band observed above the major band in stroma is not clear at present. Thus, multiple protein complexes common to any of the three nitrogen assimilation enzymes were not observed also by this Native PAGE analysis. Since the profile of the bands detected by the anti-GS antibody remained basically the same between the analyses of BN-PAGE and Native PAGE, an in-gel GS activity assay using Native PAGE was performed in order to evaluate whether the molecular species in each fraction are enzymatically active or not (Fig. 5E). After the run of the Native PAGE, chlorophyll-binding proteins were detected as a green color in the lanes of thylakoids (T) and whole chloroplasts (W). Subsequent GS activity staining caused yellowish background of the whole gel and reddish staining of only the 420 kDa protein in the lanes of stroma (lane S) and whole chloroplasts (lane W), in agreement with the results of the leaf GS activity staining of *M. medicago* plants described in the previous section. The GS transferase activity in solution was also measured to be 2.1 μmol min^{-1} mg^{-1} total protein in the stromal extract. The activity in the thylakoid extract was about 0.07 μmol min^{-1} mg^{-1} total protein, which was barely above the background level. Taken together, the 420 kDa GS holoenzyme present in stroma was shown to be a catalytically active form while the slower migrating protein complex (560 kDa) detected by the anti-GS antibody in thylakoids appeared to have no significant GS activity. The 560 kDa protein in thylakoids may represent an inactive form caused by a post-translational modification as described in the previous section. In this regard, the higher molecular weight form could result from the covalent binding of ubiquitin or ubiquitin-like proteins of approx. 10 kDa;

Arabidopsis GS2 has been identified as a potential substrate for SUMO (small ubiquitin-like modifier) by the yeast two-hybrid system [27] although SUMOylation has not been tested in the plants. Another possibility has been postulated for the slower migrating form of the root nodule GS2 protein to be a non-processed precursor form which contains the transit peptide [25]. Analytical studies such as a proteomic analysis of the isolated protein are required to verify these possibilities. Reactivity of the anti-GS antibody we used was examined by additional western analyses (Figure S1 and S2), but we can't totally deny the possibility that the higher molecular weight forms (the 560 kDa protein and the 55 kDa polypeptide) detected by the GS antibody are a non-GS protein that may contain some antigenic structures common to GS proteins.

Fd-binding ability of multiple forms of GOGAT and NiR proteins

GOGAT and NiR are known to interact with Fd as a physiological electron donor. In order to evaluate the function of the multiple forms of GOGAT and NiR, the ability of each form for binding with Fd was investigated using the stromal extracts prepared under different pH conditions (Fig. 6). FNR as another major Fd-dependent enzyme and GS as a non-Fd dependent enzyme were also analyzed as controls. We routinely use an affinity resin with recombinant Fd from the cyanobacteria *L. boryana* [19] for the purification of Fd-dependent enzymes because of its stability on an affinity resin and because *L. boryana* Fd functions as an electron donor for Fd-dependent enzymes in higher plants to an extent comparable to that achieved by plant Fds. At pH 8, where GOGAT, NiR and FNR are mostly present as monomers in the stromal extracts, profiles of the Fd-binding and elution were quite similar among these three enzymes (lanes 6–10 in panels A, B and D); part of the enzymes were bound and released with first one-column volume of the elution buffer containing 0.2 M NaCl. The relative intensity of the signals in the elution fraction (lane 9) was somewhat higher for NiR than for FNR and GOGAT. This is consistent with the binding strength of these enzymes in the leaf extracts of maize and potato on Fd-affinity chromatography; in the order of GOGAT<FNR<NiR (T. Hase and Y. Chikuma, personal communications). In contrast, no detectable band was seen in the elution fraction of GS (lane 9 in

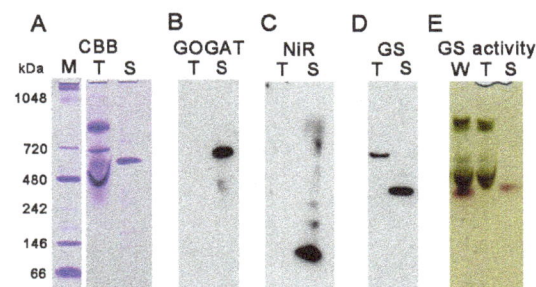

Figure 5. Protein complexes in spinach chloroplasts analyzed by using Native PAGE. CBB staining (A), western blots (B–D) and in-gel GS activity staining (E) after Native PAGE analysis of thylakoid (T) and stroma (S) proteins extracted from spinach chloroplasts. For in-gel GS activity assay, proteins extracted from whole chloroplasts (W) were also used. Lane M stands for the protein standard markers, but the migration distance does not necessarily correlate with the molecular mass in this Native PAGE. Samples loaded were derived from 10 μg (A–D) and 20 μg (E) on a chlorophyll basis of chloroplasts whose proteins were extracted.

panel C), indicating no significant binding. Thus, monomer forms of GOGAT and NiR, as well as FNR, at least partly bind with Fd in the stromal extracts under the conditions of pH 8. On the other hand, at pH 6, higher molecular weight forms of GOGAT (730 kDa) and NiR (170 kDa) in addition to the monomer forms were clearly observed in the extracts (lanes 1 in panels A and B). For NiR protein, the 170 kDa band together with the 70 kDa monomer band was greatly diminished in the unbound fraction (lane 2 in panel B), indicating their efficient binding with Fd. The reason for the detection of only the monomer band in the elution fraction (lanes 4,5 in panel B) is assumed to be due to the dissociation of the 170 kDa molecule under the conditions of 0.2 M NaCl (as implied from the results in lane 2 of Fig. 2B). FNR as a control also showed similar binding profile with NiR, although only the monomer was detected under this condition (lanes 1–5, panel D). On the other hand in GOGAT, significant reduction in the intensity of the bands was not observed in the unbound fraction (lane 2 in panel A), but partial binding was indicated from the band in the elution fraction (lane 5 in panel A). Because the 730 kDa molecule appears to dissociate into the 190 kDa monomer under 0.1 M NaCl condition (lane 2 in Fig. 2A), it is not clear which molecules, either or both of the 730 kDa and the 190 kDa proteins, actually bound with Fd. Under pH 6 conditions, GS protein also appears to partly bind with Fd resin (lane 5 in panel C); whether this is due to non-specific binding or transient complex formation of GS with other Fd-binding proteins is not clear at present.

Taken together, both the monomer (70 kDa) and the higher molecular weight complex (170 kDa) of NiR exhibited the binding ability with Fd; oligomerization of NiR may be related to its other

function such as enzyme activity and protein stability. On the other hand in GOGAT, the monomer binds with Fd at least pH 8, and the 730 kDa complex appears to have either negligible or low Fd-binding ability under pH 6 conditions. As for Fd-GOGAT from *Synechococcus* sp. PCC 6301, there is a report that two protein bands appeared after non-denaturing PAGE of purified GOGAT proteins, both presenting GOGAT activity *in situ* using methyl viologen as a small artificial electron donor although the data were not shown [28]. Thus, the higher molecular weight complex of GOGAT in spinach chloroplasts may itself retain catalytic activity but may assume conformation which reduces the binding with Fd for the purpose to regulate the GOGAT reaction in the stromal fraction.

To sum up, individual complexes with larger molecular sizes of holoenzyme were detected for all the three nitrogen assimilatory enzymes, GOGAT, GS and NiR, studied in this article. Figure 7 shows schematic models for the pH-dependent conversion of monomers and oligomers (or multi-protein complexes) of GOGAT and NiR, and possible localization of GS inferred from this study, together with a previously presented model for FNR [7]. The pH-dependent conversion of monomers and multimeric complexes of GOGAT and NiR, which appears to be accompanied by differential binding to Fd and possibly by differential enzyme activity or protein stability, indicates the physiological significance of these complexes for the regulation of enzyme function in response to the conditions such as redox state in stroma and light availability. To our surprise, GS signal was observed as another inactive complex in thylakoid fraction (or some other plastid membranes) with a larger polypeptide, postulating a different regulatory mechanism from those of the above two enzymes, which involves different localization, similarly to the case of FNR, and possibly a novel post-translational modification.

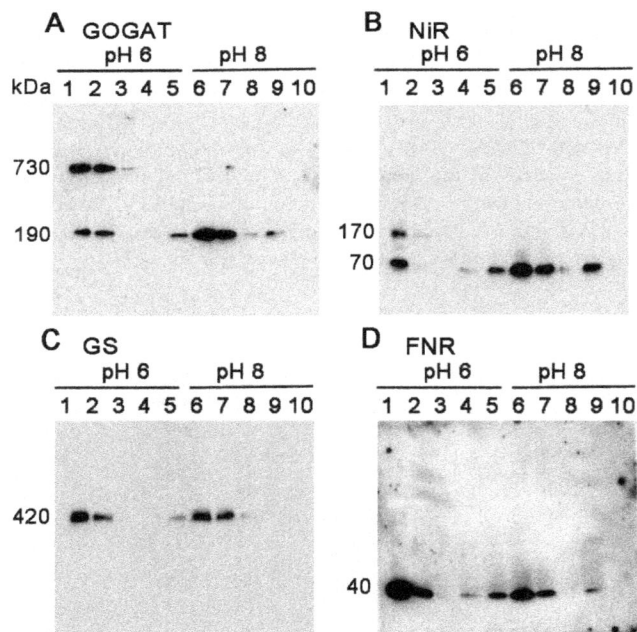

Figure 6. Fd-binding assay of protein complexes in the stromal fraction of spinach chloroplasts. Stromal proteins extracted from spinach chloroplasts by using potassium phosphate buffer at pH 6 (lane 1) or pH 8 (lane 6) were mixed with Fd resin, and the resulting unbound fractions (lanes 2&7), washed fraction (lanes 3&8) and fractions eluted with first (lanes 4&9) and second (lanes 5&10) one-column volume of the same buffer containing 0.2 M NaCl were collected and analyzed by western blots after BN-PAGE.

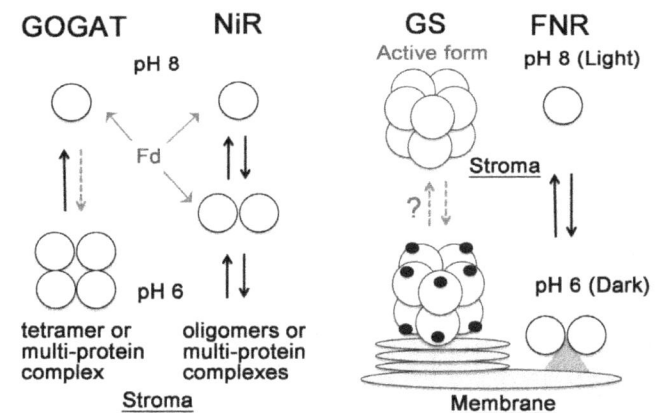

Figure 7. Models of pH-dependent complex formation and relocation of carbon- and nitrogen assimilatory enzymes in chloroplasts. Complex formation (possibly oligomerization) of GOGAT and NiR and localization of GS signals observed in this study are depicted together with a previously presented model for FNR [7]. GOGAT and NiR display pH-dependent conversion of monomers and oligomers (or protein complexes with unknown proteins) in the stromal fraction, which appears to be accompanied by differential binding to Fd (Fig. 6) and possibly by differential enzyme activity or protein stability. GS is postulated to exist both as an active enzyme of decamer in stroma and as a modified, probably inactive form (depicted with black dots) in the plastid membranes (such as thylakoids and envelop), which may interconvert by an unknown mechanism. FNR is known to relocate between stroma and thylakoids (possibly as a dimer for the stabilization of this enzyme) in the light, redox and pH-dependent manners [7,8].

Supporting Information

Figure S1   Western analysis of spinach proteins by SDS-PAGE (A) and BN-PAGE (B). A: Total proteins from spinach leaves (lane L), whole chloroplasts (W), stroma (S) and thylakoid fraction (T) were analyzed by western blot after 7.5% SDS-PAGE probed with anti-GS antibody (GS) and preimune serum (PI). In spinach leaves, only GS2 isoprotein is considered to be expressed [ref. S1 in Figure S1] and is thought to be detected by the polyclonal antibody against maize GS1a proteins [ref. S2 in Figure S1] that shares 75% identity with spinach GS2, through the consequence of the cross-reactivity of the antibody by recognizing the epitopes common to the two GS isozymes. As expected from the western analysis of the whole chloroplasts by 2D BN/SDS-PAGE (Figure 1H), stroma and thylakoid fractions contained the polypeptides with distinct sizes of approx. 41 kDa and 51 kDa, respectively. The deduced MWs are somewhat different from those in the 2D analysis (45 kDa and 55 kDa), but the size in the stromal fraction of the 1D SDS-PAGE (41 kDa) appears to be closer to the estimated size of spinach GS2 protein (372 residues) and would be more accurate than the 2D analysis. B: Proteins extracted from whole chloroplasts were analyzed by western blot after BN-PAGE, probed with anti-GS antibody (GS) and preimune serum (PI). The numbers beside each band stand for the estimated molecular weights. All the samples loaded were derived from 10 μg on a chlorophyll basis of chloroplasts whose proteins were extracted.

Figure S2   Western analysis of maize proteins by BN-PAGE (A) and SDS-PAGE (B). A: Thylakoid (T) and stroma (S) proteins were extracted from maize chloroplasts using the same protocol as described for spinach in the 'Materials and Methods'

section, and were analyzed by western blot after BN-PAGE, probed with anti-GS antibody (GS). The pattern is quite similar to that of spinach leaves (Figure 1D), although the mobility of the major band detected in the lane of thylakoid proteins is slightly faster (approx. 540 kDa) than that of spinach (560 kDa). B: Total proteins from maize leaves (lane L), stroma (S) and thylakoid fraction (T) were analyzed by western blot after 7.5% SDS-PAGE, probed with anti-GS antibody (GS). The analysis showed the bands, 41 kDa for stroma and 49 kDa for thylakoids, which are similar to those of spinach (Figure S1A). The relative content of the maize GS2 protein in the stromal fraction appears to be lower compared to that of spinach. Additional lowest band detected in the lane of the whole leaves is thought to be a cytosolic GS1 protein (39 kDa), which is known to be absent in spinach leaves, leading to the confirmation that the anti-GS antibody reacts with both GS1 and GS2. Thus, the patterns of the GS signals detected in the stroma and thylakoid fractions of maize chloroplasts were basically the same as those of spinach on both BN-PAGE and SDS-PAGE analyses. The numbers beside each band stand for the estimated molecular weights.

Acknowledgments

The authors thank Dr. Shingo Kikuchi for helpful advices with experimental methods and techniques.

Author Contributions

Conceived and designed the experiments: YKA. Performed the experiments: YKA. Analyzed the data: YKA TH. Contributed reagents/materials/analysis tools: YKA. Wrote the paper: YKA TH.

References

1. Kuriyan J, Eisenberg D (2007) The origin of protein interactions and allostery in colocalization. Nature 450: 983–90.
2. Winkel BSJ (2004) Metabolic channeling in plants. Annu Rev Plant Biol 55: 85–107.
3. Buchanan BB (1984) The ferredoxin/thioredoxin system: a key element of the regulatory function of light in photosynthesis. Bioscience 34: 378–383.
4. Scheibe R (1991) Redox-modulation of chloroplast enzymes. A common principle for individual control. Plant Physiol 96: 1–3.
5. Wedel N, Soll J (1998) Evolutionary conserved light regulation of Calvin cycle activity by NADPH-mediated reversible phosphoribulokinase/CP12/glyceraldehyde-3-phosphate dehydrogenase complex dissociation. Proc Natl Acad Sci U S A 95(16): 9699–704.
6. Howard TP, Metodiev M, Lloyd JC, Raines CA (2008) Thioredoxin-mediated reversible dissociation of a stromal multiprotein complex in response to changes in light availability. Proc Natl Acad Sci U S A 105(10): 4056–61.
7. Benz JP, Lintala M, Soll J, Mulo P, Bolter B (2010) A new concept for ferredoxin–NADP(H) oxidoreductase binding to plant thylakoids. Trends Plant Sci 15: 608–613.
8. Juric S, Hazler-Pilepic K, Tomasic A, Lepedus H, Jelicic B, et al. (2009) Tethering of ferredoxin: NADP+ oxidoreductase to thylakoid membranes is mediated by novel chloroplast protein TROL. Plant J 60: 783–794.
9. Xu G, Fan X, Miller AJ (2012) Plant nitrogen assimilation and use efficiency. Annu Rev Plant Biol 63: 153–82.
10. Masclaux-Daubresse C, Daniel-Vedele F, Dechorgnat J, Chardon F, Gaufichon L, et al. (2010) Nitrogen uptake, assimilation and remobilization in plants: challenges for sustainable and productive agriculture. Ann Bot 105(7): 1141–57.
11. Seabra AR, Silva LS, Carvalho HG (2013) Novel aspects of glutamine synthetase (GS) regulation revealed by a detailed expression analysis of the entire GS gene family of Medicago truncatula under different physiological conditions. BMC Plant Biol 13: 137.
12. Schägger H, von Jagow G (1991) Blue native electrophoresis for isolation of membrane protein complexes in enzymatically active form. Anal Biochem 199(2): 223–31.
13. Swamy U, Wang M, Tripathy JN, Kim SK, Hirasawa M, et al. (2005) Structure of spinach nitrite reductase: implications for multi-electron reactions by the iron-sulfur: siroheme cofactor. Biochemistry 44(49): 16054–63.
14. Suzuki A, Knaff DB (2005) Glutamate synthase: structural, mechanistic and regulatory properties, and role in the amino acid metabolism. Photosynth Res 83(2): 191–217.
15. Forde BG, Cullimore JV (1989) The molecular biology of glutamine synthetase in higher plants. In: Miflin, BJ, editor.Oxford Surv Plant Mol Cell Biol Vol. 6.London: Oxford Univ Press. pp. 247–296.
16. Unno H, Uchida T, Sugawara H, Kurisu G, Sugiyama T, et al. (2006) Atomic structure of plant glutamine synthetase: a key enzyme for plant productivity. J Biol Chem 281(39): 29287–96.
17. Torreira E, Seabra AR, Marriott H, Zhou M, Llorca O, et al. (2014) The structures of cytosolic and plastid-located glutamine synthetases from Medicago truncatula reveal a common and dynamic architecture. Acta Crystallogr D Biol Crystallogr 70(Pt 4): 981–93.
18. Mach J (2002) Organelle preparations. In: Weigel D, Glazebrook J, editors.Arabidopsis a Laboratory Manual.New York: Cold Spring Harbour Laboratory Press. pp. 217–218.
19. Sakakibara Y, Kimura H, Iwamura A, Saitoh T, Ikegami T, et al. (2012) A new structural insight into differential interaction of cyanobacterial and plant ferredoxins with nitrite reductase as revealed by NMR and X-ray crystallographic studies. J Biochem 151(5): 483–92.
20. Stengel A, Benz P, Balsera M, Soll J, Bölter B (2008) TIC62 redox-regulated translocon composition and dynamics. J Biol Chem 283(11): 6656–67.
21. Stabile H, Curti B, Vanoni MA (2000) Functional properties of recombinant Azospirillum brasilense glutamate synthase, a complex iron-sulfur flavoprotein. Eur J Biochem 267(9): 2720–30.
22. Petoukhov MV, Svergun DI, Konarev PV, Ravasio S, van den Heuvel RH, et al. (2003) Quaternary structure of Azospirillum brasilense NADPH-dependent glutamate synthase in solution as revealed by synchrotron radiation x-ray scattering. J Biol Chem 278(32): 29933–9.
23. Binda C, Bossi RT, Wakatsuki S, Arzt S, Coda A, et al. (2000) Cross-talk and ammonia channeling between active centers in the unexpected domain arrangement of glutamate synthase. Structure 8(12): 1299–308.
24. van den Heuvel RH, Ferrari D, Bossi RT, Ravasio S, Curti B, et al. (2002) Structural studies on the synchronization of catalytic centers in glutamate synthase. J Biol Chem 277(27): 24579–83.
25. Melo PM, Lima LM, Santos IM, Carvalho HG, Cullimore JV (2003) Expression of the plastid-located glutamine synthetase of Medicago truncatula. Accumula-

tion of the precursor in root nodules reveals an in vivo control at the level of protein import into plastids. Plant Physiol 132(1): 390–9.

26. Masalkar P, Wallace IS, Hwang JH, Roberts DM (2010) Interaction of cytosolic glutamine synthetase of soybean root nodules with the C-terminal domain of the symbiosome membrane nodulin 26 aquaglyceroporin. J Biol Chem 285(31): 23880–8.

27. Elrouby N, Coupland G (2010) Proteome-wide screens for small ubiquitin-like modifier (SUMO) substrates identify Arabidopsis proteins implicated in diverse biological processes. Proc Natl Acad Sci U S A. 107(40): 17415–20.

28. Marqués S, Florencio FJ, Candau P (1992) Purification and characterization of the ferredoxin-glutamate synthase from the unicellular cyanobacterium *Synechococcus* sp. PCC 6301. Eur J Biochem. 206(1): 69–77.

16

Complete Genomic Sequence and Comparative Analysis of the Genome Segments of Sweet Potato Chlorotic Stunt Virus in China

Yanhong Qin[1], Li Wang[2], Zhenchen Zhang[1]*, Qi Qiao[1], Desheng Zhang[1], Yuting Tian[1], Shuang Wang[1], Yongjiang Wang[1], Zhaoling Yan[3]

1 Key Laboratory of Crop Pest Control of Henan Province, Key Laboratory of Pest Management in South of North-China for Ministry of Agriculture of PRC, Institute of Plant Protection, Henan Academy of Agricultural Sciences, Zhengzhou, Henan, China, 2 School of Life Sciences and technology, Nanyang Normal University, Nanyang, Henan, China, 3 Institute of Agricultural Economics and Information, Henan Academy of Agricultural Sciences, Zhengzhou, Henan, China

Abstract

Background: *Sweet potato chlorotic stunt virus* (family *Closteroviridae*, genus *Crinivirus*) features a large bipartite, single-stranded, positive-sense RNA genome. To date, only three complete genomic sequences of SPCSV can be accessed through GenBank. SPCSV was first detected from China in 2011, only partial genomic sequences have been determined in the country. No report on the complete genomic sequence and genome structure of Chinese SPCSV isolates or the genetic relation between isolates from China and other countries is available.

Methodology/Principal Findings: The complete genomic sequences of five isolates from different areas in China were characterized. This study is the first to report the complete genome sequences of SPCSV from whitefly vectors. Genome structure analysis showed that isolates of WA and EA strains from China have the same coding protein as isolates Can181-9 and m2-47, respectively. Twenty *cp* genes and four RNA1 partial segments were sequenced and analyzed, and the nucleotide identities of complete genomic, *cp*, and RNA1 partial sequences were determined. Results indicated high conservation among strains and significant differences between WA and EA strains. Genetic analysis demonstrated that, except for isolates from Guangdong Province, SPCSVs from other areas belong to the WA strain. Genome organization analysis showed that the isolates in this study lack the *p22* gene.

Conclusions/Significance: We presented the complete genome sequences of SPCSV in China. Comparison of nucleotide identities and genome structures between these isolates and previously reported isolates showed slight differences. The nucleotide identities of different SPCSV isolates showed high conservation among strains and significant differences between strains. All nine isolates in this study lacked *p22* gene. WA strains were more extensively distributed than EA strains in China. These data provide important insights into the molecular variation and genomic structure of SPCSV in China as well as genetic relationships among isolates from China and other countries.

Editor: Darren P. Martin, Institute of Infectious Disease and Molecular Medicine, South Africa

Funding: Financial support provided by the Earmarked Fund for China Agricultural Research System (CARS-11-B-07) and Independent Innovation Foundation of Henan Academy of Agricultural Sciences. The funders had no role in study design, data collection and analysis, decision to publish, or preparation of the manuscript.

Competing Interests: The authors have declared that no competing interests exist.

* Email: Zhangzhenchen@126.com

Introduction

Sweet potato (*Ipomoea batatas*) is the third most important root crop after potato and cassava [1,2]. China is currently the largest producer of sweet potato; the country cultivates the crop over an average of 4.1 million hectares of planting area, which accounts for 48.29% of the worldwide total [3,4]. Over 30 viruses are known to infect sweet potato [5]. *Sweet potato chlorotic stunt virus* (SPCSV) is the most devastating virus affecting sweet potato. SPCSV, which was previously known as sweet potato sunken vein virus, belongs to genus *Crinivirus* of family *Closteroviridae* [6–8]. SPCSV was first reported in the 1970s [9]. It is phloem-limited

and transmitted in a semi-persistent manner by whitefly. The virus has the second largest genome and contains a single-stranded bipartite positive-sense RNA genome [10,11]. SPCSV is often found in co-infection with *Sweet potato feathery mottle virus* (SPFMV), a member of the genus *Potyvirus* that causes a synergistic disease called sweet potato virus disease (SPVD); SPVD is the main viral constraint affecting sweet potatoes worldwide [12–14]. Plants with SPVD exhibit severe symptoms, such as leaf strapping, vein clearing, chlorosis, stunting, leaf distortion, and even death, and the disease causes yield losses ranging from 70% to 100% [15–18]. Molecular studies have shown that co-infection of SPCSV enhances SPFMV RNA viral titers by at least 600-fold, whereas SPCSV titers remain equal or are reduced compared with single infection [12,13,17,19]. Besides SPFMV, several other viruses belonging to the genera *Potyvirus*, *Carlavirus*, *Cucumovirus*, *Ipovovirus*, and *Cavemovirus* can result in synergistic diseases and severely affect sweet potato yield upon co-infection with SPCSV [18,20].

SPCSV is distributed worldwide and has been detected in all sweet potato production areas except those in the Pacific region [3,5,21,22]. Based on serological studies, SPCSV can be divided into two distantly related strains: the East African (EA) strain and the West African (WA) strain [23,24]. A similar subdivision into two genetic strains was revealed by phylogenetic analysis of the coat protein (*cp*) and heat shock protein 70 homolog (*hsp70h*) gene sequences [3,25–27] as well as analysis of the RNA1 sequences [28]. WA strains are more widely distributed than EA strains [27,29–31]. The complete genomic sequence of SPCSV was first determined by Kreuze [11]. As SPCSV has a low titer in sweet potato plants and always co-infects sweet potatoes with other viruses in field [22,32–34], separation and purification of SPCSV is difficult to perform and determination of the SPCSV genomic sequence is greatly constrained. To date, only three SPCSV genomic sequences can be accessed through GenBank, including two complete sequences of the SPCSV EA strain and one sequence of the SPCSV WA strain [11,29,35]. Partial sequence analysis shows that not all EA strain isolates include the *p22* open reading frame (ORF) at the 3′ end of RNA1; other isolates may lack a 767 nt region of RNA1 that includes the *p22* gene [29]. The *p22* gene has only been found in isolates from Uganda [28,29,36]. Regardless of the presence of *p22*, isolates of SPCSV act synergistically with SPFMV in sweet potato plants and significantly enhance SPFMV titers. However, co-infection of SPFMV with SPCSV isolates containing *p22* causes more severe symptoms in the indicator plant *Ipomoea setosa* than co-infection of SPFMV with SPCSV isolates lacking *p22* [28].

Recent studies have focused on the synergism of SPCSV with other unrelated viruses; the incidence, distribution, and effects of SPVD on sweet potato yield; and management approaches to SPVD [18,33,37]. Limited information is available on the genetic variability of SPCSV, and this information is based on analysis of the nucleotide sequences of *hsp70h* and *cp* genes on RNA2 and *p7*, *p22*, and *RNase3* genes on RNA1 [3,25–27,36].

In China, SPCSV was first detected in 2011 [38], and SPVD was first reported in China in 2012 [39]. Subsequent to its discovery, the virus developed and caused a serious epidemic from 2012 to 2013 [27]. Despite the importance of the virus, however, knowledge of the molecular characterization and genetic diversity of SPCSV in China, which is an important sweet potato production area, remains limited. Moreover, the presence or absence of the *p22* gene in China isolates is unclear.

The current study presents the complete genome sequences of five SPCSV EA and WA isolates obtained from different areas in China. This study is the first to report the complete genome sequences of SPCSV from whitefly vectors, and we provide a simplified and convenient method for cloning SPCSV genomes. The genomic structures of the five isolates were analyzed and compared with three isolate genomes obtained from GenBank. Nucleotide and amino acid identities, including the molecular variance of different isolates, were also compared and a phylogenetic tree was constructed. To confirm the genetic variability of the *cp* gene of WA strains and investigate whether or not the *p22* gene is present in China, we also described the *cp* gene sequences of 20 other isolates and 4 genome segments of RNA1 flanking the *p22* insertion region. Genetic variability and

phylogenetic relationships among these isolates were further analyzed.

Results

Complete nucleotide sequence analyses of SPCSV from China and comparison with other isolates

The complete genomic sequences of five isolates of SPCSV belonging to WA and EA strains from different areas in China were characterized by RT-PCR using viruliferous whitefly as materials. The 5′ and 3′ UTRs of virus genome segments RNA1 and RNA2 were also determined by rapid amplification of cDNA ends (RACE). The cloning strategy and a schematic representation of the SPCSV genome organization are shown in Figure 1. The complete genome nucleotide sequences of the five SPCSV isolates were deposited in GenBank under accession numbers KC146840–KC146843 and KC888961–KC888966 (Table 1). Table 2 shows the primer sequences and their corresponding positions in the SPCSV genome.

The nucleotide sequences of the Guangdong-2011 isolate genome RNA1 and RNA2 comprised 8622 and 8217 bp, respectively. The genome segment RNA1 of Guangdong-2011 isolate had an 89 nt 5′-UTR, a 172 nt 3′-UTR, and four ORFs: nucleotides 90–6053 (*p227*), 6052–7569 (*RdRp*), 7586–8272 (*RNase3*), and 8277–8450 (*p7*). The genome segment RNA2 had an 88 nt 5′-UTR, a 192 nt 3′-UTR, and contained eight ORFs: nucleotides 89–238 (*p6.1*), 554–712 (*p6.2*), 984–2648 (*hsp70h*), 2670–4226 (*p60*), 4208–4429 (*p8*), 4462–5235 (major *cp*), 5238–7292 (minor *cp*), and 7297–8025 (*p28*). The nucleotide sequences of the Jiangsu-2011, Sichuan-12-8, Sichuan-12-12, and Chongqing-12-8 isolate genome RNA1 and RNA2 comprised 8637 and 8107 bp, respectively. The genomic segment RNA1 of these isolates had an 89 nt 5′-UTR, a 193 nt 3′-UTR, and contained four ORFs: nucleotides 90–6053 (*1a*), 6052–7569 (*RdRp*), 7583–8272 (*RNase3*), and 8277–8444 (*p7*). The genomic segment RNA2 of these isolates had a 191 nt 5′-UTR, a 192 nt 3′-UTR, and contained nine ORFs: nucleotides 192–329 (*p5.2*), 333–467 (*p5*), 406–534 (*p5.1*), 879–2543 (*hsp70h*), 2565–4121 (*p60*), 4103–4324 (*p8*), 4352–5125 (major *cp*), 5128–7182 (minor *cp*), and 7187–7915 (*p28*). Comparisons of the genomic structures of RNA1 and RNA2 between this study and the reported isolates are shown in Tables 3 and 4.

Multiple sequence comparisons showed that the RNA1 sequence of the Guangdong-2011 isolate was similar to that of the SPCSV EA strain. RNA1 and RNA2 genome segments of the Guangdong-2011 isolate were 99.27% and 99.68% identical to that of the EA m2-47 isolate and 82.44% and 69.99% identical to that of the WA Can181-9 isolate, respectively. The Jiangsu-2011, Sichuan-12-8, Sichuan-12-12, and Chongqing-12-8 isolates showed a closer genetic relationship to the WA Can181-9 isolate than to the EA m2-47 isolate. The nucleotide sequence identities of RNA1 and RNA2 segments respectively ranged from 98.90% to 99.26% and from 98.80% to 99.17% when these four isolates were compared with the Can181-9 isolate of the WA strain. By contrast, the nucleotide sequence identities of RNA1 and RNA2 segments respectively ranged from 82.33% to 82.41% and from 69.80% to 69.98% when these four isolates were compared with the m2-47 isolate of the EA strain.

The pairwise percent identity of complete genomic sequences of the five isolates in China and three isolates retrieved from GenBank was calculated in multiple alignment (Table 5). Results showed that the RNA1 and RNA2 segment nt sequence identities of all isolates ranged from 81.5% to 99.6% and from 70.9% to 99.7%, respectively. The nt sequence identities in the WA strain

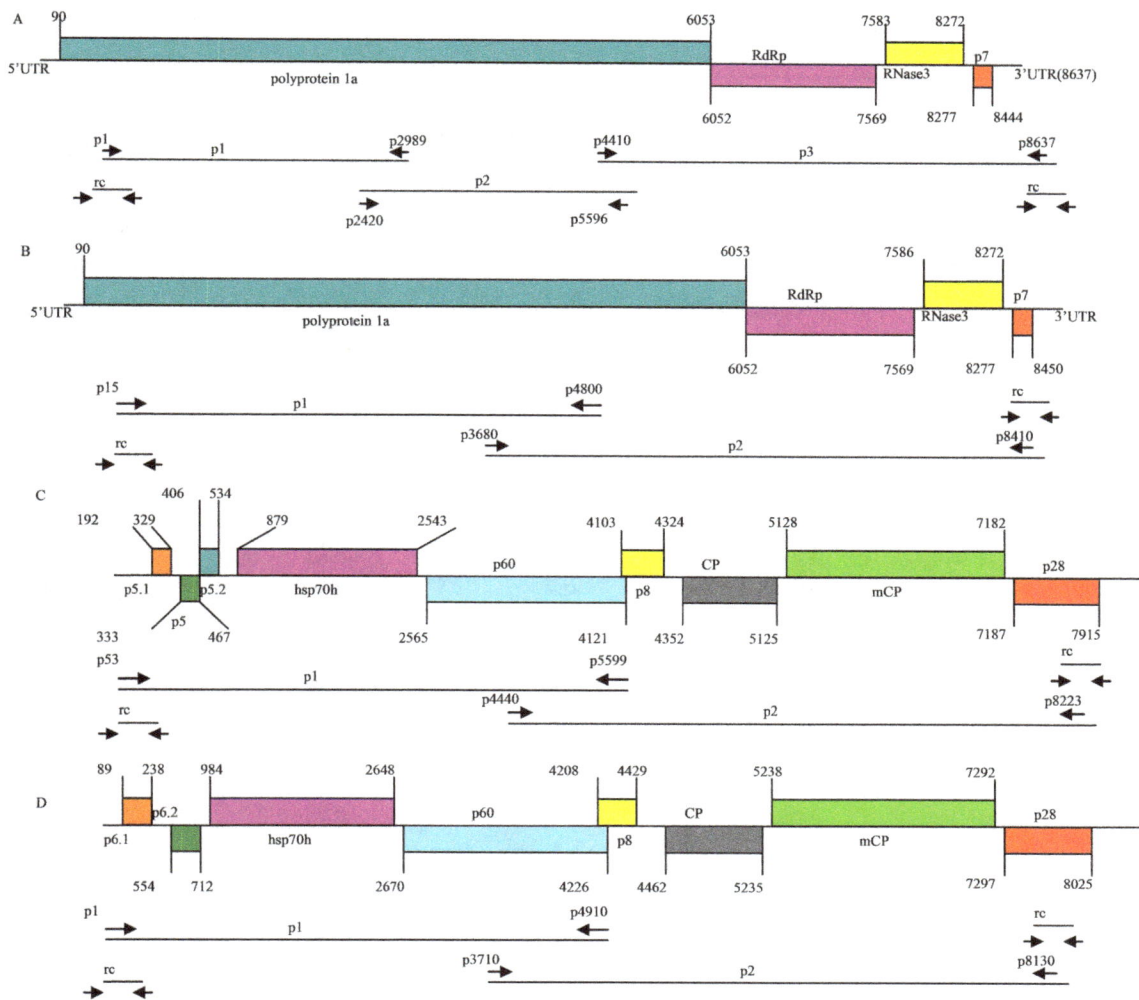

Figure 1. Schematic diagram of genomic organization and genome cloning strategy of SPCSV. Solid lines represent RNA genomes and boxes represent ORFs. Putative protein products are indicated. p1, p2, and p3 represent RT-PCR-generated sequences using specific primers. rc represents the 5- and 3-terminal clones generated by 5' and 3' RACE, respectively. Primer sequences are shown in Table 2. (A) Schematic diagram of the genomic organization and genome cloning strategy of the SPCSV WA strain RNA1 segment. (B) Schematic diagram of the genomic organization and genome cloning strategy of the SPCSV EA strain RNA1 segment. (C) Schematic diagram of the genomic organization and genome cloning strategy of the SPCSV WA strain RNA2 segment. (D) Schematic diagram of the genomic organization and genome cloning strategy of the SPCSV EA strain RNA2 segment.

were 98.8%–99.6% for RNA1 and 98.8%–99.7% for RNA2. The nt sequence identities in EA strains were 97.5%–99.4% for RNA1 and 98.3%–99.7% for RNA2. The nt identities between WA and EA strains were 81.5%–82.7% for RNA1 and 70.9%–71.2% for RNA2. This study revealed that the complete genomic sequences of the same strain group display a high degree of conservation. Compared with the reported sequences, the Chinese isolates showed high conservation and low molecular variation in their genomic sequences.

MEGA 4.0 was used to construct a phylogenetic tree of the five complete genomic sequences determined in this study and three isolates obtained from GenBank (Figure 2). Results indicated that Jiangsu-2011, Sichuan-12-8, Sichuan-12-12, and Chongqing-12-8 belong to the same branch as the Can181-9 isolate while Guangdong-2011 belongs to the same branch as the m2-47 and Uganda isolates.

Comparison of the length and nucleotide identity of the 5' and 3' UTRs of RNA1 and RNA2

Analysis of RNA1 5' UTRs indicated 89 nt in Guangdong-2011, m2-47, Uganda (EA strain), Jiangsu-2011, Sichuan-12-8, Sichuan-12-12, Chongqing-12-8 and Can181-9 (WA strain), a nucleotide identity of over 98% between different isolates in the WA and EA strains, and a nucleotide identity of 85.4%–87.6% between WA and EA strains (Table 6).

Analysis of RNA1 3' UTRs showed that this UTR, being 193 nt long (Table 3), is conserved between different isolates in WA strains and that its nucleotide identity is 98.4%–100% (Table 6). Isolates in the EA strain differed from each other significantly: the length of the 3' UTR of isolate Uganda was 226 nt, that of isolate m2-47 was 187 nt long, and that of isolate Guangdong-2011 was 172 nt long (Table 3). The nt identity between Guangdong-2011 and m2-47 isolates was 100%, that between Uganda and m2-47 was 81.8%, and that between Guangdong-2011 and Uganda was 80.2% (Table 6).

Table 1. Name, strain assignment, geographic origin, segment length, reference, and GenBank accession number of all isolates or samples used in this study.

Isolate names	accession numbers	Serotype	Segments	Segment length (bp)	Geographical origin	Reference
Jiangsu-2011	KC146840	WA	RNA1	8637	Jiangsu	This study
Jiangsu-2011	KC146841	WA	RNA2	8107	Jiangsu	This study
Guangdong-2011	KC146842	EA	RNA1	8622	Guangdong	This study
Guangdong-2011	KC146843	EA	RNA2	8217	Guangdong	This study
Sichuan-12-8	KC888964	WA	RNA1	8637	Sichuan	This study
Sichuan-12-8	KC888961	WA	RNA2	8107	Sichuan	This study
Sichuan-12-12	KC888965	WA	RNA1	8637	Sichuan	This study
Sichuan-12-12	KC888962	WA	RNA2	8107	Sichuan	This study
Chongqing-12-8	KC888966	WA	RNA1	8637	Chongqing	This study
Chongqing-12-8	KC888963	WA	RNA2	8107	Chongqing	This study
m2-47	HQ291259	EA	RNA1	8637	Peru	Cuellar et al. 2011
m2-47	HQ291260	EA	RNA2	8219	Peru	Cuellar et al. 2011
Uganda	AJ428554	EA	RNA1	9407	Uganda	Kreuze et al. 2002
Uganda	AJ428555	EA	RNA2	8223	Uganda	Kreuze et al. 2002
Can181-9	FJ807784	WA	RNA1	8637	Spain	Trenado et al. 2012
Can181-9	FJ807785	WA	RNA2	8108	Spain	Trenado et al. 2012
Jiangsu-11-19	KC243096	WA	RNA1	1356	Jiangsu	This study
Anhui-11-2	KC243097	WA	RNA1	1356	Anhui	This study
Chongqing-11-8	KC243098	WA	RNA1	1356	Chongqing	This study
Zhejiang-11-4	KC243099	EA	RNA1	1356	Zhejiang	This study
Sichuan-11-28	KC243079	WA	RNA2	774	Sichuan	This study
Jiangxi-11-3	KC243080	WA	RNA2	774	Jiangxi	This study
Guangxi-11-5	KC243081	WA	RNA2	774	Guangxi	This study
Guangxi-11-7	KC243082	WA	RNA2	774	Guangxi	This study
Hebei-11-5	KC243083	WA	RNA2	774	Hebei	This study
Shanxi-11-6	KC243084	WA	RNA2	774	Shanxi	This study
Shandong-11-7	KC243085	WA	RNA2	774	Shandong	This study
Shaanxi-11-1	KC243086	WA	RNA2	774	Shaanxi	This study
Jiangsu-11-17	KC243087	WA	RNA2	774	Jiangsu	This study
Jiangsu-11-19	KC243088	WA	RNA2	774	Jiangsu	This study
Anhui-11-2	KC243089	WA	RNA2	774	Anhui	This study
Hunan-11-4	KC243090	WA	RNA2	774	Hunan	This study
Chongqing-11-3	KC243091	WA	RNA2	774	Chongqing	This study
Henan-11-28	KC243092	WA	RNA2	774	Henan	This study
Guangdong-11-9	KC243093	WA	RNA2	774	Guangdong	This study
Zhejiang-11-3	KC243094	WA	RNA2	774	Zhejiang	This study
Zhejiang-11-4	KC243095	WA	RNA2	774	Zhejiang	This study
Guangdong-2009	HM773432	EA	RNA2	774	Guangdong	Qiao et al. 2011
Sichuan-10-77	KC146844	WA	RNA2	774	Sichuan	This study

Comparison of RNA2 5'UTRs showed that the WA strain length is 191 nt (Table 4) and that the nucleotide identity between different isolates is 99.5%–100% (Table 6). The nucleotide identity between the three isolates in the EA strain was 96.6%–98.9% (Table 6), the length of isolate Guangdong-2011 was 88 nt, and the lengths of the two other isolates were both 90 nt. Despite differences in the 5' UTR lengths among WA and EA strains, the nucleotide sequences of these strains were rather conserved; nt identities between the eight isolates were 93.2%–100% (Table 6).

Comparison of RNA2 3' UTRs suggested that the length of both WA and EA strains is 192 nt (Table 4) and that the nt identities between different isolates in the WA and EA strains exceed 99%. By contrast, the nt identity between WA and EA strains was 79.7%, which is rather low (Table 6).

Comparison of the 5' and 3' UTRs of RNA1 and RNA2 indicated that the nt identity of 5'UTRs between RNA1 and RNA2 is low (around 40%) whereas that of 3'UTRs of RNA1 and

Table 2. Polymerase chain reaction primers and PCR conditions used to amplify the SPCSV genome, CP gene, and RNA1 partial sequences.

Primer name	Sequence (5' to 3')	Nucleotide position	Size of amplicons (bp)	polarity	PCR conditions
WA-RNA1-1	GAAATACTTCCAGCTATCCAAATTTGGTG	1–29	2989	Sense	94°C, 3 minutes, 1 cycle; 94°C, 30 seconds, 55°C, 30 seconds,
WA-RNA1-2989	ATACGTCTCTCTCCAACGACAAC	2967–2989		Antisense	72°C, 3 minutes, 30 cycles; 72°C, 10 minutes, 1 cycle
WA-RNA1-2420	AGAGAACAACTTCATTTCTACTCAATTGT	2421–2449	3205	Sense	94°C, 3 minutes, 1 cycle; 94°C, 30 seconds, 55°C, 30 seconds,
WA-RNA1-5596	GATAAATAAGTTTACCTGTATTGTCGGTC	5597–5625		Antisense	72°C, 3.5 minutes, 30 cycles; 72°C, 10 minutes, 1 cycle
WA-RNA1-4410	GAGCATCAACTGTGGACGGCGAACCAAGC	4411–4439	4227	Sense	94°C, 3 minutes, 1 cycle; 94°C, 30 seconds, 55°C, 30 seconds,
WA-RNA1-8637	AACCTAGTTATTTAAATACTAGGTTTTCC	8609–8637		Antisense	72°C, 4.5 minutes, 30 cycles; 72°C, 10 minutes, 1 cycle
WA-RNA2-53	CATTGGTTGTCGTCATGACTCGCAT	53–77	5574	Sense	94°C, 3 minutes, 1 cycle; 94°C, 30 seconds, 55°C, 30 seconds,
WA-RNA2-5599	CCAACTTACCAGATTTCGAGAACTGTAC	5599–5626		Antisense	72°C, 6 minutes, 30 cycles; 72°C, 10 minutes, 1 cycle
WA-RNA2-4440	ATGCGTCTCGTCGTGACGTCCAGACTG	4440–4466	3668	Sense	94°C, 3 minutes, 1 cycle; 94°C, 30 seconds, 55°C, 30 seconds,
WA-RNA2-8223	GGCCTAGTTATTTAAATACTAGGTTTTCC	8079–8107		Antisense	72°C, 4 minutes, 30 cycles; 72°C, 10 minutes, 1 cycle
EA-RNA1-15	TATCCAAATTTGGTGTGTTCTGCAG	15–39	4815	Sense	94°C, 3 minutes, 1 cycle; 94°C, 30 seconds, 55°C, 30 seconds,
EA-RNA1-4800	TTCCAATTGTGATAGATAAAGATCTACC	4802–4829		Antisense	72°C, 5 minutes, 30 cycles; 72°C, 10 minutes, 1 cycle
EA-RNA1-3680	GGTGAGTAAGAAGGATAAATTGATCTCG	3680–3707	4754	Sense	94°C, 3 minutes, 1 cycle; 94°C, 30 seconds, 55°C, 30 seconds,
EA-RNA1-8410	TTCTAATACTCAAAAGGCAATATACAAAC	8405–8433		Antisense	72°C, 5 minutes, 30 cycles; 72°C, 10 minutes, 1 cycle
EA-RNA2-1	GAAATACTACCCAGGTTTTTCCATGAGT	1–28	4933	Sense	94°C, 3 minutes, 1 cycle; 94°C, 30 seconds, 55°C, 30 seconds,
EA-RNA2-4910	GCATGTGATTGATGAAACTATGAGTTC	4907–4933		Antisense	72°C, 5 minutes, 30 cycles; 72°C, 10 minutes, 1 cycle
EA-RNA2-3710	TTACCTAGGGACGTTGACGAATTGGT	3705–3730	4452	Sense	94°C, 3 minutes, 1 cycle; 94°C, 30 seconds, 55°C, 30 seconds,
EA-RNA2-8130	TTCATACACACACTCTAAATAGAAATACG	8128–8156		Antisense	72°C, 4.5 minutes, 30 cycles; 72°C, 10 minutes, 1 cycle
RdRp-F	CAANACNAANGAATTGAACAT	7176–7196	1356 or 2081	Sense	94°C, 3 minutes, 1 cycle; 94°C, 30 seconds, 53°C, 30 seconds,
SVV-R3	TTTTTGAGNTTTTANAATACACAC	8508–8531		Antisense	72°C, 2 minutes, 30 cycles; 72°C, 10 minutes, 1 cycle
CP-F	ATGGCTGATAGCACTAAAGTCGA	4352–4374	774	Sense	94°C, 3 minutes, 1 cycle; 94°C, 30 seconds, 58°C, 30 seconds,
CP-R	TCAACAGTGAAGACCTGTTCCAG	5103–5125		Antisense	72°C, 45 seconds, 30 cycles; 72°C, 10 minutes, 1 cycle
Hsp70h-F	AGTGGTGAYGTAATAGTCGGTGG	1008–1030	365	Sense	94°C, 3 minutes, 1 cycle; 94°C, 30 seconds, 58°C, 30 seconds,
Hsp70h-R	GCTAACGATTCACADACAGACTTCA	1348–1372		Antisense	72°C, 30 seconds, 30 cycles; 72°C, 10 minutes, 1 cycle

RNA2 is high (around 80%). The nt identity of 3′ UTRs of RNA1 and RNA2 of isolate Uganda was about 99%.

Genome structure analysis and nucleotide/amino acid sequence comparison of proteins

Analysis of genome structures suggested that the RNA1 segment contained four ORFs: 1a (p227), RdRp, RNase3, and p7. The genome positions of these ORFs were listed in Table 5. Between protein p7 and the 3′UTR, the Uganda isolate contains a p22 ORF (located at 8606–9181 nt of the genome); none of the seven other isolates had the *p22* gene. Nucleotide and amino acid identities between these proteins are listed in Table 6.

Analysis of the nucleotide and amino acid identity of protein 1a (p227), RdRp, RNase3, and p7 encoded by eight isolates in RNA1

Table 3. Comparison of genomic structures of RNA1.

	Can181-9	Chongqing-12-8	Jiangsu-2011	Sichuan-12-8	Sichuan-12-12	Guangdong-2011	m2-47	Uganda
5'UTR	1–89	1–89	1–89	1–89	1–89	1–89	1–89	1–89
polyprotein 1a	90–6053	90–6053	90–6053	90–6053	90–6053	90–6053	90–6053	90–6053
RdRP	6052–7569	6052–7569	6052–7569	6052–7569	6052–7569	6052–7569	6004–7569	6052–7569
RNase3	7583–8272	7583–8272	7583–8272	7583–8272	7583–8272	7586–8272	7586–8272	7586–8272
p7	8277–8444	8277–8444	8277–8444	8277–8444	8277–8444	8277–8450	8277–8450	8277–8450
p22								8606–9181
3'UTR	8445–8637	8445–8637	8445–8637	8445–8637	8445–8637	8451–8622	8451–8637	9182–9407

Table 4. Comparison of genomic structures of RNA2.

	Can181-9	Chongqing-12-8	Jiangsu-2011	Sichuan-12-8	Sichuan-12-12	Guangdong-2011	m2-47	Uganda
5'UTR	1–191	1–191	1–191	1–191	1–191	1–88	1–90	1–90
p5.2/p6.1	192–329	192–329	192–329	192–329	192–329	89–238	91–240	91–240
p5	333–467	333–467	333–467	333–467	333–467			
p5.1/p6.2	406–534	406–534	406–534	406–534	406–534	554–712	556–714	
hsp70h	880–2544	879–2543	879–2543	879–2543	879–2543	984–2648	986–2650	987–2651
p60	2566–4122	2565–4121	2565–4121	2565–4121	2565–4121	2670–4226	2672–4228	2673–4229
p8	4104–4325	4103–4324	4103–4324	4103–4324	4103–4324	4208–4429	4210–4431	4211–4432
CP	4353–5126	4352–5125	4352–5125	4352–5125	4352–5125	4462–5235	4464–5237	4465–5241
mCP	5129–7183	5128–7182	5128–7182	5128–7182	5128–7182	5238–7292	5240–7294	5244–7298
p28	7188–7916	7187–7915	7187–7915	7187–7915	7187–7915	7297–8025	7299–8027	7303–8031
3'UTR	7917–8108	7916–8107	7916–8107	7916–8107	7916–8107	8026–8217	8028–8217	8032–8223

Table 5. Analysis of nucleotide sequence identities (%) of RNA1 (lower diagonal) and RNA2 (upper diagonal) segments of SPCSV isolates in this study and those retrieved from the database.

Isolates	1	2	3	4	5	6	7	8
1. Can181-9		99.0	98.8	99.2	99.2	71.0	70.9	71.0
2. Chongqing-12-8	98.9		99.1	99.1	99.1	71.0	70.9	71.0
3. Jiangsu-2011	98.9	99.6		98.8	98.8	71.1	71.0	71.2
4. Sichuan-12-8	99.3	98.8	98.8		99.7	71.0	70.9	71.1
5. Sichuan-12-12	99.3	98.8	98.8	99.6		71.0	70.9	71.0
6. Guangdong-2011	82.6	82.5	82.5	82.5	82.6		99.7	98.3
7. m2-47	82.7	82.6	82.5	82.6	82.6	99.4		98.4
8. Uganda	81.6	81.5	81.5	81.6	81.6	97.5	97.6	

showed that the nucleotide identities of encoding proteins between the eight isolates are as follows: protein 1a (81.0%–99.6%), RdRp (87.4%–99.8%), RNase3 (82.7%–99.6%) and p7(75.0%–100%). The amino acid identities of encoding proteins between the eight isolates are as follows: protein 1a (87.2%–99.5%), RdRp (95.8%–99.8%), RNase3 (79.8%–99.6%), and p7 (60.0%–100%). The nt/aa identities between different isolates in the WA strain are as follows: protein 1a (98.7%–99.6%/98.6%–99.5%), RdRp (99.1%–99.8%/98.8%–99.6%), RNase3 (98.3%–99.6%/97.4%–99.6%), and p7 (98.2%–100%/96.4%–100%). The nt/aa identities between different isolates in the EA strain are as follows: protein 1a (98.3%–99.4%/97.6%–98.7%), RdRp (99.4%–99.6%/99.2%–99.8%), RNase3 (98.7%–99.4%/98.2%–99.1%), and p7 (98.3%–100%/98.2%–100%). The nt/aa identities between WA and EA strains are as follows: protein 1a (81.0%–81.3%/87.2%–87.6%), RdRp (87.4%–87.6%/95.8%–96.6%), RNase3 (82.7%–83.6%/79.8%–81.6%), and p7 (75.0%–78.0%/60.0%–63.6%). Analysis of RdRp protein sequences demonstrated that the 5′ end of isolate m2-47 is 16 aa longer than that of the other isolates. RNase3 sequence analysis showed that the ORF for RNase3 is 684 nt (228 aa) in all EA strain isolates but 687 nt (229 aa) in WA strain isolates. The 5′ end of the RNase3 ORF in EA strains was 1 aa longer than that of WA strains. P7 sequence analysis showed that the ORF for p7 is 171 nt (57 aa) in all EA strain isolates but 165 nt (55 aa) in WA strain isolates. The 3′ end of the p7 ORF in EA strains was 2 aa longer than that of WA strains.

Analysis of genomic structures indicated that the segment RNA2 contains at least six ORFs: Hsp70h, p60, p8, CP, mCP, and p28. The locations of these ORFs in the genome are listed in Table 4. The nucleotide and amino acid identities between these proteins are listed in Table 6.

Isolates Guangdong-2011, m2-47, and Uganda of the EA strain contained an ORF for p6 (named p6.1 in this study) in the upstream region of hsp70h, and the nt/aa identity of the p6.1 protein between these three isolates was 100%. Between p6.1 and hsp70h, similar to isolate m2-47, the Guangdong-2011 isolate contained one ORF for the p6 protein (named p6.2 in this study) more than isolate Uganda; the nt and aa identities of this protein were 98.7% and 96.2%, respectively. Similar to Can181-9, isolates Jiangsu-2011, Sichuan-12-8, Sichuan-12-12, and Chongqing-12-8 of the WA strain contained three ORFs in the upstream region of hsp70h, namely p5.2, p5, and p5.1. The nt and aa identities of the five isolates are as follows: p5.2 (97.8%–100% and 100%), p5 (97.0%–100% and 90.9%–100%), and p5.1 (95.3%–100% and 90.5%–100%).

Analysis showed that the nucleotide identities of the encoded proteins between the eight isolates are as follows: Hsp70h (76.5%–99.9%), p60 (72.4%–99.8%), p8 (72.1%–100%), CP (72.8%–99.7%), mCP (62.9%–99.8%), and p28 (77.4%–99.7%). The aa identities of these proteins are as follows: Hsp70h (90.1%–100%), p60 (78.8%–99.4%), p8 (68.5%–100%), CP (76.7%–99.6%), mCP (61.1%–99.4%), and p28 (85.1%–99.2%). The nt/aa identities between different isolates of WA strain are as follows: Hsp70h (98.8%–99.6%/98.4%–99.3%), p60 (98.3%–99.7%/97.3%–99.4%), p8 (98.2%–99.1%/95.9%–100%), CP (98.7%–99.6%/97.7%–99.6%), mCP (98.5%–99.8%/98.1%–99.4%), and p28 (98.6%–99.7%/97.9%–99.2%). The nt/aa identities between different isolates within the EA strain are as follows: Hsp70h (98.7%–99.9%/99.8%–100%), p60 (98.8%–99.8%/98.6%–99.4%), p8 (95.5%–100%/93.2%–100%), CP (98.6%–99.7%/95.7%–99.2%), mCP (98.2%–99.7%/98.1%–99.1%), and p28 (97.9%–99.7%/97.9%–99.2%). The nt/aa identities between the WA and EA strains are as follows: Hsp70h (76.5%–77.0%/90.1%–90.6%), p60 (72.4%–73.0%/78.8%–80.5%), p8 (72.1%–

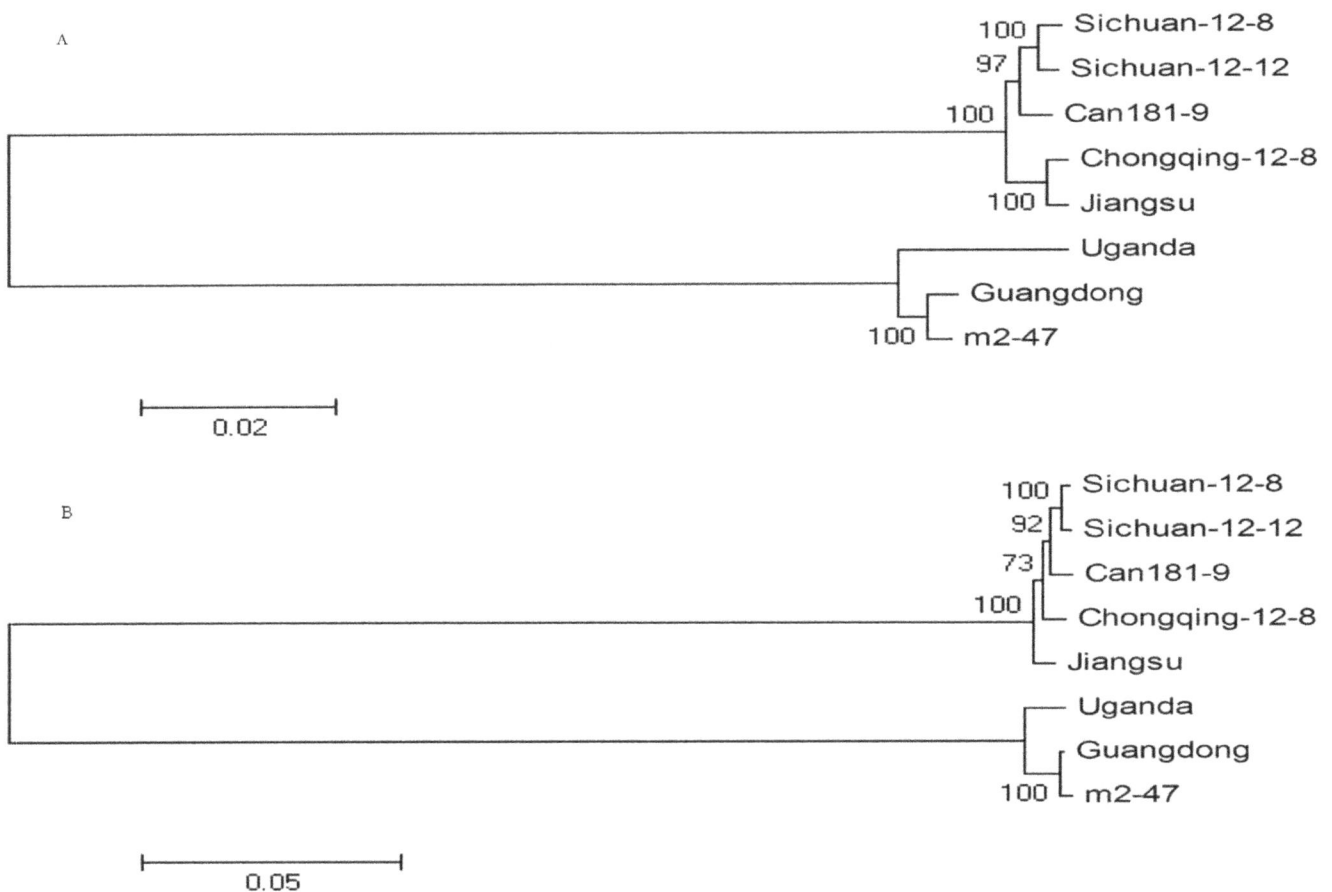

Figure 2. Phylogenetic analysis of genomic RNA1 (A) and RNA2 (B) segments of five isolates determined in China and three isolates retrieved from GenBank. Neighbor-joining trees were constructed via the maximum composite likelihood substitution model using MEGA (version 4.0). Numbers at branches represent bootstrap values of 1000 replicates. The scale bar shows the number of nucleotide substitutions per site.

73.4%/68.5%–69.9%), CP (72.8%–74.0%/76.7%–80.2%), mCP (62.9%–63.6%/61.1%–62.3%), and p28 (77.4%–78.5%/85.1%–86.4%). CP protein sequence analysis also showed that isolate Uganda has one methionine residue more than the seven other isolates in locus 231.

Absence of the *p22* gene in RNA1

Genome structure analysis showed that the *p22* gene was not present in any of the aforementioned isolates from China. To investigate whether or not *p22* is present in other Chinese regions, the primers RdRp-F and SVV-R3 were used to amplify the region from the *RdRp* gene to the 3′-UTR of RNA1; these primers were designed according to the reference [28]. Sweet potato samples infected by SPCSV were collected from Jiangsu, Zhejiang, Anhui, and Chongqing, China. Using RT-PCR, the partial RNA1 sequences from four different samples were obtained (relevant data, such as name, strain assignment, geographic origin, segment length, reference, and GenBank accession numbers, are listed in Table 1). Sequence analysis indicated that Zhejiang-11-4 shares nucleotide identities of 99.6% with Guangdong-2011 and 84.4% with isolate Jiangsu-2011. Isolates Jiangsu-11-19, Anhui-11-2, and Chongqing-11-8 shared nucleotide identities of 84.4% with isolate Guangdong-2011 and 99.4%–99.6% with isolate Jiangsu-2011. Overall, genome organization analysis showed that *p22* is missing in all nine isolates from China.

The nt sequences of 9 SPCSV isolates characterized in this study and 14 isolates deposited in GenBank were subjected to phylogenetic analysis, including the previously characterized WA strain Can181-9 isolate from Spain, EA strain m2-47 isolate from Peru, and 12 EA strain isolates from Uganda. Phylogenetic analysis (Figure 3) of partial genomic sequences of RNA1 was used to assign isolates of SPCSV from China to the two relatively distantly related strains EA and WA. Isolates Zhejiang-11-4 and Guangdong-2011 in this study belonged to the EA strain. Isolates Jiangsu-11-19, Anhui-11-2, Chongqing-11-8, Jiangsu-2011, Sichuan-12-8, Sichuan-12-12, and Chongqing-12-8 belonged to the WA strain.

Molecular variability and phylogenetic analysis of the *cp* gene in WA strains

To study the genetic variation of SPCSV in different Chinese regions, we used the nucleotide sequence of the *cp* gene of the SPCSV WA strain in GenBank (GenBank accession number FJ807785) to design primers named CP-F and CP-R to amplify the *cp* gene of SPCSV WA isolates from different regions. A total of 20 *cp* gene sequences from 14 Chinese provinces or cities were obtained (name, strain assignment, geographic origin, segment length, reference, and GenBank accession numbers are listed in Table 1). Sequencing results suggested that the *cp* gene sequence length of 20 isolates is 774 bp, as expected.

Table 6. Nucleotide (lower diagonal) and amino acid (upper diagonal) (%) sequence identities of UTRs, polyprotein 1a (p227), RdRp, RNase3, p7, hsp70h, p60, p8, CP, mCP, and p28 among SPCSV isolates.

	Can181-9	Chongqing-12-8	Jiangsu-2011	Sichuan-12-8	Sichuan-12-12	Guangdong-2011	m2-47	Uganda
RNA1-5'UTR								
Can181-9								
Chongqing-12-8	98.9							
Jiangsu-2011	97.8	98.9						
Sichuan-12-8	98.9	100	98.9					
Sichuan-12-12	98.9	100	98.9	100				
Guangdong-2011	85.4	86.5	87.6	86.5	86.5			
m2-47	85.4	86.5	87.6	86.5	86.5	100		
Uganda	85.4	86.5	87.6	86.5	86.5	98.9	98.9	
polyprotein 1a								
Can181-9		98.8	98.7	99.3	99.5	87.2	87.4	87.5
Chongqing-12-8	99.0		99.2	98.7	98.9	87.3	87.5	87.6
Jiangsu-2011	98.9	99.6		98.6	98.8	87.4	87.6	87.6
Sichuan-12-8	99.3	98.8	98.7		99.5	87.2	87.4	87.5
Sichuan-12-12	99.3	98.8	98.7	99.5		87.3	87.5	87.6
Guangdong-2011	81.2	81.2	81.0	81.2	81.2		98.7	97.6
m2-47	81.3	81.3	81.1	81.2	81.3	99.4		98.0
Uganda	81.3	81.2	81.1	81.2	81.3	99.3	98.6	
RdRp								
Can181-9		99.2	99.6	99.4	99.2	96.2	96.4	96.6
Chongqing-12-8	99.1		99.6	99.0	98.8	96.0	96.2	96.4
Jiangsu-2011	99.3	99.8		99.4	99.2	96.2	96.4	96.6
Sichuan-12-8	99.4	99.1	99.3		99.4	96.0	96.2	96.4
Sichuan-12-12	99.4	99.1	99.3	99.7		95.8	96.0	96.2
Guangdong-2011	87.6	87.4	87.5	87.4	87.4		99.2	99.4
m2-47	87.6	87.4	87.5	87.4	87.4	99.5		99.8
Uganda	87.7	87.5	87.5	87.4	87.4	99.4	99.8	
RNase3								
Can181-9		97.4	97.8	98.3	97.8	80.7	79.8	81.1
Chongqing-12-8	98.3		99.6	97.8	97.4	80.7	79.8	81.1
Jiangsu-2011	98.8	99.4		98.3	97.8	81.1	80.3	81.6
Sichuan-12-8	98.8	98.6	99.1		98.7	81.1	80.3	81.6
Sichuan-12-12	98.7	98.4	99.0	99.6		80.7	79.8	81.1
Guangdong-2011	83.0	82.8	83.4	83.3	83.1		99.1	99.1

Table 6. Cont.

	Can181-9	Chongqing-12-8	Jiangsu-2011	Sichuan-12-8	Sichuan-12-12	Guangdong-2011	m2-47	Uganda
m2-47	82.8	82.7	83.3	83.1	83.0	99.4		98.2
Uganda	83.1	83.0	83.6	83.4	83.3	99.1	98.7	
p7								
Can181-9	Can181-9	98.2	98.2	98.2	100	63.6	63.6	61.8
Chongqing-12-8	98.8	Chongqing-12-8	100	96.4	98.8	63.6	63.6	61.8
Jiangsu-2011	98.8	100	Jiangsu-2011	96.4	98.8	63.6	63.6	61.8
Sichuan-12-8	99.4	98.2	96.4	Sichuan-12-8	99.4	61.8	61.8	60.0
Sichuan-12-12	100	98.8	98.8	99.4	Sichuan-12-12	63.6	63.6	61.8
Guangdong-2011	78.0	76.8	76.8	77.4	78.0	Guangdong-2011	100	98.2
m2-47	78.0	76.8	76.8	77.4	78.0	100	m2-47	98.2
Uganda	76.2	75.0	75.0	75.6	76.2	98.3	98.2	Uganda
RNA1-3'UTR								
Can181-9	Can181-9	99.0	100	99.5	100	90.7	90.9	80.3
Chongqing-12-8	99.0	Chongqing-12-8	99.0	98.4	99.0	90.1	90.4	80.3
Jiangsu-2011	100	99.0	Jiangsu-2011	99.5	99.5	90.7	90.9	80.3
Sichuan-12-8	99.5	98.4	99.5	Sichuan-12-8	99.5	90.1	90.4	79.8
Sichuan-12-12	100	99.0	99.5	99.5	Sichuan-12-12	90.7	90.9	80.3
Guangdong-2011	90.7	90.1	90.7	90.1	90.7	Guangdong-2011	100	80.2
m2-47	90.9	90.4	90.9	90.4	90.9	100	m2-47	81.8
Uganda	80.3	80.3	80.3	79.8	80.3	80.2	81.8	Uganda
RNA2-5'UTR								
Can181-9	Can181-9	100	100	99.5	99.5	96.6	93.2	95.5
Chongqing-12-8	100	Chongqing-12-8	100	99.5	99.5	96.6	93.2	95.5
Jiangsu-2011	100	100	Jiangsu-2011	99.5	99.5	96.6	93.2	95.5
Sichuan-12-8	99.5	99.5	99.5	Sichuan-12-8	100	96.6	93.2	95.5
Sichuan-12-12	99.5	99.5	99.5	100	Sichuan-12-12	96.6	93.2	95.5
Guangdong-2011	96.6	96.6	96.6	96.6	96.6	Guangdong-2011	96.6	98.9
m2-47	93.2	93.2	93.2	93.2	93.2	96.6	m2-47	97.8
Uganda	95.5	95.5	95.5	95.5	95.5	98.9	97.8	Uganda
Hsp70h								
Can181-9	Can181-9	99.1	98.9	99.1	99.3	90.4	90.4	90.6
Chongqing-12-8	99.0	Chongqing-12-8	99.1	98.6	98.7	90.1	90.1	90.3
Jiangsu-2011	98.9	99.5	Jiangsu-2011	98.4	98.6	90.1	90.1	90.3
Sichuan-12-8	99.4	98.8	98.9	Sichuan-12-8	99.1	90.1	90.1	90.3
Sichuan-12-12	99.4	98.8	98.9	99.6	Sichuan-12-12	90.1	90.1	90.3

Table 6. Cont.

	Can181-9	Chongqing-12-8	Jiangsu-2011	Sichuan-12-8	Sichuan-12-12	Guangdong-2011	m2-47	Uganda
Guangdong-2011	76.9	76.5	76.6	76.9	76.9		100	99.8
m2-47	77.0	76.5	76.7	76.9	77.0	99.9		99.8
Uganda	76.9	76.6	76.8	76.9	76.9	98.7	98.9	
p60	Can181-9	Chongqing-12-8	Jiangsu-2011	Sichuan-12-8	Sichuan-12-12	Guangdong-2011	m2-47	Uganda
Can181-9		97.5	97.3	97.5	97.5	78.8	78.8	79.5
Chongqing-12-8	98.7		99.4	98.5	98.5	79.7	79.7	80.5
Jiangsu-2011	98.3	99.4		98.6	98.6	79.3	79.3	80.1
Sichuan-12-8	99.0	98.7	99.0		99.2	79.5	79.5	80.3
Sichuan-12-12	98.9	98.7	98.9	99.7		79.5	79.5	80.3
Guangdong-2011	72.4	72.7	72.4	72.4	72.4		99.4	98.6
m2-47	72.5	72.8	72.5	72.5	72.4	99.8		98.8
Uganda	72.7	73.0	72.7	72.7	72.6	98.8	98.9	
p8	Can181-9	Chongqing-12-8	Jiangsu-2011	Sichuan-12-8	Sichuan-12-12	Guangdong-2011	m2-47	Uganda
Can181-9		98.6	95.9	97.3	98.6	68.5	68.5	68.5
Chongqing-12-8	98.2		97.3	98.6	100	69.9	69.9	69.9
Jiangsu-2011	98.2	98.2		95.9	97.3	68.5	68.5	68.5
Sichuan-12-8	98.6	98.6	98.6		98.6	69.9	69.9	69.9
Sichuan-12-12	98.6	98.6	98.6	99.1		69.9	69.9	69.9
Guangdong-2011	72.1	72.5	71.6	72.1	72.1		100	93.2
m2-47	72.1	72.5	71.6	72.1	72.1	100		93.2
Uganda	73.0	73.4	72.5	73.0	73.0	95.5	95.5	
CP	Can181-9	Chongqing-12-8	Jiangsu-2011	Sichuan-12-8	Sichuan-12-12	Guangdong-2011	m2-47	Uganda
Can181-9		99.2	99.2	99.6	98.4	79.8	79.4	77.8
Chongqing-12-8	99.5		98.4	99.6	98.4	79.4	79.0	77.4
Jiangsu-2011	99.4	98.8		98.8	97.7	80.2	79.8	78.2
Sichuan-12-8	99.6	99.6	99.0		98.8	79.8	79.4	77.8
Sichuan-12-12	99.4	99.4	98.7	99.5		78.6	78.2	76.7
Guangdong-2011	73.3	73.3	74.0	73.3	72.9		99.2	95.7
m2-47	73.3	73.3	74.0	73.3	72.9	99.7		96.5
Uganda	73.2	73.2	73.8	73.2	72.8	98.6	98.8	
mCP	Can181-9	Chongqing-12-8	Jiangsu-2011	Sichuan-12-8	Sichuan-12-12	Guangdong-2011	m2-47	Uganda
Can181-9		98.4	98.5	98.7	99.0	61.5	61.3	61.4
Chongqing-12-8	99.2		98.1	98.8	99.1	61.4	61.1	61.3
Jiangsu-2011	98.9	98.5		98.4	98.7	62.3	62.0	62.1
Sichuan-12-8	99.4	99.5	98.7		99.4	61.5	61.3	61.4

Table 6. Cont.

	Can181-9	Chongqing-12-8	Jiangsu-2011	Sichuan-12-8	Sichuan-12-12	Guangdong-2011	m2-47	Uganda
Sichuan-12-12	99.5	99.6	98.8	99.8		61.7	61.4	61.5
Guangdong-2011	63.5	63.2	63.6	63.3	63.4		99.1	98.7
m2-47	63.2	62.9	63.3	63.0	63.1	99.7		98.1
Uganda	63.4	63.1	63.6	63.1	63.2	98.4	98.2	
p28	Can181-9	Chongqing-12-8	Jiangsu-2011	Sichuan-12-8	Sichuan-12-12	Guangdong-2011	m2-47	Uganda
Can181-9		98.9	98.3	97.9	97.9	85.1	85.1	85.5
Chongqing-12-8	98.9		98.8	99.2	99.2	86.0	86.0	86.4
Jiangsu-2011	98.9	98.6		98.8	98.8	86.0	86.0	86.4
Sichuan-12-8	99.0	99.6	98.8		99.2	86.0	86.0	86.4
Sichuan-12-12	99.0	99.6	98.8	99.7		86.0	86.0	86.4
Guangdong-2011	77.4	77.6	77.9	77.5	77.5		99.2	97.9
m2-47	77.4	77.6	77.9	77.5	77.5	99.7		98.8
Uganda	77.9	78.2	78.5	78.1	78.1	97.9	98.2	
RNA2-3'UTR	Can181-9	Chongqing-12-8	Jiangsu-2011	Sichuan-12-8	Sichuan-12-12	Guangdong-2011	m2-47	Uganda
Can181-9								
Chongqing-12-8	99.5							
Jiangsu-2011	100	99.5						
Sichuan-12-8	99.5	100	99.5					
Sichuan-12-12	99.5	100	99.5	100				
Guangdong-2011	79.7	79.7	79.7	79.7	79.7			
m2-47	79.7	79.7	79.7	79.7	79.7	99.0		100
Uganda	79.7	79.7	79.7	79.7	79.7	99.0		

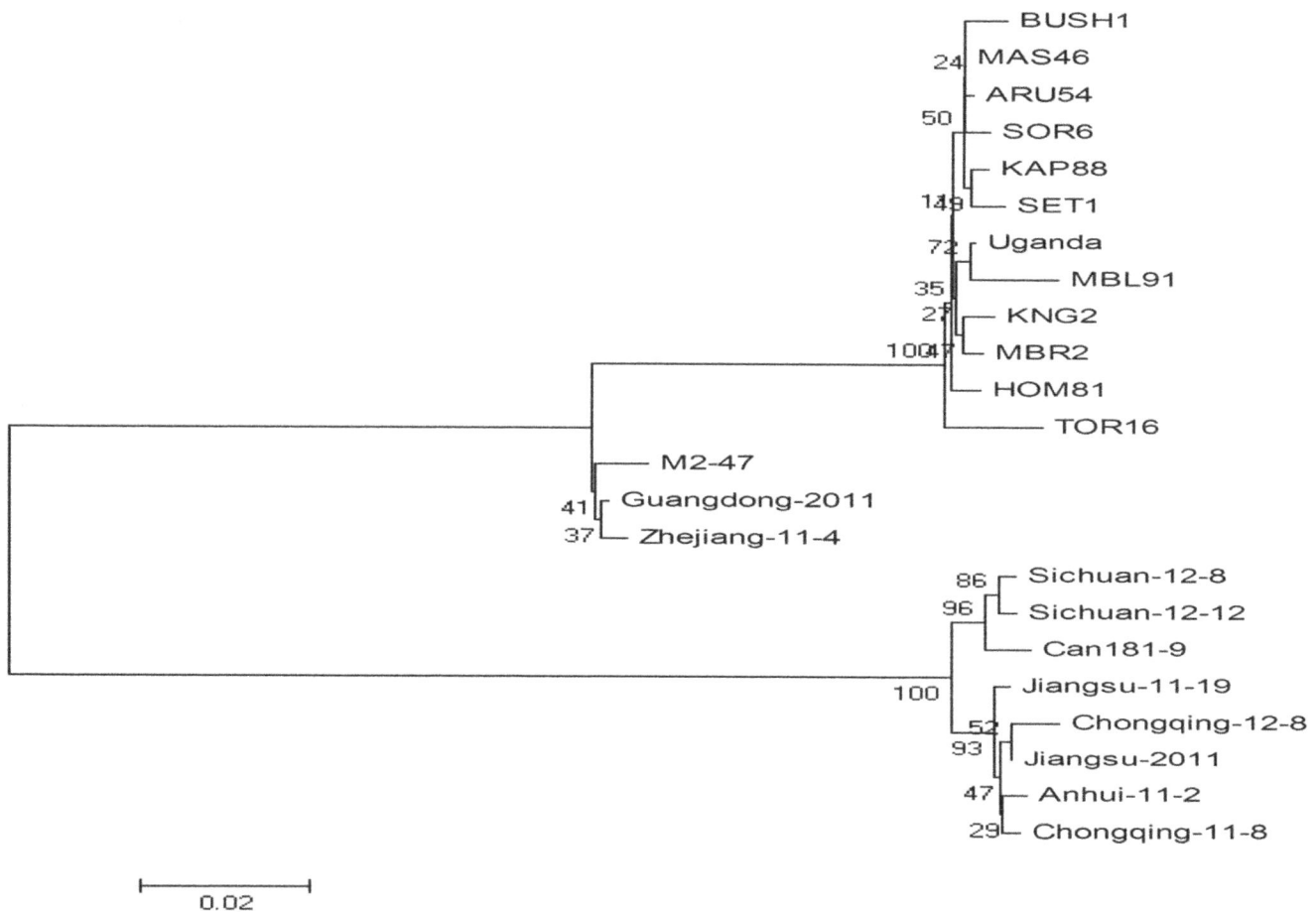

Figure 3. Phylogenetic tree based on partial sequences of RNA1. Phylogenetic tree illustrating relationships between isolates obtained from China and representative isolates of sweet potato chlorotic stunt virus (SPCSV) deposited in GenBank. Neighbor-joining trees were constructed via the maximum composite likelihood substitution model using MEGA (version 4.0). Numbers at branches represent bootstrap values of 1000 replicates. The scale bar shows the number of nucleotide substitutions per site.

Sequence alignment and phylogenetic analysis using the neighbor-joining method were performed with MEGA software (version 4.0). Nucleotide and amino acid sequence identity analysis of 25 isolates acquired in this research showed that, except for isolate Guangdong-2011 (EA strain), SPCSV isolates belonging to the WA strain exhibit 98.4%–100% nt sequence identity and 97.7%–100% deduced aa sequence identity. These findings demonstrate that isolates of the SPCSV WA strain from China show nearly identical. A phylogenetic tree was constructed (Figure 4) based on CP sequences to illustrate the probable genetic relationships between the 25 isolates and Guangdong-2009, m2-47, Uganda, and Can181-9. Results showed that Chinese SPCSV isolates could be divided into two branches. Except for those obtained from the Guangdong region, SPCSV isolates from other Chinese regions belonged to the WA strain. This finding indicates that at least two different strains may be observed among Chinese SPCSV isolates and that the WA strain is more prevalent than the EA strain.

Analysis of the phylogenetic tree based on partial RNA1 sequences showed that Zhejiang-11-4 from Zhejiang Province belongs to the EA strain. However, when the tree was constructed using the CP gene, the isolate appeared to belong to the WA

strain. This finding suggests that Zhejiang-11-4 occurs through co-infection by WA and EA.

This study reveals that nucleotide and amino acid identities are highly conserved between different isolates of Chinese strains and that the WA strain is distributed more extensively than the EA strain.

Discussion

The complete genomic sequences of five different SPCSV isolates were acquired from main sweet potato production areas in China by RT-PCR and 5′ and 3′-RACE using virus-transmitting vector whitefly as a material. Guangdong-2011, which was isolated from Guangdong Province, belongs to the EA strain, whereas Jiangsu-2011, Sichuan-12-8, Sichuan-12-8, and Chongqing-12-8, which were isolated from Jiangsu, Sichuan, and Chongqing, belong to the WA strain. To the best of our knowledge, this study is the first to report the complete genomic sequences of Chinese SPCSV isolates as well as the complete SPCSV genomic sequence acquired from whitefly.

Factors such as low titers of SPCSV, heterogeneous distribution of the virus in sweet potato, and phenol and polysaccharide enrichment in sweet potato leaves [40,41] significantly decrease the quality of extracted RNA from sweet potato leaves and the

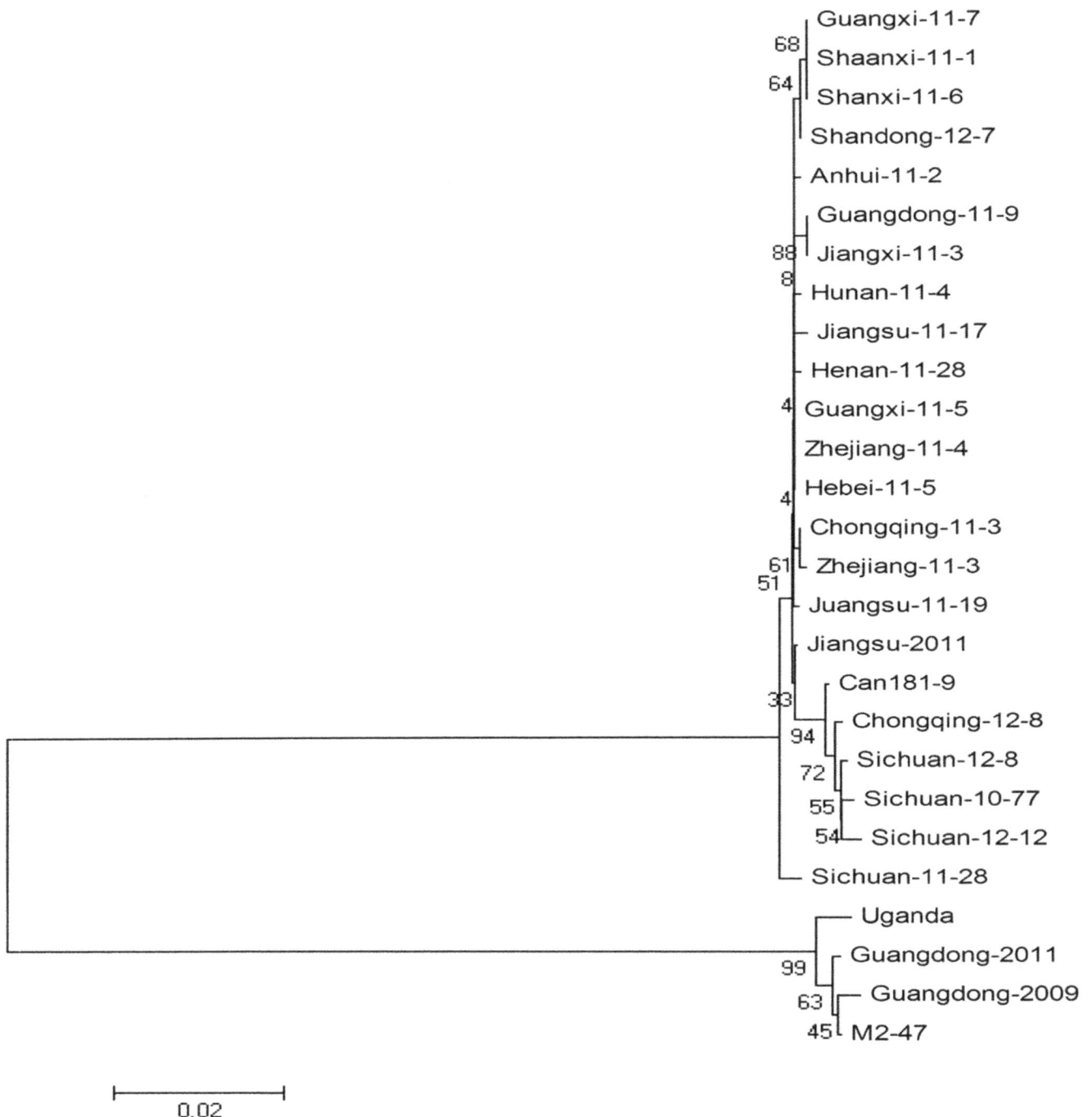

Figure 4. Phylogenetic tree based on CP gene sequences. Phylogenetic tree illustrating relationships between the CP genes of isolates obtained from China and representative isolates of sweet potato chlorotic stunt virus (SPCSV) deposited in GenBank. Neighbor-joining trees were constructed via the maximum composite likelihood substitution model using MEGA (version 4.0). The bar in this figure represents 0.02 Kimura nucleotide unites. Numbers at branches represent bootstrap values of 1000 replicates.

amplification efficiency of conventional RT-PCR. SPCSV usually co-infects sweet potato with other viruses in the field; thus SPCSV virions are difficult to separate and purify. These factors severely restrict SPCSV genome sequencing [22,32–34]. Only three SPCSV complete genomic sequences can be accessed via GenBank today, and only one complete sequence of SPCSV WA strain is available [11,29,35]. In the present study, the SPCSV

titer in whitefly was found to be significantly higher than that in sweet potato and *I. setosa*. This feature of the virus-transmitting vector whitefly is advantageous for acquiring the complete SPCSV genomic sequence. In addition, using viruliferous whitefly as an experimental material can preclude the interference of other viruses, phenols, and polysaccharides in sweet potato leaves. Our research provides a simplified and convenient method for cloning SPCSV genomic sequences.

The complete genomic sequences of the five isolates from different Chinese areas were used to analyze molecular variations and phylogenesis compared with the genomic sequences of three isolates obtained from GenBank. Analysis of the nucleotide identity of complete genomic sequences indicated that the nt identities in SPCSV strain are highly conserved and significant differences were found between WA and EA strains. Analyses of nt and aa identities based on the *cp* gene and partial RNA1 sequences are consistent with aforementioned results. High conservation was observed between different isolates in the same strain, but significant differences were found between WA and EA strains, which are inferred to be caused by the short time since SPCSV was transmitted into China, so the virus shows only limited genetic variability. This result is consistent with previous reports [3,27,28,36]. Tugume [36] analyzed the genetic variability of *p7*, *RNase3*, and *p22* genes of different SPCSV isolates infecting sweet potato and wild species in Uganda and suggested that the three genes show only limited genetic variability among strains. The EA and WA strain isolates showed nt (aa) sequence identities of 83.8%–84.3% (81.6%–82.5%) for RNase3 and 76.2%–79.8% (60.7%–66.7%) for p7 [36]. The nt and deduced aa sequence identities of *RNase3*, *p7* and *hsp70h* of EA strain isolates Ug, Tug2, Unj2, Mis1, m2-47, and WA strain isolate Is were analyzed. The nt and aa sequences of *RNase3* and *hsp70h* in the EA strain isolates were nearly identical but the nt and aa of *p7* showed more variation [28]. We previously analyzed the genetic variability of partial *hsp70h* gene of different SPCSV isolates in China, and our results demonstrated that the *hsp70* gene sequences of the same strain group display a high degree of conservation and that strain group WA has a wider geographic distribution in China than the EA strain [27].

The genome organization of SPCSV shares many similarities with other criniviruses [10]. However, the genome of SPCSV possesses unique features particularly concerning the gene content of RNA1. Downstream from the ORF for the replicase, RNA1 contains ORFs for Class 1 RNase III enzyme, a putative hydrophobic protein (p7), and a 22 kDa protein that shows no significant similarity to known proteins from any organism [11,28,42]. The 3′-proximal part of RNA1 constitutes an interesting genomic region for study owing to its unique gene functions. RNase3 contains a single endoribonuclease domain and a dsRNA-binding domain [43,44]. RNase3 inhibits posttranscriptional gene silencing and exerts a key function in the development of severe diseases in sweet potato plants co-infected with other viruses [42,43]. The *p22* gene was identified as a SPCSV RNA silencing suppressor and it could suppress silencing induced by dsRNA [42]. Similar activities were exhibited by homologs of *p22* encoded by other members of the family *Closteroviridae*. Cañizares identified *Tomato chlorosis virus* (ToCV) RNA1-encoded p22 protein as an effective silencing suppressor by using an agrobacterium co-infiltration assay. ToCV p22 suppressed local RNA silencing induced either by sense RNA or dsRNA very efficiently but did not interfere with short or long-distance systemic spread of silencing [45]. Data showed that the *Beet yellow virus* p21, *Beet yellow stunt virus* p22, *Citrus tristeza virus* p20, and *Grapevine leafroll-associated virus-2* p24, which are homologs of p22, are silencing suppressors [46–48]. Recent studies have revealed that many SPCSV isolates lacking *p22* still synergize with unrelated viruses [18,28,43], which indicates that p22 is dispensable for synergy between SPCSV and other viruses. While *p22* is a pathogenicity enhancer of SPCSV, co-infection of SPFMV with SPCSV isolates containing *p22* causes more severe symptoms than co-infection with SPCSV isolates lacking *p22* in the indicator plant *I. setosa* [28]. The *p22* gene is present only in Ugandan isolates of SPCSV, and SPCSV isolates from other areas do not contain the *p22* gene [36]. The partial RNA1 sequences of nine isolates from different Chinese regions were determined in this study, and no isolate (neither EA nor WA strains) containing the *p22* gene was observed in our research.

A novel virus isolate related to viruses in genus *Crinivirus* carrying predicted ORFs for proteins homologous to the RNase3 and p7 of SPCSV was detected and designated as KML33b. The sequences of KML33b were highly divergent from SPCSV isolates and showed <60% sequence identities for both nt and aa [36]. According to the species demarcation criteria from the Ninth Report of the International Committee on Taxonomy of Viruses (ICTV), the criteria for demarcating species in the genus *Crinivirus* are: (a) genome structure and organization (number and relative location of the ORFs) and (b) amino acid sequences of relevant gene products (polymerase, CP, Hsp70h) differing by more than 25% [49]. In the present study, comparison of RNA1 and RNA2 between WA and EA strains showed that the nucleotide identities exhibit significant differences between strains. Genomic structure analysis suggested that the quantity of proteins encoded by the RNA2 segment between SPCSV WA and EA strains differ. Before *hsp70h*, WA encodes three but EA encodes one or two hypothetical proteins. Comparison of the 3′ and 5′ UTRs of the genome shows that the lengths of RNA1 3′UTR and RNA2 5′UTR are different between WA and EA strains. The nucleotide identity of the RNA1 3′UTR and RNA2 5′UTR between WA and EA strains is low. Although the RNA1 5′ and RNA2 3′UTRs of the WA and EA strains showed the same length, their nucleotide identity was low. Comparison of aa identities suggested that proteins with an aa identity <75% between the WA and EA strains include p7 in RNA1, p8, and mCP in RNA2. Tairo previously suggested that EA and WA of SPCSV may belong to different species in the genus *Crinivirus* [3]. As mentioned above, the low sequence identity and differences in genomic structures between the EA and WA strains of SPCSV support this proposal. However, as the biological and serological relationships between the strains remain incompletely understood, a systematic assessment of these differences should be given the top priority for future research.

Materials and Methods

Ethics statement

Samples were collected from private land with the owner's permission. No specific permissions were required for sampling from any other location. Field studies did not involve endangered or protected species.

Collection of virus isolates

From 2010 to 2012, five sweet potato vine cuttings were collected from the main sweet potato-producing areas (Jiangsu, Guangdong, Sichuan Provinces, and Chongqing City) of China (Table 1) and grown in an insect-proof greenhouse. These cuttings were proven to be infected with SPCSV by nitrocellulose membrane (NCM)-ELISA and RT-PCR. The SPCSV-infected sweet potato was placed in an insect cage to feed the whiteflies. Non-viruliferous whiteflies were separately fed in SPCSV-infected sweet potato plants; here, whiteflies that had been fed for more than 3 d were considered viruliferous. Adult viruliferous whiteflies were collected and quick-frozen in liquid nitrogen, and then stored at −70°C to amplify complete genomic sequence.

From 2010 to 2012, a total of 20 vine cuttings were collected from the main sweet potato-producing areas in 14 provinces of China (Table 1). *I. setosa* was inoculated by side-grafting with infected sweet potato scions and grown in an insect-proof greenhouse (temperature, 25–30°C; relative humidity, 70%) under natural daylight. *I. setosa*-infecting SPCSV were used to clone the *cp* gene and partial RNA1 sequences.

Design and synthesis of primers used in this study

The primers used in this study are shown in Table 2. Primers for amplification of the RNA1 and RNA2 regions of WA strains were designed according to the sequences of SPCSV and deposited in GenBank (GenBank accession numbers FJ807784 and FJ807785). The primers were named WA-RNA1-P1, WA-RNA1-P2989, WA-RNA1-P2420, WA-RNA1-P5596, WA-RNA1-P4410, WA-RNA1-P8637, WA-RNA2-P53, WA-RNA2-P5599, WA-RNA2-P4440, and WA-RNA2-P8223. Primers for amplification of the RNA1 and RNA2 regions of EA strains were designed according to the sequences of SPCSV and deposited in GenBank (GenBank accession numbers HQ291259, HQ291260, AJ428554, and AJ428555). The primers were named EA-RNA1-P15, EA-RNA1-P4800, EA-RNA1-P3680, EA-RNA1-P8410, EA-RNA2-P1, EA-RNA2-P4910, EA-RNA2-P3710, and EA-RNA2-P8130. Primers for amplification of the *cp* gene were designed according to the sequences of SPCSV and deposited in GenBank (GenBank accession number FJ807785). These primers were named CP-F and CP-R. Primers for amplification of the 3′ genomic region of RNA1 were synthesized according to the reference [28] and named RdRp-F and SVV-R3.

Serological detection of SPCSV

SPCSV was detected using an NCM-ELISA kit from the International Potato Center (CIP). Briefly, 150 mg of leaf material was ground in a mortar with 1 mL of extraction buffer (1 M Tris-HCl containing 0.2% Na_2SO_3). The homogenate was transferred to a 1.5 mL Eppendorf tube and spun at $6000 \times g$ for 5 min. Aliquots of the supernatant (20 µL) were placed on membranes. The membranes were blocked with blocking solution at room temperature and then incubated for 60 min. After washing twice for 3 min each time, the membranes were incubated in virus-specific antibodies to the SPCSV coat protein (provided by CIP) at 4°C overnight. After washing twice for 3 min each time, the membranes were incubated in the conjugated anti-SPCSV antibody at room temperature for 1 h and then washed again twice for 3 min each time. The color reaction was developed using NBT/BCIP as the substrate. Color development was ceased by discarding the substrate solution and immersing the membranes in tap water. NCMs were washed in distilled water for 10 min.

RNA isolation

Total RNA was isolated from 100 mg of viruliferous whiteflies or *I. setosa* leaves as templates using the Total Plant RNA Extraction Miniprep System (Sangon, Shanghai, China). The amount and quality of the RNA were verified using agarose gel electrophoresis.

RT-PCR detection of SPCSV

RNA was used for reverse transcription using Moloney murine leukemia virus (M-MLV) reverse transcriptase (TaKaRa, Shiga, Japan) according to the manufacturer's instructions. Synthesized cDNA was amplified using Ex *Taq* DNA polymerase (TaKaRa). The partial sequence of *hsp70h* gene was amplified to confirm samples infected SPCSV. Degenerate primers for amplifying the *hsp70h* gene were designed according to the sequences of the SPCSV WA and EA strains and deposited in GenBank. These primers were respectively named Hsp70h-F and Hsp70h-R (Table 2).

Cloning and sequence analysis

RNA was used for reverse transcription using M-MLV reverse transcriptase (TaKaRa) according to the manufacturer's instructions. Synthesized cDNA was amplified using LA *Taq* DNA polymerase (TaKaRa). Genome RNA1 or RNA2 sequences of the WA strain were acquired by two or three overlapping RT-PCRs, whereas genome RNA1 or RNA2 sequences of the EA strain were obtained by two overlapping RT-PCRs. All PCR conditions in this study are shown in Table 2. Rapid amplification of cDNA ends (RACE) was used to determine the 5′ and 3′ ends of the viral genomic segments RNA1 and RNA2. 5′ and 3′ RACE was conducted by TaKaRa. The *cp* gene and the partial sequence of RNA1 obtained from the 3′-promixmal region of the RdRp gene to the middle of the 3′UTR were determined by cloning and sequencing obtained amplicons by standard RT-PCR.

The amplified products were purified from agarose gels using an AxyPrep DNA Gel Extraction Kit (Axygen, Hangzhou, China) and cloned into the PMD19-T vector (TaKaRa). Recombinant plasmids were transformed into *Escherichia coli* strain TG1 competent cells, purified using Plasmid Miniprep Kits (Bioteke, Beijing, China), and sequenced by Sangon Biotech Company (Sangon, Shanghai, China). Sequencing was conducted in both directions for each of the two independent amplicons of all isolates.

The sequences obtained were compared by BLAST search with the existing sequences in the NCBI database. Sequences were analyzed using DNAMAN Version 6.0. SPCSV Can181-9, m2-47 and Uganda isolates were used to explore the genome organization of the SPCSV WA and EA strains in China. Phylogenetic and molecular evolutionary analyses were conducted using MEGA version 4.0 [50]. Multiple alignments of protein-coding sequences were obtained using the default options in Clustal W [51]. Phylogenetic trees were constructed based on the aligned protein-coding sequences using the neighbor-joining method. The statistical significance of tree branching was tested by performing 1000 bootstrap replications.

Author Contributions

Conceived and designed the experiments: YHQ ZCZ. Performed the experiments: YHQ LW. Analyzed the data: YHQ ZCZ. Contributed reagents/materials/analysis tools: ZCZ QQ DSZ YTT SW YJW. Contributed to the writing of the manuscript: YHQ ZCZ ZLY.

References

1. FAOSTAT (2008) FAO Statistical Database. http://faostat.fao.org/.
2. Rännäli M, Czekaj V, Jones RAC, Fletcher JD, Davis RI, et al. (2008) Molecular genetic characterization of *Sweet potato virus G* (SPVG) isolates from areas of the Pacific Ocean and southern Africa. Plant Dis 92: 1313–1320. doi:10.1094/PDIS-92-9-1313.
3. Tairo F, Mukasa SB, Jones RAC, Kullaya A, Rubaihayo PR, et al. (2005) Unraveling the genetic diversity of the three main viruses involved in sweetpotato virus disease (SPVD), and its practical implications. Mol Plant Pathol 6: 199–211. doi:10.1111/j. 1364-3703.2005.00267.x.
4. Ma D, Li H, Tang J, Xie Y, Li Q, et al. (2010) Current status and future prospects of development of sweet potato industry in China. Sweet potato in food and energy security: Proceedings of China Xuzhou 4th International Sweet potato Symposium & 4th China-Japan-Korea Sweet potato Workshop. Xuzhou, China, pp. 3–10.

5. Clark CA, Davis JA, Abad JA, Cuellar W, Fuentes S, et al. (2012) Sweet potato viruses: 15 years of progress on understanding and managing complex diseases. Plant Dis 96: 168–185. doi: 10.1094/PDIS-07-11-0550.

6. Cohen J, Franck A, Vetten HJ, Lesemann DE, Loebenstein G (1992) Purification and properties of closterovirus-like particles associated with a whitefly-transmitted disease of sweet potato. Ann. Appl. Biol. 121, 257–268. doi:10.1111/j.1744-7348.1992.tb03438.x.

7. Gibson RW, Mpembe I, Alicai T, Carey EE, Mwanga ROM, et al. (1998b) Symptoms, aetiology and serological analysis of sweet potato virus disease in Uganda. Plant Pathol 47: 95–102. doi:10.1046/j.1365-3059.1998.00196.x.

8. Van Regenmortel MHV, Fauquet CM, Bishop DHL, Carstens EB, Estes MK, et al. (2000) The Seventh Report of the International Committee on Taxonomy of Viruses. San Diego: Academic Press.

9. Schaefers GA, Terry ER (1976) Insect transmission of sweet potato disease agents in Nigeria. Phytopathology 66: 642–645. doi: 10.1094/Phyto-66-642.

10. Dolja VV, Kreuze JF, Valkonen JPT (2006) Comparative and functional genomics of closteroviruses. Virus Res 117: 38–51. doi: 10.1016/j.virusres.2006.02.002.

11. Kreuze JF, Savenkov E, Valkonen JPT (2002) Complete genome sequence and analyses of the subgenomic RNAs of *Sweet potato chlorotic stunt virus* reveal several new features for the genus Crinivirus. J Virol 76: 9260–9270. doi: 10.1128/JVI.76.18.9260-9270.

12. Karyeija RF, Kreuze JF, Gibson RW, Valkonen JPT (2000) Synergistic interactions of a potyvirus and a phloem limited crinivirus in sweetpotato plants. Virology 269: 26–36. doi:10.1006/viro.1999.0169.

13. Mukasa SB, Rubaihayo PR, Valkonen JPT (2006) Interactions between a crinivirus, an ipomovirus and a potyvirus in coinfected sweetpoato plants. Plant Pathol 55: 458–467. doi: 10.1111/j.1365-3059.2006.01350.x.

14. Untiveros M, Quispe D, Kreuze J (2010) Analysis of complete genomic sequences of isolates of the Sweet potato feathery mottle virus strains C and EA: molecular evidence for two distinct potyvirus species and two P1 protein domains. Arch Virol 155: 2059–2063. doi: 10.1007/s00705-010-0805-y.

15. Salazar LE, Fuentes S (2001) Current knowledge on major virus diseases of sweetpotatoes. Pages 14-19 in: Proc. Int. Workshop Sweetpotato Cultivar Decline study, Sept. 8–9, 2000, Miyakonojo, Japan.

16. Njeru RW, Mburu MWK, Cheramgoi E, Gibson RG, Kiburi ZM, et al. (2004) Studies on the physiological effects of viruses on sweet potato yield in Kenya. Ann Appl Biol 145: 71–76. doi: 10.1111/j.1744-7348.2004.tb00360.x.

17. Kokkinos CD, Clark CA (2006) Interactions among Sweet potato chlorotic stunt virus and different potyviruses and potyvirus strains infecting sweetpotato in the United States. Plant Dis 90: 1347–1352. doi: 10.1094/PD-90-1347.

18. Untiveros M, Fuentes S, Salazar LF (2007) Synergistic interaction of Sweet potato chlorotic stunt virus (Crinivirus) with carla-, cucumo-, ipomo-, and potyviruses infecting sweetpotato. Plant Dis 91: 669–676. doi: 10.1094/PDIS-91-6-0669.

19. Untiveros M, Fuentes S, Kreuze J (2008) Molecular variability of sweet potato feathery mottle virus and other potyviruses infecting sweet potato in Peru. Arch Virol 153: 473–483. doi: 10.1007/s00705-007-0019-0.

20. Cuellar WJ, De Souza J, Barrantes I, Fuentes S, Kreuze JF (2011b) Distinct cavemoviruses interact synergistically with *Sweet potato chlorotic stunt virus* (genus *Crinivirus*) in cultivated sweet potato. J Gen Virol 92: 1233–1243. doi: 10.1099/vir.0.029975-0.

21. Loebenstein G, Thottappilly G, Fuentes S, Cohen J (2009) Virus and phytoplasma diseases. Pages 105–134 in; The Sweetpotato, Loebenstein G, Totthappilly G, eds. Springer Sciences Business Media BV, Dordrecht, The Netherlands. doi: 10.1007/978-1-4020-9475-0.8.

22. Mukasa SB, Rubaihayo PR, Valkonen JPT (2003) Incidence of viruses and viruslike diseases of sweetpotato in Uganda. Plant Dis. 87: 329–335. doi: 10.1094/PDIS.2003.87.4.329.

23. Hoyer U, Maiss E, Jelkmann W, Lesemann DE, Vetten HJ (1996) Identification of the coat protein gene of a Sweet potato sunken vein closterovirus isolate from Kenya and evidence for a serological relationship among geographically diverse closterovirus isolates from sweetpotato. Phytopathology 86: 744–750. doi:10.1094/Phyto-86-744.

24. Vetten HJ, Hoyer U, Maiss E, Lesemann DE, Jelkmann W (1996) Serological detection and discrimination of geographically diverse isolates of sweet potato sunken vein closterovirus. Phytopathology 100 (Suppl, abstract no 891A).

25. Alicai T, Fenby NS, Gibson RW, Adipala E, Vetten HJ, et al. (1999) Occurrence of two serotypes of sweetpotato chlorotic stunt virus in East Africa and their associated differences in coat protein and Hsp70 gene sequences. Plant Pathol 48: 718–726. doi:10.1046/j. 1365-3059.1999.00402.x.

26. Aritua V, Barg E, Adipala E, Gibson RW, Vetten HJ (2008) Further evidence for limited genetic diversity among East African isolates of *Sweet potato chlorotic stunt virus.* J Phytopathol 156: 181–189. doi: 0.1111/j.1439-0434.2007.01338.x.

27. Qin Y, Zhang Z, Qiao Q, Zhang D, Tian Y, et al. (2013) Molecular variability of sweet potato chlorotic stunt virus (SPCSV) and five potyviruses infecting sweet potato in China. Arch Virol 158: 491–495. doi:10.1007/s00705-012-1503-8.

28. Cuellar WJ, Tairo F, Kreuze JF, Valkonen JPT (2008) Analysis of gene content in Sweet potato chlorotic stunt virus RNA1 reveals the presence of p22 RNA silencing suppressor in only few isolates: implications to viral evolution and synergism. J Gen Virol 89: 573–582. doi:10.1099/vir.0.83471-0.

29. Cuellar WJ, Cruzado KR, Fuentes S, Untiveros M, Soto M, et al. (2011a) Sequence characterization of a Peruvian isolate of *Sweet potato chlorotic stunt virus*: further variability and a model for *p22* acquisition. Virus Res 157: 111–115. doi:10.1016/j.viruses.2011.01.010.

30. Fenby NS, Foster GD, Gibson RW, Seal SE (2002) Partial sequence of HSP70 homologue gene shows diversity between West African and East African isolates of *Sweet potato chlorotic stunt virus.* Tropical Agric 79: 26–30.

31. Sivparsad BJ, Gubba A (2012) Molecular resolution of the genetic variability of major viruses infecting sweet potato (*Ipomoea batatas* L.) in the province of KwaZulu-Natal in the Republic of South Africa. Crop Protect 41: 49–56. doi: 10.1016/j.cropro.2012.04.020.

32. Rukarwa RJ, Mashingaidze AB, Kyamanywa S, Mukasa SB (2010) Detection and elimination of sweetpotato viruses. African Crop Sci J 18: pp. 223–233. doi:10.4314/acsj.v18i4.68651.

33. Valverde RA, Clark CA, Valkonen JPT (2007) Viruses and Virus Disease Complexes of Sweetpotato. Plant Viruses 1: 116–126.

34. Tairo F, Kullaya A, Valkonen JPT (2004) Incidence of Viruses Infecting Sweetpotato in Tanzania. Plant Dis 88: 916–920. doi:10.1094/PDIS.2004.88.9.916.

35. Trenado HP, Franco AO, Navas-Castillo J (2009) http://www.ncbi.nlm.nih.gov/nuccore/FJ807784 and http://www.ncbi.nlm.nih.gov/nuccore/FJ807785.

36. Tugume AK, Amayo R, Weinheimer I, Mukasa SB, Rubaihayo PR (2013) Genetic Variability and Evolutionary Implications of RNA Silencing Suppressor Genes in RNA1 of *Sweet Potato Chlorotic Stunt Virus* Isolates Infecting Sweetpotato and Related Wild Species. PLoS ONE 8: e81479. doi:10.1371/journal.pone.0081479.

37. Gutiérrez DL, Fuentes S, Salazar LF (2003) Sweet potato Virus Disease (SPVD): Distribution, Incidence, and Effect on Sweetpotato Yield in Peru. Plant Dis 87: 297–302. doi: 10.1094/PDIS.2003.87.3.297.

38. Qiao Q, Zhang ZC, Qin YH, Zhang DS, Tian YT, et al. (2011) First report of Sweet potato chlorotic stunt virus infecting sweet potato in China. Plant Dis 95: 356. doi: 10.1094/PDIS-09-10-0675.

39. Zhang ZC, Qiao Q, Qin YH, Zhang DS, Tian YT, et al. (2012) First Evidence for occurrence of Sweet potato virus disease (SPVD) caused by dual infection of plants with *Sweet Potato Feathery Mottle Virus* and *Sweet potato chlorotic stunt virus* in China. Acta Phytopathologica Sinica 42: 10–16.

40. Cali BB, Moyer JW (1981) Purification, serology, and particle morphology of two russet crack strains of sweet potato feathery mottle virus. Phytopathology 71: 302–305. doi: 10.1094/Phyto-71-302.

41. Moyer JW, Kennedy GG (1978) Purification and properties of sweet potato feathery mottle virus. Phytopathology, 68: 998–1004. doi:10.1094/Phyto-68-998.

42. Kreuze JF, Savenkov EI, Cuellar W, Li X, Valkonen JPT (2005) Viral class 1 RNase III involved in suppression of RNA silencing. J Virol 79: 7227–7238. doi:10.1128/JVI.79.11.7227-7238.2005.

43. Cuellar WJ, Kreuze JF, Rajamäki ML, Cruzado KR, Untiveros M, et al. (2009) Elimination of antiviral defense by viral RNase III. Proc Natl Acad Sci U S A 106: 10354–10358. doi:10.1073/pnas. 0806042106.

44. Weinheimer I, Boonrod K, Moser M, Wassenegger M, Krczal G, et al. (2013) Binding and processing of dsRNA molecules by the Class 1 RNase III protein encoded by sweet potato chlorotic stunt virus. J Gen Virol doi:10.1099/vir.0.058693-0.

45. Cañizares MC, Navas-Castillo J, Moriones E (2008) Multiple suppressors of RNA silencing encoded by both genomic RNAs of the crinivirus, Tomato chlorosis virus. Virology 379: 168–174. doi:10.1016/j.virol.2008.06.020.

46. Chiba M, Reed JC, Prokhnevsky AI, Chapman EJ, Mawassi M, et al. (2006) Diverse suppressors of RNA silencing enhance agroinfection by a viral replicon. Virology 346: 7–14. doi:10.1016/j.virol.2005.09.068.

47. Lu R, Folimonov A, Shintaku M, Li WX, Falk BW, et al. (2004) Three distinct suppressors of RNA silencing encoded by a 20-kb viral RNA genome. Proc Natl Acad Sci U S A 101: 15742–15747. doi: 10.1073/pnas. 0404940101.

48. Reed JC, Kasschau KD, Prokhnevsky AI, Gopinath K, Pogue GP, et al. (2003) Suppressor of RNA silencing encoded by *Beet yellows virus*. Virology 306: 203–209. doi:10.1006/S0042-6822(02)00051-X.

49. King A, Lefkowitz E, Adams MJ (2011) Virus Taxonomy: Ninth Report of the International Committee on Taxonomy of Viruses: An Elsevier Title.

50. Tamura K, Dudley J, Nei M, Kumar S (2007) MEGA4: molecular evolutionary genetics analysis (MEGA) software version 4.0. Mol Biol Evol 24: 1596–1599. doi:10.1093/molbev/msm092.

51. Thompson JD, Higgins DG, Gibson TJ (1994) CLUSTAL W: improving the sensitivity of progressive multiple sequence alignment through sequence weighting, position-specific gap penalties and weight matrix choice. Nuc Acids Res 22: 4673–4680. doi:10.1093/nar/22.22.4673.

17

Detection of the Virulent Form of AVR3a from *Phytophthora infestans* following Artificial Evolution of Potato Resistance Gene *R3a*

Sean Chapman[1◊]**, Laura J. Stevens**[1,2,3◊]**, Petra C. Boevink**[1,3]**, Stefan Engelhardt**[2,3]**, Colin J. Alexander**[4]**, Brian Harrower**[1]**, Nicolas Champouret**[5]**, Kara McGeachy**[1]**, Pauline S. M. Van Weymers**[1,2,3]**, Xinwei Chen**[1,3]**, Paul R. J. Birch**[1,2,3]**, Ingo Hein**[1,3]*

1 Cell and Molecular Sciences, James Hutton Institute, Invergowrie-Dundee, United Kingdom, **2** Division of Plant Sciences, University of Dundee at James Hutton Institute, Invergowrie-Dundee, United Kingdom, **3** Dundee Effector Consortium, Invergowrie-Dundee, United Kingdom, **4** Biomathematics and Statistics Scotland, Invergowrie-Dundee, United Kingdom, **5** J.R. Simplot Company, Simplot Plant Sciences, Boise, Idaho, United States of America

Abstract

Engineering resistance genes to gain effector recognition is emerging as an important step in attaining broad, durable resistance. We engineered potato resistance gene *R3a* to gain recognition of the virulent AVR3a^{EM} effector form of *Phytophthora infestans*. Random mutagenesis, gene shuffling and site-directed mutagenesis of *R3a* were conducted to produce R3a* variants with gain of recognition towards AVR3a^{EM}. Programmed cell death following gain of recognition was enhanced in iterative rounds of artificial evolution and neared levels observed for recognition of AVR3a^{KI} by R3a. We demonstrated that R3a*-mediated recognition responses, like for R3a, are dependent on SGT1 and HSP90. In addition, this gain of response is associated with re-localisation of R3a* variants from the cytoplasm to late endosomes when co-expressed with either AVR3a^{KI} or AVR3a^{EM} a mechanism that was previously only seen for R3a upon co-infiltration with AVR3a^{KI}. Similarly, AVR3a^{EM} specifically re-localised to the same vesicles upon recognition by R3a* variants, but not with R3a. R3a and R3a* provide resistance to *P. infestans* isolates expressing AVR3a^{KI} but not those homozygous for AVR3a^{EM}.

Editor: Frederik Börnke, Leibniz-Institute for Vegetable and Ornamental Crops, Germany

Funding: LS is supported by The James Hutton Institute and the University of Dundee through a joint PhD student bursary. PSMVW is supported by the U.S. Department of Agriculture through National Institute of Food and Agriculture (NIFA) project 2011-68004-30154 (http://www.csrees.usda.gov/). NC is employed by J.R. Simplot Company, Simplot Plant Sciences, 5369 West Irving Street, Boise, ID 83706, USA. This work was funded by the Rural & Environment Science & Analytical Services (RESAS) Division of the Scottish Government (http://www.scotland.gov.uk/Topics/Research/About/EBAR/research-providers) and the Biotechnology and Biological Sciences Research Council (BBSRC) (http://www.bbsrc.ac.uk/home/home.aspx) through the joint projects CRF/2009/SCRI/SOP & BB/H018441/1 (IH), BB/K018299/1 and RESAS funded work package 6.4. The funders had no role in study design, data collection and analysis, decision to publish, or preparation of the manuscript.

* Email: Ingo.Hein@hutton.ac.uk

◊ These authors contributed equally to this work.

Introduction

In a process known as effector triggered immunity (ETI), plant disease resistance (*R*) genes can facilitate immunity to phylogenetically diverse and unrelated pathogens that express cognate effector molecules [1]. Effectors that are recognised by *R* genes, either directly or indirectly, and provoke successful plant defences are genetically defined as avirulence (*Avr*) genes. The best described group of plant *R* gene products contains a nucleotide-binding (NB) domain and leucine-rich repeats (LRRs), collectively known as NB-LRRs [2]. NB-LRRs are strictly regulated by plants as, upon their activation, many elicit programmed cell death (PCD) as part of the hypersensitive response (HR) which prevents further spread of disease in plant tissues [3]. Together with effectors, *R* genes are at the forefront of host/pathogen co-evolution [4]. NB-LRRs are one of the largest gene families in

plants and more than 750 members have recently been described in potato [5–6]. The organisation of many NB-LRRs into physically-linked clusters is providing insight into their evolution, which can involve duplication followed by diversification.

In agriculture, successful deployment of *R* genes to control important diseases in crop plants has so far been hampered by the ability of pathogens often to rapidly circumvent detection by the host plant's innate immune system. Advances in studying pathogen effector diversity coupled with the ability to engineer *R* genes offers the opportunity to develop more durable resistances that specifically target essential effectors and known variants [7–8].

The *Phytophthora infestans* effector AVR3a is an essential effector for this pathogen to cause late blight on potato. Stable silencing of *Avr3a* in the *P. infestans* isolate 88069 significantly reduces infection in susceptible *Solanum tuberosum* (potato) cv. Bintje and in the model solanaceous plant species *Nicotiana benthamiana* [9–10]. Two forms of AVR3a are prevalent in current *P. infestans* isolates and differ in only two amino acids within the mature protein; $AVR3aE^{80}M^{103}$ ($AVR3a^{EM}$) and $AVR3aK^{80}I^{103}$ ($AVR3a^{KI}$) [11–12]. $AVR3a^{KI}$ elicits ETI upon recognition by the potato resistance protein R3a, a member of the coiled-coil (CC) NB-LRR gene family [13]. This response is evaded by $AVR3a^{EM}$ which consequently is free to promote virulence [9,11,13]. However, a weak R3a-dependent response to $AVR3a^{EM}$ can be observed under UV light [14]. The mechanism of R3a-mediated recognition of AVR3a has been investigated recently [15]. Upon activation by $AVR3a^{KI}$, both the effector protein and R3a rapidly re-localize from the host cytoplasm to late endosomes, components of the endocytic pathway, which is thought to be a prerequisite for subsequent HR development. The un-recognised $AVR3a^{EM}$ form of the effector does not cause this re-localisation. There is no evidence of direct interaction between $AVR3a^{KI}$ and R3a, but bimolecular fluorescence complementation (BiFC) assays reveal that the two proteins are in close proximity at late endosomes [15].

Artificial evolution has previously been used to alter the recognition specificity of the potato CC-NB-LRR resistance protein Rx to gain recognition of different strains of *Potato virus X* (PVX) and a distantly related virus, *Poplar mosaic virus* (PopMV) [16–17]. Random mutagenesis, screening for beneficial mutations and designed amalgamation of these mutations has been used to generate transgenic plants of the model species *N. benthamiana* that are resistant to previously virulent strains. DNA shuffling, also known as directed evolution, was first developed in the early 1990 s and has since been used to generate a wide variety of novel genes and proteins [18]. DNA shuffling has previously been used in the functional analysis of the resistance gene *Pto* [19]. In this study, the LRR of *R3a* has been subjected to error-prone PCR and iterative rounds of DNA shuffling to identify gain-of-recognition variants (R3a*) through functional screening in *N. benthamiana*. *R3a** gene products with varying degrees of $AVR3a^{EM}$ recognition were generated in three rounds of DNA shuffling and a subsequent round of site-directed mutagenesis. The best-performing clones from each round were taken forward for further analysis and compared to wild-type R3a. R3a* variants demonstrated significantly improved gain-of-recognition towards $AVR3a^{EM}$ that manifested itself as a gain of re-localisation to late endosomes, but not yet as a gain of resistance.

Materials and Methods

Construction of plasmid vectors

The plasmid pBinPlus.R3a [13] was amplified with the primer pairs R3a-5-Asc/R3a-1564-Bam-M and R3a-1564-Bam-P/R3a-3-Not (Table S1) and the *Asc*I/*Bam*HI and *Bam*HI/*Not*I digested amplification products cloned in a three way ligation in to *Asc*I/*Not*I digested binary vector pGRAB [20] to produce pGRAB.35S::R3a. A derivative of the former plasmid, pGRAB.-R3a::R3a, was produced in which the *Cauliflower mosaic virus* 35S (35S) promoter sequence was replaced with the *R3a* promoter sequence. This derivative was produced by digestion of pBin-Plus.R3a with *Pme*I and *Xho*I, and insertion of the released fragment, containing 2358 bases upstream of the translational start site, in to pGRAB.35S::R3a that had been treated with *Bsp*EI, T4 DNA polymerase and *Xho*I in order.

Mutagenesis and DNA shuffling

Shuffling of 2283 bp of the R3a LRR region was performed as described by Stemmer [18]. First round PCRs were carried out with the primer pair R3a-1564-Bam-P/R3a-3-Not in the presence

of 0.5 mM $MnCl_2$ as described by Leung *et al.* [21] with *Taq* DNA polymerase (Roche, Mannheim, Germany). Following DNaseI treatment, fragments of circa 500 bp were reassembled through forty rounds of primer-less thermo-cycling. Gel-purified products of circa 2.3 kb were amplified through thirty cycles of PCR and, following gel-purification, digested with *Bam*HI and *Not*I prior to cloning in to pGRAB.35S::R3a digested with the same enzymes. The ligated population was amplified in *E. coli* resulting in a population with a complexity of circa 125 K, a vector background of circa 14% and a base mutation rate of 0.43%. Aliquots of plasmid, gel-purified on account of plasmid instability, were transformed in to *A. tumefaciens* cells. In the second and third rounds *Pfu*Ultra II Fusion HS DNA polymerase (Stratagene, La Jolla, CA, USA) was used in thermo-cycling reactions to produce shuffled populations with lower mutation rates, less than 0.05%. The second and third round populations had complexities of circa 200 K and 150 K, respectively, and both had a vector background of circa 10%.

Site-directed mutagenesis

A mixture of two templates, the wild-type gene and clone Rd1-2 from the first round of shuffling containing the Q931R codon substitution, was used in PCR amplifications with the primer pairs R3a-1564-Bam-P/R3a-1740W-M, R3a-1740W-P/R3a-1841W-M, R3a-1841W-P/R3a-2743W-M, R3a-2743W-P/R3a-3028-M. The products of the primary PCRs were used in an overlap PCR reaction with the flanking primer pair R3a-1564-Bam-P/R3a-3028-M. The product of the secondary PCR reaction was digested with *Bam*HI and *Aat*II and cloned between the same unique sites of pGRAB.R3a::R3a. A population of circa 50K clones, with a vector background of 1% and random base mutation rate of 0.05% was produced.

Plant growth conditions

N. benthamiana plants were grown in a glasshouse with a 16 h day period at 22°C and an 8 h night period at 18°C. Supplementary lighting was provided below 200 W m^{-2} and screening above 450 W m^{-2}.

Screening of mutated R3a clones

DNA populations prepared from *E. coli* were transformed in to *Agrobacterium tumefaciens* strain AGL1 [22] carrying the helper plasmids pSoup and $pBBR1MCS1.VirG_{N54D}$ [23] for screening. *Agrobacterium* cultures grown from single transformed colonies were co-infiltrated with cultures of *Agrobacterium* transformed with $pGRAB.35S::AVR3a^{EM}$ according to the method of Engelhardt *et al.* [15] in to *N. benthamiana* leaves with each of the components at the same final OD_{600} of 0.1–0.5. Reference mixtures were infiltrated in to opposing half-leaves and between two and seven days post infiltration leaves were inspected to assess visible symptoms and plant auto-fluorescence under day-light and 365 nm illumination from a Blak-Ray lamp (UVP, Upland, CA, USA), respectively.

Symptom scoring

Circa five week old plants were used for symptom scoring with two adjacent, expanded leaves, both circa 90 mm in length, being used for infiltrations. Symptoms were scored on an arbitrary scale for nine days after infiltration. Symptom scores were plotted; the areas under the curve determined and mean scores calculated by dividing by the duration. A linear mixed modelling approach was adopted using GenStat for Windows, 16th edition (VSN International Ltd., Hemel Hempstead, UK). The data were analysed in two stages: first, the stability of the phenotypes over repeated experiments was examined by fitting a model with experiment as a fixed effect. Infiltration mixture, leaf age, position of infiltration site on leaf, experiment and their interactions were set as fixed effects with plant and leaf within an individual plant as the random effects. This allowed for terms in the model which specifically tested for significant interactions with experiment. Lack of any significant interaction with the infiltration mixtures would provide evidence that the relative responses of the phenotypes were consistent. Having determined that the infiltration mixtures behaved consistently over the experiments, the second stage fitted a model with experiment as a random effect with plant and leaf within plant nested below this. Multiple comparison tests then examined the differences in response amongst the infiltration mixtures.

Virus-induced gene silencing (VIGS) of SGT1 and HSP90

Tobacco rattle virus (TRV)-induced gene silencing in *N. benthamiana* was performed as described previously [14]. *Agrobacterium* cultures transformed with the binary TRV RNA1 construct, pBINTRA6, or the TRV RNA2 vector constructs PTV00, PTV:eGFP, PTV:HSP90 or PTV:SGT1 were re-suspended to OD_{600} = 0.5 for the RNA1 construct and OD_{600} = 1.0 for the RNA2 constructs. Re-suspended RNA1 and RNA2 cultures were mixed in a 1:1 ratio and infiltrated into non-cotyledonous leaves of *N. benthamiana* plants at the 5-leaf stage. For each of the biological replicates, six plants per treatment were used and six plants were used as non-TRV controls. Three weeks after treatment with the VIGS constructs, plants were infiltrated with culture mixtures (OD_{600} = 0.5) designed to express R3a, Rd2-1, Rd3-1 or Rd4-1 and $AVR3a^{KI}$ or $AVR3a^{EM}$. HRs were scored at 6dpi.

Confocal laser scanning microscopy

Imaging was performed on a Leica TCS-SP2 AOBS microscope (Leica Microsystems) using HCX APO L, 40x/0.8, and 63x/0.9 water dipping lenses or a Zeiss 710 using a Plan APO 40x/1.0 water dipping lens. Images were collected using line by line sequential scanning. The optimal pinhole diameter and the same gain levels were used within experiments. YFP and CFP were imaged using 514 nm and 405 nm excitation, respectively, and emissions were collected between 520–563 nm and 455–490 nm, respectively. Photoshop CS5.1 software (Adobe Systems) was used for post-acquisition image processing.

Agrobacterium tumefaciens transient assays (ATTAs)

Functional *Agrobacterium tumefaciens* transient assays (ATTAs) were carried out in *N. benthamiana*. Cultures carrying pGRAB.-R3a::R3a, pGRAB.R3a::Rd2-1, pGRAB.R3a::Rd3-1, pGRAB.-R3a::Rd4-1 or pGRAB empty vector were re-suspended as described before to OD600 = 0.1 for each construct. Each of the five resuspensions was infiltrated into separate areas of leaves. Four leaves on each of sixteen plants were infiltrated in each replicate. Two days post infiltration, leaves were detached and infiltration sites inoculated with $AVR3a^{KI}$ homozygous *P. infestans* isolate 7804.b or $AVR3a^{EM}$ homozygous isolate 88069. Leaves were incubated in transparent sealed boxes at 100% humidity in a cool room and covered for the first 12 hours. Lesion sizes were measured up to 15dpi.

Production of transgenic potato plants

R3a wild-type gene and the three modified versions Rd2-1, Rd3-1 and Rd4-1 were cloned under *R3a* native regulatory elements in a pCambia-based binary vector with kanamycin resistance as a selectable marker, using standard restriction enzyme methods to create pSIM2093, pSIM3027, pSIM3028 and pSIM3029 respectively. *Rpi-vnt1* under its native regulatory elements was cloned in to pSIM401 to generate pSIM1620. The binary vectors were then transformed into *Agrobacterium* strains AGL1 and LBA4404 for plant transformation. Ranger Russet was transformed as described in Duan *et al.* [24]. For each construct twenty to thirty lines were regenerated with kanamycin selection. These were tested for late blight resistance and assessed at 7dpi.

Results

Identification of R3a mutants with enhanced $AVR3a^{EM}$ recognition

A binary vector, pGRAB.35S::R3a, containing the R3a open reading frame (Accession number AY849382.1) under the control of *Cauliflower mosaic virus* 35S promoter and terminator sequences was produced to allow specific mutation and shuffling of the R3a LRR domain. In this plasmid a unique *Bam*HI site was introduced silently at nucleotide position 1567 in the ORF, ninety nucleotides upstream of the sequence encoding the LRR domain as denoted by Huang *et al.* [13]. Following mutagenesis and DNA shuffling of the LRR region, regenerated recombinant clones were screened for enhanced $AVR3a^{EM}$ recognition through *Agrobacterium*-mediated transient expression in co-infiltrations with binaries expressing the virulent elicitor, $AVR3a^{EM}$. Clones were screened for enhanced recognition through comparison of visible symptoms of PCD and induced auto-fluorescence with reference to the responses produced by the wild-type R3a gene and $AVR3a^{EM}$. Clones with putatively improved recognition isolated from primary screens were screened again to confirm the phenotypic improvement and to eliminate auto-activators. In the first cycle, screening of approximately three thousand clones identified eleven R3a* clones with improved phenotypes. The eleven clones from the first round are referred to as Rd1-1 to Rd1-11 and contained, in addition to synonymous changes, base changes that resulted in between one and four amino acid substitutions (Fig. 1a; Table S2). Three of the clones contained single amino acid substitutions (Rd1-1 [K920E], Rd1-2 [Q931R], Rd1-3 [R618Q]) identifying these changes as being responsible for enhanced recognition. Interestingly, the single amino acid substitution K920E in R3a was recently also identified in a complementary study by Segretin *et al.* [25] as a substitution that enhances recognition towards $AVR3a^{EM}$. The largest numbers of amino acid substitutions were found in LRRs #3 and #15 and included those found in the three clones with single substitutions (Fig. 1a).

A second round of shuffling was carried out to combine beneficial changes and remove deleterious substitutions using the eleven isolated clones, Rd1-1 to Rd1-11, and the wild-type gene as starting material. Screening of approximately six hundred recombinant clones identified four which gave responses comparable to or greater than the best performing clone from the first round, Rd1-1 [K920E]. All of these R3a* clones from the second round, Rd2-1 to Rd2-4, contained the amino acid substitution E620D in LRR #3 in addition to at least one other coding change (Fig. 1a; Table S2).

As recognition of $AVR3a^{EM}$ improved and gave stronger responses it became more difficult to discriminate the differences in responses between modified clones with $AVR3a^{EM}$ and also in comparison to the response of the wild-type *R3a* gene with $AVR3a^{KI}$. Therefore, a third round of shuffling, using the clones from the first and second rounds of shuffling and the wild-type

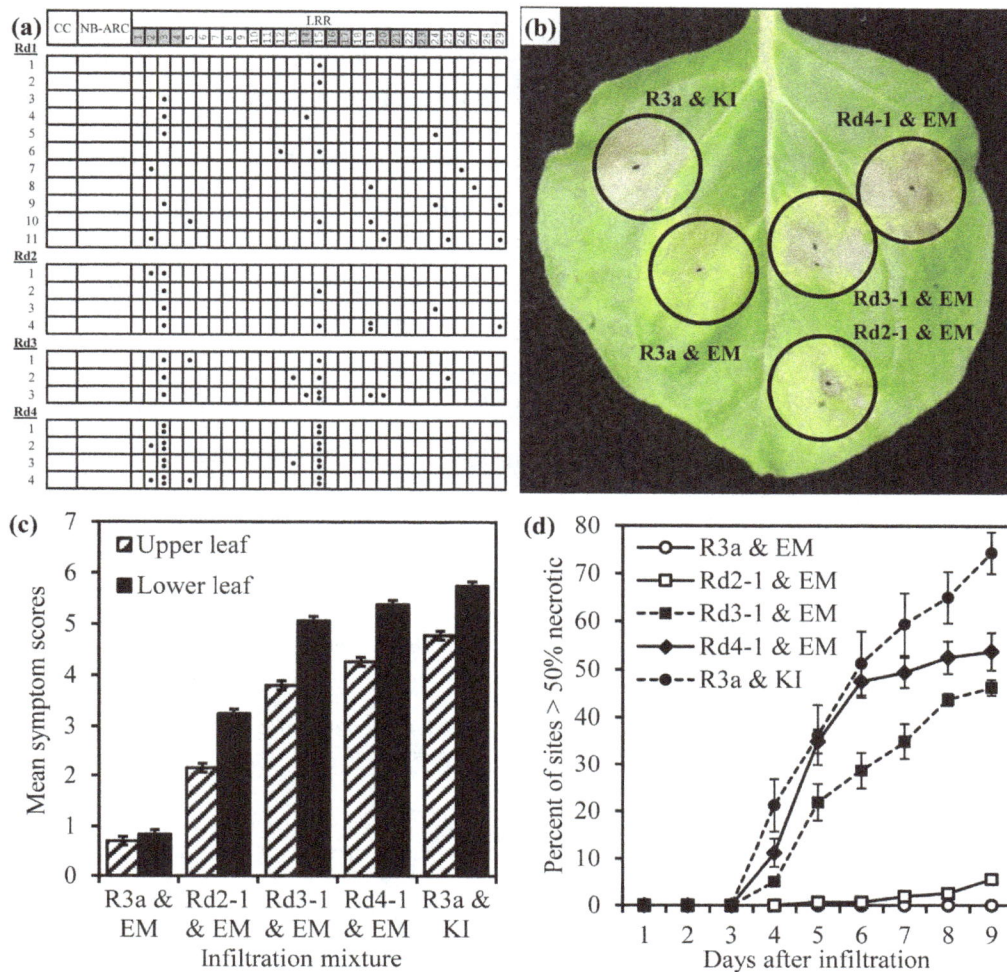

Figure 1. Four rounds of mutagenesis and shuffling identified R3a mutants with enhanced recognition of AVR3a^{EM} and disease responses (R3a*). (a) Schematic showing locations of non-synonymous mutations found in the LRRs of R3a* clones isolated from the four rounds of mutagenesis and shuffling (Rd1 to Rd4). LRRs containing amino acids under diversifying selection are shaded above. (b) Representative *N. benthamiana* leaf showing responses of best performing clones from second, third and fourth rounds (Rd2-1, Rd3-1 & Rd4-1) to AVR3a^{EM} (EM), compared to responses of wild-type R3a to AVR3a^{KI} (KI) and AVR3a^{EM} five days after co-infiltration with resistance genes under transcriptional control of *R3a* promoter. (c) Mean disease scores from the four experiments, each of nine days duration, for different infiltration mixtures in upper (hatched) and lower (solid) paired leaves. Error bars show +/− standard error. (d) Time-course of percentage of sites showing necrosis development, greater than 50% necrosis of individual infiltrated sites, for the five infiltrated mixtures. Mean percentages of the four experiments. Each experiment includes data for 40 infiltration sites (upper and lower leaves combined) and error bars show +/− standard errors.

gene as starting material, was performed. However, the shuffled LRR domains were cloned in to a binary vector, pGRAB.-R3a::R3a, containing the *R3a* gene or *R3a** variants under the wild-type *R3a* promoter, rather than the strong 35S promoter. The purpose of this was to protract the timing of the cell death responses and facilitate the discrimination of differences (Fig. 1b). Screening of approximately 300 clones from this third round population identified three R3a* clones, Rd3-1 to Rd3-3, that gave responses greater than the best performing clone from the second round of shuffling, Rd2-1 [T585A, E620D]. All of these clones contained the E620D change found in the second round clones and a number of other amino acid changes including at least one change in LRR #15. A previously conducted comparison of functional R3a with three paralogous, non-functional resistance gene analogues revealed 13 positions under diversification [13]. These positions, with the exception of one in the CC-domain, are located in LRRs 1 to 4 and 14 to 23.

Intriguingly, the majority of mutations present in the clones from the second and third rounds are within these LRRs (Fig. 1a).

One limitation of shuffling is in recombining beneficial mutations that are in close proximity due to the limited frequency of crossing-over. Therefore, as multiple mutations in LRRs #3 and #15 had been found to enhance AVR3a^{EM} recognition, an alternate approach was adopted. A population of site-directed mutants was produced using degenerate oligonucleotides that encoded different pairs of amino acids at positions 585 (T/A), 618 (R/Q), 620 (E/D), 918 (R/G), 920 (K/E), 923 (D/G) and 931 (Q/R) with the potential for 128 different permutations. Screening of approximately 400 clones from this population with the best performing clone from the third round of shuffling (Rd3-1 [E620D, L668P, Q931R]) as a reference, identified eight clones that produced responses with AVR3a^{EM} comparable to or greater than the reference. Five of these clones were found to have identical sequences. Representative unique clones, Rd4-1 to Rd4-4, contain the amino acid substitutions R618Q, E620D, K920E and Q931R, but lacked either of the designed substitutions R918G or D923G (Table S2).

Iterative rounds of shuffling have progressively improved AVR3a^{EM} recognition by R3a* variants, producing faster and stronger PCD responses upon co-infiltration

The enhanced recognition of AVR3a^{EM} was assessed more accurately in single leaf comparisons. The R3a* constructs Rd2-1, Rd3-1 and Rd4-1 were transiently expressed under the control of the native *R3a* promoter with AVR3a^{EM} and the responses compared with those produced by the wild-type gene with AVR3a^{EM} and AVR3a^{KI}. The best performing clone from the first round was not included in this analysis on account of the relatively poor response produced when it was under the control of the *R3a* promoter.

Symptom development on *N. benthamiana* was monitored for 9 days after infiltration in four independent experiments. In each experiment two adjacent, expanded leaves on each of twenty plants were infiltrated with the five infiltration mixtures in a circularly permuted arrangement to account for possible intra-leaf position effects. Symptoms were scored on an arbitrary scale ranging from 0 (no symptoms) to 10 (complete necrosis of the infiltrated area). A progressive increase in the recognition of AVR3a^{EM} was observed for the three R3a* clones with the necrotic response produced by Rd4-1 being close to that produced by the wild-type gene with AVR3a^{KI} (Fig. 1b).

A mixed model with experiment as a fixed effect was fitted to test for consistent responses of the infiltration mixtures over repeated experiments. The experiments showed significant differences in mean response (p = 0.001), but there was no significant interaction (p = 0.306) between experiment and infiltration mixture indicating that the relative responses of the phenotypes were stable over repeated experiments. In addition, there were no significant interactions between experiment and the other fixed effects.

Since the phenotypes were determined to be stable, a second analysis of the data with experiment as a random effect was now fitted. This showed highly significant effects (p<0.001) from infiltration mixture and leaf age (upper vs lower leaf). Further, there was a highly significant interaction (p<0.001) between infiltration mixture and leaf age because the combination of R3a and AVR3a^{EM} did not show a difference in mean scores between younger and older leaves, in contrast to the other combinations which showed significant differences (Fig. 1c). Position of infiltration site on leaf was non-significant and all other interactions of fixed effect were also non-significant. Multiple comparisons using Bonferroni correction with an experiment-wise significance level of 5% showed significant differences between the mean symptom scores for all infiltration mixtures within either younger or older leaves with a comparison-wise significance level of 0.0011 (Table S3). Ordering of the responses was the same for both younger and older leaves with R3a and AVR3a^{KI}> Rd4-1 & AVR3a^{EM}> Rd3-1 & AVR3a^{EM}> Rd2-1 & AVR3a^{EM}> R3a and AVR3a^{EM}. While the clone Rd2-1 from the second round gave symptoms when co-expressed with AVR3a^{EM}, it rarely produced an HR phenotype as shown in figure 1d, which shows the proportion of sites with more than 50% of the infiltrated area necrotic. Despite improved recognition of AVR3a^{EM} by the R3a* variants relative to wild-type R3a, their recognition of AVR3a^{KI} was not impaired (data not shown), indicating that this specificity had not been significantly attenuated. Further, when expressed from the strong 35S promoter in the absence of AVR3a^{EM} or AVR3a^{KI} none of the clones produced necrosis, indicating that they maintained

appropriate control mechanisms and were not auto-activators (Fig. S1).

R3a* recognition of Avr3a^{EM} is dependent on HSP90 and SGT1

Previous studies by Bos *et al.* [14] demonstrated that R3a-dependent recognition of AVR3a^{KI} involves both SGT1 (suppressor of the G2 allele of *skp1*) and HSP90 (heat shock protein 90) that are required for the activation of other R proteins [26–27]. Their involvement in the AVR3a^{EM}-dependent responses was tested through *Tobacco rattle virus* (TRV)-based gene silencing of *SGT1* and *HSP90* in *N. benthamiana* with TRV-based expression of truncated GFP (eGFP) as a control. Three biological replicates for R3a, Rd2-1, Rd3-1 and one for Rd4-1, with infiltrations in two leaves of each of six plants per TRV-based silencing construct, revealed that both SGT1 and HSP90 are required to mediate an HR upon R3a*-based recognition of AVR3a^{KI} and AVR3a^{EM} (Fig. 2, Fig. S2). HRs were abolished for all infiltrations on TRV:SGT1 inoculated plants and there were almost no HRs recorded on TRV:HSP90 inoculated plants (Fig. 2). The HRs were not affected on plants inoculated with TRV:eGFP.

Compared to TRV:eGFP inoculated plants, *SGT1* and *HSP90* silenced plants were morphologically stunted, a phenotype that had been reported previously [14]. Nevertheless, upon infection with the bacterial pathogen *Erwinia amylovora* that produces a SGT1 and HSP90 independent non-host response in *N. benthamiana* [28], all plants were able to mount the expected HR response (Fig. 2, Fig. S2).

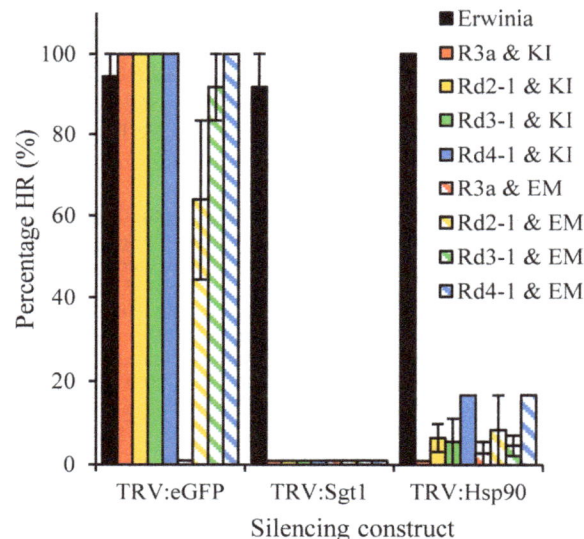

Figure 2. HR responses resulting from R3a* recognition of AVR3a^{EM} and AVR3a^{KI}, like those caused by wild-type R3a recognition of AVR3a^{KI}, are dependent on SGT1 and HSP90. SGT1- and HSP90-silenced plants were produced using TRV-based vectors. These plants and control plants inoculated with TRV:eGFP were infiltrated with different combinations of *Agrobacterium* cultures designed to express R3a, R3a* variants, AVR3a^{KI} (KI) or AVR3a^{EM} (EM). The percentage of sites (N = 12) showing HR responses six days after infiltration was recorded. The graph shows the mean percentages from three independent experiments with the exception that the dependence on HSP90 of Rd4-1 responses was only tested in a single experiment. The non-host bacterial pathogen *Erwinia amylovora* was used as a control for an SGT1- and HSP90-independnet HR response. Error bars show +/− standard error. Zero values have been transformed to 1% to facilitate their observation.

In a gain of mechanism, R3a* variants re-localize to late endosomes upon co-infiltration with Avr3a^{KI} or Avr3a^{EM}

In a previous study, Engelhardt *et al.* [15] demonstrated that, upon recognition of AVR3a^{KI} but not AVR3a^{EM}, wild-type R3a re-localizes from the host cytoplasm to specific late endosomes that can be labelled with the cyan fluorescent protein marker PS1-CFP [29]. This re-localization was found to be a pre-requisite for subsequent HR development for untagged R3a co-expressed with AVR3a^{KI} [15]. To study if R3a* variants with enhanced recognition of AVR3a^{EM} had gained the capacity to re-localize upon detection of AVR3a^{EM} and continued to exhibit this phenotype following detection of AVR3a^{KI}, N-terminal fusions of R3a* variants Rd2-1, Rd3-1 and Rd4-1 with yellow fluorescent protein (YFP) were generated as described previously with expression of these constructs driven by a 35S promoter. Western-blot analysis of protein extracts from inoculated tissue demonstrated the integrity of the fusion proteins (Fig. S3). As demonstrated for YFP-R3a wild-type fusions by Engelhardt *et al.* [15], YFP-R3a* fusions did not elicit HRs alone or in the presence of AVR3a^{KI} or AVR3a^{EM}, probably due to steric hindrance of the signalling domains of R3a (data not shown).

As anticipated, following transient expression in *N. benthamiana*, all YFP-R3a/R3a* fusions when expressed by themselves displayed cytoplasmic localizations (Fig. S4). In accord with the observations described by Engelhardt *et al.* [15], the localisation of YFP-R3a remained cytoplasmic upon co-infiltration with AVR3a^{EM} (Fig. 3), but changed to fast moving, PS1-CFP labelled vesicles, following recognition of AVR3a^{KI}. The YFP-R3a* fusions of Rd2-1, Rd3-1 and Rd4-1 proteins maintained this mechanistically characteristic re-localisation following co-expression with AVR3a^{KI} (Fig. 3). However, in contrast to YFP-R3a, all selected YFP-R3a* variants also displayed highly reproducible re-localization to PS1-CFP labelled vesicles after the perception of AVR3a^{EM} (Fig. 3; Fig S5).

AVR3a^{KI} and AVR3a^{EM} re-localize to endosomes upon co-infiltration with R3a* variants but not, in the case of AVR3a^{EM}, with wild-type R3a

As shown by Engelhardt *et al.* [15] AVR3a^{KI}, but not AVR3a^{EM}, also re-localizes from the cytoplasm to endosomes upon co-expression with R3a. This was demonstrated by N-terminal fusions of AVR3a^{KI} and AVR3a^{EM} to green fluorescent protein as well as by BiFC, also known as split-YFP, assays [9,15,30]. The latter revealed that wild-type R3a and AVR3a^{KI} are found in close proximity at PS1-CFP labelled vesicles [15].

To investigate if the vesicular co-association of R3a and AVR3a^{KI} was extended to the R3a* variants, BiFC was used to analyse and localize protein–protein interactions *in planta*. As described previously for wild-type R3a, the N-terminal portion of YFP, YN, was fused to the N-terminal end of the R3a* variants Rd2-1, Rd3-1 and Rd4-1. The constructs used to express the C-terminal portion of YFP, YC, fused to AVR3a^{KI} and AVR3a^{EM} were as described previously [15] with all constructs being transiently expressed in *N. benthamiana* from the 35S promoter.

In accord with previous findings [15], co-expression of YN-R3a with YC-AVR3a^{KI} gave strong YFP fluorescence, whereas co-expression with YC-AVR3a^{EM} did not give detectable YFP fluorescence (Fig. 4). Like the YN fusion to wild-type R3a, all the YN-R3a* fusions when co-expressed with AVR3a^{KI} gave strong, punctate, YFP signals (Fig. 4), but unlike the wild-type fusion also gave YFP fluorescence signals at PS1-CFP labelled vesicles when co-expressed with AVR3a^{EM} (Fig. 4; Fig. S6). This indicates that AVR3a^{EM} is also within close proximity of the re-localized R3a*

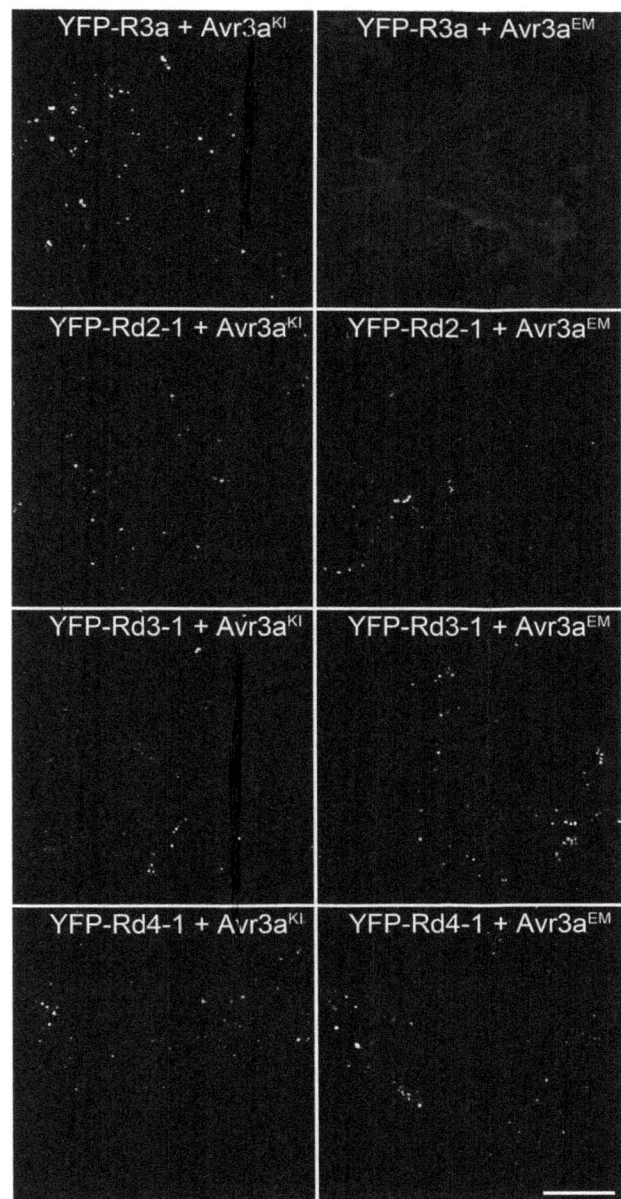

Figure 3. YFP fusions to R3a* variants re-localize to vesicles after the perception of both of AVR3a^{KI} and AVR3a^{EM}, whereas YFP-R3a remains cytoplasmic in the presence of AVR3a^{EM}. Two days after infiltration of mixtures of *Agrobacterium* cultures designed to express AVR3a^{KI}, AVR3a^{EM}, YFP-R3a or YFP fusions to the R3a* variants, infiltrated *N. benthamiana* leaf tissue was examined under a confocal laser scanning microscope. Scale bar = 50 μm.

gene products. Thus, in line with the gain of recognition of AVR3a^{EM} by the R3a* variants and subsequent necrosis responses, the R3a* variants and AVR3a^{EM} show the same mechanistic re-localization as observed for R3a and AVR3a^{KI}.

R3a* variants maintain resistance towards AVR3a^{KI}-expressing *P. infestans* isolates but have not gained resistance towards AVR3a^{EM} homozygous isolates

To evaluate if R3a* variants with gain of AVR3a^{EM} recognition and re-localisation mechanism yield effective disease resistance,

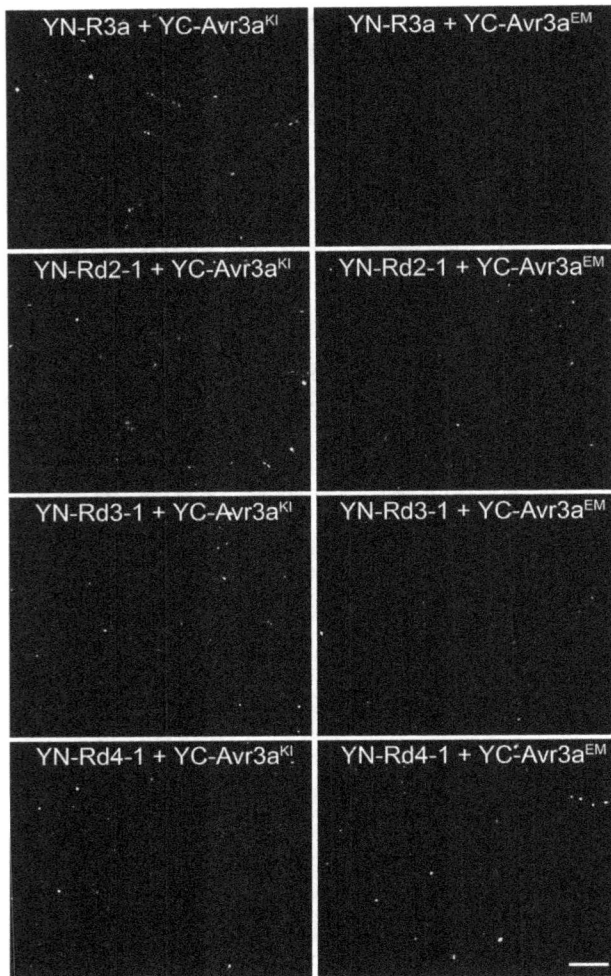

Figure 4. Both YC-AVR3a^KI and YC-AVR3a^EM when co-expressed with YN-R3a* fusions give vesicle associated YFP fluorescence like YC-AVR3a^KI and YN-R3a, whereas YC-AVR3a^EM and YN-R3a do not. Two days after infiltration of mixtures of *Agrobacterium* cultures designed to express YC-AVR3a^KI, YC-AVR3^EM, YN-R3a or YN fusions to the R3a* variants, infiltrated *N. benthamiana* leaf tissue was examined under a confocal laser scanning microscope. Representative images from two experiments. Scale bar = 50 μm.

transient and stable expression systems were utilised. *Agrobacterium tumefaciens* transient assays (ATTAs) in *N. benthamiana* have successfully been used to demonstrate function for late blight resistance gene products such as R2, Rpi_STO1 [31] and R3b [32]. Selected R3a* clones Rd2-1, Rd3-1 and Rd4-1 were transiently expressed in *N. benthamiana* using the *R3a* promoter in ATTAs alongside wild-type R3a and an empty vector control. Infiltrated leaf areas were challenged two days after infiltration with AVR3a^KI or AVR3a^EM homozygous *P. infestans* isolates via drop inoculation. Disease progression was monitored by measuring visible lesion diameters in multiple independent experiments and analysis of variance was carried out using GenStat on the data from individual experiments to test for significant differences. Multiple comparisons were performed using Bonferroni correction with a significance level of 5% and a comparison-wise error rate of 0.005.

In three experiments ATTA sites were inoculated with the AVR3a^KI homozygous *P. infestans* isolate 7804.b. In all three experiments the wild-type R3a and the R3a* variants significantly reduced spread of *P. infestans* relative to the empty vector control and there were no significant differences between the different R3a forms (Fig. 5a). This result indicates that the selected mutations in the LRR do not impair the resistance induced by AVR3a^KI. Likewise ATTA sites were inoculated with the AVR3a^EM homozygous isolate 88069 [9] in five experiments. In four of the five experiments there were no significant differences in *P. infestans* spread between any of the R3a forms and the empty vector control (Fig. 5b). In the fifth experiment there was significantly increased spread with the empty vector control, but the R3a* variants showed no significant differences from the wild-type gene which does not provide resistance against AVR3a^EM homozygous isolates. Co-infiltrations of R3a, R3a*, AVR3a^KI and AVR3a^EM constructs were carried out contemporaneously in all experiments to confirm that the conditions were conducive to HR development (data not shown).

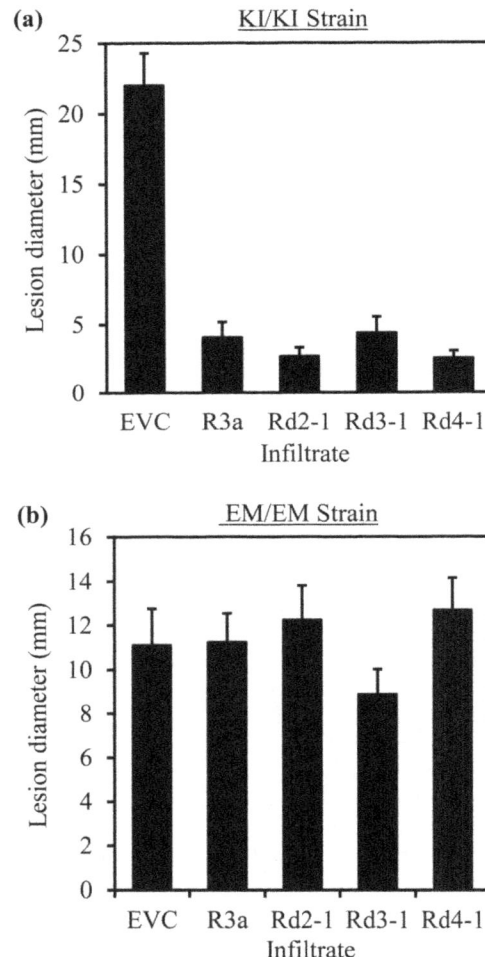

Figure 5. R3a and R3a* variants expressed from *Agrobacterium* reduce the spread of a *P. infestans* strain expressing AVR3a^KI, but not the spread of a strain expressing only AVR3a^EM. (a) Means of lesion diameters measured 12 days after drop inoculation of agro-infiltrated areas with strain 7804.b (KI/KI). (b) Means of lesion diameters measured 8 days after drop inoculation of agro-infiltrated areas with strain 88069 (EM/EM). (a) and (b) show representative experiments from sets of three and five repeated experiments, respectively. For both (a) and (b), error bars show +/− standard errors, N = 30. EVC indicates empty vector control.

To confirm these results and to rule out potential adverse effects and limitations of the ATTA system in *N. benthamiana*, transgenic potato plants were generated. The wild-type *R3a* gene and the *R3a** genes for Rd2-1, Rd3-1 and Rd4-1 were transformed into the potato cultivar Ranger Russet using the *R3a* promoter and terminator to regulate gene expression and stability. Transgenic Ranger Russet lines expressing R3a and the three R3a* variants were compared to Ranger Russet lines containing the *Rpi_vnt1* gene [33] and non-transgenic Ranger Russet plants as positive and negative controls, respectively. Transgenic lines were challenged with the Mexican isolate P6752, which is heterozygous for AVR3a^{KI} and AVR3a^{EM}, and the US isolate US-8 BF-6, which is homozygous for AVR3a^{EM}. The transgenic potato plants expressing R3a, Rpi_vnt1 and the R3a* variants, but not the non-transgenic Ranger Russet, demonstrated high levels of resistance towards the heterozygous isolate P6752 (Fig. 6). Thus, the transient ATTA data and the transgenic plants corroborate the conclusion that the selected mutations in the LRR do not negatively impact on resistance towards *P. infestans* isolates expressing Avr3a^{KI}. The transgenic Ranger Russet lines expressing Rpi_vnt1 provided resistance towards the AVR3a^{EM} homozygous isolate US-8 BF-6. However, none of the transgenic lines expressing the *R3a** or *R3a* genes, which had been shown to be resistant to isolate P6752, were able to control disease development of isolate US-8 BF-6 (Fig. 6).

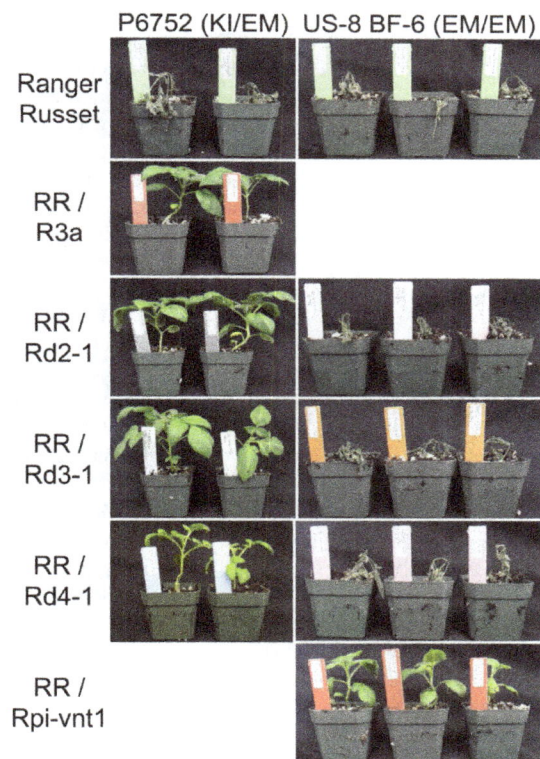

Figure 6. R3a and R3a* variants expressed via the *R3a* promoter in transgenic plants protect the susceptible cultivar Ranger Russet from *P. infestans* strain P6752, which is heterozygous for AVR3a^{KI} and AVR3a^{EM}, but not from strain US-8 BF-6, which is homozygous for AVR3a^{EM}. Non-transgenic plants were used as a control for susceptibility. Transgenic plants expressing R3a or Rpi_vnt1 were used as positive controls for resistance to P6752 or US-8 BF-6, respectively. Representative plants were photographed at 11dpi.

Discussion

The relatively narrow genetic basis of clonal potato cultivars in agriculture provides pathogens such as *P. infestans* with sufficient opportunity to adapt and overcome inducible host resistant responses and to thus cause disease on a global scale. Resistance responses rely on the direct or indirect recognition of modified-self or pathogen-derived molecules [1]. For example, in the first layer of inducible resistance, also referred to as PAMP triggered immunity (PTI), conserved pathogen-associated molecular patterns (PAMPs) and/or damage-associated molecular patterns (DAMPs) are recognised [1,34,35]. Successful pathogens circumvent this recognition with the help of effectors that perturb host resistance responses and promote effector-triggered susceptibility (ETS). However, by being in close proximity to the host, pathogen effectors provide the innate plant immune system with another opportunity for detection that is dependent on the presence of cognate plant R proteins that subsequently yield ETI [1,36]. In nature, this closely entwined co-evolution between hosts and pathogens is evident in the diversification observed for effectors and *R* genes [4].

Indeed, *P. infestans* is known as a pathogen with 'high evolutionary potential' [37] and more than 560 RXLR-type candidate effectors have been described within the late blight pathogen genome [38]. The genomic organisation of RXLR effector genes, which are often in gene-poor and repeat-rich regions, is thought to facilitate their enhanced diversification by enabling non-homologous recombination. Oomycete RXLR-type effectors have been shown to evade detection by R gene products via transcriptional regulation [39], utilising functionally equivalent effectors that allow loss of recognised effectors [40], suppressor activity of unrelated effectors [41] and/or sequence diversity [11,42]. Sequence diversity underpins virulence or avirulence behaviour for the essential *P. infestans* effector AVR3a. AVR3a^{KI} determines avirulence on plants carrying *R3a* whereas AVR3a^{EM} promotes virulence. It is thought that only AVR3a^{KI} was present in the *P. infestans* strain responsible for the outbreak of late blight disease leading to the Irish Potato Famine in the 1840 s [43]. The AVR3a^{EM} allele may have come to dominate in *P. infestans* populations once the resistance gene *R3a* was deployed in potato crops, quickly usurping the AVR3a^{KI} allele. By using clonal potato varieties in current agriculture, the diversification of *R* genes is undermined and novel, naturally occurring resistances can only slowly be introgressed via breeding.

Functional *R* genes are often found in clusters that show evidence of duplication followed by diversification [44]. The phylogenetic NB-LRR gene grouping that contains homologs of the functional *R3a* gene in the potato genome has previously been described as CNL-8 [5]. This group is, after the *R2* cluster (CNL-5), the second largest *R* gene cluster in potato where more than 750 NB-LRR-like genes have been identified [5–6]. Functional *R3a*, a homolog of the tomato *I2* resistance gene that controls races of the fungus *Fusarium oxysporum* [45], was cloned alongside three paralogous sequences that provided insight into amino acid positions under diversification [13]. With one exception, positions under diversification reside in the LRR and cluster around two regions spanning LRRs 1 to 4 and 14 to 23 that were also identified as being important in this study.

Here we used, in addition to random mutagenesis and screening, DNA shuffling and targeted mutagenesis to enhance AVR3a^{EM} recognition by R3a*. DNA shuffling, which emulates the natural evolutionary processes of mutation, recombination and selection, has proven a highly effective method for evolving new specificities/properties for a wide range of proteins that cannot be rationally designed and is of particular use in identifying mutations that are beneficial in combination [18]. Artificial evolution of a resistance gene to broaden its specificity has previously been performed on the gene Rx that protects potato plants from strains of PVX. In an initial Rx study, mutagenesis of the LRR region was performed on the basis that this region is the primary determinant of recognition specificity. Screening identified four mutations in the LRR region that affected elicitor recognition and activation functions [16]. Introduction of the mutated Rx genes in to the model host $N.$ $benthamiana$ as transgenes extended resistance to a normally resistance-breaking strain of PVX, HB, and a distantly related carlavirus, PopMV [16].

Our primary screen identified eleven mutants with enhanced AVR3a^{EM} recognition containing in total 23 amino acid substitutions. However, only three of these (R618Q, K920E, Q931R), found in the clones with single amino acid substitutions, are known to be causative to the improved phenotype. The previous study performed by Segretin et $al.$ [25] identified six amino acid substitutions in the LRR domain that enhanced AVR3a^{EM} recognition: two of these were also found in our primary screen (L668P, K920E). The fact that we did not identify all the LRR mutations identified by Segretin et $al.$ [25] and that they did not identify more mutations in the LRR region, demonstrates that neither screen was exhaustive.

The initial screen identified mutants with enhanced recognition of AVR3a^{EM} when expressed from the strong 35S promoter. Amino acid changes in LRRs #3 and #15, and thus within regions known to be under diversifying selection [13], were prevalent in the clones from the first round of screening, suggesting these regions might be of particular importance. This suggestion was supported by the fact that all the clones from the second round of shuffling contained an amino acid change, E620D, in LRR #3 and sometimes additional changes in LRR #15, while all of the clones from the third round contained the E620D change and one or two amino acid changes in LRR #15. Using DNA shuffling it is more difficult to bring together combinatorially beneficial mutations that are in close proximity. Furthermore, random mutagenesis as a source of diversity has limitations in that single nucleotide changes can only convert a codon for one amino acid to a limited set of codons for other amino acids rather than all twenty possible amino acids. To circumvent the former problem a more directed approach was adopted in which a library that contained all permutations of the amino acid changes thought important (one in LRR #2, two in LRR #3 and four in LRR #15) was produced for screening. The aptness of this approach was shown by the fact that clones were obtained with enhanced recognition responses to AVR3a^{EM} compared to the best performing clone from the third round of shuffling, Rd3-1, and that one of the possible 128 forms was prevalent. This form contained two amino acid changes in close proximity in each of LRRs #3 and #15; a combination that would have been difficult to obtain through DNA shuffling. Interestingly, an amino acid change at one of these positions, Q931, was also found in one of the clones recovered by Segretin et $al.$ [25], though their "de-convolution" did not show their substitution at this position, proline instead of arginine, to improve AVR3a^{EM} recognition. As shown by Segretin et $al.$ [25] for the R3a mutants with enhanced AVR3a^{EM} recognition, we did not note any reduction in the AVR3a^{KI} recognition responses of the clones we isolated.

Our studies show that the recognition of AVR3a^{EM} by the R3a* mutants we have isolated recapitulates the mechanistic processes of recognition of AVR3a^{KI} by the wild-type $R3a$ gene. It has previously been reported [14] that the HR triggered by R3a recognition of AVR3a^{KI} is dependent on the ubiquitin ligase-associated protein SGT1 and HSP90. VIGS of $SGT1$ and, to a lesser degree, of $HSP90$ in our experiments inhibited the cell death responses induced by recognition of AVR3a^{EM} by our R3a* mutants. Similarly, it has been shown that wild-type R3a re-localises from the cytoplasm to late endosomal compartments when co-expressed with AVR3a^{KI}, but not when co-expressed with AVR3a^{EM} [15]. We found that the mutants from the three later rounds still re-localised to endosomal compartments when co-expressed with AVR3a^{KI} and, importantly, also re-localised to the same vesicles when co-expressed with AVR3a^{EM}. The earlier study by Engelhardt et $al.$ [15] also showed that the effector AVR3a^{KI} itself relocalises from the cytoplasm to endosomes when co-expressed with wild-type R3a and is in close physical proximity to R3a, whereas AVR3a^{EM} remains distributed through the cytoplasm. Our BiFC experiments show that AVR3a^{KI} and the normally unrecognized form, AVR3a^{EM} both traffic from the cytoplasm to vesicles when co-expressed with the R3a* forms. This re-localisation of R3a and AVR3^{KI} was shown to be a prerequisite for the development of the HR [15]. Thus, the gain of AVR3a^{EM} recognition R3a* variants have gained many aspects of the mechanism of the wild-type R3a response to AVR3a^{KI}.

Although the R3a* variants produced in this study responded to AVR3a^{EM} and produced HR responses when the elicitor was transiently expressed via $Agrobacterium$, critically they only provided resistance to $P.$ $infestans$ isolates that express AVR3a^{KI} and not to isolates that express only AVR3a^{EM}. Both transient expression via $Agrobacterium$ in $N.$ $benthamiana$ and stable transgenic expression in potato corroborated this finding. Failure to protect from the pathogen itself was also reported for the R3a mutants identified by Segretin et $al.$ [25]. That this was the case for mutants with single amino acid changes is perhaps not surprising given the large differences from the wild-type R3a/AVR3a^{KI} response we observed when first and second round clones were expressed from the $R3a$ promoter with AVR3a^{EM}. For some pathogen/R gene combinations, e.g. PVX and Rx, the resistance responses can be separated from the HR [46] though the induction of necrotic responses by transient expression of the elicitor protein has been used to identify Rx mutations that, when expressed transgenically in the model host $N.$ $benthamiana$, provide resistance to the pathogen itself. However, for $P.$ $infestans$ it has been suggested that the strength of the HR correlates with resistance levels [47].

A recent, secondary mutation study of Rx provides some evidence that stepwise artificial evolution could be required to obtain an optimum combination between effector recognition and subsequent R gene activation and signal transduction [17]. In the Rx study, the resistance provided by one of the mutations in the LRR domain to PopMV was improved by random mutagenesis of the CC-NB-ARC1-ARC2 domains [17]. In addition to constitutively active mutants that by themselves gave necrotic responses, four mutants with enhanced responses to PopMV were isolated. For three of these mutants the improved phenotype was conferred by a single amino acid change, while for the fourth a pair of amino acid substitutions was required. The mutations, which affect activation sensitivity, were found to be located around the nucleotide-binding pocket of Rx. As mentioned previously, we have evidence that neither this screening nor the efforts from Segretin et $al.$ [25] were exhaustive as both approaches yielded novel beneficial mutations. Whereas our study of $R3a$ focused solely on the LRRs, in Segretin et $al.$ [25] the entire $R3a$ gene was subjected to mutagenesis. The latter study identified eight single amino acid changes that enhanced responses to AVR3a^{EM}. Out of these, six occurred in the LRR domain and one in the CC domain. This substitution enhanced the response to AVR3a^{EM} but also showed some auto-activation. The final substitution, found in the NB-ARC domain, sensitised the AVR3a^{EM} response and broadened specificity to include an elicitor from another *Phytophthora* species [25]. Interestingly, this change occurred in the nucleotide-binding pocket and is adjacent to one of the sensitizing mutations found in *Rx* [17]. Broadening resistance gene specificity merely by introducing sensitizing mutations without improving recognition may have detrimental consequences in the field. However, a natural precedent for this has been found in PM3 resistance protein alleles in which the substitution of two amino acids in the NB domain enhances the HR and broadens the spectrum of resistance to wheat powdery mildew isolates [48]. Thus, additional efforts to combine novel mutations in R3a domains responsible for AVR3a^{EM} recognition (LRR) and response (CC-NB-ARC1/ARC2) could further improve R3a* variants that already display gain of recognition and mechanistic re-localisation to ultimately yield genes that provide effective resistance in potato against isolates expressing AVR3a^{EM}. Considering the importance of AVR3a to *P. infestans*, such a resistance, combined with other, mechanistically distinct *R* genes could provide a step towards more durable late blight control.

Supporting Information

Figure S1 R3a* variants are not auto-activators. *N. benthamiana* leaves were infiltrated with *Agrobacterium* cultures designed to express R3a or the R3a* variants from the strong 35S promoter. Mixtures of cultures designed to co-express AVR3a^{KI} (KI) were used as positive controls for the induction of cell death. Leaves were examined under white-light and UV-B illumination. Photograph of representative leaf was taken five days after infiltration. In the absence of elicitor the R3a* variants, like R3a, do not produce visible cell death.

Figure S2 HR responses resulting from R3a* recognition of AVR3a^{EM} and AVR3a^{KI}, like those caused by wild-type R3a recognition of AVR3a^{KI}, are dependent on SGT1 and HSP90. SGT1- and HSP90-silenced plants were produced using TRV-based vectors. These plants and control plants inoculated with TRV:*eGFP* were infiltrated with different combinations of *Agrobacterium* cultures designed to express R3a, R3a* variants, AVR3a^{KI} (KI) or AVR3a^{EM} (EM). Photographs show representative HR responses induced by each of the different mixtures on control TRV:eGFP inoculated plants, SGT1-silenced plants and HSP90-silenced plants. The non-host bacterial pathogen *Erwinia amylovora* was used as a control for an SGT1- and HSP90-independnet HR response.

Figure S3 Western blot analysis showing integrity of YFP fusion proteins. Soluble protein extracts were prepared from *N. benthamiana* leaf tissue two days after infiltration with cultures designed to express YFP fusions to R3a, Rd2-1, Rd3-1 or Rd4-1. The blot was probed with anti-GFP antibodies as described by Engelhardt *et al.* (2012). Protein sizes are indicated

in kilodaltons (kD) and protein loading is shown by Ponceau S (PS) staining.

Figure S4 YFP fusions to R3a and the R3a* variants localize to the cytoplasm in the absence of AVR3a. *N. benthamiana* leaves were infiltrated with cultures designed to express YFP fusions to R3a, Rd2-1, Rd3-1 or Rd4-1. Leaf tissue was examined two days after infiltration under a confocal laser scanning microscope. Representative images are from five independent experiments. Scale bar = 20 μm.

Figure S5 In the presence of AVR3a^{EM} YFP fusions to R3a* variants, but not YFP-R3a, re-localize to vesicles labelled by the prevacuolar compartment marker PS1-CFP. *N. benthamiana* leaves were infiltrated with mixtures of cultures designed to express PS1-CFP, AVR3a^{EM} and YFP fusions to R3a, Rd2-1, Rd3-1 or Rd4-1. Tissue was examined two days after infiltration under a confocal laser scanning microscope. The left-hand panel shows YFP signal, the right-hand panel CFP signal and the central panel displays the merged signals. Representative images are from three independent experiments. Scale bar = 10 μm.

Figure S6 YC-AVR3a^{EM} reconstitutes YFP fluorescence with YN fusions to the R3a* variants at vesicles labelled by the prevacuolar compartment marker PS1-CFP. Generation of the YFP signal indicates that AVR3aEM and the R3a* variants are in close proximity at the vesicles. *N. benthamiana* leaves were infiltrated with mixtures of cultures designed to express PS1-CFP, YC-AVR3a^{EM} and YN fusions to Rd2-1, Rd3-1 or Rd4-1. Tissue was examined 2 d after infiltration under a confocal laser scanning microscope. Left-hand panel, YFP signal; right-hand panel, CFP signal; central panel, merged signals. Representative images from three experiments. Scale bar = 20 μm.

Acknowledgments

The authors acknowledge Dr Sanwen Huang for kindly providing sequences of wild type R3a and the native regulatory elements.

Author Contributions

Conceived and designed the experiments: SC PCB PRJB IH. Performed the experiments: SC LS PCB SE BH NC KM PSMVW XC. Analyzed the data: SC LS PCB CA NC PRJB IH. Wrote the paper: IH SC LS CA PRJB.

References

1. Jones JDG, Dangl JL (2006) The plant immune system. Nature 444: 323–329.
2. van der Biezen EA, Jones JDG (1998) Plant disease-resistance proteins and the gene-for-gene concept. Trends Biochem Sci 23: 454–456.
3. Heath MC (2000) Hypersensitive response-related death. Plant Mol Biol 44: 321–334.
4. Hein I, Gilroy EM, Armstrong MR, Birch PRJ (2009) The zig-zag-zig in oomycete–plant interactions. Mol Plant Pathol 10: 547–562.

5. Jupe F, Pritchard L, Etherington GJ, MacKenzie K, Cock PJA, et al. (2012) Identification and localisation of the NB-LRR gene family within the potato genome. BMC Genomics 13: 75.

6. Jupe F, Witek K, Verweij W, Śliwka J, Pritchard L, et al. (2013) Resistance gene enrichment sequencing (RenSeq) enables reannotation of the NB-LRR gene family from sequenced plant genomes and rapid mapping of resistance loci in segregating populations. Plant J 76: 530–544.

7. Birch PRJ, Boevink PC, Gilroy EM, Hein I, Pritchard L, et al. (2008) Oomycete RXLR effectors: delivery, functional redundancy and durable disease resistance. Curr Opin Plant Biol 11: 373–379.

8. Vleeshouwers VG, Raffaele S, Vossen JH, Champouret N, Oliva R, et al. (2011) Understanding and exploiting late blight resistance in the age of effectors. Annu Rev Phytopathol. 49:507–531.

9. Bos JIB, Armstrong MR, Gilroy EM, Boevink PC, Hein I, et al. (2010) *Phytophthora infestans* effector AVR3a is essential for virulence and manipulates plant immunity by stabilizing host E3 ligase CMPG1. Proc Natl Acad Sci USA 107: 9909–9914.

10. Vetukuri RR, Tian Z, Avrova AO, Savenkov EI, Dixelius C, et al. (2011) Silencing of the PiAvr3a effector-encoding gene from *Phytophthora infestans* by transcriptional fusion to a short interspersed element. Fungal Biol 115: 1225–1233.

11. Armstrong MR, Whisson SC, Pritchard L, Bos JIB, Venter E, et al. (2005) An ancestral oomycete locus contains late blight avirulence gene Avr3a, encoding a protein that is recognized in the host cytoplasm. Proc Natl Acad Sci USA 102: 7766–7771.

12. Cárdenas M, Grajales A, Sierra R, Rojas A, González-Almario A, et al. (2011) Genetic diversity of Phytophthora infestans in the Northern Andean region. BMC Genet 12: 23.

13. Huang S, van der Vossen EAG, Kuang H, Vleeshouwers VGAA, Zhang N, et al. (2005) Comparative genomics enabled the isolation of the R3a late blight resistance gene in potato. Plant J 42: 251–261.

14. Bos JIB, Kanneganti T-D, Young C, Cakir C, Huitema E, et al. (2006) The C-terminal half of *Phytophthora infestans* RXLR effector AVR3a is sufficient to trigger R3a-mediated hypersensitivity and suppress INF1-induced cell death in *Nicotiana benthamiana*. Plant J 48: 165–176.

15. Engelhardt S, Boevink PC, Armstrong MR, Ramos MB, Hein I, et al. (2012) Relocalization of late blight resistance protein R3a to endosomal compartments is associated with effector recognition and required for the immune response. Plant Cell 24: 5142–5158.

16. Farnham G, Baulcombe DC (2006) Artificial evolution extends the spectrum of viruses that are targeted by a disease-resistance gene from potato. Proc Natl Acad Sci USA 103: 18828–18833.

17. Harris CJ, Slootweg EJ, Goverse A, Baulcombe DC (2013) Stepwise artificial evolution of a plant disease resistance gene. Proc Natl Acad Sci USA, 110: 21189–21194.

18. Stemmer WP (1994) DNA shuffling by random fragmentation and reassembly: in vitro recombination for molecular evolution. Proc Natl Acad Sci USA 91: 10747–10751.

19. Bernal A, Pan Q, Pollack J, Rose L, Willets N, et al. (2005) Functional dissection of the Pto resistance gene using DNA shuffling. J Biol Chem 280: 23073–23083

20. Simpson CG, Lewandowska D, Liney M, Davidson DDavidson D, Chapman S, et al. (2014) Arabidopsis PTB1 and PTB2 proteins negatively regulate splicing of a mini-exon splicing reporter and affect alternative splicing of endogenous genes differentially. New Phytol 203: 424–436.

21. Leung DW, Chen E, Goeddel DV (1989) A method for random mutagenesis of a defined DNA segment using a modified polymerase chain reaction. Technique, 1: 11–15.

22. Lazo GR, Stein PA, Ludwig RA (1991) A DNA transformation-competent Arabidopsis genomic library in Agrobacterium. Biotechnology 9: 963–967.

23. van der Fits L, Deakin EA, Hoge JH, Memelink J (2000) The ternary transformation system: constitutive virG on a compatible plasmid dramatically increases Agrobacterium-mediated plant transformation. Plant Mol Biol 43: 495–502.

24. Duan H, Richael C, Rommens CM (2012) Overexpression of the wild potato eIF4E-1 variant Eva1 elicits Potato virus Y resistance in plants silenced for native eIF4E-1. Transgenic Res 21: 929–938.

25. Segretin ME, Pais M, Franceschetti M, Chaparro-Garcia A, Bos JIB, et al. (2014) Single amino acid mutations in the potato immune receptor R3a expand response to *Phytophthora* effectors. Mol Plant Microbe Interact 27: 624–637.

26. Liu Y, Burch-Smith T, Schiff M, Feng S, Dinesh-Kumar SP (2004) Molecular chaperone Hsp90 associates with resistance protein N and its signaling proteins

27. SGT1 and Rar1 to modulate an innate immune response in plants. J Biol Chem 279: 2101–2108.

27. Azevedo C, Betsuyaku S, Peart J, Takahashi A, Noel L, et al. (2006) Role of SGT1 in resistance protein accumulation in plant immunity. EMBO J, 25, 2007–2016.

28. Gilroy EM, Hein I, van der Hoorn R, Boevink PC, Venter E, et al. (2007) Involvement of cathepsin B in the plant disease resistance hypersensitive response. Plant J 52: 1–13.

29. Saint-Jean B, Seveno-Carpentier E, Alcon C, Neuhaus JM, Paris N (2010) The cytosolic tail dipeptide Ile-Met of the pea receptor BP80 is required for recycling from the prevacuole and for endocytosis. Plant Cell, 22: 2825–2837.

30. Walter M, Chaban C, Schütze K, Batistic O, Weckermann K, et al. (2004) Visualization of protein interactions in living plant cells using bimolecular fluorescence complementation. Plant J 40: 428–438.

31. Saunders DG, Breen S, Win J, Schornack S, Hein I, et al. (2012) Host protein BSL1 associates with Phytophthora infestans RXLR effector AVR2 and the Solanum demissum Immune receptor R2 to mediate disease resistance. Plant Cell 24: 3420–3434.

32. Li G, Huang S, Guo X, Li Y, Yang Y, et al. (2011) Cloning and characterization of R3b; members of the R3 superfamily of late blight resistance genes show sequence and functional divergence. Mol Plant Microbe Interact 24: 1132–1142.

33. Foster SJ, Park TH, Pel M, Brigneti G, Sliwka J, et al. (2009) Rpi-vnt1.1, a Tm-2(2) homolog from Solanum venturii, confers resistance to potato late blight. Mol Plant Microbe Interact 22: 589–600.

34. Maffei ME, Arimura G, Mithöfer A (2012) Natural elicitors, effectors and modulators of plant responses. Nat Prod Rep 29: 1288–1303.

35. Newman MA, Sundelin T, Nielsen JT, Erbs GI (2013). MAMP (microbe-associated molecular pattern) triggered immunity in plants. Front Plant Sci 16;4 139

36. Deslandes L, Rivas S (2012) Catch me if you can: bacterial effectors and plant targets. Trends Plant Sci. 17: 644–655

37. Raffaele S, Win J, Cano LM, Kamoun S (2010) Analyses of genome architecture and gene expression reveal novel candidate virulence factors in the secretome of Phytophthora infestans. BMC Genomics. 16: 637.

38. Haas BJ, Kamoun S, Zody MC, Jiang RHY, Handsaker RE, et al. (2009) Genome sequence and analysis of the Irish potato famine pathogen *Phytophthora infestans*. Nature 461: 393–398.

39. Rietman H, Bijsterbosch G, Cano LM, Lee HR, Vossen JH, et al. (2012) Qualitative and quantitative late blight resistance in the potato cultivar Sarpo Mira is determined by the perception of five distinct RXLR effectors. Mol Plant Microbe Interact 25: 910–919.

40. Van Poppel PM, Guo J, van de Vondervoort PJ, Jung MW, Birch PR, et al. (2008) The Phytophthora infestans avirulence gene Avr4 encodes an RXLR-dEER effector. Mol Plant Microbe Interact 21: 1460–1470.

41. Wang Q, Han C, Ferreira AO, Yu X, Ye W, et al. (2011) Transcriptional programming and functional interactions within the Phytophthora sojae RXLR effector repertoire. Plant Cell. 23: 2064–2086

42. Gilroy EM, Tayor RM, Hein I, Boevink P, Sadanandom A, et al. (2011) CMPG1-dependent cell death follows perception of diverse pathogen elicitors at the host plasma membrane and is suppressed by *Phytophthora infestans* RXLR effector AVR3a. New Phytol 190: 653–666.

43. Yoshida K, Schuenemann VJ, Cano LM, Pais M, Mishra B, et al. (2013) The rise and fall of the *Phytophthora infestans* lineage that triggered the Irish potato famine. eLife, 2: e00731.

44. McDowell JM, Simon SA (2006) Recent insights into R gene evolution. Mol Plant Pathol 7, 437–448.

45. Ori N1, Eshed Y, Paran I, Presting G, Aviv D, et al. (1997) The I2C family from the wilt disease resistance locus I2 belongs to the nucleotide binding, leucine-rich repeat superfamily of plant resistance genes. Plant Cell. 9: 521–532.

46. Bendahmane A, Kanyuka K, Baulcombe DC (1999) The Rx gene from potato controls separate virus resistance and cell death responses. Plant Cell, 11: 781–792.

47. Vleeshouwers VGAA, van Dooijeweert W, Govers F, Kamoun S, Colon LT (2000) The hypersensitive response is associated with host and nonhost resistance to *Phytophthora infestans*. Planta 210: 853–864.

48. Stirnweiss D, Milani SD, Jordan T, Keller B, Brunner S (2014) Substitutions of two amino acids in the nucleotide-binding site domain of a resistance protein enhance the hypersensitive response and enlarge the PM3F resistance spectrum in wheat. Mol Plant Microbe Interact 27: 265–276.

18

Inter-Plant Vibrational Communication in a Leafhopper Insect

Anna Eriksson[1,2], Gianfranco Anfora[1]*, Andrea Lucchi[2], Meta Virant-Doberlet[3], Valerio Mazzoni[1]

1 Research and Innovation Centre, Fondazione Edmund Mach, San Michele all'Adige, Italy, **2** Department of Coltivazione e Difesa delle Specie Legnose, University of Pisa, Pisa, Italy, **3** Department of Entomology, National Institute of Biology, Ljubljana, Slovenia

Abstract

Vibrational communication is one of the least understood channels of communication. Most studies have focused on the role of substrate-borne signals in insect mating behavior, where a male and a female establish a stereotyped duet that enables partner recognition and localization. While the effective communication range of substrate-borne signals may be up to several meters, it is generally accepted that insect vibrational communication is limited to a continuous substrate. Until now, interplant communication in absence of physical contact between plants has never been demonstrated in a vibrational communicating insect. With a laser vibrometer we investigated transmission of natural and played back vibrational signals of a grapevine leafhopper, *Scaphoideus titanus*, when being transmitted between leaves of different cuttings without physical contact. Partners established a vibrational duet up to 6 cm gap width between leaves. Ablation of the antennae showed that antennal mechanoreceptors are not essential in detection of mating signals. Our results demonstrate for the first time that substrate discontinuity does not impose a limitation on communication range of vibrational signals. We also suggest that the behavioral response may depend on the signal intensity.

Editor: Wulfila Gronenberg, University of Arizona, United States of America

Funding: This research was supported by the Autonomous Province of Trento (Accordo di Programma 2010, http://www.uniricerca.provincia.tn.it/accordi_programma/), Funding Research Programme P1-0255 and Research Project V4-0525 by Slovenian Research Agency, Fondi Ateneo of Pisa University. The funders had no role in study design, data collection and analysis, decision to publish, or preparation of the manuscript.

Competing Interests: The authors have declared that no competing interests exist.

* E-mail: gianfranco.anfora@iasma.it

Introduction

Substrate-borne vibrational signaling is a widespread form of animal communication, not only in arthropods [1,2] but also among vertebrates [3,4]. Although it has been recognized for centuries, its importance has long been overlooked [1,2,5]. As with any communication channel, the effective communication range of vibrational signals depends on the amplitude of the emitted signals, on attenuation and degradation during propagation [6–8] and on the sensitivity of the receiver's receptors [9]. Depending on the size, the communication range of vibrational signals can extend up to eight meters [6,10–13]. At any rate, it is generally assumed to be limited to one plant or neighboring plants with interconnected roots or touching leaves [2,10,14,15].

Until recently most studies on vibrational communication have been made within the range of few centimeters and have primarily focused on the species-specific vibrational repertoire (reviewed in [10,16]). The ability of conspecifics to recognize and locate each other in the environment depends on the efficacy of their communication. In particular, species-specific signals used in sexual communication enable identification of the sender (species and sex) and provide information necessary to determine its location [17,18]. In order to efficiently localize a conspecific partner, receivers should, in principle, determine not only a direction of the signal source, but also estimate its distance and adjust searching strategy accordingly. Currently there is no evidence of determination of source distance in plant-dwelling insects [19]. However, it has been hypothesized

that on plants, insects may be able to roughly estimate the distance by the extent of distortion and degradation due to differences in attenuation and filtering of different frequency components in the signal [6].

Signals that are perceived by insects as substrate-borne vibrations usually have a low intensity air-borne component [10,20,21] that potentially may be detected over few centimeters by antennal receptors (e.g. [22]) or even by vibration receptors in the legs [23]. Antennal receptors suggested to be involved in perception of air-borne and substrate-borne vibrations have been described in *Oncopsis flavicollis* [24–25], *Nezara viridula* [26], and *Hyalesthes obsoletus* [27]. Therefore, we investigated whether continuity of the substrate is essential in the transmission of vibrational signals for successful communication between sexes.

As a model species we chose the leafhopper *Scaphoideus titanus* Ball (Hemiptera: Cicadellidae), a major pest of grapevine, that transmits the phytoplasma responsible for the grapevine yellow disease "Flavescence dorée" in Europe [28]. The role of vibrational signals in intraspecific communication and pair formation of *S. titanus* on a single grapevine leaf has been described in detail. Pair formation begins with a spontaneous emission of a male calling signal (MCS) which in response to female reply may extend into a courtship phrase (MCrP). Females don't emit vibrational signals spontaneously [29]. In absence of female reply males may perform the "call-fly" behavior [30], by alternating emissions of MCS with jumps from the plant [29].

We show here that discontinuity of substrate is not a barrier for communication in a vibrational communicating insect and that antennal receptors are not essential for detecting mating signals when partners are placed on discontinuous substrates. The results are discussed with regard to mate searching behavior associated with different levels of signal intensity.

Results

Test 1. Male-female inter-plant communication

We placed *S. titanus* male and female on different grapevine leaves separated by a gap of varying widths. In all trials males initiated communication behavior with emission of MCS and females were observed to reply to male calls up to a 6-cm gap distance (Figure 1). As a result of female responses, most males established a duet with the female that ended either with female location or "call-fly" behavior. Few males did not show any reaction to female responses. When mating duets were observed, they were composed of short series of male pulses alternated with one or more female pulses. Within the 5-cm gap distance, most females replied to male calls, although mate locations - achieved by the short jump from the upper leaf to the lower one with the female - were observed only at shorter distance. At 7-cm distance between leaves, none of the females responded to MCS.

Test 2. Signal transmission

We studied transmission of male vibrational signals between grapevine leaves that were separated by a gap of varying distance. In playback experiments (Figure 2), the mean substrate velocity progressively decreased with the distance (i.e. width of the gap) (Jonckheere test: $J_0 = 5.93$, P<0.001). In contrast, the dominant frequency increased ($J_0 = 2.29$, P = 0.011). Compared with the signal recorded from the lower leaf, at 0.5 cm gap distance the decrease in vibration velocity was on average of $91.6\pm7.1\%$ and at 11 cm gap distance the velocity was further reduced of $7.3\pm5.6\%$. Values of velocity measured between 0.5–1 cm were over

0.001 mm/s, whereas from 2 cm gap the mean velocity was constantly lower.

Test 3. The role of antennae in perception of vibrational signals

When ten pairs of intact males and females with surgically removed antennae were tested on the same leaf, all females responded to the MCS. When pairs were tested on two leaves not connected via the common substrate and separated by 5 cm gap, seven out of ten females responded. This result is identical to test 1, when leaves were separated by 5 cm gap and females had intact antennae.

Discussion

Contrary to general belief, our findings demonstrate that the communication range of vibrational signals emitted by small insects is not limited to physically interconnected substrates. Production of low-frequency acoustic signals that are perceived by receivers as substrate-borne vibrations usually also results in emission of a low-intensity air-borne component [6,20,21]. Efficient radiation of acoustic sources in the air is possible only when emitter is bigger than 1/3 of the wavelength of the emitted sound [31,32]. For an insect of the size of *S. titanus* (4–5 mm), the optimal frequency of air-borne sound would be above 10 kHz. The effective air-borne range of low frequency vibrational signals with dominant frequencies in the range between 80–300 Hz emitted by *S. titanus* is short and we never heard air-borne sounds during their calling. Nevertheless, while communication at distances larger than a few cm is mediated by vibrations of the substrate, at closer range the role of air-borne component cannot be excluded. At a range of a few cm, such signals may be detected by mechanosensory hairs [33] or the Johnston's organ in the antennae [22]. Our results show that in *S. titanus* mechanorecep-tors in the antennae are not involved in detection of air-borne component of vibrational signals. Heteropteran insects possess hairs that may be used for detecting air-particle displacement [34]

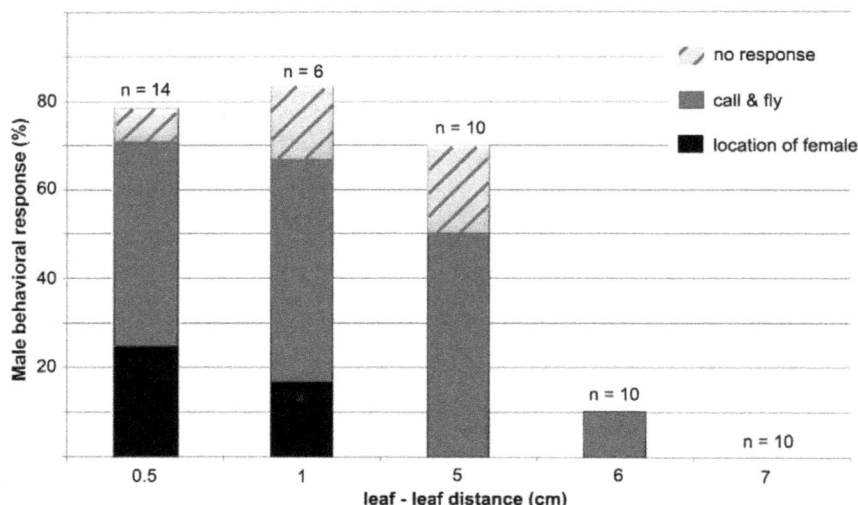

Figure 1. Male-female communication in *Scaphoideus titanus* recorded on leaves without direct contact (Test 1). Distances between upper and lower leaf were from 0.5 cm to 7 cm. The percentage of females that responded to the male calling signal (total column height) is divided according to the subsequent male behavioral response: mating duet, followed either by female location (black) or by call-fly (gray), and no male reaction (striped). n indicates the number of insect pairs tested.

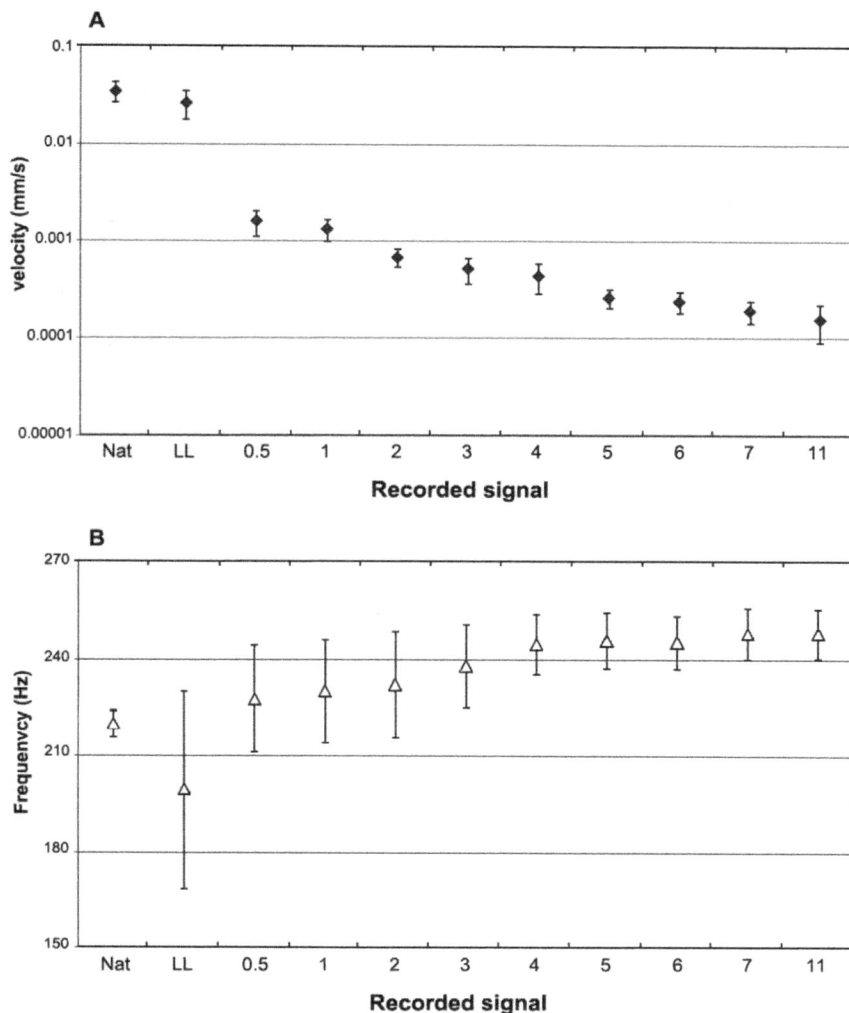

Figure 2. Signal properties measured on leaves with discontinuous substrate (Test 2). Mean (±SE) values of maximum substrate vibration velocity (mm/s) (A, logarithmic scale) and frequency (Hz) (B) of pulses from MCS (Male calling signal) are shown. While substrate velocity progressively decreased (Jonckheere test: $J_0 = 5.93$, $P < 0.001$) with the distance between leaves, the frequency increased ($J_0 = 2.29$, $P = 0.011$). Nat: MCS emitted by natural male recorded on the same leaf; LL: MCS emitted by playback recorded on the same leaf; 0.5–11: MCS emitted on the lower leaf and recorded from the upper leaf with a progressive gap width of 0.5–11 cm.

however, a systematic survey of sensilla on the leafhopper body is lacking.

Our measurements showed that vibrations are transmitted from one leaf to another even when they were separated by a gap of 11 cm and that females responded to males up to a gap width of 6 cm. From our results it was not possible to determine explicitly whether the vibrational signals were detected as air-borne sound or as substrate vibrations induced in the leaf. However, some observations, indicate the latter as the more probable hypothesis. In some cases male and female leafhoppers were not positioned within the gap between leaves, but on external sides of leaf laminae. In such situation two leaves would represent severe obstacle to any low intensity air-borne sounds. On the other hand, it has previously been shown that leaf vibrations are transmitted through the air beyond the boundary layer of the leaf and that air particle displacement triggered by leaf vibrations has the same temporal pattern as substrate vibrations [35]. The fact that in our experimental set-up we used two partly overlapping leaves with relatively large surface may also explain why in other studies in

which only the tips of the leaves were in close proximity, concluded that vibrational communication was limited to a continuous substrate. Situations in which leaves are separated by a gap but partly overlapping probably represent a more natural case for insects that communicate in a dense vegetation habitat.

The maximum intensity of vibrational signals on a leaf without any contact with the vibrated leaf, measured directly as velocity at gap distances at which females were still responding, was in the velocity range between 10^{-6} and 10^{-7} m/sec at dominant frequencies between 220–250 Hz. These values translate to displacement values between 10^{-9} and 10^{-10} m. The lowest neurobiologically determined velocity threshold values for subgenual organs in various insect groups are all in the range between 10^{-5} and 10^{-6} m/sec (Heteroptera: [36,37]; Neuroptera: [38]; Orthopteroids: [39–41]). However, in all these insects conversion of velocity threshold values into displacement values results in threshold values below 10^{-9} m. In particular, in another hemipteran insect, the southern green stink bug *Nezara viridula*, threshold values of receptor cells in the subgenual organ follow the line of equal displacement [36]. This suggests that, although displacements induced in a leaf by vibrational signals emitted on another leaf nearby are low, they are not below the threshold values of the subgenual organ. In leafhoppers nothing is known about vibration receptors in the legs [10]. However, it is likely that leafhoppers possess subgenual organs on all six legs. In insects this is the most sensitive organ to detect substrate vibrations and it was described also in closely related insect groups such as froghoppers (Cercopidae) and bugs (Heteroptera) [36,42,43]. Our measurements also revealed a significant increase in dominant frequency (from 200 to 250 Hz) when vibrational signals were transmitted through air from one leaf to another. It is interesting to note that resonant frequencies of sound-induced vibrations in bean leaves are in the frequency range between 190 and 290 Hz [44]. In the pentatomid bug *N. viridula*, for which bean is a preferred host plant, resonant frequencies correspond to best frequency sensitivity of one of the two cells in the subgenual organ [37]. We argue that transmission of vibrational signals from one leaf to another via air may be a common phenomenon. High receptor sensitivity, together with potential tuning of plant resonant frequencies with spectral properties of vibrational signals may enable the insect to extend the communication range beyond the limit of one plant.

In addition, our results suggest that the intensity of the perceived vibrational signals may have crucial effects on the leafhopper behavior. Mating duet followed by female location was observed only at the two shortest gaps, while call-fly behavior prevailed at longer distances. Although the role of shifts in dominant frequency cannot be excluded, the observed differences are small (between 20 and 40 Hz) in comparison with the 20 dB difference in intensity. When male and female were positioned on the same leaf at the beginning of our observations, MCS was immediately extended into a courtship phrase without the intermediate stage observed at other distances [29,45]. It is conceivable that leafhoppers are able to compare the intensity of their own signals and perceived signals emitted by the duetting partner. Below a certain threshold the intensity may provide information that the female is not located on the same leaf as the male and that the male therefore needs to adjust the searching strategy accordingly. Since most studies on planthopper and leafhopper mating behavior have been conducted in short range situations, the information about patterns of long-range communication is lacking.

The call-fly behavior observed in males is usually associated with a strategy to increase effective signaling space [30,46]. However, when the position of the source of low intensity female reply is unpredictable for the courting male, call-fly strategy may enable a faster localization of the leaf hosting the female. In addition, numerous changes of the position of the signaling male may reduce predation risk from eavesdropping predators like spiders [47].

In conclusion, we showed that the communication range of vibrational signals is not limited by substrate continuity and that in this situation antennal receptors are not essential in detection of vibrational mating signals. Moreover, our behavioral observations together with measurements of signal transmission between grapevine leaves suggest that behavioral responses of *S. titanus* may depend on the signal intensity.

Materials and Methods

Rearing of insects

S. titanus eggs originated from two-year-old grapevine (*Vitis vinifera*) canes collected from organic farms in Northern Italy (Povo, Trento, Italy). Egg hatching occurred in a climate chamber (24±1°C, 16L:8D photoperiod, 75% R.H.). Nymphs were removed daily into rearing boxes, consisting of plastic beakers (height 10 cm; 5 cm i.d.) with a moistened grapevine leaf laid on top of a 1-cm-layer of technical agar solution (0.8%) that was replaced twice a week. At emergence, adults were separated by sex and age (day of emergence), and kept in the rearing boxes. All experiments were made with virgin, sexually mature males and females at least 8 days old [29].

Terminology and recording of vibrational signals

In the current study we used terminology established by Mazzoni et al. [29].

The experiments were performed in an enclosed room of the Entomology Section (Pisa University) at 23±1°C from June to August, between 5 pm and 9 pm which is the peak in sexual activity in *S. titanus* [29]. The signals were recorded with a laser vibrometer (Ometron VQ-500-D-V, Brüel and Kjær Sound & Vibration A/S, Nærum, Denmark) and digitized with 48 kHz sample rate and 16-bit resolution, then stored directly onto a hard drive through Plug.n.DAQ (Roga Instruments, Waldalgesheim, Germany). Signal spectral analysis was performed by means of Pulse 14 (Brüel and Kjær Sound & Vibration A/S). Recorded signals were analyzed with a FFT window length of 400 points.

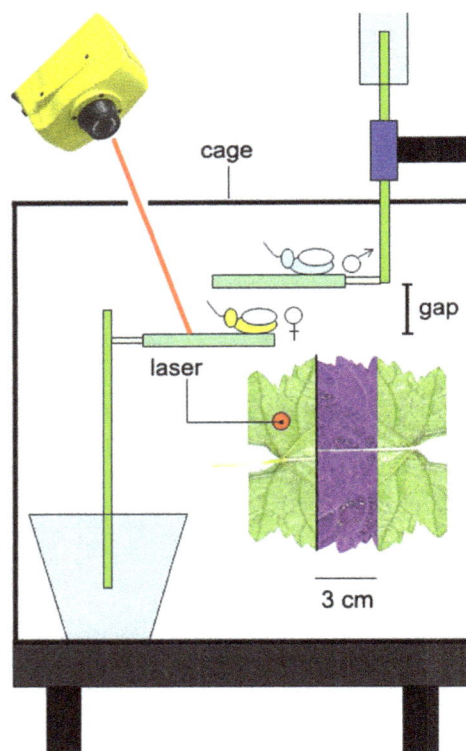

Figure 3. A schematic drawing of experimental setup. A male and a female were placed on leaves (surface 6×10 cm) of two separate grapevine cuttings. The bottom of the stem was put in a glass vial filled with water to prevent withering. One cutting was put on an anti-vibration table (Astel S.a.s., Ivrea, Italy). The second cutting was attached to a metal arm suspended from above – without any contact with the table - and positioned in parallel over half the surface of the lower leaf (as shown in the inset as viewed from above). The laser beam was focused on the lamina of the lower leaf with the female. To prevent the insects from escaping, recordings were made within a Plexiglas cylinder (50×30 cm), provided of two openings for the laser beam and the metal arm. Not drawn to the scale.

The leafhopper behavior was recorded with a Canon MV1 miniDV camera. The communication between males and females was observed for 20 minutes or until the male reached the female.

Test 1. Inter-plant communication

We placed a male and a female on leaves of two separate grapevine cuttings with one leaf (surface 6×10 cm) (see Figure 3). The gap width between the upper and lower leaf surface ranged from 0.5 cm to 7 cm. For each distance we recorded whether the female responded to the MCS emitted by male with the prompt emission of pulses. Then, we categorized and counted the male behavioral reactions to the female reply: (1) no reaction; (2) mating duet followed by call-fly; (3) mating duet with male search and location of the female.

Test 2. Signal transmission

Transmission of MCS between grapevine leaves that were not connected by a common substrate was studied by playback of pre-recorded MCS. The spectral structure of *S. titanus* MCS is characterized by a series of several prominent frequency peaks in the range between 80 and 300 Hz and maximum substrate vibration velocity above 10^{-2} mm/s. We recorded MCS at a close range on the grapevine leaf with a laser vibrometer as described above, from three different males. Since variability between spectral parameters among males was negligible we used a single randomly chosen MCS (composed of 27 pulses). Five pairs of leaves were tested from different cuttings, in the same experimental set up of Figure 3, in absence of real insects and cage. The lower grapevine leaf was vibrated by a minishaker (Type 4810; Brüel and Kjær Sound & Vibration A/S) with a conical tip attached onto the leaf surface, 2 cm distant from the anterior border. The minishaker was driven from a computer via Adobe Audition 3.0 (Adobe Systems Incorporated). The amplitude of playback signal was adjusted to the natural emitted signal. The measurements were taken from the leaf lamina in two different randomly chosen points at least 2 cm distant from the border both of the lower and upper leaf by laser vibrometer. The gap between parallel leaf surfaces was 0.5, 1, 2, 3, 4, 5, 6, 7 and 11 cm. Spectral components and velocity of leaf vibration were analyzed along the distance by taking the average of nine randomly chosen recorded pulses from each distance and each leaf. To assess the velocity and frequency differences the Jonckheere test was performed [48].

Test 3. The role of antennae in perception of vibrational signals

Females were put in a freezer ($-25°$C) for 30 seconds to cool them and prevent them from moving when placed under a stereomicroscope. Both antennae were cut off with microscissors. After ablation, females were kept separately in the rearing boxes for 24 hrs before they were used in experiments.

For the experiments, ten pairs consisting of intact males and of females whose antennae had been removed were first tested at close range on a single grapevine leaf to determine the female responsiveness after the ablation. In case of female response, they were subsequently tested on two leaves not connected via the common substrate and separated by a 5 cm gap as described above. The laser was focused on the leaf of the female.

Acknowledgments

We thank Elisabetta Leonardelli for help with insect rearing and Paolo Giannotti for assistance in the experimental set-up. At last we are grateful to the anonymous reviewers for their comments on the earlier versions of the manuscript.

Author Contributions

Conceived and designed the experiments: AE GA VM. Performed the experiments: AE VM. Analyzed the data: AE VM. Contributed reagents/materials/analysis tools: GA AL VM. Wrote the paper: AE GA AL MVD VM.

References

1. Virant-Doberlet M, Čokl A (2004) Vibrational communication in insects. Neotrop Entomol 33: 121–134.
2. Cocroft RB, Rodríguez RL (2005) The behavioral ecology of insect vibrational communication. Bio Science 55: 323–334.
3. Hill PS (2009) How do animals use substrate-borne vibrations as an information source. Naturwissenschaften 96: 1355–1371.
4. Caldwell MS, Johnston GR, McDaniel JG, Warkentin KM (2010) Vibrational signaling in the agonistic interactions of red-eyed treefrogs. Curr Biol 20: 1012–1017.
5. Hill PS (2008) Vibrational Communication in Animals. Cambridge (MA): Harvard University Press. 261 p.
6. Michelsen A, Fink F, Gogala M, Traue D (1982) Plants as transmission channel for insect vibrational songs. Behav Ecol Sociobiol 11: 269–281.
7. Cocroft RB, Shugart HJ, Konrad KT, Tibbs K (2006) Variation in plant substrates and its consequences for insect vibrational communication. Ethology 112: 779–789.
8. Miklas N, Stritih N, Čokl A, Virant-Doberlet M, Renou M (2001) The influence of substrate on male responsiveness to the female calling song in *Nezara viridula*. J Insect Behav 14: 313–332.
9. Endler JA (1993) Some general comments on the evolution and design of animal communication systems. Philos T R Soc B 340: 215–225.
10. Čokl A, Virant-Doberlet M (2003) Communication with substrate-borne signals in small plant-dwelling species. Annu Rev Entomol 48: 29–50.
11. Stewart KW, Zeigler DD (1984) The use of larval morphology and drumming in Plecoptera systematic and further studies of drumming behavior. Ann Limnol 20: 105–114.
12. McVean A, Field LH (1996) Communication by substratum vibration in the New Zealand tree weta *Hemideina femorata* (Stenipelmatidae: Orthoptera). J Zool 239: 101–122.
13. Barth FG (2002) Spider senses-technical perfection in biology. Zoology 105: 271–285.
14. Ichikawa T, Ishii S (1974) Mating signal of the brown planthopper *Nilaparvata lugens* Stål (Homoptera: Delphacidae): vibration of the substrate. Appl Entomol Zool 9: 196–198.
15. Hunt RE (1993) Role of vibrational signals in mating behavior of *Spissistilus festinus* (Homoptera: Membracidae). Ann Entomol Soc Am 86: 356–361.
16. Claridge MF (1985) Acoustic signals in the Homoptera: behavior, taxonomy and evolution. Annu Rev Entomol 30: 297–317.
17. Bradbury JW, Vehrencamp SL (1998) Principles of Animal Communication. Sunderland: Sinauer Associates. 882 p.
18. Gerhardt HC, Huber F (2002) Acoustic Communication in Insects and Anurans. Chicago: University of Chicago Press. 531 p.
19. Virant-Doberlet M, Čokl A, Zorović M (2006) Use of substrate vibrations for orientation: from behavior to physiology. In: Drosopoulos S, Claridge MF, eds. Insect Sounds and Communication. Boca Raton: Taylor & Francis. pp 81–97.
20. Ossiannilsson F (1949) Insect drummers. A study on morphology and function of sound-producing organ of Swedish Homoptera Auchenorrhyncha with notes on their sound-production. Opusc Entomol Suppl. 10: 1–145.
21. Percy DM, Taylor GS, Kennedy M (2006) Psyllid communication: acoustic diversity, mate recognition and phylogenetic signal. Invertebr Syst 20: 431–445.
22. Kirchner WH (1994) Hearing in honeybees: the mechanical response of the bee's antenna to near filed sound. J Comp Physiol A 175: 261–265.
23. Shaw SR (1994) Detection of airborne sound by a cockroach "vibration detector": a possible missing link in insect auditory evolution. J Exp Biol 193: 13–47.
24. Howse PE, Claridge MF (1970) The fine structure of Johnston's organ of the leafhopper *Oncopsis flavicollis*. J Insect Physiol 16: 1665–1675.
25. Claridge MF, Nixon GA (1986) *Oncopsis flavicollis* (L.) associated with tree birches (*Betula*): a complex of biological species or host plant utilization polymorphism? Biol J Linn Soc 27: 381–397.
26. Jeram A, Pabst MA (1996) Johnston's organ and central organ in *Nezara viridula* (L.) (Heteroptera, Pentatomidae). Tissue Cell 28: 227–235.
27. Romani R, Rossi Stacconi MV, Riolo P, Isidoro N (2009) The sensory structures of the antennal flagellum in *Hyalesthes obsoletus* (Hemiptera: Fulgoromorpha: Cixiidae): a functional reduction? Arthropod Struct Dev 38: 473–483.
28. Schvester D, Carle P, Motous G (1963) Transmission de la flavescence dorée de la vigne par *Scaphoideus littoralis* Ball. Annales des Epiphyties 14: 175–198.

29. Mazzoni V, Prešern J, Lucchi A, Virant-Doberlet M (2009) Reproductive strategy of the Nearctic leafhopper *Scaphoideus titanus* Ball (Hemiptera: Cicadellidae). Bull Entomol Res 99: 401–413.

30. Hunt RE, Nault LR (1991) Roles of interplant movement, acoustic communication, and phototaxis in mate-location behavior of the leafhopper *Graminella nigrifrons*. Behav Ecol Sociobiol 28: 315–320.

31. Markl H (1983) Vibrational communication. In: Huber F, Markl H, eds. Neuroethology and Behavioral Physiology. Heidelberg: Springer-Verlag. pp 332–353.

32. Bennet-Clark HC (1998) Size and scale effects as constraints in insect sound communication. Philos T R Soc B 353: 407–419.

33. Keil TA (1997) Functional morphology of insect mechanoreceptors. Microsc Res Tech 39: 506–531.

34. Drašlar K (1973) Functional properties of trichobotria in the bug *Pyrrhocoris apterus* (L.). J Comp Physiol 84: 175–184.

35. Casas J, Bacher S, Tautz J, Meyhöfer R, Pierre D (1998) Leaf vibrations and air movements in a leafminer-parasitoid system. Biol Cont 11: 147–153.

36. Čokl A, Virant-Doberlet M, Zorović M (2006) Sense organs involved in the vibratory communication of bugs. In: Drosopoulos S, Claridge MF, eds. Insect Sounds and Communication. Boca Raton: Taylor & Francis. pp 71–80.

37. Čokl A (1983) Functional properties of vibroreceptors in the legs of *Nezara viridula* (L.) (Heteroptera: Pentatomidae). J Comp Physiol A 150: 261–269.

38. Devetak D, Gogala M, Čokl A (1978) A contribution to physiology of the vibration receptors in bugs of the family Cydnidae. Biol Vestn 36: 131–139.

39. Shaw SR (1994) Re-evaluation of absolute threshold and response mode of the most sensitive known "vibration" detector, the cockroach's subgenual organ: a cochlea-like displacement threshold and a direct response to sound. J Neurobiol 25: 1167–1185.

40. Čokl A, Kalmring K, Rössler W (1995) Physiology of a tympanate tibial organs in forelegs and midlegs of the cave-living Ensifera *Troglophilus neglectus* (Raphidophoridae, Gyllacridoidea). J Exp Zool 273: 376–388.

41. Čokl A, Virant-Doberlet M (1997) Tuning of tibial receptor cells in *Periplaneta americana* L. J Exp Zool 278: 395–404.

42. Debaisieux P (1938) Organes scolopidiaux des pattes d'Insectes. Cellule 47: 77–202.

43. Michel K, Amon T, Čokl A (1983) The morphology of the leg scolopidial organs in *Nezara viridula* (L.) (Heteroptera, Pentatomidae). Rev Can Biol Exp 42: 139–150.

44. Čokl A, Zorović M, Žunič A, Virant-Doberlet M (2005) Tuning of host plants with vibratory songs of *Nezara viridula* L. (Heteroptera: Pentatomidae). J Exp Biol 208: 1481–1488.

45. Mazzoni V, Lucchi A, Čokl A, Prešern J, Virant-Doberlet M (2009) Disruption of the reproductive behavior of *Scaphoideus titanus* by playback of vibrational signals. Entomol Exp Appl 133: 174–185.

46. Gwynne DT (1987) Sex-biased predation and the risky mate-locating behavior of male tick-tock cicadas (Homoptera: Cicadidae). Anim Behav 35: 571–576.

47. Virant-Doberlet M, King AR, Polajnar J, Symondson WOC (2011) Molecular diagnostics reveal spiders that exploit prey vibrational signals used in sexual communication. Mol Ecol, doi: 10.1111/j.1365-294X.2011.05038.x.

48. Siegel S, Castellan NJ (1988) Nonparametric statistics for the behavioral sciences, 2nd ed. New York: McGraw-Hill. 399 p.

19

Phytoavailability of Cadmium (Cd) to Pak Choi (*Brassica chinensis* L.) Grown in Chinese Soils: A Model to Evaluate the Impact of Soil Cd Pollution on Potential Dietary Toxicity

Muhammad Tariq Rafiq[1,2]ᵍ, Rukhsanda Aziz[1]ᵍ, Xiaoe Yang[1], Wendan Xiao[1], Peter J. Stoffella[3], Aamir Saghir[4], Muhammad Azam[5], Tingqiang Li[1]*

1 Ministry of Education Key Laboratory of Environmental Remediation and Ecological Health, College of Environmental and Resource Sciences, Zhejiang University, Hangzhou, China, 2 Department of Environmental Science, International Islamic University, Islamabad, Pakistan, 3 Indian River Research and Education Center, Institute of Food and Agricultural Sciences, University of Florida, Fort Pierce, Florida, United States of America, 4 Institute of Statistics, Zhejiang University, Hangzhou, China, 5 College of Agriculture and Biotechnology, Zhejiang University, Hangzhou, China

Abstract

Food chain contamination by soil cadmium (Cd) through vegetable consumption poses a threat to human health. Therefore, an understanding is needed on the relationship between the phytoavailability of Cd in soils and its uptake in edible tissues of vegetables. The purpose of this study was to establish soil Cd thresholds of representative Chinese soils based on dietary toxicity to humans and develop a model to evaluate the phytoavailability of Cd to Pak choi (*Brassica chinensis* L.) based on soil properties. Mehlich-3 extractable Cd thresholds were more suitable for Stagnic Anthrosols, Calcareous, Ustic Cambosols, Typic Haplustalfs, Udic Ferrisols and Periudic Argosols with values of 0.30, 0.25, 0.18, 0.16, 0.15 and 0.03 mg kg^{-1}, respectively, while total Cd is adequate threshold for Mollisols with a value of 0.86 mg kg^{-1}. A stepwise regression model indicated that Cd phytoavailability to Pak choi was significantly influenced by soil pH, organic matter, total Zinc and Cd concentrations in soil. Therefore, since Cd accumulation in Pak choi varied with soil characteristics, they should be considered while assessing the environmental quality of soils to ensure the hygienically safe food production.

Editor: Wenju Liang, Chinese Academy of Sciences, China

Funding: This study was supported by the Ministry of Environmental Protection of China (grant no. 2011467057), the ministry of Science and Technology of China (grant no. 2012AA100605), and by the Fundamental Research Funds for the Central Universities of China. The funders had no role in study design, data collection and analysis, decision to publish, or preparation of the manuscript.

Competing Interests: The authors have declared that no competing interests exist.

* Email: litq@zju.edu.cn

ᵍ These authors contributed equally to this work.

Introduction

Cadmium (Cd) is an important environmental pollutant toxic to animals and human beings. It is one of the most mobile elements, among all the toxic heavy metals [1]. Cadmium is not required for plants growth or reproduction, however its bioaccumulation and subsequent accrual in the food chain surpasses all other trace elements due to its high mobility in soil [2]. It is the most toxic element in the environment and even at low concentrations is very toxic to living cells and considered as carcinogenic [3]. In humans, Cd exposure can result in multiple adverse effects, such as testicular damage, renal and hepatic dysfunction, etc. [3]. Moreover, Cd is implicated in the development of cancer, phytotoxic at higher levels of concentrations [4] and classified as a type I carcinogen by the International Agency for Cancer Research [5]. Significant quantities of Cd can be transferred from contaminated soil to plants [6]. Therefore, crops produced from Cd contaminated soils may be unsuitable or even detrimental for animal and human consumption [7].

Vegetables are an important component of human diet since they contain proteins, carbohydrates as well as minerals and vitamins [8]. The proportion of vegetables consumed in the total diet has been increased with the improvement of living standards. However, vegetables are also one of the most important pathways through which heavy metals enter the food chain and affect human health. Leafy vegetables can accumulate higher concentrations of Cd than other crops [9,10]. Leafy vegetables are known to accumulate higher concentrations of Cd in the edible parts even when grown in soils containing low concentrations of Cd [11]. Pak choi (*Brassica chinensis* L.), also known as Chinese cabbage, is a popular leafy vegetable, grown and consumed worldwide. Therefore, it is imperative to control Cd concentrations in Pak choi, especially in its edible parts to ensure food safety. To limit the transfer of soil Cd into the edible parts of Pak choi, an understanding of its accumulation characteristics is required. Currently, there is an elevated concern over Cd accumulation in food and its potential risks to human health [12]. Cadmium accumulation and distribution varies among vegetable cultivars and tissues [13]. However, the accumulation and distribution of Cd in vegetables grown in a diversity of soil types were rarely studied [14].

About one fifth of agricultural land is contaminated by Cd, lead (Pb) and arsenic (As) in China [15]. Moreover, it was reported that about 20% of farm lands in China are contaminated with heavy metals and Cd contamination accounts for more than 1.3×10^5 ha of the total affected area [16,17]. Cadmium uptake by rice (*Oryza sativa* L.) and vegetables from soil is the initial source of exposure for human beings [18,19]. Therefore, there are environmental concerns of soils, food safety and human health for the present and future agricultural and environmental sustainability of world vegetable supplies. As, only a small fraction of total trace metals in soil is available for plant absorption, it is widely accepted that the total metal content in soils is neither a viable indicator of phytoavailability nor an adequate tool to assess the potential risk of dietary toxicity [12]. Tracy and Sheila [20] reported that extractable Cd content in soil may be an improved indicator of bioavailability and toxicity than the total contents and toxicity and availability of metals differed among soils types. Metal uptake and translocation studies were conducted for different crops under varying soil conditions, to further understanding uptake and the transport mechanisms [21,22].

To ensure the food safety and environmental quality of soils, guidelines for permissible concentrations of Cd in agricultural soils need to be established. Due to limited number of studies, the soil environmental quality guidelines for heavy metals in farmland soils developed and applied in the world are still based on total metal contents of soil. Minimal attention has been focused on metal accumulation differences among the edible parts of crops, and the relationship between total concentration and phytoavailability of heavy metals in different soil types [4]. Developing the linkage between the bioavailability of Cd in soil and its transfer into the edible plant parts is a key to improving existing soil environmental quality standards. Information is vital on the degree of transloca-tion of heavy metals from soils to plants, which are used as food crops, and absorption of metals in food plants to concentration that does not cause phytotoxicity symptoms [23]. This study was conducted in seven Chinese soil types to establish direct relationship of Cd level in such contaminated soils and Cd uptake in Pak choi. The main objectives were to establish soil Cd thresholds for representative Chinese soils based on human dietary toxicity and to determine the relationships between several soil properties and Cd accumulation in Pak choi. This information will be useful in establishing soil protection guidelines to produce hygienically safe vegetables.

Materials and Methods

Ethics Statement

The soils used in this study were agricultural soils. No specific permissions were required for the described locations. We confirm that the field studies did not involve endangered or protected species.

Soil Collection and Analysis

Seven Chinese soils were selected for this study. Udic Ferrisols, Mollisols, Periudic Argosols, Typic Haplustalfs, Ustic Cambosols, Calcaric Regosols and Stagnic Anthrosols were collected from Chinese cities of Guilin (104°40'–119°45'E, 24°18'– 25°41'N), Harbin (125°42'– 130°10'E, 44°04'–46°40'N), Huzhou (119°14'–120°29'E, 30°22'–31°11'N), Zhanjiang (110°08'–110°77'E, 20°33'–21°62'N), Qufu (116°51'–117°13'E, 35°29'–35°49'N), Ya'an (102°37'–103°12'E, 29°23'–30°37'N) and Jiaxing

(120°7'–121°02'E, 30°5'–30°77'N), respectively. Soils samples were taken at a depth of up to 20 cm from the upper horizon. Each sample was air-dried, ground, and screened through two mm sieve before laboratory analysis. Soil pH, cation exchange capacity, organic matter contents, and particle size density were measured by using previously described methods [24–27]. Physicochemical properties of these soils are reported (Table 1)

Cadmium Spiking and Aging

Soil samples of Mollisols, Periudic Argosols, Stagnic Anthrosols and Ustic Cambosols were spiked with Cd as $Cd(NO_3)_2$ in an aqueous solution at loading rates of 1.0, 2.0, 4.0, 6.0 and 8.0 mg Cd kg^{-1} soil along with an untreated control (Ck), the background values of Cd concentration was below 0.50 mg kg^{-1} in these soil. However, the Udic Ferrisols, Typic Haplustalfs and Calcaric Regosols soil samples, with the background values of Cd concentration above 0.50 mg kg^{-1}, were spiked with Cd to establish the contamination levels of 2.0, 4.0, 6.0 and 8.0 mg Cd kg^{-1} soil along with the untreated control (Ck). Soil moisture was maintained up to 70% of its water-holding capacity by using distilled water. All the spiked soils were aged for one year subsequent to greenhouse experimentation. After one year aging period, the concentrations of total Cd, and Mehlich-3 extractable Cd were determined in each of the spiked soils.

Containerized Experiment

A containerized experiment was performed in greenhouse by growing Pak choi (*Brassica chinensis* L.) during March – April, 2012 at Zhejiang University, Hangzhou, China. Seed of Pak choi was obtained from the Zhejiang Seed Co. Hangzhou, China. Seeds were washed with distilled water and air-dried prior to sowing. Seeds were germinated in dark at 25°C and transplanted into quartz sand bed to establish seedlings. Four healthy, uniform and 21-day-old seedlings were transplanted into plastic containers with a diameter of 18 cm and height of 17 cm. Each container had 3 kg of soil. Fertilizers were applied at the rates of 0.4 g of N as CO $(NH_2)_2$ and 0.2 g P as KH_2PO_4 per kg of soil. The experiment was carried out in a completely randomized design (CRD). Treatments were established in triplicate, and the containers were randomly arranged in a greenhouse bench under controlled conditions of 16 h of light at 30°C and 8 h of dark at 22°C. Plants were monitored daily and watered as necessary.

Plant Sample Collection

Pak choi was harvested after 30 days from transplanting. The plants of Pak choi were removed from each container and separated into root and shoots (including stems and leaves). Roots and shoots of Pak choi were first washed with tap water and then with ultrapure distilled water, to remove all visible soil particles. Clean plant samples were first blotted dry, and then dried at 70°C for 72 h in an oven. Dry shoot weight of samples was recorded. Dry plant samples were ground to pass through a 60 mm sieve using an agate mill prior to Cd concentration analysis.

Total Cd of Soil and Plant

For determination of total Cd concentration in soil, 0.20 g of soil samples was digested with HNO_3–HF –$HClO_4$ (5:1:1) [4]. For plant samples, 0.20 g of shoots for each treatment was digested with HNO_3–H_2O_2 (5:1). After cooling the digest was transferred to a volumetric flask, diluted with distilled water to 50 mL [28]. The concentrations of Cd in the filtrate were determined using inductively coupled plasma–mass spectrometry (ICP-MS, Agilent, 7500a, CA, USA). The ICP-MS was operated at the following

Table 1. Basic Chemical and Physical Characteristics of Seven Chinese soils.

Soil Types	Mollisols	Ustic Cambosols	Stagnic Anthrosols	Periudic Argosols	Typic Haplustalfs	Udic Ferrisols	Calcaric Regosols
pH	7.23±0.08	7.80±0.02	6.49±0.03	4.85±0.06	5.16±0.05	4.43±0.07	8.02±0.04
OM (g kg⁻¹)	32.2±0.32	7.54±0.20	21.4±0.34	11.6±0.17	6.37±0.56	19.1±0.15	21.8±0.14
CEC (cmol kg⁻¹)	34.0±2.51	15.8±1.62	20.2±1.41	12.6±1.52	8.33±2.14	17.3±1.96	25.5±1.46
Total Cd (mg kg⁻¹)	0.51±0.02	0.59±0.06	0.79±0.04	0.47±0.02	0.92±0.01	1.06±0.09	0.96±0.06
Total Zn (mg kg⁻¹)	31.18±1.47	26.93±0.43	41.36±1.71	15.17±0.88	25.3±1.44	24.6±0.23	28.59±1.38
Sand (%)	20.6±1.54	21.6±1.29	11.4±0.26	24.8±0.65	37.4±0.96	32.75±1.65	31.6±0.57
Silt (%)	60.2±2.21	65.4±2.62	73.0±2.41	58.2±1.04	40.8±1.66	39.8±1.26	44.0±1.26
Clay (%)	19.2±1.24	13.0±1.05	15.6±1.17	17.0±0.34	21.8±0.82	49.6±1.19	24.4±1.32

conditions: the radio frequency power at the torch 1.2 kW, the plasma gas flow 15 L min⁻¹, the auxiliary gas flow 0.89 L min⁻¹, and the carrier gas flow 0.95 L min⁻¹ [28]. The same procedure without samples was used as control and three replications were conducted for each sample.

Mehlich-3 Extractable Cd in Soils

Mehlich-3 extractable Cd in soils was determined following the extraction method described by Mehlich [29]. Briefly, 5 g (0.2 mm sieved) of dry soil was shaken with 50 mL of Mehlich-3 solution (0.2 mol L⁻¹ CH_3COOH, 0.25 mol L⁻¹ NH_4NO_3, 0.015 mol L⁻¹ NH_4F, 0.013 mol L⁻¹ HNO_3, 0.001 mol L⁻¹ EDTA) for 5 min (200 rpm) at 25°C. The suspension was centrifuged at 5000 rpm for 10 min and filtered through 0.45 μm filter paper. The same procedure without samples was used as control and three replications were conducted for each soil sample. The Cd concentration in the filtrate was analyzed by inductively coupled plasma–mass spectrometry (ICP-MS, Agilent 7500a, CA, USA).

Quality Control for Cd Analysis

Quality assurance and quality control (QA/QC) for Cd in soil and Pak choi were conducted by determining Cd contents in the certified reference materials (soil GSBZ 50013-88 and plant GBW-07402) respectively, approved by General Administration of Quality Supervision, Inspection and Quarantine of the People's Republic of China (AQSIQ) and National Center for Reference Materials. The analytical results showed a recovery rate of 97.3% and 102.1% respectively.

Derivation of Soil Cd Thresholds for Potential Dietary Toxicity in Pak choi

For ensuring the environmental and food safety for human beings, an effort has been made to develop guidelines for acceptable concentrations of potentially harmful Cd in seven agricultural soils types of China. In this context, the amounts of Cd in Pak choi above than threshold level of food safety are adversely affecting humans are critical. Since, Cd bioavailability differed among soil types, the focus was on the development of soil Cd thresholds for representative Chinese soils based on food safety, Provisional Tolerable Weekly Intake (PTWI) of Cd recommended by FAO/WHO Joint Expert Committee on Food Additives, is 7 μg kg⁻¹ of body weight [30]. Estimated daily intake of metal (EDIM) was determined by the following equation.

$$\text{EDIM} = \frac{C_{cadmium} \times C_{factor} \times D_{daily\ intake}}{B_{average\ weight}}$$

Where, Cc_{admium}, C_{factor}, $D_{food\ intake}$ and $B_{average\ weight}$ represent average Cd concentration in Pak choi (mg kg⁻¹), conversion factor, daily consumption of Pak choi (g) and average body weight (kg) of the adult consumers, respectively. Average daily consumption of Pak choi for adults was considered to be 0.345 kg person⁻¹ d⁻¹ [31] and a conversion factor 0.085 was used to convert fresh Pak choi weight to dry weight [32]. Average body weight of adult was considered to be 60 kg as motioned in previous reports [30]. According to the above equation of EDIM, the provisional tolerable daily tolerable intake of Cd for Pak choi was 2.04 mg kg⁻¹ on a dry weight basis. Soil Cd threshold levels for potential dietary toxicity from Pak choi were calculated according to the tolerable daily dietary intake level of Cd (2.04 mg kg⁻¹) and the regression equations.

Statistical Analysis

Stepwise multiple regression analysis, single linear regression and one-way analysis of variance (ANOVA) were performed using the statistical software package SPSS (version 18.0). All values reported in this work are means of three independent replications. Treatment means were separated by least significant difference (LSD) test, at 5% level.

Results

Characteristics of Soils

Soils evaluated were representative of most of Chinese soil types, pH range of soils were strongly acidic to mild alkaline. Chemical and physical characteristics varied among the seven soils. Highest total Cd and Zn concentrations (background value) were observed in Udic Ferrisols and Stagnic Anthrosols respectively. Mollisols contained the highest amount of organic matter and exhibited an elevated cation exchange capacity as well (Table 1).

Mehlich-3 Extractable Cd in Soils after Aging of 1 Year

Mehlich-3 extractable Cd content increased significantly with increasing Cd spiking levels in all the seven soils. Mehlich-3 extractable Cd ranged from $0.16-3.95$ mg kg^{-1} in these soils under different Cd levels (Table 2). The Cd contents varied significantly among these soils, decreasing in order: Periudic Argosols> Typic Haplustalfs> Udic Ferrisols> Stagnic Anthrosols> Mollisols> Ustic Cambosols> Calcaric Regosols. Mehlich-3 extractable Cd concentration was greater at higher rates of Cd spiking in each soil. These results indicated that minimum and maximum extractability of Cd was found in Calcaric Regosols and Periudic Argosols, respectively under the highest (8 mg kg^{-1}) level of Cd spiked. Mehlich-3 extractable concentrations were dependent on total Cd in each soil, however the extractability was significantly higher in low pH soils as compared to the medium and high pH soils (Table 2).

Biomass Yield of Pak choi

Generally, Pak choi had tolerance to Cd toxicity in Mollisols, Stagnic Anthrosols and Calcaric Regosols soils, indicating low phytoavailability of Cd in these soils. Shoot biomass of Pak choi under different Cd treatments of these soils did not decrease significantly as compared with their respective controls. However, the shoot biomass of Pak choi grown in Ustic Cambosols, Udic Ferrisols, Periudic Argosols and Typic Haplustalfs decreased significantly as compared with the control indicating higher phytoavailability of Cd in these soils (Table 3). The stimulating effect of Cd on shoot biomass of Pak choi occurred at 1 mg kg^{-1} and 2 mg kg^{-1} in Ustic Cambosols and Mollisols respectively, whereas in Stagnic Anthrosols, it occurred at 4 mg kg^{-1}. The dry weight of Pak choi shoots at 8 mg kg^{-1} Cd generally decreased in order of: Calcaric Regosols> Mollisols, Stagnic Anthrosols> Ustic Cambosols> Udic Ferrisols> Periudic Argosols> Typic Haplustalfs (Table 3).

Accumulation and Distribution of Cadmium in Pak choi

Cadmium concentration in the shoots and roots of Pak choi varied significantly among soils at different Cd levels and soil types. Roots exhibited the higher Cd contents as compared with Pak choi shoots. The content of Cd enhanced with increasing Cd loading rate in the soils. Cd concentration was high in the roots (2.42 to 169. 95 mg kg^{-1} DW), while low in the shoot (0.48 to 89.21 mg kg^{-1} DW) (Table 4). Cd uptake in Pak choi tissues was affected by soil type, primarily due to the variation in Cd

Table 2. Mehlich-3 Extractable Cd Contents (mg kg^{-1}) in Seven Chinese Soils at the onset of Containerized Experiment after Aging of 1 year.

Cd conc. (mg kg^{-1})	Mollisols	Ustic Cambosols	Stagnic Anthrosols	Periudic Argosols	Typic Haplustalfs	Udic Ferrisols	Calcaric Regosols
Ck	0.19±0.05c	0.17±0.07c	0.22±0.08d	0.16±0.06e	0.31±0.12d	0.42±0.09d	0.29±0.07c
1	0.21±0.04c	0.19±0.06c	0.30± 0.04d	0.41±0.13e	-	-	-
2	0.62±0.04bc	0.48±0.10c	0.56±0.27d	0.99±0.29d	0.69±0.17d	0.65±0.11d	0.35±0.06c
4	1.19±0.08b	1.05±0.90bc	1.17±0.24c	1.79±0.54c	1.52±0.26c	1.46±0.60c	0.71±0.10bc
6	2.21±0.91a	2.13±0.90b	2.01±0.47b	2.98±0.46b	2.44±0.64b	2.39±0.90b	1.38±0.29ab
8	2.89±1.04a	3.28±0.84a	3.29±0.73a	3.95±0.41a	3.58±0.53a	3.47 ±0.70a	2.01±0.94a

Mean values followed by different letters within the same column are significantly different at $P < 0.05$.

Table 3. Dry Biomass (g plant^{-1}) of Pak choi Shoots Grown on Seven Chinese Soils with Different Loading Rates of Cd.

Cd conc. (mg kg^{-1})	Mollisols	Ustic Cambosols	Stagnic Anthrosols	Periudic Argosols	Typic Haplustalfs	Udic Ferrisols	Calcaric Regosols
Ck	1.95±0.18a	1.54±0.13a	1.13±0.31a	0.84±0.29a	0.73±0.20a	0.95±0.11a	2.07±0.46a
1	1.91±0.37a	1.66±0.59a	1.10±0.30a	0.63±0.35ab	-	-	-
2	2.22±0.58a	1.02±0.49a	1.14±0.53a	0.59±0.47ab	0.54±0.41ab	0.79±0.23ab	1.98±0.11a
4	1.89±0.49a	1.47±0.08a	1.25±0.28a	0.51±0.03ab	0.43±0.38ab	0.58±0.06abc	1.95±0.06a
6	1.86±0.48a	1.35±0.13ab	1.09±0.09a	0.36±0.02ab	0.38±0.11ab	0.50±0.24bc	1.90±0.15a
8	1.54±0.28a	0.95±0.15b	1.11±0.51a	0.17±0.06b	0.10±0.01b	0.39±0.29c	1.75±0.17a

Mean values followed by different letters within the same column are significantly different at $P <0.05$.

bioavailability. The lowest and highest Cd concentrations in the Pak choi tissues were at the highest (8 mg kg^{-1}) level of Cd in Calcaric Regosols and Periudic Argosols, respectively. Cadmium concentrations in Pak choi followed an order of: Periudic Argosols> Typic Haplustalfs> Udic Ferrisols> Ustic Cambosols> Mollisols> Stagnic Anthrosols> Calcaric Regosols at 8 mg kg^{-1} soil Cd level (Table 4).

Relationship between Mehlich-3 Extractable Cd in Soils and Pak choi Cd Content

Cadmium concentrations in shoots of Pak choi were significantly correlated to total Cd and Mehlich-3 extractable Cd contents in soils ($R^2 = 0.95$ to 0. 99, and 0.97 to 0.99 respectively). Cadmium concentrations in Pak choi shoots were best related to total Cd content in Mollisols ($R^2 = 0.99$). Whereas, the Cd concentrations of Pak choi shoots were best correlated to Mehlich-3 extractable Cd in Ustic Cambosols, Stagnic Anthrosols Periudic Argosols, Udic Ferrisols, Typic Haplustalfs and Calcaric Regosols with $R^2 = 0.97$, 0.99, 0.99,0.98,0.99 and 0.98, respectively (Table 5).

Soil Cd Thresholds for Potential Dietary Toxicity in Pak choi

Total soil Cd thresholds for potential dietary toxicity from the consumption of Pak choi conformed to an order of: Calcaric Regosols> Stagnic Anthrosols> Ustic Cambosols> Mollisols> Udic Ferrisols> Typic Haplustalfs> Periudic Argosols, and were 1.25, 1.16, 1.02, 0.86, 0.72, 0.70 and 0.12 mg kg^{-1}, respectively. Mehlich-3 extractable Cd thresholds were 0.30, 0.25, 0.23, 0.18, 0.16, 0.15 and 0.03 mg kg^{-1} and decreased in the following order of: Stagnic Anthrosols> Calcareous> Mollisols> Ustic Cambosols> Typic Haplustalfs> Udic Ferrisols>Periudic Argosols, respectively (Table 6).

Discussion

Biomass Yield of Pak choi

Dry weight of Pak choi did not decrease significantly under different Cd levels (Ck to 8.0 mg kg^{-1}) in Mollisols, Stagnic Anthrosols and Calcaric Regosols and even increased at 1, 2 and 4.0 mg kg^{-1} of treatment levels. Similar stimulatory responses of biomass to Cd exposure have also been reported in several plant species [33,34]. The stimulatory effect of Cd on plant biomass may be explained by various mechanisms, for examples, metal ions can serve as enzyme activators in cytokinins metabolism, which stimulates the growth of plants, [35] and a low dose of metal exposure may cause changes in cytokinins and plant hormones that regulate growth and development of plants [36]. Kaminek [36] reported that cytokinins may delay senescence by maintaining chlorophyll production and photosynthetic activity in plant leaves.

Cd exposure may cause changes to various physiological and biochemical processes in plant tissues, such as, reduction in dry biomass may be due to the negative effects of Cd on the roots, and plants could not take up nutrients to continue their normal activities. It has been well reported that Cd can reduce plant growth and development by interfering in various metabolic processes, such as, inhibition of the proton pump, reduction in root elongation, and damage to photosynthetic activity [37,38]. The excess amount of Cd in soil may be responsible for causing disturbances in mineral nutrition and carbohydrate metabolism [39].

Shoot biomass of Pak choi grown in Ustic Cambosols, Udic Ferrisols, Periudic Argosols and Typic Haplustalfs decreased significantly as compared with the control. The inhibitory effect of

Table 4. Cd Concentration (mg kg^{-1} DW) in Pak choi Grown under Different Cd Levels in Seven Chinese Soils.

Cd (mg kg^{-1})	Mollisols		Ustic Cambosols		Stagnic Anthrosols		Periudic Argosols		Typic Haplustalfs		Udic Ferrisols		Calcaric Regosols	
	Root	Shoot	Root	Shoot	Root	Shoot	Root	Shoot	Root	Shoot	Root	Shoot	Root	Shoot
Ck	3.41±0.64e	1.00±0.28e	2.42±0.45f	0.48±0.22e	2.73±0.28e	0.85±0.12e	7.26±1.10f	1.84±0.56f	12.35±1.68e	4.37±1.02e	11.81±0.78e	6.77±0.92e	3.11±0.66e	1.92±0.58b
1	5.09±0.91e	2.53±0.76e	6.69±0.65e	2.06±0.98e	4.12±0.75e	1.92±0.94e	21.3±2.01e	11.00±1.57e	-	-	-	-	-	-
2	8.92±1.09d	4.31±0.54d	9.30±1.29d	4.77±0.22d	8.34±1.11d	3.94±0.83d	39.33±3.86d	28.51±2.61d	25.41±1.79d	12.09±1.34d	19.41±1.27d	10.06±1.10d	4.24±0.99d	2.70±0.99b
4	14.21±0.90c	8.34±1.11c	14.00±1.43c	9.01±1.06c	13.21±1.55c	7.06±1.06c	68.43±3.71c	39.12±3.06c	39.11±2.28c	32.51±2.51c	33.55±1.71c	20.89±1.46c	10.37±1.21c	3.46±1.10b
6	21.24±1.20b	11.75±1.17b	24.36±1.11b	17.23±1.50b	20.21±1.80b	11.63±1.54b	117.76±6.53b	68.21±4.08b	68.00±3.96b	41.11±3.01b	55.53±2.67b	31.45±2.48b	14.41±1.03b	6.57±0.93a
8	38.96±1.39a	18.11±1.12a	43.21±2.21a	22.22±1.51a	28.56±1.42a	17.63±1.37a	169.95±8.11a	89.21±5.70a	118.21±6.24a	69.21±4.11a	85.32±4.28a	53.73±2.88a	18.73±1.54a	8.13±0.87a

Mean values followed by different letters within the same column are significantly different at P <0.05.

Cd on shoot growth is consistent with earlier reports of three Chinese cabbage cultivars exposed to different soil Cd levels of 1, 2.5 and 5 mg kg^{-1}. A significant decrease in the shoot biomass was observed at 2.5 and 5 mg kg^{-1} levels of Cd as compared to their respective controls [40]. Shentu et al. [4] found a 46% reduction in root dry weight of radish at 6.31 mg kg^{-1} Cd exposure in red yellow soil, which is in accordance with our results as we also noticed a shoot dry weight reduction of 58.9%, 79.7%, and 86.3% in Udic Ferrisols, Periudic Argosols and Typic Haplustalfs respectively at 8 mg kg^{-1} level of soil Cd as compared to their respective controls.

Accumulation and Distribution of Cadmium in Pak choi

Variations of Cd accumulation in Pak choi grown in different soils with different pH may be due to the difference in bioavailability of Cd in each soil. Liang et al. [13] stated that Cd content of spinach plants was highly dependent upon the soil pH being highest at pH 5.3. Lai and Chen [41] reported that Cd concentration in Pak choi shoots was up to 85 mg kg^{-1} DW with an application of soil Cd up to 20 mg kg^{-1}. Moreover, it was observed that accumulation of the Cd in rice shoot ranged from 67.9 to 241.7 µg/pot in different rice genotypes at 5 mg kg^{-1} soil Cd level [42].

Relationship between Mehlich-3 Extractable Cd in Soils and Pak choi Cd Content

Mehlich-3 extraction technique appeared efficient to assess Cd phytoavailability to Pak choi, grown in seven textured soils, as evidenced by high correlation coefficients (R^2>0.97). This is in agreement with our previous studies, [43,28] which reflected a high linear correlation (R^2>0.98) between Mehlich-3 Cr and Cr contents in Pak choi and rice grown under six different textured soils. These results are similar to those reported in which extractable soil metal was an improved indicator for Cd phytoavailability in several vegetable crops [4]. Mehlich-3 extraction method is applicable to a large range of soil types, from acidic to alkaline, and makes it ideal for application at a wide scale [44]. Generally, the extraction techniques are assumed to have a relationship between the extractable fraction of metals and the phytoavailability of the metals to plants, and these metals such as exchangeable, soluble, and loosely adsorbed metals are labile and thus readily available to plants [12,45]. The efficiency of Mehlich-3 extraction method was compared with the EPA 3050 B method (a strong acid digestion method) to assess the predictive capabilities through a lettuce (green specie) bioassay. Mehlich-3 extraction was positively correlated with the more costly EPA test, and could be developed as a less expensive and easily conduct able technique [46].

Soil Cd Thresholds for Potential Dietary Toxicity in Pak choi

Cadmium concentrations in the shoots of Pak choi were significantly correlated to total Cd and Mehlich-3 extractable Cd contents in soils, with R^2 values of 0.95 to 0.99, and 0.97 to 0.99, respectively. From this investigation, Cd contents in Pak choi shoots were correlated to total Cd content in Mollisols (R^2 values of 0.99). Cadmium concentrations of Pak choi shoots were highly correlated to Mehlich-3 extractable Cd in Ustic Cambosols, Stagnic Anthrosols, Periudic Argosols, Udic Ferrisols, Typic Haplustalfs and Calcaric Regosols with R^2 values of 0.97, 0.99, 0.99, 0.98, 0.99 and 0.98, respectively. Total Cd threshold levels for potential dietary toxicity conformed to an order of: Calcaric Regosols> Stagnic Anthrosols> Mollisols> Ustic Cambosols>

Table 5. Regression Correlation between Cd Contents in the Edible Shoots of Pak choi and Different Forms of Cd in Various Soils.

Soil type	Form of Soil Cd	Regression equation	R^2
Mollisols	Total Cd	$y = 2.1706x +0.1669$	0.99
	Mehlich-3 extractable Cd	$y = 5.7207x +0.7037$	0.98
Ustic Cambosols	Total Cd	$y = 2.9358x -0.9586$	0.96
	Mehlich-3 extractable Cd	$y = 7.0409x +0.7648$	0.97
Stagnic Anthrosols	Total Cd	$y = 2.2344x -0.5651$	0.98
	Mehlich-3 extractable Cd	$y = 5.3848x +0.4458$	0.99
Periudic Argosols	Total Cd	$y = 11.061x +0.6961$	0.98
	Mehlich-3 extractable Cd	$y = 22.326x +1.3968$	0.99
Typic Haplustalfs	Total Cd	$y = 8.7318x -4.0825$	0.97
	Mehlich-3 extractable Cd	$y = 19.123x -0.8046$	0.98
Udic Ferrisols	Total Cd	$y = 6.6697x -2.7793$	0.97
	Mehlich-3 extractable Cd	$y = 14.873x -0.3773$	0.99
Calcaric Regosols	Total Cd	$y = 0.8936x +0.924$	0.95
	Mehlich-3 extractable Cd	$y = 3.5937x +1.1512$	0.98

Typic Haplustalfs> Udic Ferrisols> Periudic Argosols and were 1.25, 1.16, 1.02, 0.86, 0.72, 0.70 and 0.12 mg kg^{-1}, respectively. Mehlich-3 extractable Cd thresholds decreased in the following order of: Stagnic Anthrosols> Calcareous> Mollisols> Ustic Cambosols> Udic Ferrisols>Typic Haplustalfs> Periudic Argosols and were 0.30, 0.25, 0.23, 0.18, 0.16, 0.15 and 0.03 mg kg^{-1}, respectively (Table 6).

Cadmium concentrations in Pak choi shoots, were highly correlated to total Cd content in Mollisols with the threshold levels of 0.86 mg kg^{-1} with a $R^2 = 0.99$. However, the Cd concentrations of Pak choi shoots were best related to Mehlich-3 extractable Cd in Stagnic Anthrosols, Calcaric Regosols, Mollisols, Ustic Cambosols, Typic Haplustalfs and Periudic Argosols with thresholds values of 0.30, 0.25, 0.23, 0.18, 0.16, 0.15 and 0.03 mg kg^{-1}, (R^2 values of 0.97, 0.99, 0.99, 0.98, 0.98 and 0.98), respectively. Based on the wide range of applicability and the simplicity of extraction method, it is proposed that Mehlich-3 extractable Cd is more suitable to be used as soil Cd thresholds for potential dietary toxicity in Pak choi. Our previous study evaluated the phytoavailability of Cd to rice, and demonstrated the suitability of Mehlich-3 extraction method in different textured soils [47]. Similar to our results, Murakami et al. [48] reported that Mehlich-3 extractable Cd was an improved indicator than total soil Cd and HCl-

extractable Cd to predict the grain Cd content of japonica rice varieties. Our results are also in agreement with Shentu et al. [4] who also concluded that extractable Cd was a better soil test index for Cd phytoavailability of several vegetables and could be used as soil Cd thresholds for food safety. Among the predicted thresholds (total soil Cd) the lowest value (0.12 mg kg^{-1}) was observed for the Periudic Argosols, an acidic soil. Bioavailability and uptake of Cd are very high in this soil. The leafy vegetables like Pak choi can accumulate large quantities of Cd as compared to other crops [9,10]. Therefore the predicted threshold is even lower than background value of Cd in soil; it means that there is a risk for dietary toxicity from Pak choi grown on it, even with the background value of total Cd in soil. This kind of information has been reported in our previous study. The threshold of total soil Cd for rice was 0.21 mg kg^{-1} which was also lower than background value of total Cd in soil [47].

Cd levels (Ck, 1, 2, 4, 6, 8 mg kg^{-1}) used in this investigation represented uncontaminated, lightly contaminated, and moderately Cd polluted soils. Therefore, these levels of Cd contamination are realistic, comparable to those applied in other soil safety risk assessment studies, and thus, the results are applicable in field conditions as well.

Table 6. Soil Cd Threshold Levels for Potential Dietary Toxicity in Edible Part of Pak choi Calculated from the Permissible Limit of Cd (2.04 mg kg^{-1} DW) in Leafy Vegetables and Regression Equations.

Soil Type	Total Cd (mg kg^{-1})	Mehlich-3 extractable Cd (mg kg^{-1})
Mollisols	0.86	0.23
Ustic Cambosols	1.02	0.18
Stagnic Anthrosols	1.16	0.30
Periudic Argosols	0.12	0.03
Udic Ferrisols	0.72	0.16
Typic Haplustalfs	0.70	0.15
Calcaric Regosols	1.25	0.25

Table 7. Stepwise Regression Model for Predicting Cd Concentration (Y) in Edible Part of Pak choi based on Soil Properties.

Stepwise regression model	R^2	F value		T value and R^2 of partial regression coefficient	
				T value	R^2
$Y = -39.256-15.516 \text{ pH}-0.944 \text{ Zn}_T-0.379 \text{ OM} +26.752 \text{ Cd}_T$	0.977	138.808*	pH	−18.682**	0.964
			Zn_T	−3.788**	0.524
			OM	−3.652*	0.376
			Cd_T	2.67*	0.354

[a]Cd_T and Zn_T refer to the total Cadmium and Zinc concentrations.
[b]Superscripts * and ** indicate significant levels of probability at 0.05 and 0.01, respectively.

Stepwise Regression Model for Predicting Cd Phytoavailability to Pak choi

Many physicochemical properties of soils can influence the heavy metal accumulation in vegetables. For example, the amount of heavy metal uptake from soils was influenced by soil pH, organic matter (OM) content, cation exchange capacity (CEC) and soil texture [49]. The combinations of basic soil properties may explain Cd uptake by plants [50]. By considering this aspect, soil pH, OM, CEC, total soil Cd, total Zn and clay contents were integrated to simulate the combined effects of soil environment on Cd phytoavailability to Pak choi. Stepwise linear regression was conducted and four independent variables pH, total Zn, OM and total Cd significantly influenced the accumulation of Cd in Pak choi plants (Table 7). Both the multiple correlation and partial regression coefficients reached the statistically significant levels at least the 0.05. For multiple linear regression analyses, R^2 values could be used to explain variation of the dependents [12]. It was found that R^2 value was above 0.97, which means that more than 97% of variation in Cd concentration in Pak choi shoots could be attributed to soil pH, total Zinc, OM and total Cd contents in soils (Table 7).

The influence of each factor on Cd concentration of Pak choi (Y) shoots could be further explained by the values of each coefficient [12]. Stepwise regression model revealed that Cd concentration in the Pak choi was enhanced by lower soil pH (negative coefficients showed negative effect and vice versa), total Zinc, OM contents and higher total soil Cd. Lower soil pH, zinc, OM and higher soil total Cd are among the factors which enhance the bioavailability Cd contents in soils. Therefore, these three variables had the contradictory effect on Cd phytoavailability to Pak choi. Wang et al. [12] reported that soil characteristics (e.g. pH, CEC and OM) affected the phytoavailability of different heavy metals in soils, and such influences could be considered in the assessment of phytoavailability of heavy metals. There are four parameters involved in this model and then interactions between them were obvious (e.g. Cd concentration in the extractable fraction was correlated with lower soil pH, soil zinc and OM content). Furthermore, the coefficients obtained in the present model can regulate these cross effects and result in an improved model fitting. For example, there was a negative correlation between the soil pH, Zinc and OM, these factors had an inverse effect on Cd phytoavailability and soil Cd was the leading factor influencing Cd phytoavailability to Pak choi (coefficient of soil Cd was greater than those of pH, Zinc and OM). Our results about soil Cd and pH are in accordance with our recent study which developed an empirical model to correlate the Cd phytoavailability to rice with several soil properties. Soil pH and bioavailable soil Cd were major influencing factors which (pH negatively and soil Cd positively) correlate with the Cd phytoavailability, however total Zn and OM were not included in our previously developed model [47]. Eriksson and Soderstrom [51] reported that the Cd concentration of wheat grain grown on non-calcareous soils of Sweden was positively correlated to soil total Cd and negatively to extractable Zn. A study was conducted on Cd contaminated soils in Taiwan, whereas regression equation was developed to predict Cd concentrations in rice roots by available fractions of Cd and Zn in soil [52]. The negative coefficient of Zn indicated that soil Zn suppressed the uptake of Cd by rice roots in all varieties as Zn has an antagonistic effect on Cd uptake by root [53]. Oliver et al. [54] also observed a significant decrease of Cd up to 50% in wheat grain when $2.5-5.0$ kg Zn ha^{-1} was applied to Cd contaminated Australian soils.

Organic matter content was negatively correlated with the accumulation of Cd in Pak choi shoots (Table 7). Organic matter plays an important role in determining the bioavailability and mobility of heavy metals in soils. Organic matter is involved in supplying organic chemicals to the soil solution, which may act as chelates and increase metal bioavailability to plants [55]. However, OM could reduce the bioavailability of heavy metals in soils by adsorption or forming stable complexes with humic substances [56]. Halim et al. [57] reported that addition of humic acid demonstrated a decrease in extractable heavy metal fraction in metal contaminated soils. This could partially explain the negative correlation of organic matter contents and Cd uptake observed in our present study.

Conclusions

The present study concludes that Cd concentration in Pak choi tissues was dependent on soil type. To establish the soil Cd thresholds of potential dietary toxicity from Pak choi, both Cd bioavailability in garden soils and Pak choi tissues should be taken into consideration. The selection of proper soil types for vegetable production can help us to avoid the toxicity of Cd in our daily diet. Stepwise regression model demonstrated that soil pH, organic matter, total Cd and Zinc contents may be the major factors having influence on the phytoavailability of Cd in different textured soils.

Acknowledgments

M.T. Rafiq acknowledges International Islamic University Islamabad Pakistan for providing a PhD scholarship under HEC-IIUI faculty development program.

Author Contributions

Conceived and designed the experiments: MTR XY TL. Performed the experiments: MTR RA WX. Analyzed the data: RA AS MA. Contributed reagents/materials/analysis tools: XY. Wrote the paper: MTR RA PJS.

References

1. Liu JG, Qian M, Cai GL, Yang JC, Zhu QS (2007) Uptake and translocation of Cd in different rice cultivars and the relation with Cd accumulation in rice grain. J Hazard Mater 143: 443–447.

2. Mahler RJ, Bingham FT, Page AL (1978) Cadmium-enriched sewage sludge application to acid and calcareous soils: Effect on yield and cadmium uptake by lettuce and Swiss chard. J Environ Qual 7: 274–281.

3. Stohs SJ, Bagchi D, Hassoun E, Bagchi M (2000) Oxidative mechanisms in the toxicity of chromium and cadmium ions. J Environ Pathol Toxicol Oncol 19: 201–213.

4. Shentu J, He Z, Yang XE, Li TQ (2008) Accumulation properties of cadmium in a selected vegetable –rotation system of south eastern China. J Agric Food Chem 56: 6382–6388.

5. IARC (1993) (International agency for research on cancer), Monographs on the evaluation of the carcinogenic risks to humans beryllium, cadmium, mercury and exposures in the glass manufacturing industry. IARC, Scientific Publications. Lyon, France. 119–238.

6. Li ST, Liu RL, Wang M, Wang XB, Shan H, et al. (2006) Phytoavailability of cadmium to cherry-red radish in soil applied composted chicken or pig manure. Geoderma 136: 260–271.

7. Lebeau T, Bagot D, Jezequel K, Fabr B (2002) Cadmium biosorption by free and immobilized microorganisms cultivated in a liquid soil extract medium: Effects of Cd, pH, and techniques of culture. Sci Total Environ 291: 73–83.

8. Abdola M, Chmtelnicka J (1990) New aspects on the distribution and metabolism of essential trace elements after dietary exposure to toxic metals. Biol Trace Element Res 23: 25–53.

9. Yang JX, Guo HT, Ma YB, Wang LQ, Wei DP, et al. (2010) Genotypic variations in the accumulation of exhibited by different vegetables. J Environ Sci 22: 1246–1252.

10. Yang Y, Zhang FS, Li HF, Jiang RF (2009) Accumulation of cadmium in the edible parts of six vegetable species grown in Cd-contaminated soils. J Environ Manag 90: 1117–1122.

11. Chen HS, Huang QY, Liu LN, Cai P, Liang W, et al. (2010) Poultry manure compost alleviates the phytotoxicity of soil cadmium: Influence on growth of Pak choi (Brassica chinensis L.). Pedosphere 20: 63–70.

12. Wang XP, Shan XQ, Zhang SZ, Wen B (2004) A model for evaluation of the phytoavailability of trace elements to vegetables under field conditions. Chemosphere 55: 811–822.

13. Liang Z, Ding Q, Wei D, Li J, Chen S, et al. (2013) Major controlling factors and predictable equations for Cd transfer factor involved in soil-spinach system. Ecotox Environ Saf 93: 180–185.

14. Ge Y, Murray P, Hendershot WH (2000) Trace metal speciation and bioavailability in urban soils. Environ. Pollut. 107, 137–144.

15. Gu JG, Zhou QX, Wang X (2003) Reused path of heavy metal pollution in soils and its research advance. J Basic Sci Eng 11: 143–151 (in Chinese).

16. Gu JG, Zhou QX (2002) Cleaning up through phytoremediation: A review of Cd contaminated soils. Ecol Sci 21: 352–356. (in Chinese with English abstract)

17. Du TP (2005) Food safety and strategy in China. Productivity Res: 6, 139–141. (in Chinese with English abstract)

18. Franz E, Römkens P, Van Raamsdonk L, Van Der Fels-Klerx I (2008) A chain modeling approach to estimate the impact of soil cadmium pollution on human dietary exposure. J Food Protect: 71, 2504–13.

19. Kobayashi E, Suwazono Y, Dochi M, Honda R, Nishijo M, et al. (2008) Estimation of benchmark doses as threshold levels of urinary cadmium, based on excretion of β2-microglobulin in cadmium polluted and non-polluted regions in Japan. Toxicol Lett 179: 108–12.

20. Tracy S, Sheila M (2006) Cadmium and zinc accumulation in soybean: A threat to food safety. Sci Total Environ 371: 63–73.

21. Ide G, Becker B (1995) Relationship between the arsenic concentration in soil and in the cropped vegetables. International Conference on Heavy Metal in the Environment, Hamburg, vol. 2, CEP Consultants Ltd, Norwich, UK 302–304

22. Szteke B, Jedrzejczak R (1995) The variability of heavy metal contents in plant and soil samples from fields of one farm. International Conference on Heavy Metal in the Environment, Hamburg, vol. 2, CEP Consultants Ltd., Norwich, UK 228–231.

23. Salvatore M, Carratù G, Carafa A (2009) Assessment of heavy metals transfer from a moderately polluted soil into the edible parts of vegetables. J Food Agric Environ 7: 683–688.

24. Chaturvedi R, Sankar K, (2006) Laboratory manual for the physicochemical analysis of soil, water and plant. Wildlife Institute of India, Dehradun, India.

25. Hendershot WH, Duquette M (1986) A simple barium chloride method for determining cation exchange capacity and exchangeable cations. Soil Sci Soc Am J 50: 605–608.

26. Rashid A, Ryan J, Estefan G (2001) Soil and plant analysis laboratory manual. International center for agricultural research in the dry areas (ICARDA), Aleppo, Syria.

27. Day PR (1965) Particle fractionation and particle-size analysis. In: Klute, A. (eds.), Methods of soil analysis. ASA and SSSA, Madison, WI, pp. 545–567.

28. Xiao W, Yang XE, He Z, Rafiq MT, Hou D, et al. (2013) Model for evaluation of the phytoavailability of chromium (Cr) to rice (Oryza sativa L.) in representative Chinese soils. J Agric Food Chem 61: 2925–2932.

29. Mehlich A (1984) Mehlich-3 soil test extractant a modification of Mehlich-2 extractant. Commun Soil Sci Plan 15: 1409–1416.

30. FAO/WHO (2003) Report of the sixty first meeting of Joint FAO/WHO expert committee on food additives. Rome.

31. Wang XL, Sato T, Xing BS, Tao S (2005) Health risks of heavy metals to the general public in Tianjin, China via consumption of vegetables and fish. Sci Total Environ 350: 28–37.

32. Rattan R, Datta S, Chhonkar P, Suribabu K, Singh A (2005) A Long-term impact of irrigation with sewage effluents on heavy metal content in soils, crops and groundwater: A case study. Agric Ecosyst Environ 109: 310–322.

33. Peter MC (2002) Ecological risk assessment (ERA) and hormesis. Sci Total Environ 288: 131–140.

34. Liu X, Peng K, Wang A, Lian CL, Shen ZG (2010) Cadmium accumulation and distribution in populations of Phytolacca Americana L. and the role of transpiration. Chemosphere 78: 1136–1141.

35. Peter N, Karoly B, Laszlo G (2003) Characterization of the stimulating effect of low-dose stressors in maize and bean seedlings. J Plant Physiol 160: 1175–1183.

36. Kaminek M (1992) Progress in cytokinin research. TIBTECH 10: 159–162.

37. Ali B, Tao QJ, Zhou YF, Gill RA, Ali S, et al. (2013) 5-aminolevolinic acid mitigates the cadmium-induced changes in Brassica napus as revealed by the biochemical and ultra-structural evaluation of roots. Ecotox Environ Safe 92: 271–280.

38. Ali B, Wang B, Ali S, Ghani MA, Hayat MT, et al. (2013) 5-Aminolevulinic acid ameliorates the growth, photosynthetic gas exchange capacity and ultrastructural changes under cadmium stress in Brassica napus L. J Plant Growth Regul 32: 604–614.

39. Moya JL, Ros R, Picazo I (1993) Influence of cadmium and nickel on growth, net photosynthesis and carbohydrate distribution in rice plants. Photosyn Res 36: 75–80.

40. Weitao L, Zhou Q, Ana J Suna Y, Liu R (2010) Variations in cadmium accumulation among Chinese cabbage cultivars and screening for Cd-safe cultivars. J Hazard Mater 173: 737–743.

41. Lai HY, Chen BC (2013) The dynamic growth exhibition and accumulation of cadmium of Pak choi (Brassica campestris L. ssp. chinensis) grown in contaminated soils. Int J Environ Res Public Health 10: 5284–5298.

42. Cui YJ, Zhu YG, Smith SA, Smith SE (2004) Cadmium uptake by different rice genotypes that produce white or dark grains. J Environ Sci 16: 962–967.

43. Xiao W, Yang XE, Zhang Y, Rafiq MT, He Z, et al. (2013) Accumulation of chromium in Pak choi (Brassica chinensis L.) grown on representative Chinese soils. J Agric Qual 42: 758–765.

44. De Villiers S, Thiart C, Basson NC (2010) Identification of sources of environmental lead in South Africa from surface soil geochemical maps. Environ Geochem Health 32: 451–459.

45. Kabata-Pendias A, Pendias H (2001) Trace Elements in Soils and Plants. Boca Raton, FloridaCRC Press.

46. Laura W, Wander M, Phillips E (2011) Testing and educating on urban soil lead: A case of Chicago community gardens. J Agric Food Sys Comm Develop, ISSN: 2152–0801 online.

47. Rafiq MT, Aziz R, Yang XE, Wendan X, Rafiq MK, et al. (2014) Cadmium phytoavailability to rice (Oryza sativa L.) grown in representative Chinese soils. A model to improve soil environmental quality guidelines for food safety. Ecotox Environ Safe 103:101–107.

48. Murakami M, Nakagawa F, Ae N, Ito M, Arao T (2009) Phytoextraction by rice capable of accumulating Cd at high levels: Reduction of Cd content of rice grain. Environ Sci Technol 43: 5878–5883.

49. Jung MC, Thornton I (1996) Heavy metal contamination of soils and plants in the vicinity of a lead-zinc mine, Korea. Appl Geochem 11: 53–59.

50. McBride M (2002) Cadmium uptake by crops estimated from soil total Cd and pH. Soil Sci 15: 84–92.

51. Eriksson JE, Sderstrom M (1996) Cadmium in soil and winter wheat grain in southern Sweden. Factors influencing Cd levels in soils and grain. Acta Agric Scand Sect B 46: 240–248.

52. Romkens PFAM, Guo HY, Chu CL, Liu TS, Chiang CF, et al. (2009) Prediction of cadmium uptake by brown rice and derivation of soil-plant transfer models to improve soil protection guidelines. Environ Pollut 157: 2435–2444.

53. Giordano PM, Mays DA, Behel AD (1979) Soil temperature effects on the uptake of cadmium and zinc by vegetables grown on sludge amended soil. J Environ Qual 8: 232–236.

54. Oliver DP, Hannam R, Tiller KG, Wilhelm NS, Merry RH, et al. (1994) The effects of zinc fertilization on cadmium concentration in wheat grain. J Environ Qual 23: 705–711.

55. McCauley A, Jones C, Jacobsen J (2009) Soil pH and Organic Matter. Nutrient management modules 8, #4449-8. Montana State University Extension Service, Bozeman, Montana, pp. 1–12.

56. Liu LN, Chen HS, Cai P, Liang W, Huang QY (2009) Immobilization and phytotoxicity of Cd in contaminated soil amended with chicken manure compost. J Hazard Mater 163: 563–567.

57. Halim M, Conte P, Piccolo A (2003) Potential availability of heavy metals to phytoextraction from contaminated soils induced by exogenous humic substances. Chemosphere 52: 265–275.

20

Cropping Systems and Cultural Practices Determine the *Rhizoctonia* Anastomosis Groups Associated with *Brassica* spp. in Vietnam

Gia Khuong Hoang Hua[ꝰ]**, Lien Bertier**[ꝰ]**, Saman Soltaninejad, Monica Höfte***

Laboratory of Phytopathology, Department of Crop Protection, Faculty of Bioscience Engineering, Ghent University, Gent, Belgium

Abstract

Ninety seven *Rhizoctonia* isolates were collected from different *Brassica* species with typical *Rhizoctonia* symptoms in different provinces of Vietnam. The isolates were identified using staining of nuclei and sequencing of the rDNA-ITS barcoding gene. The majority of the isolates were multinucleate *R. solani* and four isolates were binucleate *Rhizoctonia* belonging to anastomosis groups (AGs) AG-A and a new subgroup of A-F that we introduce here as AG-Fc on the basis of differences in rDNA-ITS sequence. The most prevalent multinucleate AG was AG 1-IA (45.4% of isolates), followed by AG 1-ID (17.5%), AG 1-IB (13.4%), AG 4-HGI (12.4%), AG 2-2 (5.2%), AG 7 (1.0%) and an unknown AG related to AG 1-IA and AG 1-IE that we introduce here as AG 1-IG (1.0%) on the basis of differences in rDNA-ITS sequence. AG 1-IA and AG 1-ID have not been reported before on *Brassica* spp. Pathogenicity tests revealed that isolates from all AGs, except AG-A, induced symptoms on detached leaves of several cabbage species. In *in vitro* tests on white cabbage and Chinese cabbage, both hosts were severely infected by AG 1-IB, AG 2-2, AG 4-HGI, AG 1-IG and AG-Fc isolates, while under greenhouse conditions, only AG 4-HGI, AG 2-2 and AG-Fc isolates could cause severe disease symptoms. The occurrence of the different AGs seems to be correlated with the cropping systems and cultural practices in different sampling areas suggesting that agricultural practices determine the AGs associated with *Brassica* plants in Vietnam.

Editor: Mark Gijzen, Agriculture and Agri-Food Canada, Canada

Funding: This work was supported by a scholarship from the Special Research Fund of Ghent University (BOF) given to Gia Khuong Hoang Hua. Lien Bertier was funded by a PhD grant of the Agency of Innovation by Science and Technology in Flanders (IWT). The funders had no role in study design, data collection and analysis, decision to publish, or preparation of the manuscript.

Competing Interests: The authors have declared that no competing interests exist.

* Email: monica.hofte@ugent.be

ꝰ These authors contributed equally to this work.

Introduction

Vietnam is a country in Southeast Asia in which the agricultural sector accounts for more than 22% of the GDP, 30% of export and 52% of all employment. Vietnam is not only one of the world leaders in rice and coffee export, but also the third world's largest vegetable producer. *Brassicas* are among the main vegetables produced for both local consumption and export [1]. Vegetables in Vietnam are mainly produced by poor households living in the Red River and Mekong River delta (see Figure 1) in intensive cultivation systems or in rotation with other crops. Due to the lack of knowledge in crop management, limited availability of technology and land fragmentation, farmers are suffering heavy yield losses year after year. Among the limiting factors in vegetable production is the occurrence of *Rhizoctonia* diseases, which has been recognized as one of the most important threats.

Rhizoctonia is a genus of basidiomycete fungi causing many important plant diseases. Based on differences in the number of nuclei per cell, *Rhizoctonia* isolates have been differentiated into uninucleate *Rhizoctonia*, binucleate *Rhizoctonia* (teleomorphs: *Ceratobasidium* spp. and *Tulasnella* spp.) and multinucleate *Rhizoctonia* (teleomorphs: *Thanatephorus* spp. and *Waitea* spp.) [2]. *Rhizoctonia* species can also be classified using biochemical and molecular techniques. Among those, rDNA-ITS sequence analysis appears to be the most convenient and reliable method [3]. Currently, isolates of *R. solani*, the most widely recognized species within the multinucleate *Rhizoctonia* group, have been divided into 13 anastomosis groups (AGs), while 16 AGs of binucleate *Rhizoctonia* have been recognized [2–7]. Due to the considerable genetic diversity, several AGs have been further divided into subgroups based on phylogenetic differences. These phylogenetic differences can be associated with differences in morphology, ecology, pathogenicity and biochemical characteristics, although this is not necessarily the case [8].

Compared to binucleate *Rhizoctonia*, multinucleate *R. solani* AGs usually have a wider host range and higher virulence. *R. solani* can survive for a long period in plant debris, contaminated seeds, or infested soils as mycelium or sclerotia [9,10]. Under favorable conditions, sclerotia geminate and form delicate hyphae that will grow toward the host plants [11]. *Brassica* vegetables can be attacked by several different AGs of *R. solani* resulting in the development of various diseases such as foliar blight, wirestem and damping off. In previous studies, AGs 1-IB, 1-IC, 2-1, 2-2 IIIB, 3, 4-HGI, 4-HGII, 4-HGIII, 5, 7, 9 and 10 were shown to be pathogenic on *Brassica* crops grown in Canada [12,13], Australia [14], Japan [15,16], North America [17–19], Brazil [20], China [21], Belgium [22], and the UK [23]. Although *Rhizoctonia*

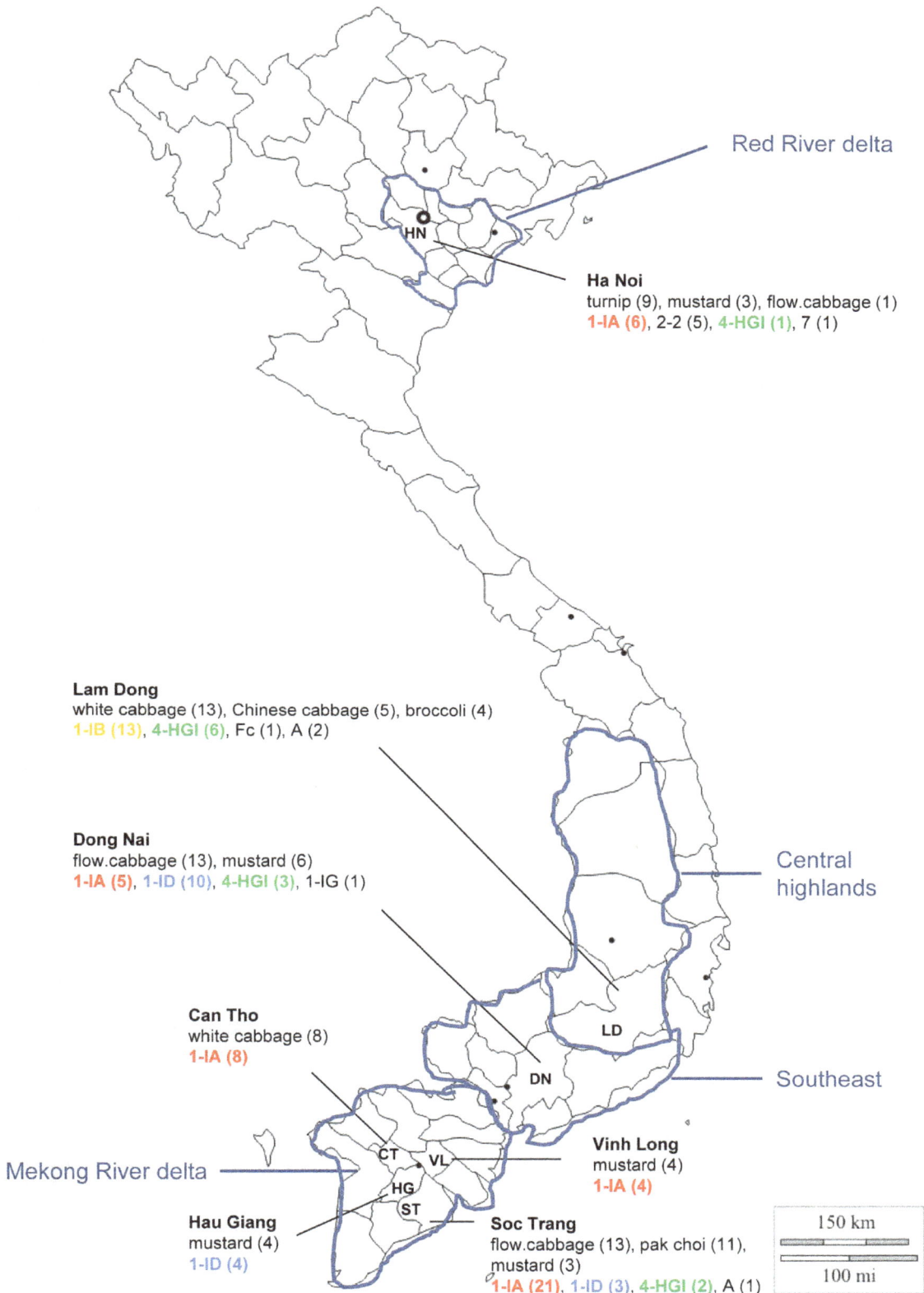

Figure 1. Location of sites for collection of *Rhizoctonia* isolates from *Brassica* spp. in Vietnam. The seven provinces sampled are: Ha Noi (districts of Gia Lam, Thanh Tri and Dong Anh), Lam Dong (Da Lat city and Duc Trong district), Dong Nai (Bien Hoa city), Vinh Long (Binh Tan district), Can Tho (Cai Rang district), Hau Giang (Phung Hiep district) and Soc Trang (Soc Trang city and My Xuyen district). In each city or district of one province, one to two wards were surveyed and these wards are marked with a start+. Different colors are used to highlight the most important AGs found in our survey including AG 1-IA, AG 1-IB, AG 1-ID and AG 4-HGI.

diseases occur severely and frequently on leafy vegetables cultivated in Vietnam, there have been no reports about the AGs and subgroups that attack *Brassica* crops. Therefore, our research aimed at (i) identifying the species and AGs of *Rhizoctonia* present on *Brassica* plants in different vegetable producing regions in Vietnam, and (ii) verifying the susceptibility of *Brassica* vegetables to *Rhizoctonia* isolates collected. Our story revealed that *Rhizoctonia* AGs such as AG 1-IA and AG 1-ID, which have not been reported before on *Brassica* spp., are predominant in Vietnam, which is presumably linked with the cultural practices and cropping systems in the different sampling areas. Moreover, we describe two new AGs that were previously unknown.

Materials and Methods

Field sampling and pathogen isolation

Sampling was done on private farms by Gia Khuong Hoang Hua (Vietnamese citizen) with permission from the farmer. In general, permission by the authorities was not required since the sampling studies were carried out on private farms. Only cabbage plants showing symptoms of the undesired *Rhizoctonia* fungus were sampled, hence the field studies did not involve endangered or protected species. The GPS coordinates of the locations where samples were collected are presented in Table S1 in File S1.

The survey was conducted from September to October 2011 on various fields in the Red River delta (Ha Noi; an important vegetable production area of the North), the Mekong River delta (Vinh Long, Can Tho, Hau Giang and Soc Trang; main vegetable production areas of the South), the Central Highlands (Lam Dong; vegetables are mainly produced for export) and the Southeast (Dong Nai; vegetables are mainly produced for domestic consumption) (Figure 1). These regions were chosen because of their importance in vegetable production and because they offer a good representation of the different climatic and topographic conditions in Vietnam and, therefore, a good representation of the distribution of *Rhizoctonia* spp. on *Brassica* spp. in Vietnam can be obtained. Total production area, climatic conditions and main agricultural activities of these regions are listed in Table 1.

A total of 142 *Brassica* plants with *Rhizoctonia*-like symptoms were sampled. Infected root and leaf tissues were washed in running tap water, surface-disinfected in 1% sodium hypochlorite solution for two min and then rinsed twice in sterile water before placing on 1% water agar medium supplemented with strepto-mycin (0.05 g L^{-1}). After 24 h of incubation, *Rhizoctonia*-like hyphal tips growing out of these tissues were transferred to fresh potato dextrose agar (PDA; Difco) plates and incubated for two to four days at 28°C.

Ninety seven *Rhizoctonia* isolates were recovered and subjected to nuclei staining and sequencing of the ITS-rDNA region. All isolates are listed in Table 2.

Nuclei staining

Rhizoctonia isolates were cultured on sterile glass slides covered by PDA for two days at 28°C. Actively growing fungal hyphae were stained with 10 µg mL^{-1} 4, 6-diamino-2-phenyl indole (DAPI; Sigma-Aldrich) and the number of nuclei per hyphal cell was determined using an Olympus BX51 microscope [22].

DNA extraction, PCR and sequencing of the rDNA-ITS region

The rDNA-ITS region of all collected isolates was sequenced for identification to the AG and subgroup level. The usefulness of the rDNA-ITS region for identification of unknown *Rhizoctonia* spp. has been clearly shown by Sharon et al. [2,3].

Rhizoctonia isolates were grown on potato dextrose broth at 28°C for one week. Mycelial mats were harvested by filtration and ground in liquid nitrogen to produce a fine powder. Total genomic DNA was extracted using the DNeasy Plant Mini Kit (Qiagen). The rDNA-ITS fragment including the 5.8 S gene was amplified using primers ITS4 (5′-TCCTCCGCTTATTGATATGC-3′) and ITS5 (5′-GGAAGTAAAAGTCGTAACAAGG-3′) [24]. The PCR amplification reactions were performed by adding 2 µL genomic DNA (5–10 ng $µL^{-1}$) to 23 µL of reaction mixture containing 2.5 µL PCR buffer (10×; Qiagen), 5 µL Q-solution (Qiagen), 0.5 µL dNTPs (10 mM; Fermentas GmbH), 1.75 µL of each primer (10 µM), 0.15 µL Taq DNA polymerase (5 units $µL^{-1}$; Fermentas GmbH) and 11.35 µL ultrapure sterile water. Amplification was performed using a Flexcycler PCR Thermal Cycler (Analytik Jena) programmed for an initial denaturation step at 94°C for 10 min followed by 35 cycles at 94°C for 1 min, 55°C for 1 min and 72°C for 1 min. Cycling ended with a final extension step at 72°C for 10 min. Amplification products were separated in 1% agarose gels in TAE-buffer at 100 V for 30 min and visualized by ethidium bromide staining on a UV transillu-minator. The sequences of both strands were determined by LGC Genomics GmbH (Berlin, Germany) using Sanger sequencing.

Identification using BLAST and phylogenetic analysis

Consensus sequences for all 97 isolates were created with BioEdit version 7.1.11. To determine the AG of the isolates, the rDNA-ITS consensus sequences obtained were compared to those in Genbank using the BLASTn tool.

However, since Genbank is an uncurated database, it can contain inaccurately designated *Rhizoctonia* spp., as has been previously shown by Sharon et al. [3]. Therefore, comparison of rDNA-ITS sequences of unknown isolates to a curated database of sequences containing representative rDNA-ITS sequences of all known uninucleate, binucleate and multinucleate *Rhizoctonia* AG and subgroups provides a more reliable identification. Such a database of representative sequences is available from Sharon et al. [2] and was provided by Michal Sharon to us. However, not all known AGs were present in this database, therefore we added representative isolates of the following AGs: AG 1-1E, AG 1-1F, AG 2-2 WB [25] and AG 13 [4]. The total number of sequences in the database was 129 and the Genbank accession numbers of all these sequences can be found in Table S2 in File S2.

Multiple alignments for multinucleate and binucleate *Rhizocto-nia* isolates were constructed using MUSCLE which is imple-mented in MEGA 6 [26] and checked manually afterwards. The resulting alignments had a length of 717 bp (multinucleate *Rhizoctonia* isolates) and 768 bp (binucleate *Rhizoctonia* isolates).

Separate phylogenetic trees were constructed for multinucleate *Rhizoctonia* and binucleate *Rhizoctonia* isolates. For the binucleate tree, one representative isolate for each known binucleate AG was included together with the binucleate *Rhizoctonia* isolates obtained in this work.

For the multinucleate tree, reference isolates from the curated database (Table S2 in File S2) were added for all AGs present in our collection from Vietnam. Another 32 isolates from a characterization study in Vietnam that has not been published (Thuan et al., unpublished; Genbank accession numbers with prefix 'EF' in Table 3), and found on a range of crops and belonging to AGs AG 1-IA, AG 1-ID and AG 4-HGI were also added to the multinucleate *Rhizoctonia* alignment for phylogenetic analysis. Phylogenetic trees were built using the neighbour joining algorithm with 1000 bootstrap repeats using MEGA 6 [26]. Model

Table 1. Sampling locations and their relevant characteristics [53–59].

Sampling location	Agricultural area (1000 ha)	Average temperature (°C)	Main crop
Red River Delta			
Ha Noi	152.24	24	- Cabbage, tomato, cucumber, radish
			- Rice
Central highlands region			
Lam Dong	279.00	High land: 14	- Lettuce, cabbage, carrot, potato
		Low land: 21	- Coffee, tea, cashew-nut tree, cotton
Southeast region			
Dong Nai	289.02	27	- Coffee, cotton, black pepper
			- Durian, grapefruit, mango
Mekong River Delta			
Vinh Long	116.18	27	- Rice
			- Mango, orange, grapefruit, durian
			- Cabbage, cucumber, bean
Can Tho	115.00	27	- Rice
			- Cabbage, cucumber
Hau Giang	139.07	27	- Durian, pineapple, grapefruit
			- Rice
Soc Trang	278.15	27	- Rice
			- Grapefruit, mango, durian

testing was done using the software implemented in MEGA 6 and the K2+ G DNA substitution model was chosen.

Aggressiveness of *Rhizoctonia* isolates towards detached leaves of cabbages, rice and water spinach

Nine isolates (representing the nine different *Rhizoctonia* AGs collected) were randomly selected and tested for virulence towards several *Brassica* crops in two independent experiments (Figures 2, 3 and 4). To confirm their pathogenicity, tests were repeated for AGs 1-IA, 1-IB, 1-ID, 2-2, 4-HGI and A since each of these AGs consists of more than one isolate. Results of additional tests are shown in Tables S3, S4 and S5 in File S1.

Leaves of white cabbage (*Brassica oleracea*), Chinese cabbage (*B. chinensis*), pak choi (*B. chinensis*), mustard cabbage (*B. juncea*) and Chinese flowering cabbage (*B. parachinensis*) were cut into pieces (3×3 cm). The soil substrate used in our experiments was a mixture (w/w) of 50% potting soil (Structural; Snebbout, Kaprijke, Belgium) and 50% sand (Cobo garden; Belgium). Sets of six leaf discs were placed in a plastic box (16×11×6 cm) containing 400 g of soil substrate. Inoculum of *Rhizoctonia* spp. was produced according to the method described by Scholten et al. [27]. Briefly, water-soaked wheat kernels were autoclaved for 25 min on two successive days and then inoculated with three fungal discs (diameter 5 mm) cut at the edge of a 3-day-old *Rhizoctonia* colony cultured on PDA. Flasks containing the inoculated kernels were incubated for 14 days at 28°C and shaken every 3–4 days to avoid coagulation. Two *Rhizoctonia*-infected kernels which were comparable in size were buried 2 cm below each leaf disc. Leaf discs inoculated with sterile wheat kernels served as a control. All boxes were incubated in a growth chamber at 22°C.

The detached leaf bio-assay was also conducted to investigate the pathogenicity of the *Rhizoctonia* isolates on rice and water spinach, two important hosts of *Rhizoctonia* spp. in tropical countries. Surface-sterilized seeds of rice (*Oryza sativa* cv. CO39)

and water spinach (*Ipomoea aquatic* cv. Trang Nong) were sown in plastic trays (45×45×10 cm) filled with 4 kg of soil substrate and kept in a growth chamber at 28°C for four weeks before their leaves were detached. Rice leaves were then cut into pieces (8 cm long) and six rice leaf pieces or six water spinach leaves were put in one square Petri dish containing a sterile filter paper moistened with sterile water. Two sterile glass slides were placed in the middle of each Petri dish to keep the leaves away from water. A 5-mm plug harvested from 3-day-old cultures of *Rhizoctonia* spp. on PDA was placed at the center of each leaf or leaf piece and the Petri dishes were incubated at 28°C.

After four days of incubation, disease severity was scored based on the following disease scale: 0 = no symptoms observed; 1 = lesions covered less than 25% of leaf surface; 2 = lesions covered 25–50% of leaf surface; 3 = lesions covered 50–75% of leaf surface; 4 = lesions covered more than 75% of leaf surface or dead leaf. The experiment had a completely randomized design. Each treatment consisted of 12 leaves or leaf pieces equally divided into two boxes or Petri dishes.

In vitro pathogenic potential of *Rhizoctonia* spp. on seedlings of white cabbage and Chinese cabbage

The same nine *Rhizoctonia* isolates were studied for their *in vitro* pathogenic potential using the method described by Keijer et al. [28]. Six surface-sterilized seeds of white cabbage (*B. oleracea* cv. TN180) or Chinese cabbage (*B. chinensis* cv. Elton) were germinated on Gamborg B5 medium (Gamborg B5 medium including vitamins; Duchefa) in a square Petri dish. Two mycelial disks (5 mm in diameter) from 3-day-old *Rhizoctonia* cultures grown on PDA were placed between seeds. In the control dishes, sterile PDA discs were used for inoculation. The Petri dishes were incubated at 22°C in the dark for two days for seed germination. Then, the second halves of the Petri dishes were covered with aluminum foil to protect the roots from light and placed in an

Table 2. Characterization of *Rhizoctonia* isolates collected from diseased *Brassica* crops grown in Vietnam by sequencing the ITS-region.

AG/Subgroup	Host plant	Isolate[ab]	Genbank accession numbers
1-IA	*B. parachinensis* (Chinese flowering cabbage)	STST03-1, STST03-3, STST03-4, STST04-2, **STMX04-1**, **STMX04-2**, **STMX04-3**, **STMX04-4**, STMX04-5	KF907702, KF907703, KF907704, KF907705
	B. juncea (Mustard cabbage)	DNBH01-1, DNBH01-2, DNBH01-3, DNBH02-2, DNBH02-3	
		HNGL01-1, HNGL01-2, **HNGL01-3**	KF907706
		STST02-1, STST02-2	
		VLBT01-1, VLBT01-2, VLBT01-3, VLBT01-4	
	B. chinensis (Pak choi)	**STMX01-1**, STMX01-2, **STMX01-4**, STMX01-5, STMX02-1, STMX02-2, **STMX02-3**, **STMX03-1**, **STMX03-2**, STMX03-3	KF907707, KF907708, KF907709, KF907710, KF907711
	B. oleraceae (Turnip cabbage)	HNDD01-1, HNDD01-2, **HNDD01-3**	KF907712
	B. oleraceae (White cabbage)	CTCR01-1, CTCR01-2, **CTCR01-3**, **CTCR02-1**, **CTCR02-2**, CTCR02-3, CTCR03-1, CTCR03-2	KF907713, KF907714, KF907715
1-IB	*B. chinensis* (Chinese cabbage)	LDDT03-1	
	B. oleraceae (Broccoli)	LDDL01-1, **LDDL01-2**, LDDL01-3, LDDL01-4	KF907716
	B. oleraceae (White cabbage)	**LDDL04-1**, LDDL04-2, LDDL04-3, LDDL04-4, LDDL04-5, **LDDL05-1**, LDDL05-2, **LDDL05-3**	KF907717, KF907718, KF907719
1-ID	*B. parachinensis* (Chinese flowering cabbage)	**DNBH03-1**, **DNBH03-2**, **DNBH03-3**, DNBH03-4, DNBH03-5, DNBH05-1-1, DNBH05-1-3, DNBH05-2-2, **DNBH05-3-1**, **DNBH05-4**	KF907720, KF907721, KF907722, KF907723, KF907724
		STST04-1, STST04-3	
	B. juncea (Mustard cabbage)	HGPH01-1, HGPH01-2, **HGPH01-3**, HGPH01-4	KF907725
		STST02-3	
1-IG	*B. parachinensis* (Chinese flowering cabbage)	**DNBH05-1-2**	KF907730
2-2	*B. parachinensis* (Chinese flowering cabbage)	**HNTT01-1**	KF907726
	B. oleraceae (Turnip cabbage)	**HNDA01-1**, HNDA01-2, **HNDA01-3**, **HNDA01-4**	KF907727, KF907728, KF907729
4-HGI	*B. chinensis* (Chinese cabbage)	LDDT01-1, LDDT01-2, LDDL02-2	KF907731
	B. parachinensis (Chinese flowering cabbage)	DNBH05-2-1, DNBH05-3-2	
		STST01-1, **STST03-2**	KF907732
	B. juncea (Mustard cabbage)	DNBH02-1	
	B. oleraceae (Turnip cabbage)	**HNDD01-4**	KF907733
	B. oleraceae (White cabbage)	LDDT02-1, LDDT02-2, LDDT02-3	
7	*B. oleraceae* (Turnip cabbage)	**HNDA02-1**	KF907734
A	*B. chinensis* (Pak choi)	STMX01-3	
	B. oleraceae (White cabbage)	**LDDL03-1**, LDDL03-2	KF907735
Fc	*B. chinensis* (Chinese cabbage)	**LDDL02-1**	KF907736

[a]The first two letters represent provinces in which the samples were collected (i.e. CT: Can Tho, VL: Vinh Long, HG: Hau Giang, ST: Soc Trang, DN: Dong Nai, LD: Lam Dong and HN: Ha Noi).
[b]Isolates in bold (unique sequences) are submitted to Genbank.

upright position in a growth chamber (22°C, 12 h light). The disease severity was recorded for root and hypocotyl or for leaves after six days of incubation using the following disease scale: 0 = healthy, no symptoms; 1 = lesions covering less than 25% of the root, hypocotyl or leaf surface; 2 = lesions covering between 25% and 50% of the root, hypocotyl or leaf surface; 3 = wilted plant with lesions covering between 50% and 75% of the root, hypocotyl or leaf surface; 4 = lesions covering more than 75% of root,

Table 3. Multinucleate and binucleate *Rhizoctonia* isolates derived from Genbank included in the phylogenetic analysis for comparison.

AG/Subgroup	Isolate	Host plant	Origin	Genbank accession number	Reference
1-IA	L31-1, L66-1, L73, L59, L38, L52, L62-1	Rice	Vietnam	EF206342, EF429208, EF429211, EF429212, EF429210, EF429209, EF429207	unpublished
	RM61	Water spinach	Vietnam	EF429216	unpublished
	LB71	Water hyacinth	Vietnam	EF429215	unpublished
	DP38	Peanut	Vietnam	EF429214	unpublished
	CLV72-2	Barnyard grass	Vietnam	EF429213	unpublished
	BV71-2, BV61-2, BV50-1	Cotton	Vietnam	EF429206, EF429205, EF206341	unpublished
	CC72	Bermuda grass	Vietnam	EF429204	unpublished
	B34-1	Corn	Vietnam	EF429203	unpublished
1-IG	RMPG28	Chickpea	India	JF701750	[29]
1-ID	BV62-1, BV61-6, BV61-5, BV61-4, BV61-1	Cotton	Vietnam	EF197803, EF197804, EF197802, EF197801, EF197800	unpublished
	SR61, SR650	Durian	Vietnam	EF197798, EF197797	[30]
	B61-1	Corn	Vietnam	EF197796	unpublished
	CCD61-1	Sugar beet	Vietnam	EF197799	unpublished
4-HGI	XL4	Cauliflower	Vietnam	EF203247	unpublished
	CB63, CB34-2	Cabbage	Vietnam	EF203251, EF203245	unpublished
	CP50-2	Coffee	Vietnam	EF203250	unpublished
	BV68-1, BV68-2	Cotton	Vietnam	EF203249, EF203248	unpublished
	KT63-1	Potato	Vietnam	EF203246	unpublished
Fc	BS-YT-06-5-14, YT, BS-J-06-6-3, DL-jiang-06-2-4, DL-YT-06-4-10, DL-YT-06-4-9, DL-YT-06-3-4	Taro, Ginger	China	HM623619, HM623631, HM623615, HM623622, HM623625, HM623624, HM623623	unpublished

AG/Subgroup	Isolate	Disease Index						
		White cabbage	Chinese cabbage	Pak choi	Mustard cabbage	Chinese flowering cabbage	Rice	Water spinach
Control		0.00 a	0.00 a	0.00 a	0.00 a	0.00 a	0.00 a	0.00 a
1-IA	STMX04-2	4.00 d	2.63 d	1.79 d	1.79 c	2.17 e	3.92 d	3.21 f
1-IB	LDDL05-3	3.83 c	0.00 a	4.00 e	3.92 e	3.96 g	1.96 c	2.83 ef
1-ID	DNBH05-4	4.00 d	0.00 a	0.42 b	1.54 c	1.67 d	2.38 c	3.54 f
1-IG	DNBH05-1-2	4.00 d	1.00 b	0.63 bc	1.04 b	1.04 c	1.04 b	2.29 de
2-2	HNDA01-1	3.83 c	0.00 a	0.33 b	0.08 a	0.04 ab	1.29 b	1.75 cd
4-HGI	STST01-1	3.88 cd	1.54 c	2.13 d	3.88 e	3.71 f	2.33 c	2.67 e
7	HNDA02-1	3.13 b	0.00 a	0.00 a	0.13 a	0.17 b	2.29 c	0.46 b
A	DNBH04-1	0.00 a	0.00 a	0.00 a	0.00 a	0.00 a	0.00 a	0.00 a
Fc	LDDL02-1	3.38 b	0.71 b	1.17 c	2.58 d	1.17 c	0.00 a	1.38 c

Figure 2. Aggressiveness of *Rhizoctonia* isolates towards detached leaves of white cabbage, Chinese cabbage, pak choi, mustard cabbage, Chinese flower cabbage, rice and water spinach. Leaves were scored using a scale ranging from 0 (no disease symptoms) to 4 (lesions covered more than 75% of leaf surface or dead leaf). For rapid visual evaluation of the data, a coloring scale with green ($0<DI\leq1$), yellow ($1<DI\leq2$), orange ($2<DI\leq3$) and red ($3<DI\leq4$) was used. The experiment was conducted twice and each treatment consisted of 12 leaves or leaf pieces. The data of the two experiments were pooled before Mann-Whitney comparisons were applied at $p=0.05$. Within columns, disease severities followed by the same letter are not significantly different.

AG/Subgroup	Isolate	Disease Index			
		Root		Leaf	
		White cabbage	Chinese cabbage	White cabbage	Chinese cabbage
Control		0.00 a	0.00 a	0.00 a	0.00 a
1-IA	STMX04-2	3.63 d	3.08 d	3.42 d	2.71 cd
1-IB	LDDL05-3	4.00 e	3.75 ef	4.00 e	3.54 d
1-ID	DNBH05-4	1.88 bc	1.04 b	2.17 c	0.63 b
1-IG	DNBH05-1-2	4.00 e	4.00 g	4.00 e	4.00 e
2-2	HNDA01-1	4.00 e	3.96 fg	4.00 e	3.21 d
4-HGI	STST01-1	4.00 e	4.00 g	4.00 e	4.00 e
7	HNDA02-1	2.50 c	1.83 c	2.29 c	1.58 c
A	DNBH04-1	1.17 b	1.42 bc	0.29 b	0.00 a
Fc	LDDL02-1	3.75 de	3.42 de	3.83 de	3.50 d

Figure 3. Pathogenic potential of *Rhizoctonia* isolates on seedlings of white cabbage and Chinese cabbage in *in vitro* bio-assays. Disease severity was assessed on a scale ranging from 0 (no symptoms) to 4 (lesions covering more than 75% of root, hypocotyl or leaf surface or dead plant). For rapid visual evaluation of the data, a coloring scale with green ($0<DI\leq1$), yellow ($1<DI\leq2$), orange ($2<DI\leq3$) and red ($3<DI\leq4$) was used. The experiment was done twice with 12 seedlings maintained in two square Petri plates for one treatment. The data of the two experiments were pooled before Mann-Whitney comparisons were applied at $p=0.05$. Within columns, disease severities followed by the same letter are not significantly different.

hypocotyl or leaf surface or dead plant. In this test, a complete randomized design was applied with two Petri dishes (six seedlings each) per treatment and this experiment was done twice.

In vivo pathogenic potential of *Rhizoctonia* spp. on roots and hypocotyls of white cabbage and Chinese cabbage

Surface-sterilized seeds of white cabbage and Chinese cabbage were germinated on wet filter paper in Petri dishes at 22°C one day before sowing into 600 g of soil substrate. Four days after sowing, each perforated plastic box ($22\times15\times6$ cm) with six

AG/Subgroup	Isolate	Disease Index	
		White cabbage	Chinese cabbage
Control		0.00 a	0.00 a
1-IA	STMX04-2	0.27 b	0.21 ab
1-IB	LDDL05-3	1.57 c	0.29 bc
1-ID	DNBH05-4	0.00 a	0.08 ab
1-IG	DNBH05-1-2	0.00 a	0.13 ab
2-2	HNDA01-1	3.50 e	1.57 d
4-HGI	STST01-1	3.63 e	3.08 e
7	HNDA02-1	0.26 b	0.79 c
A	DNBH04-1	0.00 a	0.00 a
Fc	LDDL02-1	2.33 d	2.96 e

Figure 4. Pathogenic potential of *Rhizoctonia* isolates on roots and hypocotyls of white cabbage and Chinese cabbage in *in vivo* experiment. Disease severity on roots was assessed on a scale ranging from 0 (no symptoms) to 4 (seedling dead). For rapid visual evaluation of the data, a coloring scale with green ($0<DI\leq1$), yellow ($1<DI\leq2$), orange ($2<DI\leq3$) and red ($3<DI\leq4$) was used. The experiment was performed twice with 12 seedlings cultivated in two plastic boxes per treatment. The data of the two experiments were pooled before Mann-Whitney comparisons were applied at $p=0.05$. Within columns, disease severities followed by the same letter are not significantly different.

seedlings was inoculated by placing a row of 12 *Rhizoctonia*-colonized wheat kernels in the middle of the box. The kernels used for inoculation had equivalent sizes and were produced as described previously. Control seedlings were similarly treated with sterile wheat kernels. All plants were incubated at 22°C. Disease severity on root and hypocotyl was evaluated 14 days after inoculation using the same disease scale described for the *in vitro* experiment. A completely randomized design was used with 12 seedlings cultivated in two experimental boxes per treatment and this test was performed twice with the same nine isolates that were used in the previous tests.

Statistical analysis

The severity of *Rhizoctonia* diseases on roots and leaves of *Brassica* seedlings are presented in Figures 2–4 and Tables S3–S5 in File S1 as Disease index (DI). DI was calculated using the following formula:

$$DI = \frac{\sum (\text{Disease class} \times \text{number of plants within that class})}{\text{Total number of plants within treatment}}$$

Pathogenicity data for the two experiments with nine isolates of nine different AGs were always very similar and no significant interaction was found between the experiments. Therefore, statistical analysis was done on pooled data for the different repeats. The non-parametric Kruskal-Wallis test for k independent samples was used, after which pair-wise comparisons were performed for all treatments using Mann-Whitney tests at a confidence level of $p=0.05$.

To determine the correlation between the distributions of AGs and sampling locations and between the distributions of AGs and sampled *Brassica* spp., contingency tables were constructed using Excel. Then, potential significant differences between the variables were revealed with Fisher's Exact Tests. All statistical analyses were conducted in SPSS 22.0 (SPSSinc, Illinois, USA).

Table 4. Pairwise sequence similarities of unknown isolates LDDL02-1 and DNBH05-1-2 to all known AGs from the curated database in Table S2 in File S2.

	LDDL02-1	DNBH05-1-2
AG 1-IA	0.88	0.92
AG 1-IB	0.84–0.87	0.84–0.87
AG 1-IC	0.89–0.90	0.89–0.90
AG 1-ID	0.85–0.86	0.85
AG 1-IE	0.89	0.94
AG 1-IF	0.84	0.85
AG 2-1	0.87–0.90	0.82–0.85
AG 2-2	0.86–0.87	0.84–0.85
AG 2-3	0.89–0.90	0.85
AG 3	0.89–0.90	0.84–0.85
AG 4	0.86–0.88	0.85–0.88
AG 5	0.91	0.86–0.87
AG 6	0.89–0.94	0.88–0.90
AG 7	0.91	0.89
AG 8	0.92–0.93	0.89–0.90
AG 9	0.9	0.86
AG 10	0.89–0.90	0.84–0.85
AG 11	0.88–0.89	0.84
AG 12	0.89–0.90	0.87–0.88
AG 13	0.90	0.90
AG 2-BI	0.85–0.86	0.81–0.82
AG-A	0.84	0.84
AG-K	0.84	0.84
AG-Bb	0.79–0.80	0.78–0.79
AG-Q	0.80	0.79
AG-Bo	0.83	0.83
AG-Ba	0.83–0.84	0.83–0.84
AG-C	0.80–0.82	0.82
AG-H	0.81–0.82	0.81
AG-I	0.81	0.80–0.81
AG-D	0.78–0.81	0.77–0.79
AG-G	0.85	0.85
AG-L	0.85	0.85–0.86
AG-O	0.86	0.86
AG-Fb	0.93	0.89
AG-P	0.86–0.89	0.86–0.90
AG-R	0.88	0.85
AG-S	0.88	0.88
AG-Fa	0.91	0.91
AG-E	0.90–0.91	0.89
UNR1	0.82–0.83	0.84
UNR2	0.80	0.82
AG-N	0.67	0.66
W. circinata	0.66–0.67	0.64–0.66

LDDL02-1 shows most similarity to AG 6 and AG-Fb. DNBH05-1-2 shows highest pairwise sequence similarity to AG 1-IA and AG 1-IE.

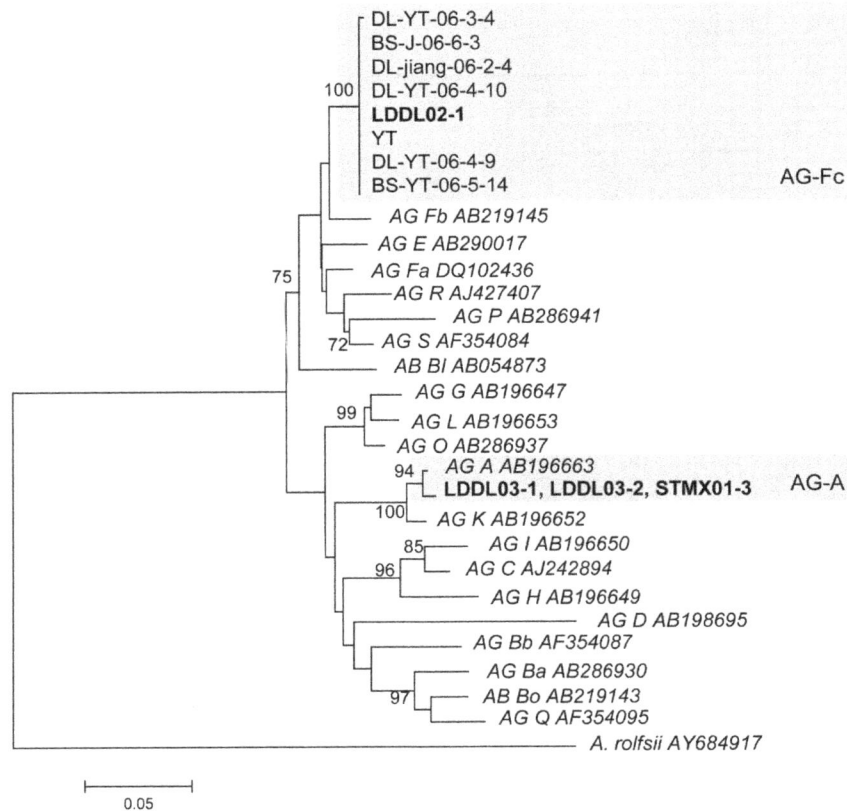

Figure 5. rDNA-ITS phylogeny of binucleate *Rhizoctonia* spp. sampled from *Brassica* spp. in Vietnam. Neighbour joining tree derived from the alignment of 31 binucleate *Rhizoctonia* isolates and the outgroup *Athelia rolfsii* (AY684917). Isolates in bold are the 4 isolates derived from *Brassica* spp. in Vietnam during this study. For each known binucleate *Rhizoctonia* AG, a representative isolate (in italics) from the curated database (Table S2 in File S2) is included. Bootstraps are only given for those branches with bootstrap support higher than 70. The tree was made using only isolates with unique sequences. Isolates with identical sequences were added afterwards on the same line.

Results

Molecular characterization and phylogenetic analysis of *Rhizoctonia* isolates

A total of 142 *Brassica* plants with *Rhizoctonia*-like symptoms were sampled in various important vegetable producing regions in Vietnam (see Figure 1). Of the 97 *Rhizoctonia* isolates recovered, only four were binucleate with two nuclei per hyphal cell. The other 93 isolates had multinucleate cells (data not shown).

Analysis of the rDNA-ITS region using the BLASTn tool (against Genbank and against the curated database) revealed that three binucleate *Rhizoctonia* isolates belonged to AG-A while the fourth isolate (LDDL02-1) could not be assigned to any known AG.

Pairwise sequence similarity scores of isolate LDDL02-1 against all isolates in the curated database (representing all known AGs) were determined. Highest pairwise sequence similarities were found with isolates of multinucleate AG 6 (94%) and binucleate AG-Fb (93%) (Table 4). When blasting to Genbank, several isolates from taro and ginger from Yunnan Province in China were found to be nearly identical to isolate LDDL02-1.

From the binucleate phylogenetic tree (Figure 5), it is clear that the unknown isolate LDDL02-1 clusters together with the isolates from Yunnan province, forming a clade with high bootstrap support that is different from any known binucleate AG, with the closest related AG being AG-Fb. So far, nothing has been

published about the Chinese isolates. Pairwise sequence similarity within AG-F is 90–100% [2], and the closest related AGs are AG-Fb (93%) and AG-Fa (91%). Therefore, we propose to assign these isolates as a new AG-F subclade, namely AG-Fc.

In the multinucleate tree, all our Vietnamese isolates form clades with high bootstrap supports together with representative isolates from the curated database (Figure 6). The largest group belonged to AG 1-IA (44 isolates), followed by AG 1-ID (17 isolates), AG 1-IB (13 isolates) and AG 4-HGI (12 isolates). Five isolates belonged to AG 2-2. AG 2-2 is further divided into four groups: AG 2-2 IV, AG 2-2 LP, AG 2-2 IIIB and AG 2-2 WB. The similarity scores between our isolates and AG 2-2 IIIB and AG 2-2 LP were comparable (between 97 and 98%). Also, our isolates show 99% similarity with isolate Barranca (DQ452119), which is assigned as AG 2-2 WB by Godoy-Lutz et al. [25]. However, with other isolates from AG 2-2 WB (DQ452111-114), pairwise sequence similarity is lower (between 96 and 97%). Therefore, we decided that it is impossible at this point to assign our isolates to any of the AG 2-2 phylogenetic subgroups. One isolate belonged to AG 7 and the last isolate (DNBH05-1-2) could not be assigned to any of the known multinucleate AG groups in our reference database. The isolate showed the highest pairwise sequence similarity with AG 1-IE (94%, see Table 4). A BLASTn search on Genbank identified an isolate from chickpea in India (RMPG28, [29]) assigned to AG 2-3 to be >99% similar (1 bp substitution) to DNBH05-1-2. However, when comparing these

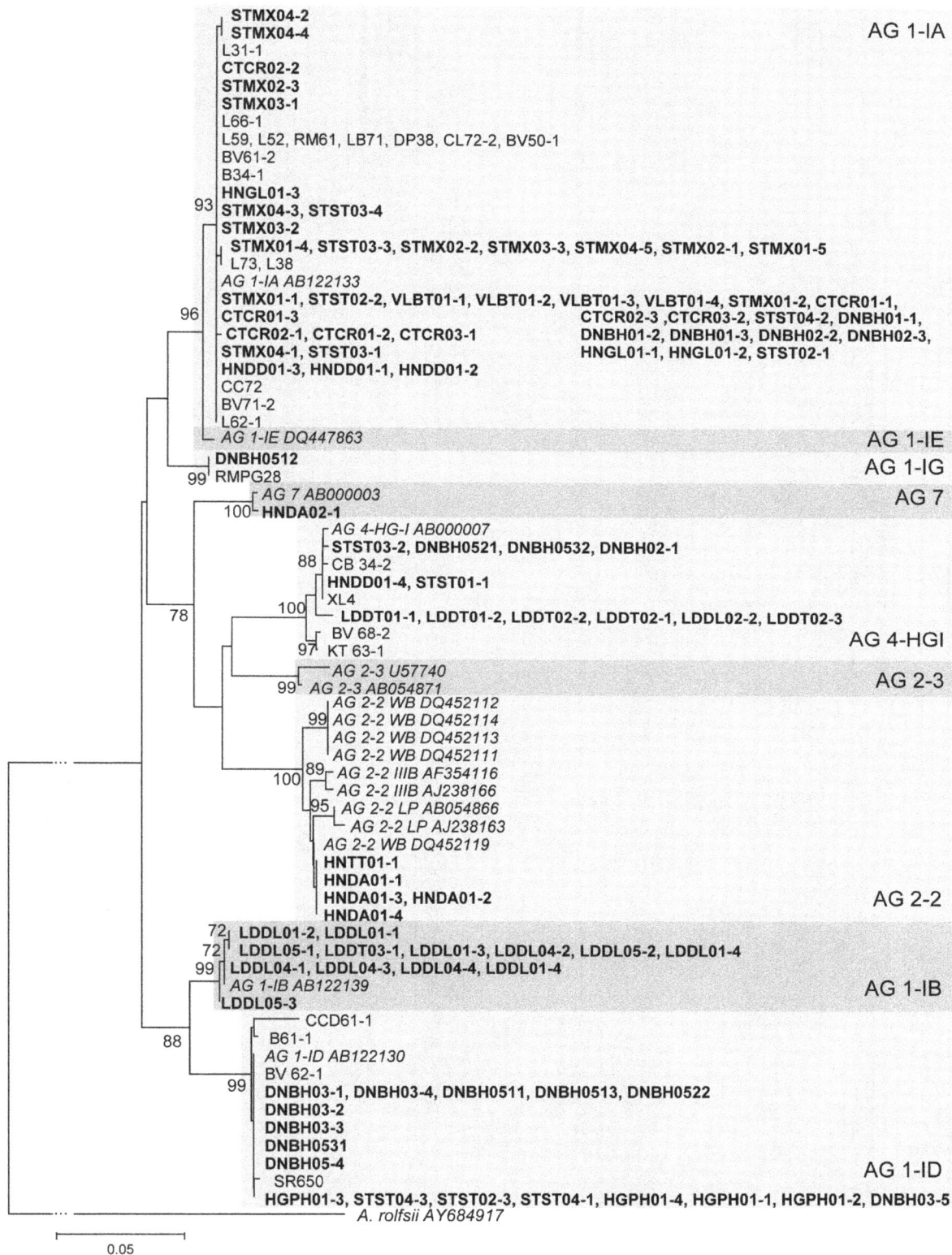

Figure 6. rDNA-ITS phylogeny of multinucleate *Rhizoctonia* spp. sampled from *Brassica* spp. in Vietnam. Neighbour joining tree derived from the alignment of 128 multinucleate *Rhizoctonia* isolates and the outgroup *Athelia rolfsii* (AY684917). Isolates in bold are the isolates derived from *Brassica* spp. in Vietnam during this study. For each of the multinucleate *Rhizoctonia* AG subgroups present in our sampling, representative isolates (in italics) from the curated database (Table S2 in File S2) are included. Bootstraps are only given for those branches with bootstrap support higher than 70. The tree was made using only isolates with unique sequences. Isolates with identical sequences were added afterwards on the same line. Only half of the length of the outgroup branch is shown to increase clarity.

Table 5. Contingency table with observed and expected frequencies of anastomosis groups (AGs) of *Rhizoctonia* spp. obtained from *Brassica* fields in different provinces of Vietnam.

Province	AG/Subset									Total/province
	Multinucleate *Rhizoctonia*							Binucleate *Rhizoctonia*		
	1-IA	1-IB	1-ID	2-2	1-IG	4-HGI	7	A	Fc	
Ha Noi	6 (5.90)	0 (1.74)	0 (2.28)	5 (0.67)	0 (0.13)	1 (1.61)	1 (0.13)	0 (0.40)	0 (0.13)	13
Lam Dong	0 (9.98)*	13 (2.95)*	0 (3.86)	0 (1.13)	0 (0.45)	6 (2.72)*	0 (0.23)	2 (0.68)	1 (0.23)	22
Dong Nai	5 (8.62)	0 (2.55)	10 (3.33)*	0 (0.98)	1 (0.20)	3 (2.35)	0 (0.20)	0 (0.59)	0 (0.20)	19
Vinh Long	4 (1.81)	0 (0.54)	0 (0.70)	0 (0.21)	0 (0.04)	0 (0.49)	0 (0.04)	0 (0.12)	0 (0.04)	4
Can Tho	8 (3.63)	0 (1.07)	0 (1.40)	0 (0.41)	0 (0.08)	0 (0.99)	0 (0.08)	0 (0.25)	0 (0.08)	8
Hau Giang	0 (1.81)	0 (0.54)	4 (0.70)*	0 (0.21)	0 (0.04)	0 (0.49)	0 (0.04)	0 (0.12)	0 (0.04)	4
Soc Trang	21 (12.25)*	0 (3.62)	3 (4.73)	0 (1.39)	0 (0.28)	2 (3.34)	0 (0.28)	1 (0.84)	0 (0.28)	27
Total/AG	44	13	17	5	1	12	1	3	1	97

Data show actual numbers of isolates collected among the different provinces. According to Fisher's exact test (p = 0.05), the AGs found are related to the sampling locations. An asterisk* indicates significant differences between observed and expected numbers. Values in parentheses represent the expected numbers.

Table 6. Contingency table with observed and expected frequencies of anastomosis groups (AGs) of *Rhizoctonia* spp. obtained from field-grown *Brassica* crops in Vietnam.

Host plant	AG/Subset[a]									Total/host plant
	Multinucleate *Rhizoctonia*							Binucleate *Rhizoctonia*		
	1-IA	1-IB	1-ID	2-2	1-IG	4-HGI	7	A	Fc	
Mustard cabbage	14 (9.07)	0 (2.68)	5 (3.51)	0 (1.03)	0 (0.21)	1 (2.47)	0 (0.21)	1 (0.62)	0 (0.21)	20
White cabbage	8 (9.53)	8 (2.81)*	0 (3.68)	0 (1.08)	0 (0.22)	3 (2.60)	0 (0.22)	2 (0.65)	0 (0.22)	21
Chinese flowering cabbage	9 (12.25)	0 (3.62)	12 (4.73)*	1 (1.39)	1 (0.28)	4 (3.34)	0 (0.28)	0 (0.84)	0 (0.28)	27
Chinese cabbage	0 (2.27)	1 (0.67)	0 (0.88)	0 (0.26)	0 (0.05)	3 (0.62)	0 (0.05)	0 (0.15)	1 (0.05)	5
Pak choi	10 (4.99)	0 (1.47)	0 (1.93)	0 (0.57)	0 (0.11)	0 (1.36)	0 (0.11)	1 (0.34)	0 (0.11)	11
Broccoli	0 (1.81)	4 (0.54)*	0 (0.70)	0 (0.21)	0 (0.04)	0 (0.49)	0 (0.04)	0 (0.12)	0 (0.04)	4
Turnip cabbage	3 (4.08)	0 (1.21)	0 (1.58)	4 (0.46)*	0 (0.09)	1 (1.11)	1 (0.09)	0 (0.28)	0 (0.09)	9
Total/AG	44	13	17	5	1	12	1	3	1	97

Data show actual numbers of isolates collected among the different *Brassica* crops. According to Fisher's exact test (p = 0.05), the AGs found are related to the crops. An asterisk* indicates significant differences between observed and expected numbers. Values in parentheses represent the expected numbers.

isolates to the AG 2–3 isolates in the curated database, we found low pairwise sequence similarity (85%) (Table 4). Also in the phylogenetic tree (see Figure 6), isolate DNBH05-1-2 clusters together with isolate RMPG28, but not with the representative AG 2–3 isolates from the curated database, indicating that isolate RMPG28 is wrongly designated as AG 2–3 in Genbank. Hence, these isolates represent a new subgroup of AG 1 and we propose the name AG 1-IG.

For most of the AGs we detected in Vietnam there was very little variation in ITS sequence, except for AG 4-HGI where six variable positions were found between the isolates from Lam Dong and the isolates from the other three provinces where this AG was detected (Soc Trang, Dong Nai and Ha Noi).

The AG 1-IA isolates showed identical or nearly identical (max. 4 SNPs) sequences to AG 1-IA isolates previously isolated from rice, water spinach, water hyacinth and other crops in Vietnam (Table 3). The AG 1-ID isolates showed high similarity (identical sequence or one SNP) to isolates detected on coffee, cotton, durian [30], corn and sugar beet (Table 3) in Vietnam.

Relationship between the AGs found and sampling locations and between AGs found and *Brassica* species

There seems to be a relationship between the AGs present and the sampling areas (Figure 1 and Table 5). Thirty three out of 44 isolates belonging to AG 1-IA were recovered from samples collected from Soc Trang, Can Tho and Vinh Long in the Mekong River delta, the main rice production region of Vietnam. On the other hand, no AG 1-IA isolates were detected in Lam Dong although the expected frequency of this group in this province was high (9.98). The observed numbers were also significantly higher than the expected numbers for AG 1-IB and AG 4-HGI in Lam Dong as well as for AG 1-ID in Hau Giang and Dong Nai.

The relationship between the occurrence of different AGs and the host plants is presented in Table 6. Significant AG-host correlations were identified in the following combinations: AG 1-IB and white cabbage, AG 1-IB and broccoli, AG 1-ID and Chinese flowering cabbage, and AG 2-2 and turnip cabbage.

Aggressiveness of *Rhizoctonia* isolates towards detached leaves

Nine *Rhizoctonia* isolates, randomly selected from each identified AG, were tested for their pathogenicity on detached leaves. Due to practical reasons, experiments with *Brassica* crops were conducted in a growth chamber specifically built for *Brassica* spp. (22°C, RH = 60%, 12 h photoperiod) and experiments with rice and water spinach were done in a growth chamber specifically built for rice (28°C, RH = 60%, 16 h photoperiod). Although the temperature and humidity in these chambers are slightly lower than those of Vietnam, they are still suitable for the growth of host plants and *Rhizoctonia* isolates involved in our study.

As shown in Figure 2, leaves of white cabbage, Chinese cabbage, pak choi, mustard cabbage, Chinese flowering cabbage, rice, and water spinach responded differently to the infection of *Rhizoctonia* spp. although each host was affected by at least four AGs. For each plant species, a wide variation in symptom severity induced by different AGs was observed and the only AG that could not cause disease on any of the plants tested was AG-A. White cabbage leaves were most severely infected by all AGs (except AG-A) with a DI varying from 3.13 to 4.00. Compared to other hosts, Chinese cabbage appeared to be most resistant to *Rhizoctonia* isolates. No disease symptoms were observed on Chinese cabbage leaves challenged with AG 1-IB, AG 1-ID, AG

2-2 and AG 7, while other AGs were weak to moderately aggressive (DI varied from 0.71 in AG-Fc to 2.63 in AG 1-IA). Inoculation of pak choi with AG 1-IB resulted in a complete decay of all leaf discs (DI = 4.00). For mustard cabbage and Chinese flowering cabbage, a strong disease pattern was obtained on leaves confronted with AG 1-IB (DI≥3.92) and AG 4-HGI (DI≥3.71). Rice leaves were very susceptible to AG 1-IA and they were severely destroyed within four days of inoculation (DI = 3.92). Disease induced by other *R. solani* isolates on rice was also significantly different from the control although the two binucleate *Rhizoctonia* isolates tested were not pathogenic on rice. For water spinach, a wide variation in aggressiveness of *R. solani* isolates was observed, resulting in DI values ranging from 0.46 to 3.54. Large lesions (DI≥3.21) developed on water spinach leaves inoculated with AG 1-IA and AG 1-ID isolates. The difference in pathogenicity of these AGs towards the different *Brassica* species was confirmed when additional isolates from AGs 1-IA, 1-IB, 1-ID, 2-2, 4-HGI and A were tested in a detached leaf assay on the same series of plants (see supplementary information, Table S3 in File S1).

In vitro pathogenic potential of *Rhizoctonia* spp. isolates on seedlings of white cabbage and Chinese cabbage

The results of the detached leaf assay revealed that white cabbage leaves were remarkably more susceptible than other cabbage species, while only some AGs could attack the leaves of Chinese cabbage. Therefore, these two plant species were selected as hosts for *in vitro* and *in vivo* pathogenicity tests using the same nine *Rhizoctonia* isolates as mentioned above. Under *in vitro* conditions (Figure 3), all AGs could cause disease and symptoms were observed on both roots and leaves. Isolates of AG 1-IA, AG 1-IB, AG 2-2, AG 1-IG, AG 4-HGI and AG-Fc were most aggressive towards these hosts (for white cabbage: DI on roots ≥ 3.63 and DI on leaves ≥3.42; for Chinese cabbage: DI on roots ≥ 3.08 and DI on leaves ≥2.71). Fewer symptoms were recorded on seedlings inoculated with isolates belonging to AG 1-ID, AG 7 and AG-A. The virulence of AG 1-IA, AG 1-IB, AG 2-2 and AG 4-HGI towards white cabbage and Chinese cabbage was confirmed when *in vitro* assays were repeated with additional isolates of these AGs (Table S4 in File S1). Towards white cabbage, DI on roots varied from 2.17 to 4.00 and DI on leaves fluctuated between 2.33 and 4.00. DI on roots and leaves of Chinese cabbage ranged from 2.08 to 4.00 and from 1.92 to 4.00, respectively.

In vivo pathogenic potential of *Rhizoctonia* spp. isolates on roots and hypocotyls of white cabbage and Chinese cabbage

The results displayed in Figure 4 demonstrate that severe *Rhizoctonia*-induced damage on white cabbage roots could only be seen for AG 2-2 (DI = 3.50) and AG 4-HGI (DI = 3.63). Moderate infection (DI = 2.33) was observed in response to AG-Fc. Towards Chinese cabbage, severe disease symptoms were incited by AG 4-HGI (DI = 3.08) and AG-Fc (DI = 2.96). No symptoms were detected on roots and hypocotyls of seedlings challenged with AG-A. In addition, the inoculation of isolates belonging to AG 1-IA, AG 1-IB, AG 1-ID, AG 1-IG and AG 7 did not result in the formation of large lesions on roots of white cabbage and Chinese cabbage seedlings (DI≤1.57). Data obtained from the tests conducted with additional *Rhizoctonia* isolates belonging to AGs 1-IA, 1-IB, 1-ID, 2-2, 4-HGI and AG-A are presented in Table S5 in File S1. These data confirm the high aggressiveness of AG 2-2 on white cabbage (DI = 4.00) and that of AG 4-HGI on both white cabbage and Chinese cabbage (DI≥3.50).

Discussion

Rhizoctonia is an important fungal 'form genus' occurring worldwide and including many important plant pathogenic strains as well as mycorrhizal fungi and hypovirulent or avirulent strains among which there are strains that are capable to protect plants against pathogenic *Rhizoctonia* and other pathogens as well as increase plant growth. Plants can be infected by different *Rhizoctonia* AGs from the time of sowing resulting in the development of both foliar and root diseases. This is the first time *Rhizoctonia* species that attack field-grown *Brassica* crops in Vietnam were isolated and characterized. Ninety seven isolates of *Rhizoctonia* were recovered from symptomatic plant tissues of seven brassicaceous hosts (mustard cabbage, white cabbage, Chinese flowering cabbage, pak choi, turnip cabbage, Chinese cabbage and broccoli). Of all the isolates collected, 4% were binucleate *Rhizoctonia* and 96% were multinucleate *Rhizoctonia*. Molecular characterization by sequencing of the rDNA-ITS region showed that the binucleate isolates found belonged to AG-A (3 isolates) and an unknown AG introduced here as AG-Fc (1 isolate). Comparison to a curated sequence database of all known *Rhizoctonia* multinucleate, binucleate and uninucleate AGs did not reveal high homology of this isolate to a known AG. Pairwise sequence similarities to all known multinucleate, binucleate and uninucleate *Rhizoctonia* AGs showed highest similarity to AG-Fb, but also to AG 6, which is a multinucleate AG. Close relationships of some binucleate groups with multinucleate groups has been previously noted by Sharon et al. [2] who did a combined ITS sequence analysis of multinucleate, binucleate and uninucleate groups. These authors stated that the clustering of binucleate and uninucleate groups close to certain multinucleate clusters may indicate a possible evolutionary bridge between multinucleate and binucleate groups and our results support this hypothesis.

We conducted pathogenicity assays on detached leaves, *in vitro* seedlings and plants grown *in vivo*. In general, pathogenicity was highest on roots and leaves under *in vitro* conditions, which is probably due to the high humidity and the young age of the plants in this system. It should be mentioned that we only checked *in vivo* pathogenicity towards roots and hypocotyls, but it should be borne in mind that some of our isolates are mainly leaf pathogens.

Although AG-A isolates were obtained from symptomatic plants, they were unable to induce disease on detached leaves or on seedlings *in vivo*, and were only slightly pathogenic on cabbage seedlings under *in vitro* conditions. This low virulence may be due to the differences in humidity between our experimental conditions and the field situation in Vietnam. Alternatively, AG-A isolates might be avirulent, but since they grow very fast *in vitro*, they may have masked the presence of slower growing, virulent AGs. In contrast to AG-A, AG-Fc appeared to be moderately to highly virulent in all experiments. This also concurs with previous studies that some binucleate *Rhizoctonia* species are highly virulent [31–33], whereas others are weakly virulent or avirulent [34–36].

Among the multinucleate AGs, AG 1-IA was the dominant group, followed by AG 1-ID, AG 1-IB, AG 4-HGI, AG 2-2, AG 7 and an unknown AG that we introduced here as AG 1-IG. The occurrence of AG 1-IB and AG 4 on *Brassica* spp. is well known from previous studies. According to Pannecoucque et al. [22], AG 1-IB is one of the causal agents of wirestem in Belgian cauliflower fields. AG 1-IB isolates are also highly virulent on lettuce in Belgium [37]. In our study, AG 1-IB isolates were only recovered from Lam Dong, a province in the cool Central Highlands of Vietnam and the main lettuce production region of Vietnam, suggesting that the presence of this AG is associated with cool climates. The presence of AG 4 has also been reported on *Brassica oleracea* [21] and *B. rapa* subsp. *chinensis* [38] in China, *B. oleracea* in the UK [23], *B. napus* L. and *B. campestris* L. in Canada [39], and *B. oleracea* [18] and *B. napus* L. in the US [40]. AG 4-HGI occurred in most regions sampled. This AG has a wide host range due to its ability to adapt to temperature variation and cropping patterns [41]. AG 4 isolates are able to induce disease on all plant parts and in our pathogenicity trials, the highest disease ratings on cabbage in all bioassays were shown for the AG 4-HGI isolates. For white cabbage, severe disease symptoms were observed on seedlings inoculated with AG 2-2 and AG 4-HGI isolates *in vivo* or on detached leaves challenged with isolates of all AGs. Although *R. solani* AG 2-1 is considered the most dominant and damaging anastomosis group attacking *Brassica* spp. [22,23,42], isolates belonging to this group were not detected in our survey.

The predominance of *R. solani* isolates of AG 1-IA and also the presence of isolates belonging to AG 1-ID in our collection were not anticipated because these AGs have not been described on *Brassica* crops before. The presence of unusual AGs in our sampling is probably due to three aspects: (i) alternative hosts of *Rhizoctonia* present in the sampling locations; (ii) poor cultural practices and (iii) high temperature. AG 1-IA isolates were mainly recovered from samples collected in provinces of the Mekong delta including Vinh Long, Can Tho and Soc Trang. This is a low-lying coastal region of Vietnam, characterized by high temperature and humidity and prone to flooding every rainy season. Due to water availability and soil type, the agricultural production in this area is dominated by rice [43,44]. As previously reported, sheath blight, caused by *R. solani* AG 1-IA, is a major disease of rice cultivated in intensive production systems [45–47]. With the ability to float and to survive in water [48], sclerotia of AG 1-IA isolates may easily spread from the rice paddy fields to vegetable fields through irrigation or flood water. The recovery of AG 1-IA in the hot region located in the South of Vietnam is consistent with the findings of Harikrishnan and Yang [41] that temperature can influence growth rate, sclerotia production of *Rhizoctonia* spp. and the distribution of *Rhizoctonia* isolates belonging to different anastomosis groups. AG 1 is a high temperature group [8] and its vegetative growth as well as sclerotia production and survival are inhibited at low temperatures. The occurrence of AG 1-IA on *Brassica* crops is probably even increased due to farmers' lack of knowledge about crop protection. AG 1 (not specific) has been considered as one of the causal agents of foliar diseases on water hyacinth (*Eichhornia crassipes*) [49], water lettuce (*Pistia stratiotes*) and anchoring hyacinth (*E. azurea*) [50]. In this study, water spinach (*Ipomoea aquatic*) was found to be susceptible to the AG 1-IA isolates that we collected from *Brassica* spp. Vietnamese farmers commonly use these aquatic plants as cover materials in vegetable production and use irrigation water from sources where these plants are present, thus bringing the fungus from these alternative hosts to *Brassica* spp. (Figure 7). Isolates of AG 1-IA appeared to be very pathogenic towards leaves of cabbages but they could not induce severe disease on roots, especially under *in vivo* conditions. This finding is in agreement with results reported previously by Yang and Li [51] that AG 1-IA isolates have a tendency to attack aerial parts of plants. These data also support our hypothesis about the spread of AG 1-IA isolates from rice and water spinach to vegetables. In other words, rice and water spinach that are infected by *R. solani* could be an important source of inoculum that may contribute to the disease caused by AG 1-IA on *Brassica* spp.

In our assays AG 1-ID was mainly pathogenic on leaves of white cabbage. *Rhizoctonia* AG 1-ID was previously found to be

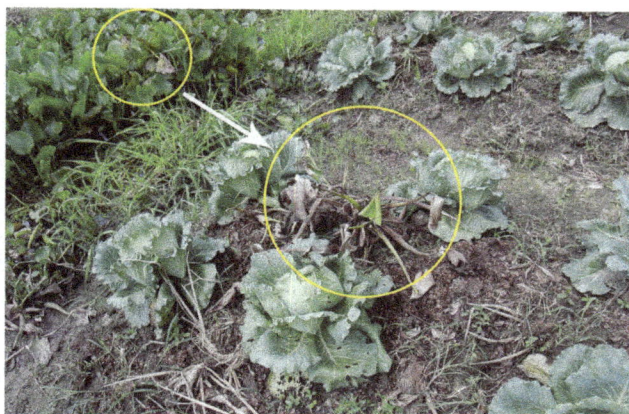

Figure 7. *Rhizoctonia*-**infected water hyacinth is introduced as cover material to a white cabbage field in Vietnam.** *Rhizoctonia*-infected water hyacinth is taken from a nearby water ditch and used as cover material on the white cabbage field. Via this common practice, Vietnamese farmers unintentionally introduce the *Rhizoctonia* fungus to their crops.

pathogenic on durian [30] and coffee [52]. Interestingly, ten out of 17 AG 1-ID isolates were collected from Dong Nai, a province where durian, coffee and cotton are widely cultivated [53], again suggesting a correlation between cropping patterns and *Rhizoctonia* distribution in Vietnam.

Collectively, it seems that the distribution of *Rhizoctonia* AGs in Vietnam is correlated with the cropping patterns and climatic conditions. However, it is difficult to draw a strong conclusion about the influence of cropping patterns and climatic conditions on the occurrence of *Rhizoctonia* spp. in Vietnam because the sampling regime for *R. solani* isolates comprises two variable parameters, namely the plant species infected by *R. solani* and the sampling site. Due to the variation in soil and climatic conditions, different *Brassica* species are grown in different geographic regions and there is a possibility that particular plant species might be more susceptible to infections by specific *R. solani* AGs than others. Therefore, another field survey with a systematic sampling regime needs to be conducted to confirm our hypothesis.

Our research also points towards the need to have good extension programs to improve the farmers' knowledge about crop protection. Additionally, knowing which AGs are responsible for *Rhizoctonia* diseases on *Brassica* spp. in Vietnam is an essential prerequisite for developing successful disease management strategies in this country.

Supporting Information

File S1   Contains the following files: Table S1. GPS co-ordinates of the wards in each province of Vietnam where *Rhizoctonia*-infected *Brassica* crops were sampled. Table S3. Aggressiveness of *Rhizoctonia* isolates towards white cabbage, Chinese cabbage, pak choi, mustard cabbage, Chinese flower cabbage, rice and water spinach in detached leaf bio-assays. Leaves were scored using a scale ranging from 0 (no disease symptoms) to 4 (lesions covered more than 75% of leaf surface or dead leaf). For rapid visual evaluation of the data, a coloring scale with green (0<DI≤1), yellow (1<DI≤2), orange (2<DI≤3) and red (3<DI≤4) was used. The test was done once with 12 leaves or leaf pieces per treatment. All data were statistically analyzed and within columns, disease severities followed by the same letter are not significantly different. Table S4. Aggressiveness of *Rhizoctonia* isolates towards roots and leaves of white cabbage and Chinese cabbage seedlings in *in vitro* bio-assays. Disease severity on roots or leaves was assessed on a scale ranging from 0 (no symptoms) to 4 (lesions covering more than 75% of root, hypocotyl or leaf surface or dead plant). For rapid visual evaluation of the data, a coloring scale with green (0<DI≤1), yellow (1<DI≤2), orange (2<DI≤3) and red (3<DI≤4) was used. Experiment was conducted once with 12 seedlings maintained in two square Petri plates for one treatment. All data were statistically analyzed and within columns, disease severities followed by the same letter are not significantly different. Table S5. Aggressiveness of *Rhizoctonia* isolates towards roots of white cabbage and Chinese cabbage seedlings in in planta experiment. Disease severity on roots was assessed on a scale ranging from 0 (no symptoms) to 4 (seedling dead). For rapid visual evaluation of the data, a coloring scale with green (0<DI≤1), yellow (1<DI≤2), orange (2<DI≤3) and red (3<DI≤4) was used. Experiment was performed once. Each treatment consisted of 12 seedlings cultivated in two plastic boxes. Data were statistically analyzed and within columns, disease severities followed by the same letter are not significantly different.

File S2   Table S2. Curated database of sequences containing representative rDNA-ITS sequences of all known uninucleate, binucleate and multinucleate *Rhizoctonia* AG and subgroups.

Acknowledgments

The authors wish to thank Dr. Tran Thi Thu Thuy and Dr. Nguyen Thi Thu Nga (Can Tho University) and Dr. Ha Viet Cuong (Ha Noi University of Agriculture) for helpful advice and Dr. Michal Sharon for providing the rDNA-ITS sequence alignment of UNR, BNR and MNR reference isolates.

Author Contributions

Conceived and designed the experiments: GKHH MH. Performed the experiments: GKHH SS. Analyzed the data: GKHH LB MH. Wrote the paper: GKHH LB MH.

References

1. Vietnam trade promotion agency (2008) Report on Vietnamese vegetable and fruit sector. Available: http://www.aseankorea.org/aseanZone/downloadFile2.asp?boa_filenum=1575. Accessed 2014 Jan 24.

2. Sharon M, Kuninaga S, Hyakumachi M, Naito S, Sneh B (2008) Classification of *Rhizoctonia* spp. using rDNA-ITS sequence analysis supports the genetic basis of the classical anastomosis grouping. Mycoscience 49: 93–114.

3. Sharon M, Kuninaga S, Hyakumachi M, Sneh B (2006) The advancing identification and classification of *Rhizoctonia* spp. using molecular and biotechnological methods compared with the classical anastomosis grouping. Mycoscience 47: 299–316.

4. Carling DE, Baird RE, Gitaitis RD, Brainard KA, Kuninaga S (2002) Characterization of AG-13, a newly reported anastomosis group of *Rhizoctonia solani*. Phytopathology 92: 893–899.

5. Carling DE, Pope EJ, Brainard KA, Carter DA (1999) Characterization of mycorrhizal isolates of *Rhizoctonia solani* from an orchid, including AG-12, a new anastomosis group. Phytopathology 89: 942–946.

6. Carling DE, Kuninaga S, Brainard KA (2002) Hyphal anastomosis reactions, rDNA-internal transcribed spacer sequences, and virulence levels among subsets of *Rhizoctonia solani* anastomosis group-2 (AG-2) and AG-BI. Phytopathology 92: 43–50.

7. Hyakumachi M, Priyatmojo A, Kubota M, Fukui H (2005) New anastomosis groups, AG-T and AG-U, of binucleate *Rhizoctonia* spp. causing root and stem rot of cut-flower and miniature roses. Phytopathology 95: 784–792.

8. Sneh B, Burpee L, Ogoshi A (1991). Identification of *Rhizoctonia* species. St. Paul Minnesota: APS Press. 133 p.

9. Schwartz HF, Gent DH, Franc GD, Harveson RM (2007) Dry bean-Rhizoctonia root rot. Available: http://www.scarab.msu.montana.edu/HpIPMSearch/AuthorSearch.exe?Complete=true&sort=asc. Accessed 2014 Jan 24.

10. Wharton P, Kirk W, Berry D, Snapp S (2007) Rhizoctonia stem canker and black scurf of potato. Available: http://www.potatodiseases.org/pdf/rhizoctonia-bulletin.pdf. Accessed 2014 Jan 24.

11. Keijer J (1996) The initial steps of the infection process in *Rhizoctonia solani*. In: Sneh B, Jabaji-Hare S, Neate S, Dijst G, editors. *Rhizoctonia* species: Taxonomy, molecular biology, ecology, pathology and disease control. Dordrecht: Springer Netherlands. pp. 149–162.

12. Verma PR (1996) Biology and control of *Rhizoctonia solani* on rapeseed: A review. Phytoprotection 77: 99–111.

13. Yang J, Kharbanda PD, Wang H (1996) Characterization, virulence, and genetic variation of *Rhizoctonia solani* AG-9 in Alberta. Plant Dis 80: 513–518.

14. Khangura RK, Barbetti MJ, Sweetingham MW (1999) Characterization and pathogenicity of *Rhizoctonia* species on canola. Plant Dis 83: 714–721.

15. Sayama A (2000) Occurrence of damping-off disease caused by *Rhizoctonia solani* AG 2-2-IIIB on cabbage plug seedlings. Annu Rep Soc Plant Protect N Jpn 51: 54–57.

16. Homma Y, Yamashita Y, Ishii M (1983) A new anastomosis group AG-7 of *Rhizoctonia solani* Kuhn from Japanese radish fields. Ann Phytopathol Soc Jpn 49: 184–190.

17. Keinath AP, Farnham MW (1997) Differential cultivars and criteria for evaluating resistance to *Rhizoctonia solani* in seedling *Brassica oleracea*. Plant Dis 81: 946–952.

18. Rollins PA, Keinath AP, Farnham MW (1999) Effect of inoculum type and anastomosis group of *Rhizoctonia solani* causing wirestem of cabbage seedlings in a controlled environment. Can J Plant Pathol 21: 119–124.

19. Paulitz TC, Okubara PA, Schillinger WF (2006) First report of damping-off of canola caused by *Rhizoctonia solani* AG 2-1 in Washington State. Plant Dis 90: 829–829.

20. Kuramae E, Buzeto A, Ciampi M, Souza N (2003) Identification of *Rhizoctonia solani* AG 1-IB in lettuce, AG 4 HG-I in tomato and melon, and AG 4 HG-III in broccoli and spinach, in Brazil. Eur J Plant Pathol 109: 391–395.

21. Yang GH, Chen JY, Pu WQ (2007) First report of head rot of cabbage and web blight of snap bean caused by *Rhizoctonia solani* AG-4 HGI. Plant Pathol 56: 351–351.

22. Pannecoucque J, Van Beneden S, Höfte M (2008) Characterization and pathogenicity of *Rhizoctonia* isolates associated with cauliflower in Belgium. Plant Pathol 57: 737–746.

23. Budge GE, Shaw MW, Lambourne C, Jennings P, Clayburn R, et al. (2009) Characterization and origin of infection of *Rhizoctonia solani* associated with *Brassica oleracea* crops in the UK. Plant Pathol 58: 1059–1070.

24. White TJ, Bruns TD, Lee S, Taylor J (1990) Amplification and direct sequencing of fungal ribosomal RNA genes for phylogenetics. In: Innis MA, Gelfland DH, Sninsky JJ, White TJ, editors. PCR protocols: A guide to methods and applications. New York: Academic Press. pp. 315–322.

25. Godoy-Lutz G, Kuninaga S, Steadman JR, Powers K (2008) Phylogenetic analysis of *Rhizoctonia solani* subgroups associated with web blight symptoms on common bean based on ITS-5.8S rDNA. J Gen Plant Pathol 74: 32–40.

26. Tamura K, Stecher G, Peterson D, Filipski A, Kumar S (2013) MEGA6: Molecular evolutionary genetics analysis version 6.0. Mol Biol Evol 30: 2725–2729.

27. Scholten OE, Panella LW, De Bock TSM, Lange W (2001) A greenhouse test for screening sugar beet (*Beta vulgaris*) for resistance to *Rhizoctonia solani*. Eur J Plant Pathol 107: 161–166.

28. Keijer J, Korsman MG, Dullemans AM, Houterman PM, De Bree J, et al. (1997) *In vitro* analysis of host plant specificity in *Rhizoctonia solani*. Plant Pathol 46: 659–669.

29. Dubey SC, Tripathi A, Upadhyay BK (2012) Molecular diversity analysis of *Rhizoctonia solani* isolates infecting various pulse crops in different agroecological regions of India. Folia Microbiol (Praha) 57: 513–524.

30. Thuan TTM, Tho N, Tuyen BC (2008) First report of *Rhizoctonia solani* subgroup AG 1-ID causing leaf blight on durian in Vietnam. Plant Dis 92: 648–648.

31. Martin B (1988) Identification, isolation, frequency, and pathogenicity of anastomosis groups of binucleate *Rhizoctonia* spp. from strawberry roots. Phytopathology 78: 379–384.

32. Demirci E, Eken C, Zengin H (2002) First report of *Rhizoctonia solani* and binucleate *Rhizoctonia* from Johnsongrass in Turkey. Plant Pathol 51: 391–391.

33. Babiker EM, Hulbert SH, Schroeder KL, Paulitz TC (2013) Evaluation of *Brassica* species for resistance to *Rhizoctonia solani* and binucleate *Rhizoctonia* (*Ceratobasidum* spp.) under controlled environment conditions. Eur J Plant Pathol 136: 763–772.

34. Herr LJ (1995) Biological control of *Rhizoctonia solani* by binucleate *Rhizoctonia* spp. and hypovirulent *R. solani* agents. Crop Prot 14: 179–186.

35. Poromarto SH, Nelson BD, Freeman TP (1998) Association of binucleate *Rhizoctonia* with soybean and mechanism of biocontrol of *Rhizoctonia solani*. Phytopathology 88: 1056–1067.

36. Ross RE, Keinath AP, Cubeta MA (1998) Biological control of wirestem on cabbage using binucleate *Rhizoctonia* spp. Crop Prot 17: 99–104.

37. Van Beneden S, Pannecoucque J, Debode J, De Backer G, Höfte M (2008) Characterisation of fungal pathogens causing basal rot of lettuce in Belgian greenhouses. Eur J Plant Pathol 124: 9–19.

38. Yang GH, Chen XQ, Chen HR, Naito S, Ogoshi A, et al. (2004) First report of foliar blight in *Brassica rapa* subsp. *chinensis* caused by *Rhizoctonia solani* AG-4. Plant Pathol 53: 260–260.

39. Yitbarek SM, Verma PR, Morrall RAA (1987) Anastomosis groups, pathogenicity, and specificity of *Rhizoctonia solani* isolates from seedling and adult rapeseed/canola plants and soils in Saskatchewan. Can J Plant Pathol 9: 6–13.

40. Baird RE (1996) First report of *Rhizoctonia solani* AG-4 on canola in Georgia. Plant Dis 80: 104–104.

41. Harikrishnan R, Yang XB (2004) Recovery of anastomosis groups of *Rhizoctonia solani* from different latitudinal positions and influence of temperatures on their growth and survival. Plant Dis 88: 817–823.

42. Ohkura M, Abawi GS, Smart CD, Hodge KT (2009) Diversity and aggressiveness of *Rhizoctonia solani* and *Rhizoctonia*-like fungi on vegetables in New York. Plant Dis 93: 615–624.

43. Ministry of Natural Resources and Environment Sub-Institute of Hydrometeorology and Environment of South Vietnam (2010) Climate change in Mekong delta: Climate scenario's, sea level rise, other effects. Available: http://wptest.partnersvoorwater.nl/wp.../CLIMATE-CHANGE-final-draft.pdf. Accessed 2014 Jan 24.

44. International Federation of Red Cross and Red Crescent Societies (2013) Emergency appeal final report Viet Nam: Mekong delta floods. Available: http://reliefweb.int/sites/reliefweb.int/files/resources/MDRVN009fr.pdf. Accessed 2014 Jan 24.

45. Lee FN, Rush MC (1983) Rice sheath blight: A major rice disease. Plant Dis 67: 829–832.

46. Ogoshi A (1987) Ecology and pathogenicity of anastomosis and intraspecific groups of *Rhizoctonia solani* Kuhn. Annu Rev Phytopathol 25: 125–143.

47. Taheri P, Gnanamanickam S, Höfte M (2007) Characterization, genetic structure, and pathogenicity of *Rhizoctonia* spp. associated with rice sheath diseases in India. Phytopathology 97: 373–383.

48. Hashiba T, Yamaguchi T, Mogi S (1972) Biological and ecological studies on the sclerotium of *Pellicularia sasakii* (Shirai) S. Ito. I. Floating on the water surface of sclerotium. Ann Phytopathol Soc Jpn 38: 414–425.

49. Freeman TE, Charudattan R, Cullen RE (1982) Rhizoctonia blight on waterhyacinth in the United States. Plant Dis 66: 861–862.

50. Zettler FW, Freeman TE (1972) Plant pathogens as biocontrols of aquatic weeds. Annu Rev Phytopathol 10: 455–470.

51. Yang G, Li C (2012) General description of *Rhizoctonia* species complex. In: Cumagun CJR, editor. Plant Pathology. Croatia: InTech. pp. 41–52.

52. Priyatmojo A, Escopalao VE, Tangonan NG, Pascual CB, Suga H, et al. (2001) Characterization of a new subgroup of *Rhizoctonia solani* anastomosis group 1 (AG-1-ID), causal agent of a necrotic leaf spot on coffee. Phytopathology 91: 1054–1061.

53. People's committee of Dong Nai province (2011) Program to develop main crops, domestic animals and build up trade name of agricultural products, duration 2011–2015. Available: http://laws.dongnai.gov.vn/2011_to_2020/2011/201109/201109260003/lawdocument_view. Accessed 2014 Jan 24.

54. Commerce (2014) Can Tho province. Available: http://www.thuongmai.vn/thuong-mai/du-an-keu-goi-dau-tu/54004-can-tho.html. Accessed 2014 Jan 24.

55. Department of Culture, Sport and Tourism of Soc Trang (2014) Overview about Soc Trang province. Available: http://www.sovhttdl.soctrang.gov.vn/wps/portal/!ut/p/c4/04_SB8K8xLLM9MSSzPy8xBz9CP0os3gLR1dvZ09LYwOL4GAnA08TRwsfvxBDI4swc_2CbEdFAM7mSnc!/. Accessed 2014 Jan 24.

56. Department of Planning and Investment of Vinh Long (2014) Overview about Vinh Long province. Available: http://www.skhdt.vinhlong.gov.vn/Default.aspx?tabid=36. Accessed 2014 Jan 24.

57. Department of Propaganda and Training of Hau Giang (2014) Overview about Hau Giang province. Available: http://www.haugiang.gov.vn/Portal/Default.aspx?pageindex=2&pageid=2979&siteid=59#. Accessed 2014 Jan 24.

58. Integrated Water Resources Management (2014) Lam Dong province. Available: http://www.iwrm.vn/index.php?page=2&id=21. Accessed 24 January 2014.

59. Viet's life (2014) Ha Noi capital. Available: http://www.cuocsongviet.com.vn/index.asp?act=dp&id_tinh=27. Accessed 2014 Jan 24.

21

Estimated Dietary Intake of Radionuclides and Health Risks for the Citizens of Fukushima City, Tokyo, and Osaka after the 2011 Nuclear Accident

Michio Murakami[1,2]*, Taikan Oki[1,2]

1 Institute of Industrial Science, The University of Tokyo, Meguro, Tokyo, Japan, **2** Japan Science and Technology Agency, Core Research for Evolutionary Science and Technology (CREST), Chiyoda, Tokyo, Japan

Abstract

The radionuclides released from the Fukushima Daiichi nuclear power plant in 2011 pose a health risk. In this study, we estimated the 1st-year average doses resulting from the intake of iodine 131 (^{131}I) and cesium 134 and 137 (^{134}Cs and ^{137}Cs) in drinking water and food ingested by citizens of Fukushima City (~50 km from the nuclear power plant; outside the evacuation zone), Tokyo (~230 km), and Osaka (~580 km) after the accident. For citizens in Fukushima City, we considered two scenarios: Case 1, citizens consumed vegetables bought from markets; Case 2, citizens consumed vegetables grown locally (conservative scenario). The estimated effective doses of ^{134}Cs and ^{137}Cs agreed well with those estimated through market basket and food-duplicate surveys. The average thyroid equivalent doses due to ingestion of ^{131}I for adults were 840 µSv (Case 1) and 2700 µSv (Case 2) in Fukushima City, 370 µSv in Tokyo, and 16 µSv in Osaka. The average effective doses due to ^{134}Cs and ^{137}Cs were 19, 120, 6.1, and 1.9 µSv, respectively. The doses estimated in this study were much lower than values reported by the World Health Organization and the United Nations Scientific Committee on the Effects of Atomic Radiation, whose assessments lacked validation and full consideration of regional trade in foods, highlighting the importance of including regional trade. The 95th percentile effective doses were 2–3 times the average values. Lifetime attributable risks (LARs) of thyroid cancers due to ingestion were 2.3–39×10^{-6} (Case 1) and 10–98×10^{-6} (Case 2) in Fukushima City, 0.95–14×10^{-6} in Tokyo, and 0.11–1.3×10^{-6} in Osaka. The contributions of LARs of thyroid cancers due to ingestion were 7.5%–12% of all exposure (Case 1) and 12%–30% (Case 2) in Fukushima City.

Editor: Bart O. Williams, Van Andel Institute, United States of America

Funding: This study was supported by a "Core Research for Evolutionary Science and Technology (CREST)" grant from the Japan Science and Technology Agency, http://www.jst.go.jp/kisoken/crest/en/index.html. MM and TO receive the funding. The funder had no role in study design, data collection and analysis, discussion to publish, or preparation of the manuscript.

Competing Interests: The authors have declared that no competing interests exist.

* Email: michio@iis.u-tokyo.ac.jp

Introduction

Radionuclides were released from the Tokyo Electric Power Company's Fukushima Daiichi nuclear power plant, mainly on 15 March 2011, after the Great East Japan Earthquake and tsunami on 11 March. They were diffused into the atmosphere, deposited mainly through precipitation, and incorporated into surface waters, drinking waters, agricultural crops, and aquatic organisms. Radionuclides have been detected in drinking water and foods in Fukushima and other prefectures in Japan, including Tokyo [1,2,3]. Radioactive iodine 131 (^{131}I), which has a short half-life (8.04 d [4]), can contaminate drinking water and foods immediately after its release, whereas radioactive cesium 134 and 137 (^{134}Cs and ^{137}Cs), which have longer half-lives (2.06 and 30 y, respectively [4]), cause contamination over longer periods of time. Although other radionuclides such as strontium 90 (^{90}Sr), ruthenium 106 (^{106}Ru), and plutonium have half-lives of >1 y, in the year after the accident ^{134}Cs and ^{137}Cs contributed an estimated 84% to 88% of the total radiation dose from radionuclides with half-lives of >1 y in the diet [5].

In response to the accident, the Japanese government announced provisional "indices relating to limits on food and drink ingestion" on 17 March 2011 [1]. The government began releasing monitoring data on radionuclide concentrations in foods on 19 March and restricted the distribution of foods collected in some municipalities from 21 March, as the foods exceeded the limits. Some municipalities in Fukushima prefecture voluntarily imposed more stringent limits. In addition, some local governments, including the Tokyo Metropolitan Government, distributed bottled water to protect the health of citizens, particularly infants, against the high concentrations of ^{131}I detected in drinking water.

Among the exposure pathways, internal exposure from dietary intake has not been fully elucidated. External exposure can be determined from personal dosimeters [6], calculated from ambient dose monitoring and shielding factors [7], or estimated by using numerical dispersion models [8]. Internal exposure from inhalation can be estimated by air monitoring [9] or by numerical dispersion models [8,10]. Internal exposure to Cs can be measured by whole-body-counter surveys; however, the Cs burdens in most inhabitants of Fukushima have been below the detection limit

[11,12]. Internal exposure from the intake of water and food can be determined from market basket surveys (i.e., foods are purchased from the market according to consumption patterns) or food-duplicate surveys (i.e., cooked foods are collected from home) [1,13], but exposure immediately after the accident could not be evaluated by these methods because of the lack of samples. On the other hand, individual foods were monitored just after the accident for restriction of the distribution of foods, as described above. On the basis of the monitoring data, working groups of the Ministry of Health, Labour and Welfare [14] and the World Health Organization (WHO) [15] approximated the nationwide average internal exposure to radionuclides from foods. The United Nations Scientific Committee on the Effects of Atomic Radiation (UNSCEAR) also recently reported that an increased risk of thyroid cancer for infants and children in particular is of concern, that most of the absorbed dose to the thyroid was received during the first year after the accident, and that the major route of exposure was ingestion; e.g., absorbed dose to thyroid in the first year for 1-year-olds in Fukushima Prefecture (districts not evacuated): external + inhalation, 0.2–19 mGy; ingestion, 33 mGy [16]. However, the regional trade in foods was not fully considered in these assessments despite the low food self-sufficiency rates in Japan (54% for major grains, 39% on basis of total energy [17]), and no validations based on the results of market basket, food-duplicate, or whole-body-counter surveys were done. More detailed estimates and validation are required on account of regional variations in concentrations in drinking water and foods, the regional trade in foods, and variations in individual exposure. Furthermore, recent screening of all children in Fukushima Prefecture identified thyroid cancer in 90 of more than 280 000 children (an incidence of 313 per 1 million) by March 2014 [18]. Although this incidence might be attributable to increased screening by advanced ultrasound techniques [18], an increase incidence of cancer in future, especially thyroid cancer in children, is a matter of concern. The the contribution of cancer risk from ingestion needs to be understood.

To provide more detail, we estimated the 1st-year average doses due to intake of drinking water and foods after the accident from radionuclide concentrations measured in water and foods and the effects of countermeasures (i.e., restriction of food distribution, voluntary withholding of rice, and distribution of bottled water for infants). We previously reported the average thyroid doses resulting from the intake of ^{131}I by citizens of Tokyo and the effects of countermeasures [19]. Here, we estimated the intake of ^{131}I, [134]Cs, and ^{137}Cs by the citizens of Fukushima City, Tokyo, and Osaka, which differ in the regional trade in foods. The results were validated against observations in market basket, food-duplicate, and whole-body-counter surveys. By comparing the results with reports by WHO and UNSCEAR, whose assessments lacked validation and full consideration of regional trade in foods, we evaluated the role of regional trade in foods. We also estimated variations in intakes by using a Monte Carlo simulation. Finally, we assessed the cancer risks due to intake. We considered all sources of exposure for citizens in Fukushima City, and evaluated the contributions of cancer risks due to ingestion to the overall risk. This information will help to address current concerns of citizens and be useful for establishing the regulation of foods and drinking water.

Methods

Intake pathway

To assess intake via the ingestion of drinking water and foods, we estimated the thyroid equivalent dose and effective dose of [131]I

and the effective dose of [134]Cs and ^{137}Cs among citizens of Fukushima City (~50 km from the nuclear power plant; capital of prefecture), Tokyo (~230 km), and Osaka (~580 km). All three cities, including Fukushima City, were outside the evacuation zone. Citizens were classified into ten groups based on age, sex, and pregnancy: <1-y-old infants, males and females aged 1–6, 7–12, 13–18, and ≥19 y old, and pregnant females.

Monitoring data on drinking water and foods were obtained from ref [1,2,3,20]. Drinking water and foods were classified into 18 categories [21]: drinking water, rice, other grains, potato, leafy vegetables, root crops, beans, fruit vegetables, milk, dairy products, formula milk, beef, pork, chicken, chicken eggs, fresh fisheries products, marine products, and others. "Fruit vegetables" included vegetable-like fruits such as tomatoes and fruits such as persimmons and apples. They were then further classified into sub-categories based on governmental restrictions [1]: leafy vegetables were divided into "spinach", "garland chrysanthemum and ging-geng-cai", "mustard spinach and non-heading lettuce", "heading leafy vegetables", "broccoli and cauliflower", and "naganegi onion, chive, and asparagus"; root crops into "turnip", "bamboo shoots", and "other root crops"; fruit vegetables into "kiwifruit", "chestnut", and "other fruit vegetables"; fresh fisheries products into "wild *ayu*, wild Japanese dace, and wild landlocked *masu* salmon" and "other fresh fisheries products"; and other into "tea", "shiitake mushroom (virgin wood)", and "other mushrooms".

Drinking water

The average doses (thyroid equivalent dose of ^{131}I and the effective dose of ^{131}I, [134]Cs, and ^{137}Cs) from drinking water were estimated as:

$$Dose_k(\mu Sv) = A_k \times B \times \sum_t C_{kt} \qquad (1)$$

where k is the radionuclide, A_k is the dose coefficient for radionuclide k (μSv/Bq) (Table S1), B is the daily consumption of drinking water per person (g/d) (Table S2), t is the number of days after the accident (consumption date), and C_{kt} is the concentration of radionuclide k in drinking water at t days after the accident (Bq/g).

So as not to underestimate the dose, we set A_k for ^{131}I on the assumption that the fractional thyroid uptake from blood is 0.3, as used in the determination of control index levels by the Nuclear Safety Commission of Japan [22]. All the dose coefficients are reference values taken from the publications of the International Commission on Radiological Protection (ICRP) (Table S1).

B was set at 710 g/d for <1 y, 1000 g/d for 1–6 y, and 1650 g/d for >6 y [22]. The average daily consumption of water in soup and rice was then added [19]: The consumption of water in soup was calculated from the consumption of soup by adults [23] and the ratio of miso (soybean paste) consumption by each age group to that by adults [24]. The consumption of water in rice was calculated from the consumption of rice [21] and the ratio of cooking water to rice [25].

The radionuclide concentrations in drinking water were the values monitored in tap water in Fukushima City [3]; Shinjuku, Tokyo [2]; and Osaka (not detected) [20]. As details of [134]Cs and ^{137}Cs concentrations in tap water in Fukushima City were not available, we estimated them from the ^{131}I concentrations and the ratios of [134]Cs and ^{137}Cs to ^{131}I in tap water in Tokyo between 19 March and 8 April. Intake was estimated from the date when radionuclides were first detected (16 March in Fukushima City, 18 March in Tokyo). As they were not detected in tap water after 4

May (except on 2 July in Tokyo), intake after then was considered to be negligible.

Foods

The average doses from foods were estimated by the following equation [19], which is modified to include ^{134}Cs and ^{137}Cs, product origins, and food categories:

$$Dose_k(\mu Sv) = A_k \times \sum_i \sum_j \sum_t \left(B_i \times C_{kijt} \times D_{ij}/100\right) \quad (2)$$

where k is the radionuclide, A_k is the dose coefficient for radionuclide k (μSv/Bq) (Table S1), i is the individual food category, j is the individual area (source prefecture or prefectural sub-group), t is the number of days after the accident (consumption date), B_i is the daily consumption of food i per person (g/d) (Table S2), C_{kijt} is the arithmetic mean concentration of radionuclide k in food i in area j at t days after the accident (Bq/g), and D_{ij} is the arrival share in an area (the fraction of food i in the market that comes from prefecture or area j) (%).

Daily consumption. For the daily consumption of each food recently reported detailed values were adopted [21]. The daily consumption of the sub-categorized foods was estimated from the daily consumption of the foods in the main categories and the ratio of the amount of each sub-category arriving at the Tokyo Metropolitan Central Wholesale Market to the total in 2010 [26]. The national fishery yield was used to apportion the daily consumption of fresh fisheries products [27]. The daily consumption of mushrooms followed reported values [24]. The daily consumption of tea by ≥19-y-old females was 439 g/d [28], and that by other groups was calculated from that and the ratio of the daily consumption of drinking water excluding water in soup and rice in each group to that in ≥19-y-old females. We assumed that 10 g of leaf is used to make 300 g of tea and that 60% of radionuclides in the leaf enter the tea [14]. The ratio of virgin wood to other substrates used to grow shiitake was used to estimate the daily consumption of "shiitake mushroom (virgin wood)" [29].

Radionuclide concentrations. The radionuclide concentrations in food were drawn from >130 000 food monitoring data [1] and rice monitoring data [3]. Details of the number of samples analyzed each month and in each prefecture are given in the Supporting Information and Tables S3 and S4. Foods were washed with water and the edible parts were then analyzed for radionuclides by Ge, NaI or CsI detectors. We considered the intake of radionuclides via the consumption of foods collected in 16 prefectures: Hokkaido, Aomori, Iwate, Miyagi, Akita, Yamagata, Fukushima, Ibaraki, Tochigi, Gunma, Saitama, Chiba, Kanagawa, Tokyo (Metropolis), Yamanashi, and Shizuoka (Figure S1). Analysis of the variations in radionuclide concentrations among foods in Fukushima Prefecture showed that spinach, broccoli and cauliflower, and marine products were heavily contaminated (Figures S2 and S3). Analysis of the regional differences in radionuclide concentrations in spinach showed that foods collected in Fukushima Prefecture and in prefectures in the Kanto region (Ibaraki, Tochigi, Gunma, Saitama, Chiba, Kanagawa, and Tokyo) had much higher levels than those in the other 8 prefectures (Figure S4). The amounts of radionuclides originating from other prefectures or from overseas were regarded as negligible. When radionuclides were not detected in a food, the contribution of radionuclides from that food to the dose was considered to be negligible. The concentrations in each food were classified by source prefecture and date. The prefectures where the distribution of the food was restricted (for example, rice in Fukushima Prefecture) were further classified into prefectural sub-

groups. The arithmetic mean concentration in each area on each day was calculated. When only the sum of ^{134}Cs and ^{137}Cs concentrations was available, individual concentrations were estimated from the average ratio of ^{134}Cs to the sum of ^{134}Cs and ^{137}Cs, namely 0.49 (March–June 2011), 0.46 (July–September 2011), 0.45 (October–December 2011), or 0.42 (January–March 2012), as calculated from foods in which both the ^{134}Cs and the ^{137}Cs concentration exceeded 25 Bq/kg [1]. The radionuclide concentrations in crude tea and manufactured tea were converted to those in tea leaf by multiplying by the weight ratio of crude tea to tea leaf (0.22) [30]. The concentrations of ^{131}I in foods on days before monitoring began or for which there were no data were approximated from the concentrations measured on the closest day and from the half-life of ^{131}I (8.04 d [4]). The concentrations of ^{134}Cs and ^{137}Cs on days before monitoring began or for which there were no data were regarded as the same as the concentrations measured on the closest day. The consumption dates and concentrations in foods were assumed to be as reported [1,3].

Arrival shares. Since citizens generally purchase food from markets, which are supplied from the whole country, the arrival share is an important parameter for estimating the doses due to ingestion. A few citizens in Fukushima City consumed vegetables produced in local fields or home gardens (4.5% of workers in Fukushima City were agricultural workers [31]). We therefore considered two scenarios: Case 1, citizens consumed foods from markets; Case 2, citizens consumed vegetables produced in Fukushima Prefecture and other foods from markets. Case 2 means the arrival shares of vegetables from Fukushima Prefecture are 100%, so this scenario can be regarded as conservative. For citizens in Tokyo and Osaka, only Case 1 was considered. The proportions of targeted foods produced in each area for the markets in each city were calculated as follows. The arrival shares of vegetables and fisheries products were based on the amounts arriving from each prefecture at each central wholesale market in 2010 [26,32,33]. We used the 2010 data so as to evaluate the effects of countermeasures. Since data on the importation of fisheries products from overseas were not available, we used the Japanese food self-sufficiency rates based on quantities of production, imports, and exports [17]. The production of marine products in Fukushima Prefecture was regarded to be nil owing to the tsunami damage and the suspension of fishing. The arrival shares of beef, pork, and chicken eggs were based on the quantities transported from each prefecture to Fukushima, Tokyo, and Osaka prefectures [34,35]. The arrival shares of beef and pork were corrected to the food self-sufficiency rates of Japan. The arrival shares of milk and dairy products were based on the production and transport of raw milk in each prefecture, the amount of milk transported from each prefecture to Fukushima, Tokyo, and Osaka prefectures [36], and the food self-sufficiency rates. For rice, other grains, tea, soybeans, and chicken, we used the nationwide production rates [30,35,37,38] and the food self-sufficiency rates of Japan. The use of these sources is reasonable, especially for dried foods. To estimate the ratios of restricted food production among the prefectural sub-groups, we used the number of cows in each municipality for milk and dairy products [39], the cultivated areas of vegetables and tea [40], the areas of rice paddies [40] and the number of rice farm households [3], and the production [41] or the number of management bodies [42] for shiitake mushrooms.

Calculation periods. It is likely that rainfall on the night of 15–16 March 2011 [43] caused heavy contamination of leafy vegetables and of milk and dairy products by radionuclides, so we estimated intake via these products before the first release of data

(19 March 2011): from 17 March 2011 in Fukushima City, on the assumption that delivery to citizens takes 1 day; and from 18 March 2011 in Tokyo and Osaka, on the assumption that it takes 2 days. The intake of other foods was estimated from 21 March 2011, when the Japanese government restricted the distribution of foods. Since radionuclide contamination in other foods was not attributable to direct deposition [22], the intake before 21 March was not included. This was supported by the lower contamination of other foods by ^{131}I and [134]Cs and ^{137}Cs than of leafy vegetables (Figures S2, S3). The intake was estimated from these days to 20 March 2012.

Effects of countermeasures

We evaluated the effects of countermeasures (restrictions on the distribution of foods, voluntary withholding of rice, and distribution of bottled water for infants) on reducing intake. We assumed the consumption of foods harvested in the areas where distribution was restricted or rice was withheld to be nil: citizens ate alternative foods containing negligible radionuclides. In this regard, however, for Case 2, the conservative scenario, we considered the effects of restrictions on the distribution of other foods except vegetables, because we assumed that some citizens consumed vegetables grown locally. Since the Tokyo Metropolitan Government distributed bottled water for infants on the morning of 24 March, we evaluated the effect of this countermeasure on 24 and 25 March. (Fukushima City distributed bottled water on the night of 11 March following the earthquake, but this was too early to consider).

Variations in individual doses

The variations in ingestion doses for ≥19-y-old males (with countermeasures in place) were estimated by Monte Carlo simulation using the commercially available software Crystal Ball (Oracle, California, USA). The drinking water and food variation data input into the Monte Carlo simulation were obtained as below.

We took into account variations in daily consumption of drinking water [44] (Table S2). The drinking water in Tokyo is derived from several sources: 38.0% of tap water comes from the Edogawa River, 41.3% from the Arakawa River, 14.4% from the Tamagawa River, and 6.3% from other sources such as ground water [45,46]. Therefore, we also took into account the radionuclide concentrations in each source. We assumed that citizens drank water from a single source. The radionuclide concentration factors were estimated from the ratios of ^{131}I concentrations in representative purification plants (Kanamachi purification plant for the Edogawa River, Asaka purification plant for the Arakawa River, Ozaku purification plant for the Tamagawa River) to those in tap water in Shinjuku [2,46]. We used the arithmetic mean concentrations on 22–28 March in the purification plants and on 23–29 March in Shinjuku, on the assumption that delivery to the tap takes 1 day. Concentrations were corrected for half-life. Concentration factors were 3.51 for the Edogawa River, 1.31 for the Arakawa River, and 0.46 for the Tamagawa River (i.e., the ^{131}I concentration in the Ozaku purification plant on the Tamagawa River was lower than that in tap water in Shinjuku). The ^{131}I concentration in other water systems was regarded as negligible. Fukushima City has only one drinking water treatment plant, and radionuclides were not detected in the drinking water in Osaka. We therefore did not need to take into account variations in the water resources in these two cities.

For input into the Monte Carlo simulation, variations in the dose from drinking water in Fukushima City were estimated from the variation in daily consumption under the assumption of a log-normal distribution. Variations in Tokyo were estimated from variations in daily consumption, the dependence on water source, and the concentration factor.

Variations in the dose from foods for input into the model were estimated under the assumption that citizens bought each food once a week, and the production area was randomly selected according to the distribution of the arrival shares. The weekly average concentrations of radionuclides in each food were calculated. The variations in the daily consumption of each food were as reported [28,47]. Since these values are based on 1-day surveys, not long-term surveys, their use could have caused us to overestimate the results. The variations in doses obtained by ingesting each food from the same area in the same week at the same daily rate of consumption were considered to indicate variations in the radionuclide concentrations. The relative standard deviations of the variations were estimated from the effective doses calculated from data measured on the same day in Tochigi Prefecture for "ging-geng-cai," in Ibaraki Prefecture for "marine products," in Shizuoka Prefecture for "tea," and in Fukushima Prefecture for other foods (n≥4 for each food; Table S2). Daily consumption of foods and the radionuclide concentrations in foods were assumed to follow a log-normal distribution. The variations in the dose from each food were estimated from the input data for the Monte Carlo simulation, namely, the variation in daily consumption, the distribution of arrival shares, and the variation in doses received by ingesting the foods in the same areas. The simulation was performed 10 000 times for drinking water and for each food.

Estimation of cancer risk

The lifetime attributable risks (LARs) of cancer incidences up to the age of 89 y were estimated in accordance with the method described in a WHO report [48] and Harada et al. [49]. We estimated those due only to the ingestion of ^{131}I, [134]Cs, and ^{137}Cs for citizens in each city and those due to three pathways–ingestion, inhalation in the radioactive cloud, and external exposure to material deposited on the ground and in the cloud–for citizens in Fukushima City. Estimation of doses due to inhalation and external exposure followed previous references [15,50] as described in the Supporting Information. The risk models were based on data from survivors of the atomic bombing in Japan; the validity of the thyroid cancer incidence was also confirmed by a cohort study performed after the Chernobyl accident [51] in the WHO report [48]. A linear-quadratic dose-response model was used for leukemia [52], and linear non-threshold (LNT) models were used for all solid cancers, breast cancer, and thyroid cancer [53]. LAR was calculated from cancer-free survival rates ($S(a, g)$), an excess absolute risk (EAR) model, and an excess relative risk (ERR) model, as follows:

$$LAR(D,e,g) = \int_{e+L}^{89} [w \times EAR(D,e,a,g) + (1-w) \times ERR(D,e,a,g) \times m(a,g)] \frac{S(a,g)}{S(e,g)} da \quad (3)$$

where D is the annual dose (Sv), e is the age at exposure, g is sex, L is the minimum latency period, w is the weight of the EAR and the ERR models combined, a is the age attained, and $m(a, g)$ is the baseline cancer incidence rate in the unexposed population.

The organ (colon, bone marrow, breast, and thyroid) doses were calculated from the effective doses and the organ dose-to-effective dose ratios for three age groups (<1 y, 10 y, and 20 y in the first year) and for the two sexes in the three cities [48]. Doses in the second and subsequent years were calculated from the doses of ^{134}Cs and ^{137}Cs in March 2012 and from the physical decay of ^{134}Cs and ^{137}Cs (half-lives of 2.06 and 30 y, respectively [4]).

The cancer-free survival rates of males and females were derived from the age- and sex-stratified all-cause mortality in Japan in 2010 [54] plus the difference between the all-cancer incidence in Japan in 2008 [55,56] and the all-cancer mortality in Japan in 2010 [54]. Cancer incidences in Japan in 2008 [55,56] were used as baseline incidence rates for all solid cancers, leukemia, breast cancer, and thyroid cancer.

The minimum latency period was set at 2 y for leukemia, 3 y for thyroid cancer, and 5 y for breast and all solid cancers. The weights of the EAR model and the ERR model combined were set at 0.5 for all solid cancers, leukemia, and thyroid cancer, and 1 for breast cancer.

The details of the EAR and ERR models and their parameters for leukemia, all solid cancers, breast cancer, and thyroid cancer are described in the Supporting Information.

The relationship between cancer risk and dose is still uncertain [57]. However, for the same dose, the risks at low dose rates are known to be lower than those at high dose rates [58]. In particular, coordination of DNA repair processes plays a critical role in allowing proper development and survival of organisms [59].To correct the risk at low dose levels, the International Commission on Radiological Protection (ICRP) and the BEIR VII committee adopted "dose and dose-rate effectiveness factors" of 2 [60] and 1.5 [61]. We followed the ICRP lead and set a factor of 2 for LNT models from the perspective of radiological protection.

Sources of uncertainty

This study included sources of uncertainty: ^{134}Cs and ^{137}Cs concentrations in tap water in Fukushima City, limited data on foods in the early stages, data that were less than detection limits, individual behaviors (such as purchasing of bottled water or not purchasing products from Fukushima Prefecture), and assessment of doses in the second and subsequent years for cancer estimation.

^{134}Cs and ^{137}Cs concentrations in tap water in Fukushima City were estimated from the ^{131}I concentrations and the ratios of ^{134}Cs and ^{137}Cs to ^{131}I in tap water in Tokyo by assuming that the ratios of ^{134}Cs and ^{137}Cs deposition to ^{131}I deposition and the removal efficiencies in drinking water treatment plants were similar between Fukushima City and Tokyo. The deposition ratios are known to be similar in the two cities [62], and drinking water treatment plants in both cities use sedimentation and rapid sand filtration, which are effective for removing ^{134}Cs and ^{137}Cs [63]. This uncertainty did not have a large effect, because the contributions of ^{134}Cs and ^{137}Cs from drinking water were minor (see "Comparison of doses among ages, effects of countermeasures, and change of intake over time).

In the early stages of monitoring, data were not available for some foods in some prefectures in the Kanto region, but these contributions were judged to be small (see Supporting Information). The number of samples was limited for some foods in the early stages even in Fukushima Prefecture. We therefore used a Monte Carlo simulation and show the variations in dose (e.g., 95th percentile value), including those resulting from radionuclide concentrations, in foods in the early stages.

We regarded radionuclide concentrations that were less than detection limits as nil. There were differences in detection limits among periods and institutions surveyed. We therefore confirmed the results through validation against observations in market basket, food-duplicate, and whole-body-counter surveys.

Differences in individual behaviors, such as not purchasing products from Fukushima Prefecture, could have influenced the

variation in estimates. However, as the volume of major crops shipped from Fukushima Prefecture did not decrease after the accident [64,65], we ignored differences in behaviors. The nationwide average of bottled soft drink consumption was 410 g/d in 2011 [66], and some citizens also purchased bottled water so as to avoid tap water. We ignored the consumption of bottled water and soft drinks: the assumption that people drank only tap water would conservatively overestimate the dose from drinking water.

The doses due to ingestion in the second and subsequent years for cancer estimation were calculated from the physical decay, although the actual doses might be lower because of tighter regulations. The doses due to external exposure from September 2014 were also calculated from the physical decay. These assumptions can be regarded as conservative.

Results and Discussion

Validation of results against market basket and food-duplicate surveys

The effective doses due to ingestion of ^{134}Cs and ^{137}Cs in the diet in Fukushima City (Case 1), Tokyo, and Osaka agreed within a factor of 2, in general, with those calculated in the market basket [1] and food-duplicate surveys [1,13,67] in five periods from July 2011 to March 2012 (Table 1). The market basket survey includes drinking water as well as foods. This good agreement supports the accuracy and reliability of our results. The doses in the diet in Fukushima City (Case 2) were higher than those in the market basket and food-duplicate surveys. This result is reasonable because Case 2 is conservative. Validation for the period from March to June, including for the thyroid equivalent doses due to ^{131}I, was not done, because no data from market basket and food-duplicate surveys were available.

Comparison of doses among cities

The average thyroid equivalent doses (and effective doses) due to ingestion of ^{131}I for a ≥19-y-old male were 840 µSv (43 µSv) in Fukushima City (Case 1), 2700 µSv (140 µSv) in Fukushima City (Case 2), 370 µSv (19 µSv) in Tokyo, and 16 µSv (0.82 µSv) in Osaka (Table 2). The average effective doses due to ^{134}Cs and ^{137}Cs were 19, 120, 6.1, and 1.9 µSv, respectively. The average effective doses due to total radionuclides were 62, 260, 25, and 2.7 µSv, respectively. There were no large differences between the average and median (within a factor of <1.5, with the exception of the thyroid equivalent dose due to ^{131}I in Osaka). The slightly lower thyroid equivalent dose due to ^{131}I for citizens of Tokyo in this study (370 µSv) than in our previous study (410 µSv) [19] is attributable to the modification of daily consumption rates.

The dose due to ingestion of ^{131}I in Case 2 was 3 times that in Case 1 and the dose due to ^{134}Cs and ^{137}Cs was 6 times that in Case 1. The doses in Fukushima City (Case 1) were lower than the values reported in the WHO preliminary assessment (adults in Fukushima City, thyroid equivalent dose, 800–8000 µSv; effective dose due to total radionuclides, 500–5000 µSv) [15] and the UNSCEAR report (absorbed dose to thyroid, 7800 µGy; effective dose due to total radionuclides, 900 µSv) [16]. Regional trade in foods was considered in this study, whereas the WHO preliminary assessment and UNSCEAR report assumed that consumers consumed mainly food produced in Fukushima and neighboring prefectures. Inclusion of regional trade in foods is a key to accurate dose assessment.

In both doses, Fukushima City had the highest values and Osaka the lowest. This was consistent with the distances from the nuclear power plant (i.e. Fukushima City, ~50 km; Tokyo,

Table 1. Comparison of effective doses of ^{134}Cs and ^{137}Cs in the diet between this study and the market basket and food-duplicate surveys (µSv/month).

	Jul.2011	Sep.–Nov. 2011	Dec. 2011	Feb.–Mar. 2012	Mar. 2012
Fukushima City (Case 1; this study)[a]	2.36	1.18	0.96	0.51	0.44
Fukushima City (Case 2; this study)[a]	4.15	2.85	3.08	6.84	2.60
Fukushima Prefecture (ref[67])[b]	0.53 ± 1.04[f]	-	-	-	-
Fukushima Prefecture (ref[1])[c]	-	1.6	-	0.33-0.55	-
Fukushima Prefecture (ref[1])[b]	-	-	-	-	0.18[g]
Fukushima Prefecture (ref[13])[b]	-	-	2.17± 1.67[f]	-	-
Tokyo (this study) [a]	-	0.44	0.24	0.18	0.19
Tokyo (ref[1])[c]	-	0.22	-	-	-
Kanto (ref[1])[c, d]	-	-	-	0.28-0.33	-
Kanto (ref[1])[b, d]	-	-	-	-	0.15[g]
Kanto (ref[13])[b, e]	-	-	0.92± 1.42[f]	-	-
Osaka (this study)[a]	-	0.15	0.09	0.07	0.07
Osaka (ref[1])[c]	-	-	-	0.13	-
Osaka (ref[1])[b]	-	-	-	-	0.1[g]

Case 1, citizens consumed vegetables bought from markets. Case 2, citizens consumed vegetables grown locally.
[a]with countermeasure (≥19 y old male).
[b]food-duplicate survey.
[c]market basket survey.
[d]including Saitama, and Kanagawa prefectures.
[e]including Tochigi, Gunma, Ibaraki, Saitama, Chiba, Tokyo, Kanagawa, and Nagano prefectures.
[f]arithmetic mean ± standard deviation.
[g]Mar.-May, 2012.

~230 km; Osaka, ~580 km). The contribution of effective dose from ^{131}I to total radionuclides was higher than that from ^{134}Cs and ^{137}Cs in Fukushima City and Tokyo, and lower in Osaka.

Except in Case 2, drinking water contributed the highest thyroid equivalent dose due to ingestion of ^{131}I in Fukushima City and Tokyo, followed by vegetables (Figures 1, S5, and S6). ^{131}I was not

Table 2. Average doses and variations due to countermeasures resulting from the intake of ^{131}I, ^{134}Cs and ^{137}Cs from drinking water and foods by a ≥19-y-old male from 16 March 2011 to 20 March 2012.

	Average	5th percentile	25th percentile	Median	75th percentile	95th percentile
^{131}I (thyroid equivalent dose: µSv)						
Fukushima City (Case 1)	840	360	520	690	940	1700
Fukushima City (Case 2)	2700	940	1500	2200	3300	6200
Tokyo	370	140	290	470	780	1400
Osaka	16	<1	1	4	13	74
^{134}Cs and ^{137}Cs (effective dose: µSv)						
Fukushima City (Case 1)	19	8.1	11	15	22	43
Fukushima City (Case 2)	120	60	86	110	150	220
Tokyo	6.1	3.0	4.4	5.8	7.7	12
Osaka	1.9	0.49	0.88	1.4	2.1	4.5
Total (effective dose: µSv)						
Fukushima City (Case 1)	62	29	39	50	66	110
Fukushima City (Case 2)	260	120	170	230	310	530
Tokyo	25	11	21	30	47	80
Osaka	2.7	0.59	1.2	1.8	3.0	7.1

Case 1, citizens consumed vegetables bought from markets. Case 2, citizens consumed vegetables grown locally.

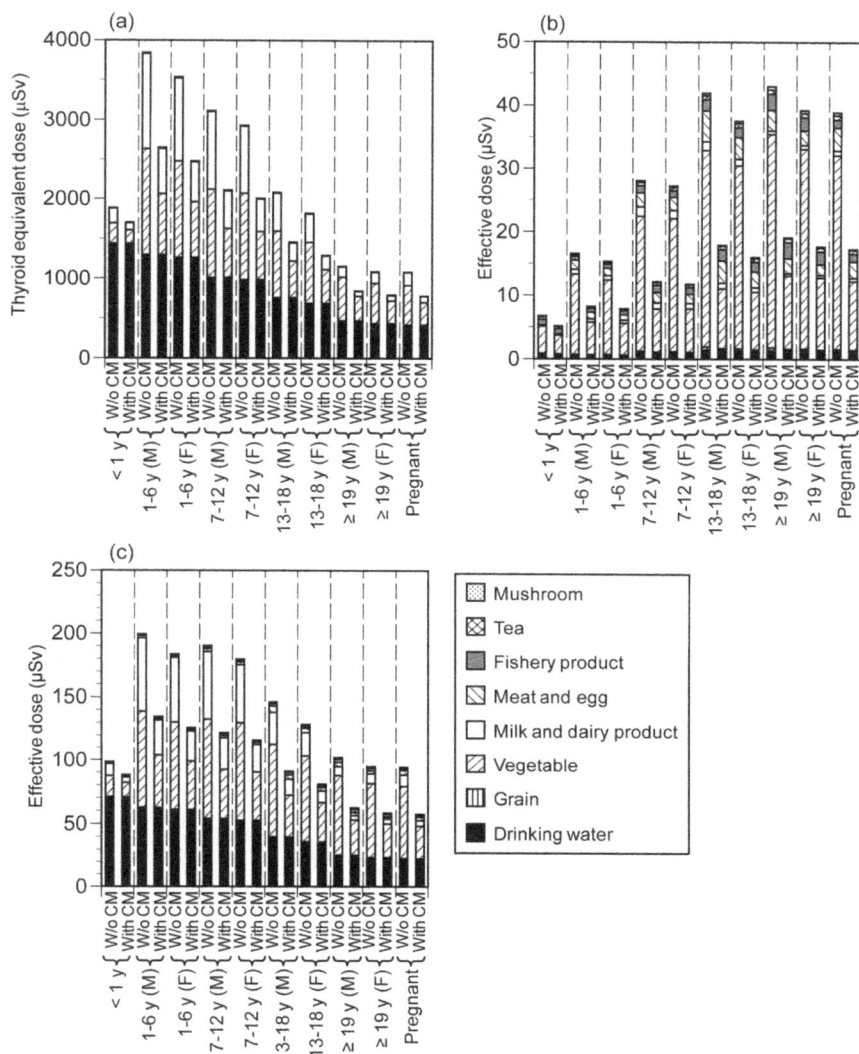

Figure 1. Average doses with and without countermeasures in Fukushima City (Case 1): (a) ^{131}I, (b) ^{134}Cs and ^{137}Cs, (c) total. CM, countermeasures; M, male; F, female. Case 1, citizens consumed vegetables bought from markets.

detected in drinking water in Osaka, and vegetables contributed the highest effective dose due to ^{134}Cs and ^{137}Cs (Figure S7). These results indicate that local contamination of drinking water caused higher intake of ^{131}I by citizens in Fukushima City and Tokyo than in Osaka, and that regional trade in foods played an important role in the intake of ^{134}Cs and ^{137}Cs by citizens in Osaka.

The 95th percentile effective dose due to ingestion of ^{134}Cs and ^{137}Cs in Fukushima City (Case 1) was 43 μSv; this was 4 times the 97th percentile internal exposure to ^{134}Cs and ^{137}Cs (i.e., 10 μSv, the value below which the exposures of 328 out of 337 individuals fell) in inhabitants of the village of Kawauchi in Fukushima Prefecture, as monitored by whole-body-counter surveys in November 2011 [12]. The effects of differences in monitoring time were small, because our average effective dose due to ^{134}Cs and ^{137}Cs in November 2011 (1.3 μSv/month) was similar to the annual average (19 μSv/y = 1.6 μSv/month). Note that the whole-body-counter surveys had sampling bias and may have underes-

timated the doses received by the general population [11]. The 95th percentile effective doses due to total radionuclides were 110 μSv in Fukushima City (Case 1), 530 μSv in Fukushima City (Case 2), 80 μSv in Tokyo, and 7.1 μSv in Osaka, 2 to 3 times the average values (Table 2). These values were much lower than the annual effective dose due to other natural radionuclides in the diet; e.g., potassium 40 (^{40}K), 130–217 μSv [68]; polonium 210 (^{210}Po), 730 μSv [69]. The provisional limits of radionuclides in drinking water and foods, which were announced by the Japanese government just after the accident, were determined from the intervention levels of 50 000 μSv of an annual thyroid equivalent dose due to ^{131}I and 5000 μSv of an annual effective dose due to ^{134}Cs, ^{137}Cs, ^{89}Sr, and ^{90}Sr [22]. Standards released in April 2012 were determined from the intervention level of 1000 μSv of annual effective dose [21]. The 95th percentile doses due to ingestion of individual and total radionuclides in Fukushima City (Cases 1 and 2) and Tokyo were greater than or equal to 1 order of magnitude lower than the provisional limits and also lower than the new limits. Those in Osaka were greater than 2 orders of magnitude lower.

Comparison of doses among ages, effects of countermeasures, and change of intake over time

The thyroid equivalent dose due to ingestion of ^{131}I in Fukushima City (Case 1) decreased with increasing age from 1 y (Figure 1, Tables S5, S7), although the daily consumption of drinking water and most foods increases with age (Table S2). This discrepancy between dose and daily consumption is attributable to the thyroid ingestion dose coefficient, which depends largely on age. In contrast, the effective doses due to ^{134}Cs and ^{137}Cs increased with age. Overall, the effective doses due to total radionuclides were higher for younger people except for <1-y-old infants (Table S6, S8). The difference in dose between sexes was small. Because the dose coefficients are the same for males and females (Table S1), the differences in doses can be attributed to differences in consumption patterns. Similar results were found in Case 2, Tokyo and Osaka, with the exception that there were no large differences in effective dose due to total radionuclides among ages in Osaka (Figures S5–S7, Tables S9–S20).

The countermeasures reduced the intake of ^{131}I by >1-y-olds by 27%–32% in Fukushima City (Case 1), by 11%–15% in Tokyo, and by 33%–37% in Osaka, and reduced the intake of ^{134}Cs and ^{137}Cs by 49%–57% in Fukushima City (Case 1), by 18%–25% in Tokyo, and by 16%–24% in Osaka (Table S21). The reductions of intake were generally lower for <1-y-old infants because the contributions of drinking water were higher for them.

The cumulative intakes of ^{131}I for ≥19-y-old males in Fukushima City (Case 1) in the presence of countermeasures were 72% of the total in week 1 and 87% in week 2 (Figure 2) and stopped increasing within 1 month. This was consistent with our previous result [19] and is attributable to the short half-life of ^{131}I and the absence of new serious releases from the nuclear power plant to the atmosphere. The intake of ^{131}I was dominant within the first 2 weeks: rapid countermeasures are therefore important in reducing intake. On the other hand, the cumulative intakes of ^{134}Cs and ^{137}Cs were 25% of the total in week 1 and 27% in week 2. The longer-term intake of ^{134}Cs and ^{137}Cs is due to the longer

half-lives. In particular, the intake via consumption of vegetables and fisheries products was continuous. Similar results were found in Case 2, Tokyo and Osaka (Figures S8–S10).

LARs of cancer incidence

The LARs of cancer from ingestion are summarized in Table 3. These values do not include risks from external exposure and inhalation. The LARs of all the solid cancer risks were 8.4–12×10^{-6} in Fukushima City (Case 1), $45–71 \times 10^{-6}$ in Fukushima City (Case 2), $3.0–5.0 \times 10^{-6}$ in Tokyo, and $1.1–2.0 \times 10^{-6}$ in Osaka, whereas the LARs of leukemia were >1 order of magnitude lower than those for all the solid cancers (i.e., 0.34–0.59×10^{-6} in Fukushima City (Case 1), $2.0–3.5 \times 10^{-6}$ in Fukushima City (Case 2), $0.14–0.25 \times 10^{-6}$ in Tokyo, and 0.05–0.09×10^{-6} in Osaka. The risk of all solid cancers combined together with leukemia is intended to provide an overall indication of the lifetime risk of cancer; however, in circumstances where the tissue doses are highly heterogeneous, such as with doses of ^{131}I to the thyroid, the risk of all solid cancers underestimates the cancer risk in specific tissues [48]. This is because the LARs of thyroid cancer in some young groups (<1 y and 10 y in the first year) in Fukushima City and Tokyo exceeded those of all solid cancers. The LARs of all solid cancers estimated from the colon dose are not suitable in the case of ingestion, although they may be still useful as indicators of the risk of deadly cancers. The LARs of thyroid cancers were $2.3–39 \times 10^{-6}$ in Fukushima City (Case 1), $10–98 \times 10^{-6}$ in Fukushima City (Case 2), $0.95–14 \times 10^{-6}$ in Tokyo, and $0.11–1.3 \times 10^{-6}$ in Osaka, and the maximum LAR in Fukushima City (Case 1) was found in females <1 y old. These values were calculated from the average doses. The 95th percentile doses were 2× the average values for Fukushima City (Cases 1 and 2) and 3× those for Tokyo and Osaka. Because of the variation of doses, a factor of at least 2–3 could be applied to the 95th percentiles of LARs. In addition, these values cannot be directly compared with the ongoing results of diagnosis of thyroid cancers in Fukushima Prefecture because of the screening effects of diagnosis and the use of advanced ultrasound techniques [18]. The contributions of ^{131}I to the lifetime-exposure thyroid doses were 92%–96% in Fukushima City (Case 1), 84%–92% in Fukushima

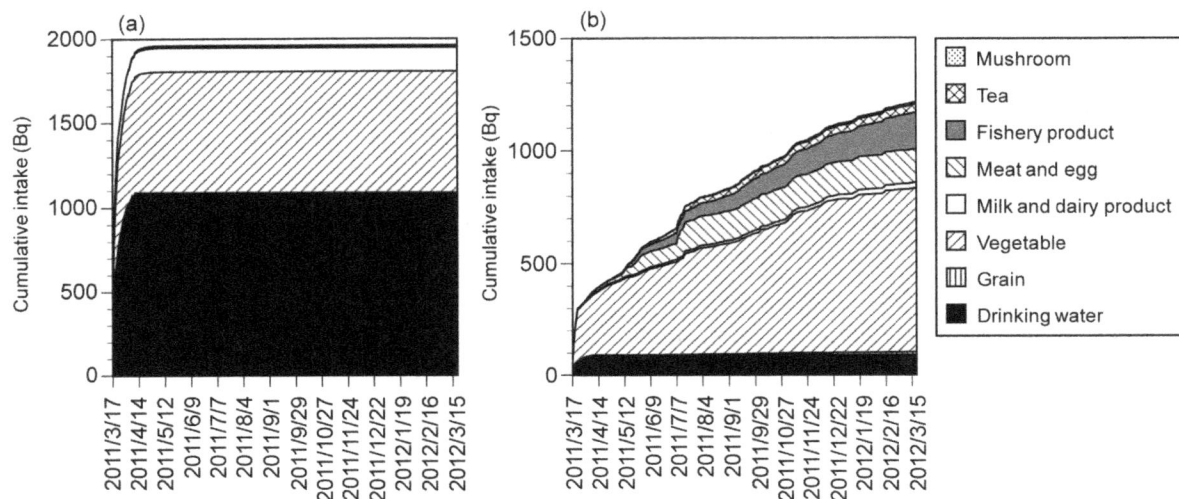

Figure 2. Cumulative intakes for ≥19-y-old male in Fukushima City (Case 1) with countermeasures: (a) ^{131}I, (b) ^{134}Cs and ^{137}Cs. Case 1, citizens consumed vegetables bought from markets.

Table 3. LARs for all solid cancers, leukemia, breast cancer, and thyroid cancer ($\times 10^{-6}$) due to ingestion.

	All solid cancers	Leukemia	Breast cancer	Thyroid cancer
Fukushima City (Case 1)				
<1 y (M)	8.6	0.57	-	9.5
<1 y (F)	12	0.36	1.5	39
10 y (M)	8.4	0.59	-	7.6
10 y (F)	11	0.36	1.5	30
20 y (M)	7.4	0.56	-	2.3
20 y (F)	9.8	0.34	1.1	9.0
Fukushima City (Case 2)				
<1 y (M)	47	3.1	-	15
<1 y (F)	66	2.0	8.4	62
10 y (M)	51	3.5	-	25
10 y (F)	71	2.2	9.3	98
20 y (M)	45	3.4	-	10
20 y (F)	61	2.1	7.1	38
Tokyo				
<1 y (M)	3.6	0.24	-	3.4
<1 y (F)	5.0	0.15	0.63	14
10 y (M)	3.6	0.25	-	3.1
10 y (F)	4.7	0.15	0.61	12
20 y (M)	3.0	0.23	-	0.95
20 y (F)	3.9	0.14	0.44	3.6
Osaka				
<1 y (M)	1.4	0.09	-	0.15
<1 y (F)	2.0	0.06	0.25	0.62
10 y (M)	1.3	0.09	-	0.20
10 y (F)	1.8	0.06	0.23	1.3
20 y (M)	1.1	0.08	-	0.11
20 y (F)	1.4	0.05	0.16	0.42

Ages represent ones in the first year. Case 1, citizens consumed vegetables bought from markets. Case 2, citizens consumed vegetables grown locally.

Table 4. LARs for all solid cancers, leukemia, breast cancer, and thyroid cancer ($\times 10^{-6}$) due to external exposure, inhalation, and ingestion.

	All solid cancers	Leukemia	Breast cancer	Thyroid cancer
Fukushima City (Case 1)				
<1 y (M)	2700	300	-	120
<1 y (F)	4100	210	1200	490
10 y (M)	2000	170	-	66
10 y (F)	3000	110	690	270
20 y (M)	1400	120	-	29
20 y (F)	2100	79	390	120
Fukushima City (Case 2)				
<1 y (M)	2700	310	-	130
<1 y (F)	4200	210	1200	520
10 y (M)	2100	170	-	83
10 y (F)	3100	120	700	340
20 y (M)	1400	120	-	37
20 y (F)	2100	81	390	150

Ages are those in the first year. Case 1, citizens consumed vegetables bought from markets. Case 2, citizens consumed vegetables grown locally.

City (Case 2), 91%–96% in Tokyo, and 62%–78% in Osaka. As described above, most of the ^{131}I was taken up within the first 2 weeks. The LARs of thyroid cancers were attributed mainly to ^{131}I in the first 2 weeks, highlighting again the fact that rapid implementation of countermeasures after a nuclear accident is important. On the other hand, Japanese dietary exposure gives lifetime cancer risks of 310×10^{-6} from inorganic arsenic (for the sum of skin, liver, and lung cancers) [70] and 1400×10^{-6} from acrylamide [71]. Cancer risks from ingestion–especially in the case of the thyroid cancer risk in young females in Fukushima City–might not be negligible, but they are 1 to 2 orders of magnitude lower than the cancer risks from ubiquitous carcinogens in the daily diet.

The total effective doses due to ingestion, inhalation, and external exposure in the first year and over a lifetime (up to 89 y) for Fukushima City are summarized in Table S22. The LARs of all solid cancers due to the three pathways were $1400–4100 \times 10^{-6}$ in Fukushima City (Case 1) and $1400–4200 \times 10^{-6}$ in Fukushima City (Case 2); those of thyroid cancer were $29–490 \times 10^{-6}$ in Fukushima City (Case 1) and $37–520 \times 10^{-6}$ in Fukushima City (Case 2) (Table 4). The contributions of LARs of all solid cancer due to ingestion were 0.3%–0.5% (Case 1) and 1.6%–3.1% (Case 2); those of thyroid cancer were 7.5%–12% (Case 1) and 12%–30% (Case 2). Even when the 95th percentile values were 2× the average, the contributions of LARs of all solid cancers due to ingestion were minor. However, for the 95th percentile values in Case 2, the contribution of LARs of thyroid cancer due to ingestion was approximately half of the total. Except in such extreme situations, the contributions of LARs of thyroid cancer due to ingestion cannot be regarded as either negligible or predominant but as minor.

As described above, the contributions of ^{131}I to the lifetime-exposure thyroid doses were >84% for Fukushima City, and the contributions of LARs from foods in the second and subsequent years were negligible. However, in Japan and other countries, some people are still concerned about artificial radionuclides in the diet. Although a greater understanding of risks alone may not lead to reasonable countermeasures, the doses and cancer risks calculated in this study are essential information.

Supporting Information

Figure S1 Locations of each prefecture in Japan.

Figure S2 ^{131}I concentrations in leafy vegetables, fruit vegetables, milk and dairy products, meat and eggs, and marine products in Fukushima Prefecture in March 2011. a: Gunma; b: April.

Figure S3 ^{134}Cs and ^{137}Cs concentrations in foods in Fukushima Prefecture in 2011. a: August; b: July; c: June; d: March; e: Gunma in March; f: March–April; g: May; h: April; i: October; j: September; k: Tochigi; December; l: April–May; m: Ibaraki in May.

Figure S4 Radionuclide concentrations in each prefecture in March 2011. (a) ^{131}I; (b) ^{134}Cs and ^{137}Cs.

Figure S5 Average doses with and without countermeasures in Fukushima City (Case 2): (a) ^{131}I, (b) ^{134}Cs and ^{137}Cs, (c) total. CM, countermeasures; M, male; F, female. Case 2, citizens consumed vegetables grown locally.

Figure S6 Average doses with and without countermeasures in Tokyo: (a) ^{131}I, (b) ^{134}Cs and ^{137}Cs, (c) total. CM, countermeasures; M, male; F, female.

Figure S7 Average doses with and without countermeasures in Osaka: (a) ^{131}I, (b) ^{134}Cs and ^{137}Cs, (c) total. CM, countermeasures; M, male; F, female.

Figure S8 Cumulative intakes for ≥ 19 y old male in Fukushima City (Case 2) with countermeasures: (a) ^{131}I, (b) ^{134}Cs and ^{137}Cs. Case 2, citizens consumed vegetables grown locally.

Figure S9 Cumulative intakes for ≥ 19 y old male in Tokyo with countermeasures: (a) ^{131}I, (b) ^{134}Cs and ^{137}Cs.

Figure S10 Cumulative intakes for ≥ 19 y old male in Osaka with countermeasures: (a) ^{131}I, (b) ^{134}Cs and ^{137}Cs.

Methods S1 Numbers of samples analyzed; Dose estimation for inhalation and external exposure; Risk models and their parameters.

Table S1 Thyroid equivalent dose coefficients for ingestion of ^{131}I and the effective dose coefficients for ingestion of ^{131}I, ^{134}Cs and ^{137}Cs (μSv/Bq).

Table S2 Daily consumption rates of drinking water and foods, and relative standard deviations (RSDs) of doses received by ingesting the foods in the same areas.

Table S3 Numbers of samples analyzed each month.

Table S4 Numbers of samples (leafy vegetables, fruit vegetables, milk and dairy products, meat and eggs, and marine products) analyzed in each prefecture in March and April 2011.

Table S5 Average thyroid equivalent doses of ^{131}I without countermeasures in Fukushima City (Case 1) in the first year after the accident (μSv). M, male; F, female. Case 1, citizens consumed vegetables bought from markets.

Table S6 Average effective doses of ^{134}Cs and ^{137}Cs without countermeasures in Fukushima City (Case 1) in the first year after the accident (μSv). M, male; F, female. Case 1, citizens consumed vegetables bought from markets.

Table S7 Average thyroid equivalent doses of ^{131}I with countermeasures in Fukushima City (Case 1) in the first year after the accident (μSv). M, male; F, female. Case 1, citizens consumed vegetables bought from markets.

Table S8 Average effective doses of ^{134}Cs and ^{137}Cs with countermeasures in Fukushima City (Case 1) in the first

year after the accident (μSv). M, male; F, female. Case 1, citizens consumed vegetables bought from markets.

Table S9 Average thyroid equivalent doses of ^{131}I without countermeasures in Fukushima City (Case 2) in the first year after the accident (μSv). M, male; F, female. Case 2, citizens consumed vegetables grown locally.

Table S10 Average effective doses of ^{134}Cs and ^{137}Cs without countermeasures in Fukushima City (Case 2) in the first year after the accident (μSv). M, male; F, female. Case 2, citizens consumed vegetables grown locally.

Table S11 Average thyroid equivalent doses of ^{131}I with countermeasures in Fukushima City (Case 2) in the first year after the accident (μSv). M, male; F, female. Case 2, citizens consumed vegetables grown locally.

Table S12 Average effective doses of ^{134}Cs and ^{137}Cs with countermeasures in Fukushima City (Case 2) in the first year after the accident (μSv). M, male; F, female. Case 2, citizens consumed vegetables grown locally.

Table S13 Average thyroid equivalent doses of ^{131}I without countermeasures in Tokyo in the first year after the accident (μSv). M, male; F, female.

Table S14 Average effective doses of ^{134}Cs and ^{137}Cs without countermeasures in Tokyo in the first year after the accident (μSv). M, male; F, female.

Table S15 Average thyroid equivalent doses of ^{131}I with countermeasures in Tokyo in the first year after the accident (μSv). M, male; F, female.

Table S16 Average effective doses of ^{134}Cs and ^{137}Cs with countermeasures in Tokyo in the first year after the accident (μSv). M, male; F, female.

Table S17 Average thyroid equivalent doses of ^{131}I without countermeasures in Osaka in the first year after the accident (μSv). M, male; F, female.

Table S18 Average effective doses of ^{134}Cs and ^{137}Cs without countermeasures in Osaka in the first year after the accident (μSv). M, male; F, female.

Table S19 Average thyroid equivalent doses of ^{131}I with countermeasures in Osaka in the first year after the accident (μSv). M, male; F, female.

Table S20 Average effective doses of ^{134}Cs and ^{137}Cs with countermeasures in Osaka in the first year after the accident (μSv). M, male; F, female.

Table S21 Reductions of doses by countermeasures (%). M, male; F, female. Case 1, citizens consumed vegetables bought from markets.

Table S22 Effective doses in the first year and over the total lifetime (up to 89 y) due to three pathways (μSv). Ages are those in the first year. M, male; F, female. Case 1, citizens consumed vegetables bought from markets. Case 2, citizens consumed vegetables grown locally.

Author Contributions

Conceived and designed the experiments: MM TO. Performed the experiments: MM. Analyzed the data: MM. Contributed reagents/materials/analysis tools: MM. Wrote the paper: MM TO.

References

1. Ministry of Health Labour and Welfare (2012) Available: http://www.mhlw.go.jp/shinsai_jouhou/shokuhin.html. Accessed: 7 Oct. 2013. [in Japanese]
2. Tokyo Metropolitan Institute of Public Health (2011) Available: http://monitoring.tokyo-eiken.go.jp/monitoring/w-past_data.html. Accessed: 7 Oct. 2013.
3. Fukushima Prefecture (2013) Available: http://www.new-fukushima.jp/monitoring/. Accessed: 13 Sep. 2013. [in Japanese]
4. ICRP (1983) Radionuclide transformations - Energy and intensity of emissions. ICRP Publication 38, Ann. ICRP 11–13.
5. Working Group on Radionuclides Pharmaceutical Affairs and Food Sanitation Council (2011) Estimation of contribution of radiocesium to internal exposure from dietary intake. [in Japanese]
6. Yoshida-Ohuchi H, Hirasawa N, Kobayashi I, Yoshizawa T (2013) Evaluation of personal dose equivalent using optically stimulated luminescent dosemeters in Marumori after the Fukushima nuclear accident. Radiat Prot Dosim 154: 385–390.
7. Yasutaka T, Iwasaki Y, Hashimoto S, Naito W, Ono K, et al. (2013) A GIS-based evaluation of the effect of decontamination on effective doses due to long-term external exposures in Fukushima. Chemosphere 93: 1222–1229.
8. Evangeliou N, Balkanski Y, Cozic A, Moller AP (2013) Global transport and deposition of Cs-137 following the Fukushima nuclear power plant accident in Japan: Emphasis on Europe and Asia using high-resolution model versions and radiological impact assessment of the human population and the environment using interactive tools. Environ Sci Technol 47: 5803–5812.
9. Amano H, Akiyama M, Bi C, Kawamura T, Kishimoto T, et al. (2012) Radiation measurements in the Chiba Metropolitan Area and radiological aspects of fallout from the Fukushima Dai-ichi nuclear power plants accident. J Environ Radioactiv 111: 42–52.
10. Christoudias T, Lelieveld J (2013) Modelling the global atmospheric transport and deposition of radionuclides from the Fukushima Dai-ichi nuclear accident. Atmos Chem Phys 13: 1425–1438.
11. Hayano RS, Tsubokura M, Miyazaki M, Satou H, Sato K, et al. (2013) Internal radiocesium contamination of adults and children in Fukushima 7 to 20 months after the Fukushima NPP accident as measured by extensive whole-body-counter surveys. P Jpn Acad B-Phys 89: 157–163.
12. Tsubokura M, Kato S, Nihei M, Sakuma Y, Furutani T, et al. (2013) Limited internal radiation exposure associated with resettlements to a radiation-contaminated homeland after the Fukushima Daiichi Nuclear Disaster. PLOS ONE 8: e81909.
13. Harada KH, Fujii Y, Adachi A, Tsukidate A, Asai F, et al. (2013) Dietary intake of radiocesium in adult residents in Fukushima prefecture and neighboring regions after the Fukushima nuclear power plant accident: 24-h food-duplicate survey in December 2011. Environ Sci Technol 47: 2520–2526.
14. Ministry of Health Labour and Welfare (2011) Available: http://www.mhlw.go.jp/stf/shingi/2r9852000001ip01-att/2r9852000001ipae.pdf. Accessed: 7 Oct. 2013. [in Japanese]
15. World Health Organization (2012) Preliminary dose estimation from the nuclear accident after the 2011 Great East Japan Earthquake and Tsunami. World Health Organization, Geneva.
16. United Nations Scientific Committee on the Effects of Atomic Radiation (2014) Sources, effects and risks of ionizing radiation. UNSCEAR 2013 Report to the General Assembly with Scientific Annexes. United Nations, New York.
17. Ministry of Agriculture Forestry and Fisheries (2013) Food demand and supply 2010: http://www.maff.go.jp/j/tokei/kouhyou/zyukyu/index.html. Accessed: 13 Sep. 2013. [in Japanese]
18. Nagataki S, Takamura N (2014) A review of the Fukushima nuclear reactor accident: radiation effects on the thyroid and strategies for prevention. Curr Opin Endocrinol Diabetes Obes 21: 384–393.
19. Murakami M, Oki T (2012) Estimation of thyroid doses and health risks resulting from the intake of radioactive iodine in foods and drinking water by the citizens of Tokyo after the Fukushima nuclear accident. Chemosphere 87: 1355–1360.

20. Osaka Water Supply Authority (2013) Available: http://www.wsa-osaka.jp/gaiyou/jigyoukanri/keikakuka/housyanou.html. Accessed: 13 Sep. 2013. [in Japanese]

21. Working Group on Radionuclides Pharmaceutical Affairs and Food Sanitation Council (2011) Revision of new standard for radionuclides in foods. [in Japanese]

22. Nuclear Safety Commission of Japan (1998) Control index levels on the ingestion of food and water. [in Japanese]

23. Kuroda M, Ohta M, Okufuji T, Takigami C, Eguchi M, et al. (2011) Frequency of soup intake and amount of dietary fiber intake are inversely associated with plasma leptin concentrations in Japanese adults. Appetite 54: 538–543.

24. Izumo Y (2001) Ministry of Health Labour and Welfare Grants 200000078A. [in Japanese]

25. Nakamura T, Yurimoto M, Matsumoto K, Kawasaki N, Tanada S (1996) Sensory taste evaluation of milled rice cooked with water different in hardness. Jpn J Food Chem 3: 141–144.

26. Tokyo Metropolitan Central Wholesale Market (2011) Available: http://www.shijou-tokei.metro.tokyo.jp/index.html. Accessed: 7 Oct. 2013. [in Japanese]

27. Ministry of Agriculture Forestry and Fisheries (2013) Fishery and aquaculture industrial production statistics 2011: http://www.e-stat.go.jp/SG1/estat/List.do?lid=000001104479. Accessed: 7 Oct. 2013. [in Japanese]

28. Suzuki K, Okuda T, HIgasine Y (2006) The Daily intakes of drinks and folate from the drinks by young women. Mem Osaka Kyoiku Univ Ser 2 Soc Sci Home Econ 54: 27–34. [in Japanese]

29. Ministry of Agriculture Forestry and Fisheries (2011) Basic data on specific forest products industry: http://www.e-stat.go.jp/SG1/estat/List.do?lid=0000010 85178. Accessed: 13 Sep. 2013. [in Japanese]

30. Ministry of Agriculture Forestry and Fisheries (2011) Crop production statistics 2009: http://www.e-stat.go.jp/SG1/estat/List.do?lid=000001069768. Accessed: 7 Oct. 2013. [in Japanese]

31. Statistics Bureau of Japan (2011) National Population Census 2010.

32. Fukushima City Central Wholesale Market (2013) Available: http://www.city.fukushima.fukushima.jp/soshiki/24/1030.html. Accessed: 13 Sep. 2013. [in Japanese]

33. Osaka Central Wholesale Market (2013) Available: http://www.pref.osaka.jp/fuichiba/toukeijouhou/nenpou22.html. Accessed: 13 Sep. 2013. [in Japanese]

34. Ministry of Agriculture Forestry and Fisheries (2011) Statistics of animal product distribution 2009: http://www.e-stat.go.jp/SG1/estat/List.do?lid=00000107 2695. Accessed: 13 Sep. 2013. [in Japanese]

35. Ministry of Agriculture Forestry and Fisheries (2011) Statistics of animal product distribution 2010: http://www.e-stat.go.jp/SG1/estat/List.do?lid=000001076341. Accessed: 7 Oct. 2013. [in Japanese]

36. Ministry of Agriculture Forestry and Fisheries (2010) Statistical survey on milk and dairy products 2008: http://www.e-stat.go.jp/SG1/estat/List.do?lid=000001063504. Accessed: 13 Sep. 2013. [in Japanese]

37. Ministry of Agriculture Forestry and Fisheries (2013) Crop statistics 2011: http://www.e-stat.go.jp/SG1/estat/List.do?lid=000001087011. Accessed: 13 Sep. 2013. [in Japanese]

38. Ministry of Agriculture Forestry and Fisheries (2012) Crop production statistics: http://www.e-stat.go.jp/SG1/estat/List.do?lid=000001081820. Accessed: 13 Sep. 2013. [in Japanese]

39. Ministry of Agriculture Forestry and Fisheries (2007) Statistical survey on livestock 2006: http://www.e-stat.go.jp/SG1/estat/List.do?lid=000001060567. Accessed: 7 Oct. 2013. [in Japanese]

40. Ministry of Agriculture Forestry and Fisheries (2011) Statistical survey on cultivated acreage (FY 2009): http://www.e-stat.go.jp/SG1/estat/List.do?lid=000001061777. Accessed: 7 Oct. 2013. [in Japanese]

41. Fukushima Prefecture (2010) Statistics of forest industry in Fukushima 2009. [in Japanese]

42. Ministry of Agriculture Forestry and Fisheries (2008) Census of agriculture and forestry 2005: http://www.maff.go.jp/j/tokei/census/afc/2010/report05_archives.html. Accessed: 13 Sep. 2013. [in Japanese]

43. Japan Meteorological Agency (2011) Available: http://www.data.jma.go.jp/obd/stats/etrn/index.php. Accessed: 7 Oct. 2013. [in Japanese]

44. Murakami M, Takeda H, Okaneya M, Kobayashi Y, Oki T (2012) Classification and motives of drink intakes based on the situations using behavioral record. SEISAM-KENKYU 64: 359–366. [in Japanese]

45. Japan Water Works Association (2011) Statistics of tap water (2009 FY). [in Japanese]

46. Bureau of Waterworks Tokyo Metropolitan Government (2013) Available: https://www.waterworks.metro.tokyo.jp/press/shinsai22/press01.html. Accessed: 13 Sep. 2013.

47. Ministry of Health and Welfare (2012) The National Health and Nutrition Survey in Japan, 2010. [in Japanese]

48. World Health Organization (2013) Health risk assessment from the nuclear accident after the 2011 Great East Japan Earthquake and Tsunami based on a preliminary dose estimation.

49. Harada KH, Niisoe T, Imanaka M, Takahashi T, Amako K, et al. (2014) Radiation dose rates now and in the future for residents neighboring restricted areas of the Fukushima Daiichi nuclear power plant. P Natl Acad Sci USA 111: E914–E923.

50. Akahane K, Yonai S, Fukuda S, Miyahara N, Yasuda H, et al. (2013) NIRS external dose estimation system for Fukushima residents after the Fukushima Dai-ichi NPP accident. Sci Rep 3: 1670.

51. Brenner AV, Tronko MD, Hatch M, Bogdanova TI, Oliynik VA, et al. (2011) I-131 dose response for incident thyroid cancers in Ukraine related to the Chornobyl accident. Environ Health Persp 119: 933–939.

52. United Nations Scientific Committee on the Effects of Atomic Radiation (2008) Annex A: Epidemiological studies of radiation and cancer. UNSCEAR 2006 Report to the General Assembly: Effects of Ionizing Radiation (United Nations Scientific Committee on the Effects of Atomic Radiation, Vienna), Vol. 1.

53. Preston DL, Ron E, Tokuoka S, Funamoto S, Nishi N, et al. (2007) Solid cancer incidence in atomic bomb survivors: 1958–1998. Radiat Res 168: 1–64.

54. Statistics Bureau of Japan (2011) General mortality in 2010: http://www.e-stat.go.jp/SG1/estat/ListE.do?lid=000001101825. Accessed: 5 Mar. 2014.

55. Center for Cancer Control and Information Services National Cancer Center Japan (2013) Cancer incidence (1975–2008): http://ganjoho.jp/professional/statistics/statistics.html. Accessed: 5 Mar. 2014. [in Japanese]

56. Matsuda A, Matsuda T, Shibata A, Katanoda K, Sobue T, et al. (2013) Cancer incidence and incidence rates in Japan in 2007: A study of 21 population-based cancer registries for the monitoring of cancer incidence in Japan (MCIJ) project. Jpn J Clin Oncol 43: 328–336.

57. Mullenders L, Atkinson M, Paretzke H, Sabatier L, Bouffler S (2009) Assessing cancer risks of low-dose radiation. Nat Rev Cancer 9: 596–604.

58. Russell WL, Kelly EM (1982) Mutation frequencies in male-mice and the estimation of genetic hazards of radiation in men. P Natl Acad Sci–Biol 79: 542–544.

59. Ciccia A, Elledge SJ (2010) The DNA damage response: Making it safe to play with knives. Mol Cell 40: 179–204.

60. ICRP (2007) The 2007 recommendations of the international commission on radiological protection. ICRP Publication 103, Ann. ICRP. 37 (2–4).

61. National Research Council of the National Academies (2006) Health risks from exposure to low levels of ionizing radiation: BEIR VII Phase 2. Board on Radiation Effects Research.

62. Kinoshita N, Sueki K, Sasa K, Kitagawa J-i, Ikarashi S, et al. (2011) Assessment of individual radionuclide distributions from the Fukushima nuclear accident covering central-east Japan. P Natl Acad Sci USA 108: 19526–19529.

63. Kosaka K, Asami M, Kobashigawa N, Ohkubo K, Teracia H, et al. (2011) Removal of radioactive iodine and cesium in water purification processes after an explosion at a nuclear power plant due to the Great East Japan Earthquake. Water Res 46: 4397–4404.

64. Diamond Online (2013) 8th July. http://diamond.jp/articles/-/38458. [in Japanese]

65. Hangui S (2012) Enormous influences on agriculture in Fukushima prefecture caused by the Great East Japan Earthquake and nuclear disaster: Focused on distribution and consumption stage. Jpn J Farm Manag 49: 93–96. [in Japanese]

66. Japan Soft Drink Association (2010) Available: http://j-sda.or.jp/statistically-information/stati01.php. Accessed: 5 Sep. 2013.

67. Koizumi A, Harada KH, Niisoe T, Adachi A, Fujii Y, et al. (2012) Preliminary assessment of ecological exposure of adult residents in Fukushima Prefecture to radioactive cesium through ingestion and inhalation. Environ Health Prev Med 17: 292–298.

68. Sugiyama H, Terada H, Takahashi M, Iijima I, Isomura K (2007) Contents and daily intakes of gamma-ray emitting nuclides, ^{90}Sr, and ^{238}U using market-basket studies in Japan. J Health Sci 53: 107–118.

69. Ota T, Sanada T, Kashiwara Y, Morimoto T, Sato K (2009) Evaluation for committed effective dose due to dietary foods by the intake for Japanese adults. Jpn J Health Phys 44: 80–88.

70. Oguri T, Yoshinaga J, Tao H, Nakazato T (2012) Daily intake of inorganic arsenic and some organic arsenic species of Japanese subjects. Food Chem Toxicol 50: 2663–2667.

71. Murakami M, Nagai T (2013) Cancer risks from micro-pollutants and their acceptable levels. J Jpn Soc Water Environ 36(A): 322–326. [in Japanese]

22

Plant-Adapted *Escherichia coli* Show Increased Lettuce Colonizing Ability, Resistance to Oxidative Stress and Chemotactic Response

Maria de los Angeles Dublan[1,2,3]*, **Juan Cesar Federico Ortiz-Marquez**[1,2], **Lina Lett**[3], **Leonardo Curatti**[1,2]*

1 Instituto de Investigaciones en Biodiversidad y Biotecnología, Consejo Nacional de Investigaciones Científicas y Técnicas, Mar del Plata, Buenos Aires, Argentina, **2** Fundación para Investigaciones Biológicas Aplicadas, Mar del Plata, Buenos Aires, Argentina, **3** Laboratorio Integrado de Microbiología Agrícola y de Alimentos, Facultad de Agronomía, Universidad Nacional del Centro de la Provincia de Buenos Aires, Azul, Buenos Aires, Argentina

Abstract

Background: *Escherichia coli* is a widespread gut commensal and often a versatile pathogen of public health concern. *E. coli* are also frequently found in different environments and/or alternative secondary hosts, such as plant tissues. The lifestyle of *E. coli* in plants is poorly understood and has potential implications for food safety.

Methods/Principal Findings: This work shows that a human commensal strain of *E. coli* K12 readily colonizes lettuce seedlings and produces large microcolony-like cell aggregates in leaves, especially in young leaves, in proximity to the vascular tissue. Our observations strongly suggest that those cell aggregates arise from multiplication of single bacterial cells that reach those spots. We showed that *E. coli* isolated from colonized leaves progressively colonize lettuce seedlings to higher titers, suggesting a fast adaptation process. *E. coli* cells isolated from leaves presented a dramatic rise in tolerance to oxidative stress and became more chemotactic responsive towards lettuce leaf extracts. Mutant strains impaired in their chemotactic response were less efficient lettuce colonizers than the chemotactic isogenic strain. However, acclimation to oxidative stress and/or minimal medium alone failed to prime *E. coli* cells for enhanced lettuce colonization efficiency.

Conclusion/Significance: These findings help to understand the physiological adaptation during the alternative lifestyle of *E. coli* in/on plant tissues.

Editor: Dawn Arnold, University of the West of England, United Kingdom

Funding: The authors received no specific funding for this work.

Competing Interests: The authors have declared that no competing interests exist.

* Email: mdublan@faa.unicen.edu.ar (MAD); lcuratti@fiba.org.ar (LC)

Introduction

Escherichia coli is a common resident of animal hosts mostly as a commensal of the lower intestine of mammals. However, some strains are versatile pathogens of vertebrates, thought to kill more than 2 million humans per year through both intraintestinal and extraintestinal diseases, some of them at infective doses as low as ten cells [1]. Among them, *E. coli* O157:H7 has been a serious public-health concern worldwide since the first outbreak report in 1982 [2].

It is presumed that about one-half of the total population of *E. coli* resides in the primary habitat of the host (lower intestine of warm blood animals) and the other half is in the external environment (secondary habitat) [3]. It has been traditionally accepted that *E. coli* grows and divides in its primary habitat but does not live in non-host environments. Thus, continuous bulk transfer from humans and animals would account for a stable population in the secondary hosts or habitats. This thought became the basis for the use of coliforms determination as an indication of fecal contamination [3].

However, several *E. coli* strains are members of the natural flora in certain tropical ecosystems in the absence of known fecal contamination [3–5]. *Escherichia coli* internalization, and often colonization, of vegetables such as lettuce [6], spinach [7], apple [8], orange [9], cress and radish [7], carrots and onions [10] constitutes a serious risk for the health of fresh produce consumers since most commonly used sanitizers are less effective under these circumstances [11]. The factors that allow *E. coli* strains to persist in a non-host environment such as plant tissues remain poorly understood [12,13]. The available datasets of genomic information from plant associated isolates and its comparison to their corresponding close relatives displaying an animal host preference has shed some valuable insights into the endophytic lifestyle of *E. coli*. Thus, it has been proposed that similar strategies comprising initial adherence, invasion, and establishment are required for bacteria to colonize any host, whether plant or animal [14].

In this work we characterized physiological adaptations during the alternative lifestyle of *E. coli* in/on plant tissues. We show that bacteria isolated from leaves are more efficient plant colonizers, more tolerant to oxidative stress and present a more stringent chemotactic response to leaf extracts. These findings support and improve the current understanding of *E. coli* lifestyle in/on plants.

Results

Escherichia coli Colonization of Lettuce

A gnotobiotic experimental system comprising lettuce seedlings transplanted onto agarized-Hoagland medium (0.6% agar-agar) inoculated with *E. coli* K12 strain MG1655 was used to study basic aspects of *E. coli* internalization and multiplication in/on plants as a non-host environment. Using the standardized colonization assay, *E. coli* colonized lettuce roots and leaves at an average of 6.9×10^9 or 7.3×10^7 CFUs per gram of tissue, respectively. The colonized roots or leaves were rinsed with sterile physiological solution until no CFUs were obtained from the washing solution. This treatment reduced the number of *E. coli* CFUs retrieved from homogenized plant samples by approximately one log.

Inspection of surface-sterilized lettuce seedlings colonized with fluorescent bacteria through the laser confocal microscope confirmed the internalization of *E. coli* cells into the lettuce leaves (Fig. S1 and video S1).

We observed *E. coli* cells in contact with lettuce roots mostly localized to the root hairs outside the plant organ (not shown), and adjacent to some root domes in zones of lateral root emergence (Fig. 1A–D). Inside the root tissue, bacterial cells tended to form clusters in rows, apparently in the apoplastic space between root cells (Fig. 1E–H).

In/on leaves, patches or aggregates of bacterial cells were observed in close proximity to the vascular tissue. These bacterial cell patches were characteristically larger at the base of leaves than in tips (Fig. 2A–F, Fig. S2A–B). This characteristic pattern of cell patches in close proximity to the vascular tissue was different from that observed when *E. coli* cells were directly spotted onto the leaves surfaces. In this case the bacterial cells remained on the surface of the leaves for a few weeks, mostly in association with the stomata (Fig. 2G–H). Furthermore, *E. coli* cells were efficiently removed by washing when using this inoculation mode (not shown).

To investigate the origin of the *E. coli* cell aggregates we inoculated the Hoagland medium containing 0.6% agar-agar with a 1:1 mixed population of GFP- and RFP-labeled bacteria. The *E. coli* cell aggregates in/on roots were looser than those of leaves and contained mostly random combinations of both GFP- and RFP-labeled bacteria (Fig. 3A). Conversely, cell aggregates in/on leaves were green or red at a roughly 1:1 ratio (Fig. 3B).

Escherichia coli Isolated from Lettuce Leaves Are More Efficient Seedlings Colonizers

When we used *E. coli*-colonized leaf macerates as a source of bacteria for subsequent rounds of seedlings colonization we observed, on average, an apparent 10-fold increase in the total number of bacteria recovered from the next round of leaves colonization (Fig. 4A). Although the level of colonization appeared to be naturally variable, the observed maximal bacterial counts from lettuce leaves were up to 100-fold higher when inoculated with bacteria isolated from leaves. To further confirm this trait, competence experiments were performed by co-inoculating an equal number of RFP-labeled *E. coli* cells cultured in LB medium (non-adapted cells) and GFP-labeled cells (plant-adapted cells) freshly recovered from a previous colonization assay. Either in this or the reciprocal experiment (GFP-bacterial cells from LB medium and RFP-bacterial cells from leaves) a 9-fold increase in the competitive index towards the plant-acclimated cells was observed.

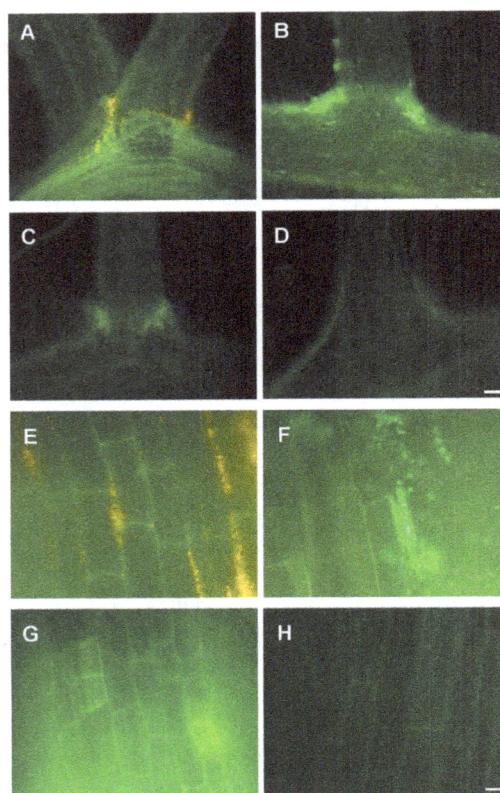

Figure 1. Representative fluorescence photomicrographs of *E. coli* K12-colonized lettuce roots. (A–D) Detail of *E. coli* localization to zones of lateral root emergence or (E–H) root elongation zone after inoculating seedlings with (A and E) RFP-, (B and F) GFP-, (C and G) non-labeled cells or (D and H) non-inoculated seedlings. Magnifications were at (A–D) 150X or (E–H) 750X. The white bar corresponds to 20 μm (A–D) or 10 μm (E–H).

These results indicated no bias according to possible differences in fitness cost due to labelling the bacteria with either fluorescent protein. Moreover, when *E. coli* populations were recovered from leaves after two consecutive cycles of seedlings colonization and then challenged against non-adapted bacteria, an even larger competitive index of 24-fold was observed for this *E. coli* population. Control experiments showed a competitive index for RFP-labeled to GFP-labeled cells of 1.0 (non-adapted) or 1.4 (plant-adapted), respectively. These results fit reasonably well with the expected value (1.0) for both non-competitive bacterial populations (Fig. 4B). When bacteria isolated from leaves were cultivated in LB medium overnight and then used for colonization assays they only partially conserved the enhanced colonization efficiency with a competitive index of 4.0 in comparison with non-adapted bacteria.

Failure to observe any considerable difference between cells cultivated in LB or M9 for non-adapted bacteria suggested that the increase in colonization ability may not be related to the nature or composition of the medium itself (not shown).

Escherichia coli Isolated from Lettuce Leaves Is More Tolerance to Oxidative Stress

We challenged plant-adapted *E. coli* cells with hydrogen peroxide and observed a dramatic increase in tolerance in comparison with non-adapted bacteria cultivated in M9 medium.

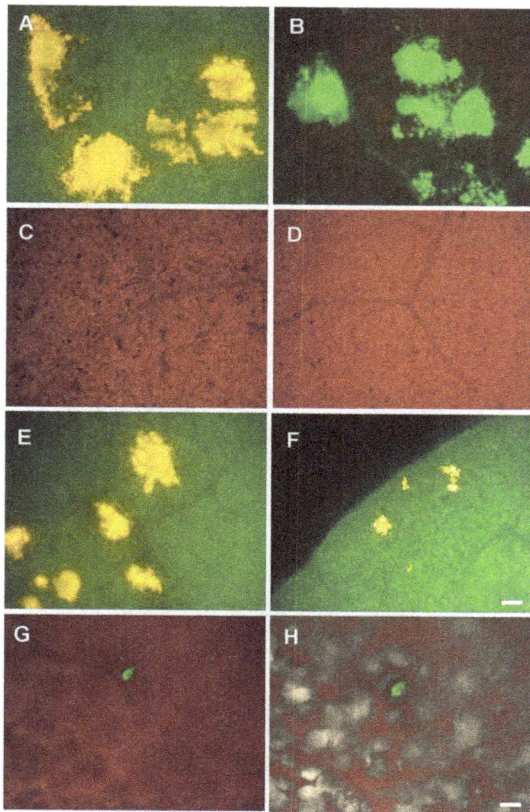

Figure 2. Representative fluorescence photomicrographs of E. coli K12-colonized lettuce leaves. Detail of E. coli cell aggregates in leaves after the inoculation of seedlings with (A) RFP-, (B) GFP-, (C) non-labeled cells or (D) non-inoculated seedlings. Detail of characteristic bacterial aggregates at (E) leaf base (close to the petiole) or (F) leaf tip (opposite to the petiole margin) of seedlings colonized with RFP-labeled E. coli. (G–H) Detail of a leaf from a lettuce seedling that had been superficially inoculated with GFP-labeled E. coli. (G) Representative fluorescence photomicrography and (H) merged images of fluorescence and bright field microscopy. Magnifications were at 150X (A–F) or 750X (G–H). White bars correspond to 20 μm (A–F) or 10 μm (G–H).

Remarkably, cells isolated from leaves still grew at hydrogen peroxide concentrations that largely compromised survival of M9 cultivated cells (Fig. 5A–B).

Since the increased colonization ability correlated with increased oxidative stress tolerance, LB cultivated cells were acclimated to oxidative stress (Fig. S3 A), and then used for lettuce colonization assays. However, these treatments failed to trigger an increased colonizing ability (Fig. S3 B).

Chemotactic Response of Escherichia coli Isolated from Lettuce Leaves

Plant-adapted E. coli cells presented a considerably more active chemotactic response as observed by migration onto TB plates (Fig. 6A–B). To confirm this result migration towards glass capillaries filled with leaf blade or vascular tissue cell-free extracts was scored as a ratio of migration towards sterile PBS buffer (control) from the same bacterial reservoir. Migration towards lettuce extracts was on average 4- to 5-fold more prominent for plant-adapted than non-adapted bacteria cultivated in LB medium. As a control we show that, conversely to its isogenic line (RP437), a mutant strain impaired in the chemotactic response

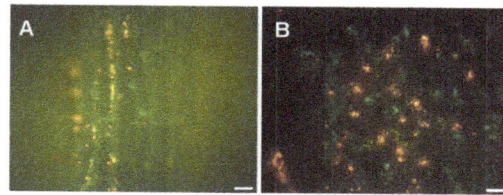

Figure 3. Origin of E. coli cell-aggregates in roots or leaves of colonized lettuce seedlings. Representative fluorescence photomicrographs of E. coli K12-colonized (A) roots or (B) leaves after the inoculation of seedlings with a 1:1 mixture of RFP- and GFP-labeled bacteria. Magnifications were at (A) 750X or (B) 150X. White bars correspond to (A) 10 μm or (B) and 20 μm.

(cheA- ΔcheA1643 strain RP9535) showed no preferential migration towards leaf-extracts (Fig. 6C–D).

Using the standard colonization assay (10^8 non-chemotactic cells$\times$mL^{-1} for 20 days), bacterial counts from roots (Fig. 7A) or leaves (Fig. 7B) were on average 15-fold lower in comparison with those reached by a similar inoculum of the chemotactic isogenic strain. Results were highly variable, similarly to those obtained for E. coli strain MG1655 colonized seedlings (Fig. 4A). It is noteworthy that the minimal colonization counts observed for the chemotactic cells were up to 1,000-fold higher than those of the non-chemotactic mutant cells (Fig. 7A–B).

In the standard assay used in this study both the inoculum size (Fig. S4 A) and incubation time (Fig. S4 B) were set up in excess to maximize the bacterial colonization of leaves. Thus, to further challenge strains impaired in their chemotactic response, we compared their performance to that of the wild type cells. We applied conditions where excess inoculum size and progression of

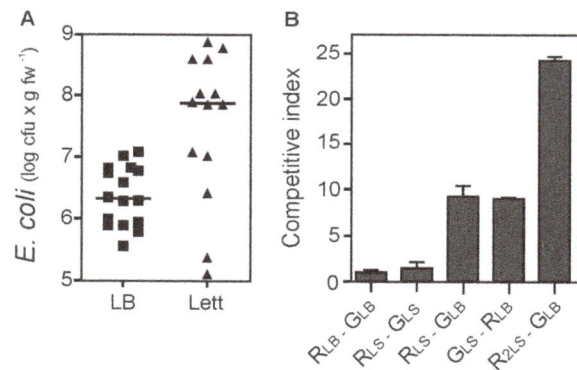

Figure 4. Escherichia coli adaptation to lettuce seedlings. (A) Data from individual assays of the colonization of lettuce leaves with non-adapted E. coli K12 strain MG1655 cultivated in LB medium (LB) or plant-adapted E. coli strain MG1655 freshly isolated from colonized lettuce leaves (Lett). The bars represent the median value of the data sets. (B) Competitive index (CI) analysis of the adaptation of E. coli K12 to colonize lettuce seedling. R_{LB} and G_{LB} are non-adapted (cultivated in LB medium) RFP- or GFP-labeled bacteria, respectively. R_{LS} and G_{LS} are RFP- or GFP-labeled, plant-adapted bacteria (isolated from colonized lettuce leaves). R_{2LS} are RFP-labeled bacteria adapted by two consecutive cycles of colonization of lettuce seedlings (isolated from leaves colonized with R_{LS} bacteria). Each data point in (A) (16 for non-adapted bacteria and 14 for adapted bacteria) corresponded to independent assays conducted at 4 to 6 different times over the course of two years. In (B) three assays per condition were conducted at two different times. Each assay (data point) consisted of three seedlings that were inoculated with bacteria in the same test tube. Differences were statistically significant (P≤0.05).

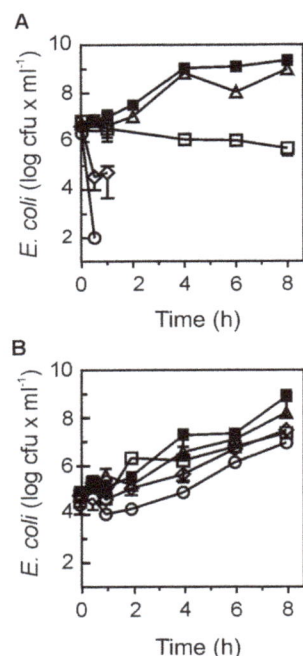

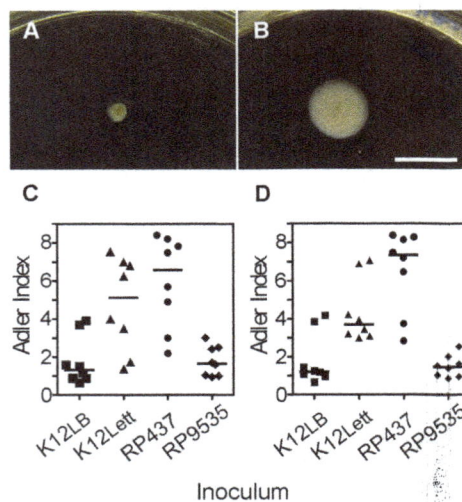

Figure 6. Chemotactic response of plant-adapted *Escherichia coli*. (A–B) Chemotactic response onto TB medium of (A) non-adapted or (B) plant-adapted bacteria. The bar in (A–B) represents 1 cm. (C–D) Chemotactic response towards (C) lettuce leaf blade or (D) lettuce-leaf main vein extracts. K12LB, non-adapted *E. coli* strain MG1655; K12, plant-adapted *E. coli* strain MG1655; RP437, non-adapted *E. coli* strain RP437 (genomic background of reference for chemotactic analysis); and RP9535, non-adapted *E. coli* strain RP9535 (ΔcheA1643 mutation in a RP437 genomic background). Strains RP437 and RP9535 were used as control of chemotactic and non-chemotactic *E. coli* K12 bacteria. Results in C–D are statistically different (P≤0.05).

Figure 5. Effect of adaptation to lettuce leaves on *Escherichia coli* K12 tolerance to oxidative stress. Survival of (A) non-adapted bacteria (cultivated in LB medium) or (B) plant-adapted bacteria (isolated from leaves) after incubation for 0 to 8 h in the presence of (■) 0; (△) 0.5; (□) 2; (◇) 8; or (○) 32 mM H_2O_2. Data represent the mean and SD of 2 independent assays.

colonization could not have compensated for some cell defect in colonization ability. Thus, *E. coli* cells were inoculated at 10^5 cells×mL^{-1} in pools of three seedlings each. After 14 days, all of them showed root colonization and one showed leaf colonization when inoculated with chemotactic bacteria. Conversely, no root or leaf colonization was observed when seedlings were inoculated with non-chemotactic bacteria in a same number of assays, confirming the results obtained by the standard assay. Inoculation with 10^2 cells of either strain×mL^{-1} failed to produce colonized seedlings in 14 days.

Discussion

This work describes basic aspects of lettuce colonization by a human commensal strain of *E. coli* K12. The colonization assays that were set up for this research produced lettuce colonization patterns that are in agreement with previous findings. Briefly, *E. coli* (among some other enterobacteria) (*i*) preferentially invade plant root tissue rather than foliage, (*ii*) about 10% of the bacteria are presumably internalized and/or tightly attached to the plant tissues, (*iii*) bacterial invasion of roots is more prominent at lateral root junctions, and in the root cortex apoplast [13]. Using an *in vitro* system similar to the one describe here, Wright et al. [15] showed that *E. coli* MG1655 was able to colonize internally only one lettuce plant out of 10. In contrast 0.5% of *E. coli* (sakai) population was internalized [15]. In field grown leafy greens, including lettuce, *E. coli* internalization and colonization was also shown to be rare [16].

Additionally, our work shows that *E. coli* K12 cells produce large microcolony-like cell aggregates in lettuce leaves that appear to arise from multiplication of single bacterial cells that reach those spots. Such a defined pattern was not observed in roots, suggesting

variations of the endophytic lifestyle of *E. coli* when internalized into different plant tissues. It was apparent that while bacteria build up their number by massive multiplication from a rather reduced number of foci (probably single cells) in leaves of colonized seedlings, direct access from the substrate into the roots might account for the observed number and distribution of bacteria in this plant organ.

The characteristic association of bacterial aggregates to the leaves vascular tissue suggests that this might be the point of access of the bacteria from the roots into the leaf tissue. Bulk transfer of bacteria along the surface of the seedlings [17] could be ruled out since surface-inoculation produced a noticeably different colonization pattern. Only very occasionally during this study fluorescent cells were found inside the vascular tissue, suggesting that in the

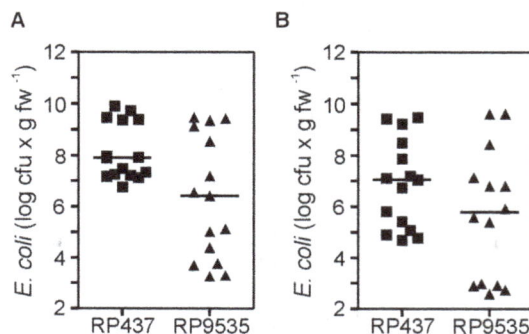

Figure 7. Effect of disrupting the chemotactic response of *Escherichia coli* on the bacterial colonization of lettuce seedlings. (A–B) *E. coli* colonization of lettuce seedlings (A) roots or (B) leaves after inoculation with non-adapted strains RP437 or PR9535. The bars represent the median value of the datasets.

experimental system used *E. coli* propagation and residency might not be prominent in this plant tissue. These results are in agreement with previous studies that show that vascular tissue and xylem cells may be invaded by comparatively lower densities of endophytic bacteria, representing a transit path for systemic spreading into leaves [12]. On the other hand, it has been shown that *Pseudomonas syringae* develops cell aggregates within bean leaves preferentially associated with veins, which has been interpreted as bacteria localizing next to a rich source of nutrients [18]. Thus, these two alternative hypotheses for the localization of cell aggregates close to vascular tissue are not mutually exclusive.

Remarkably, we showed and confirmed by means of reciprocal competitive assays that *E. coli* isolated from colonized leaves progressively colonize lettuce seedlings to higher titers. This result suggests a fast adaptation process probably mediated by the selection of more efficient genotypic/phenotypic variant(s) of the bacterium towards lettuce colonization. Similar results were obtained after sequentially passaging the type strain of *S. enterica* sv. *typhimurium* ATCC14028 through tomatoes. It was shown that spontaneous mutants that arose during tomato passages were 5- to 50-fold more competitive than the wild-type inside tomato fruits but presented a reduced fitness in laboratory medium [19]. In agreement, cultivation in laboratory medium between passages through lettuce leaves decreased the colonization competence of plant-adapted *E. coli* populations, suggesting that these populations might comprise bacteria with contrasting fitness in lettuce seedlings and laboratory medium.

We reasoned that a physiological comparison of plant-adapted vs. non-adapted bacteria would cast some insights into the physiology of the lifestyle of *E. coli* in/on lettuce as an alternative host. We observed that *E. coli* K12 adaptation to lettuce seedlings was accompanied by a dramatic increase in tolerance to oxidative stress caused by hydrogen peroxide. According to Queval et al. [20] leaf H_2O_2 concentrations in unstressed plants might range from 0.05 to 5.0 $\mu mol \times g\ FW^{-1}$. Thus, the concentration of H_2O_2 used in our study might be in the range that could be found in plant tissues and/or it might produce a level of oxidative stress on the bacterial cells comparable to that exerted by the colonized seedlings.

A microarray-based whole-genome transcriptional profiling and quantitative reverse transcription PCR of *E. coli* strain O157 exposed to lettuce lysates identified the up-regulation of genes for oxidative stress tolerance, among others. However, bacteria bearing a deleted copy of the *oxyR* gene (coordinating the response to oxidative stress) did not show reduced survival or growth over 5 h in lettuce lysates compared to the parental strain [21]. In agreement, the sole acclimation to oxidative stress appears not to be enough for *E. coli* to enhance its lettuce colonization efficiency, suggesting that adapted bacteria might display a more complex array of adaptations for enhanced fitness in/on the alternative host.

Plant-adapted *E. coli* also showed a stronger chemotactic response than non-adapted bacterial. Additionally, non-chemotactic bacteria were less competent for lettuce colonization. It had been shown before that root exudates are an important source of nutrients for the microorganisms present in the rhizosphere and participate in the colonization process through chemotaxis of soil microorganisms. It was further demonstrated that rice root exudates induce a higher chemotactic response for endophytic bacteria than for other bacterial strains present in the rice rhizosphere [22]. Mutational analysis conducted with the rhizospheric bacterium *Pseudomonas fluorescens* further demonstrated the requirement of the CheA-dependent chemotactic response for full colonization capacity of tomato roots. Although as competent

for roots colonization as the wild type strain when inoculated alone in a gnotobiotic system, the *cheA*-deficient strains were 10- to 1000-fold less competent to colonize tomato roots in either the gnotobiotic system or in non-sterile potting substrate in competition assays together with the wild type strain [23]. Thus, it appears that human commensal *E. coli* are not only genetically equipped but also remains flexible enough to activate adaptive responses to signals from alternative hosts such as plants in an apparently similar fashion as bacteria displaying a preferred rhizospheric or endophytic lifestyle.

Hence, the findings of this study help to understand the physiology of *E. coli* during the adaptation process of colonization of alternative hosts such as lettuce seedlings.

Materials and Methods

Bacterial Strains and Culture Conditions

Escherichia coli K12 strain MG1655 was the reference strain used in this work. Strain MG1655 was transformed using pKEN2-GFPmut2 [24] or pDsRed2 plasmid (Clontech) encoding Green Fluorescent Protein (GFP) or Red Fluorescent Protein (RFP) respectively. For chemotaxis assays, the strains of *E. coli* K12 RP437 [25] and RP9535 ($\Delta cheA1643$) [24] were used. *Escherichia coli* strains were cultivated overnight at 37°C in Luria Bertani broth (LB) or complete M9 medium with shaking at 150 rpm. Fluorescence-labeled strains were cultured in the presence of 100 $\mu g \times mL^{-1}$ ampicillin. For the standardized colonization assay (see below), inocula were prepared from bacterial cells cultivated in LB or M9 media and cells at the exponential phase of growth were collected by centrifugation at 6,000 rpm for 5 min at room temperature and rinsed with sterile 0.85% NaCl.

Cultivation and Bacterial Colonization of Lettuce Seedlings

Lettuce (*Lactuca sativa* cv. criolla) seeds were provided by Pro-Huerta Program of Instituto Nacional de Tecnología Agropecuaria (INTA, Argentina). Seeds were disinfected with 95% (v/v) ethanol for 1 min and 0.36% (w/v) active chlorine for 3 min and then rinsed with sterile distilled water three times. Disinfected seeds were allowed to germinate onto 0.6% agarized Hoagland solution in Petri dishes and maintained for 48 h at 22±2°C under a photoperiod of 16 h of white light at 176 μmol photons $\times m^{-2} \times s^{-1}$. The effectiveness of disinfection was tested in parallel assays by incubating a sample of the pool of disinfected seeds into LB broth overnight at 37°C with shaking. Only seeds producing no microorganisms growth from these tests were kept for colonization assays.

For the standard colonization assay bacterial cells were inoculated into 20 mL of melted agarized-Hoagland-medium at 0.5X strength containing 0.6% agar at about 40°C to obtain a final titer of 10^8 colony forming units (CFUs) $\times mL^{-1}$ in 50 mL sterile test tubes. Inoculum size was first estimated by counting bacterial cells under a microscope using a Neubauer chamber. Samples of the inocula were used for confirmatory determinations of CFUs onto solidified LB medium. Since either GFP- or RFP-expression in the fluorescent-labeled strains is directed by the *E. coli lacZ* promoter, 10 μM IPTG was incorporated into the agarized medium for consistent strong fluorescence. After allowing the inoculated medium to solidify at room temperature, three equally developed two-day-old seedlings were aseptically transferred onto the inoculated medium of each test tube inside a transfer hood aided by sterile tweezers. It was noticed in preliminary assays that pooling three seedlings together produced a similar range of variation of *E. coli* loading of colonized seedlings than scoring individual seedlings. Thus, each data point of the standard assay consisted of three seedlings in order to simplify the preparation of multiple samples. For surface inoculation assays, seedlings were transferred onto non-inoculated agarized-Hoagland medium and when the first true leaf appeared, a droplet containing bacteria at 10^8 CFU×mL^{-1} was inoculated onto this leaf. The inoculated plants were placed in a growing chamber at $22\pm2°C$ under a photoperiod of 16 h of white light at 176 µmol photons×m^{-2}×s^{-1} and analyzed 20 days after inoculation.

For each data point of CFUs from colonized seedlings, roots or leaves samples corresponding to the three seedlings from each test tube were pooled together, weighed and immediately homogenized in sterile 0.85% NaCl in a Potter-type tissue grinder. Samples were immediately spotted onto LB plates in triplicate for CFUs determination of each data point.

For fluorescence microscopy, samples were mounted directly onto glass slides and examined through a Nikon eclipse 600 fluorescence microscope. Images were captured with an Olympus DP72 camera. Samples were further examined using a Nikon confocal laser scanning microscope (Nikon C1 Plus with an Eclipse Ti Inverted Microscope).

Competitive Index Assays

GFP-labeled non-adapted bacteria (cultivated in LB medium) were allowed to compete against RFP-labeled cells isolated from leaves. For confirmatory purposes reciprocal assays were run in which non-adapted cells were RFP-labeled and plant acclimated cells were GFP-labeled. Cells were co-inoculated into test tubes at a titer of 0.5×10^8 CFU×mL^{-1} of each strain (1:1 ratio) to achieve the final titer of the standard colonization assay. Initially, bacterial doses were estimated by counting the cells of the suspensions under a microscope before inoculation of the test tubes with the mixed populations. Finally, samples of the suspensions were used to determine CFUs onto LB plates containing 10 µM IPTG, and the initial input ratio of GFP-labeled to RFP-labeled viable cells was determined. Adapted bacteria were obtained from ground colonized lettuce leaves (see standard colonization assay) and directly inoculated into the substrate of two-days-old seedlings without passages through laboratory media. After 20 days, post inoculation macerates were prepared from leaves to determine the output ratio of GFP-labeled to RFP-labeled viable cells. The competitive index (CI) was defined as the ratio between the output and the input ratios. A deviation from CI = 1 indicated that one of the strains outcompeted the other under the conditions imposed.

Oxidative Stress Tolerance

Bacterial *Escherichia coli* cells cultured in LB medium or isolated from leaves (first cycle of colonization) were inoculated at 10^5 to 10^6 CFUs×mL^{-1} into M9 complete medium supplemented with 0; 0.5; 2; 8 or 32 mM H_2O_2. Cultures were incubated at 37°C with shaking at 150 rpm and CFUs were determined over time onto LB plates. As a control for *E. coli* acclimation to oxidative stress, cells were incubated in the presence of 8 mM H_2O_2 in M9 medium for 30 min and then survival in the presence of 10 mM H_2O_2 was scored over time.

To determine the effect of acclimation to oxidative stress on the ability of *E. coli* to colonize lettuce seedlings bacteria, inocula were prepared by treating cells cultured in LB medium with 8 mM H_2O_2 (sub lethal dose) or lettuce extract for 30 min before inoculation into lettuce seedlings.

Chemotaxis Assays

The chemotaxis response of plant-adapted cells (isolated from colonized leaves) was evaluated by puncture inoculation of the cells onto TB solid medium (1 g peptone, 1 g NaCl and 0.25 g agar-agar per 1L of distilled water). Chemotaxis towards aqueous leaf extracts was assayed by the method proposed by Adler [27] with modifications [22]. To prepare leaf extracts, lettuce plants were purchased from a local store and young leaves were thoroughly rinsed with sterile distilled water. Aided by a scalpel, samples of the main vascular bundle or the remainder of the leaf (leaf blade) were taken and immediately ground in liquid nitrogen. To obtain the clarified extracts the thawed samples were subjected to 4 cycles of vortexing and incubating on ice for 5 minutes, and then centrifuged at 18,000 rpm at 4°C for 15 minutes. Finally, samples were sterilized by filtration through 0.22 µm membranes. The protein concentration in a typical preparation were 3.12 mg×mL^{-1} or 2.24 mg×mL^{-1} for leaf blade or main vein preparations, respectively. Sterile capillary tubes of 3 cm long with an internal diameter of 1 mm were plunged open end down into a 1.5 mL centrifuge-tubes containing 500 µl of lettuce extracts or phosphate buffered saline (PBS; 1.1 g K_2HPO_4, 0.32 g KH_2PO_4 and 8.5 g NaCl in 1L of distilled water, pH 7.2). Once the liquid ascended about 1.5 cm, the capillary was inserted into the 1.5 mL centrifuge-tubes containing the suspension of 10^7 bacteria×mL^{-1}. After 40 minutes of incubation, the capillary was removed and its exterior was rinsed with sterile 0.85% NaCl. The sealed end was broken off over a 1.5 mL tube containing sterile 0.85% NaCl. Serial dilutions were prepared and the CFUs were determined by counting onto LB agar plates. The plates were incubated at $36\pm1°C$ for 24 h. Finally, responses were calculated as the number of CFU per capillary containing lettuce extract, divided by the number of CFU in capillaries containing the control (PBS). *Escherichia coli* strains RP437 (wild type for chemotaxis analysis) and RP9535 ($\Delta cheA$1643) were used as positive and negative controls, respectively.

Statistical Analysis

Data are expressed as means or medians $\pm$ SD or as individual data points and the corresponding mean value. Colonization data were transformed to log$_{10}$ prior to performing statistical analysis. The data shown in Fig. 4 were subjected to the Mann-Whitney test and those shown in Figs. 6 and 7 to a Wilcoxon test using the GraphPad statistical package.

Supporting Information

Figure S1 Laser confocal images of GFP-labeled *E. coli* K12 colonization of lettuce leaves. Representative photomicrographs in the (A) XY and corresponding (B) YZ, or (C) XZ planes. In (B) and (C) data integration spans the complete width of the leaf. White bars correspond to (A) 20 µm or (B and C) 50 µm. Note that a single GFP-labeled *E. coli* aggregate spans the complete width of the leaf.

Figure S2 Representative fluorescence photomicrographs of *E. coli* K12-colonized lettuce leaves. (A) Detail of single *E. coli* cells labeled with RFP in lettuce leaves at an early stage of colonization photographed at a magnification of 750X, and the white bar represents 10 µm. (B) Detail of bacterial aggregates in the proximity of vascular tissue in lettuce leaves at advanced stages of colonization. The picture was taken at a magnification of 150X and the white bar represents 20 µm.

Figure S3 Effect of acclimation of *E. coli* K12 to H_2O_2 on lettuce colonization. (A) Survival of *E. coli* K12 with (■) or without (▲) previous incubation in the presence of 8 mM H_2O_2

for 30 min and then transferred to 10 mM H_2O_2. Data represent the mean and SD of 2 independent assays. (B) Effect of acclimation of *E. coli* K12 to 8 mM H_2O_2 or leaves lysates for 30 minutes on its performance during colonization of lettuce leaves. No addition or freshly isolated cells from leaves were used as controls. Data represent the mean and SD of 3 independent assays.

Figure S4 Effect of initial *E. coli* K12 inoculum size and time course of colonization of lettuce leaves. (A) Assays were run according to the standardized conditions at 20 days after inoculation (see materials and methods) except that the initial dose varied from 10 to 10^8 cells$\times$mL^{-1} of agarized medium. Data represent the mean and SD of 3 independent assays comprising a pool of all leaves from three seedlings from each assay. (B) Assays were run according to the standardized conditions (see materials and methods) except that data were separately collected for leaves (empty bars) or roots (full bars) at different times after inoculation. Data represent the mean and SD of 3 independent assays.

Video S1 Laser confocal video of GFP-labeled *E. coli* K12 colonization of a lettuce leaf. The video integrates confocal images at 1 μm intervals starting from the adaxial face of the leaf. Note the stomata (just a few) plane at 0.19 min at the adaxial face; two GFP-labeled *E. coli* microcolonies: one at the left

(0.23–0.33 min) and one to the right (0.29–0.42 min); a leaf vein longitudinal-section confocal plane at 0.31 min; and abaxial face stomata (more abundant) plane at 0.39 min. Both microcolonies are close to the vascular tissue of the leaf.

Acknowledgments

Authors are very thankful to Jose Angel Hernandez, John Rogers and Mauro Do Nascimento for critical reading of the Ms, to Sydney Kustu and Claudia Studdert for kind provision of *E. coli* K12 strains MG1655 or RP437 and RP9535, respectively; and to Gabriela Pagnussat and María Victoria Martín for assistance with the confocal microscope. This work will be included in the Ph.D. thesis of MA Dublan (Universidad Nacional de Mar del Plata). During part of this work MA Dublan was a Fellow at the Agencia Nacional de Promoción Científica y Tecnológica (ANPCyT) and then at the Consejo Nacional de Investigaciones Científicas y Técnicas (CONICET). L Curatti is a career researcher at the CONICET, Argentina.

Author Contributions

Conceived and designed the experiments: MAD LC. Performed the experiments: MAD. Analyzed the data: MAD JCFOM LL LC. Contributed reagents/materials/analysis tools: MAD JCFOM LL LC. Contributed to the writing of the manuscript: MAD LC.

References

1. Tenaillon O, Skurnik D, Picard B, Denamur E (2010) The population genetics of commensal *Escherichia coli*. Nat Rev Microbiol 8: 207–217.
2. Riley LW, Remis RS, Helgerson SD, McGee HB, Wells JG, et al. (1983) Hemorrhagic colitis associated with a rare *Escherichia coli* serotype. N Engl J Med 308: 681–685.
3. Winfield MD, Groisman EA (2003) Role of nonhost environments in the lifestyles of Salmonella and *Escherichia coli*. Appl Environ Microbiol 69: 3687–3694.
4. Jimenez L, Muniz I, Toranzos GA, Hazen TC (1989) Survival and activity of *Salmonella typhimurium* and *Escherichia coli* in tropical freshwater. J Appl Bacteriol 67: 61–69.
5. Walk ST, Alm EW, Gordon DM, Ram JL, Toranzos GA, et al. (2009) Cryptic lineages of the genus *Escherichia*. Appl Environ Microbiol 75: 6534–6544.
6. Franz E, Visser AA, Van Diepeningen AD, Klerks MM, Termorshuizen AJ, et al. (2007) Quantification of contamination of lettuce by GFP-expressing *Escherichia coli* O157:H7 and *Salmonella enterica* serovar *Typhimurium*. Food Microbiol 24: 106–112.
7. Jablasone J, Warriner K, Griffiths M (2005) Interactions of *Escherichia coli* O157:H7, *Salmonella typhimurium* and *Listeria monocytogenes* plants cultivated in a gnotobiotic system. Int J Food Microbiol 99: 7–18.
8. Buchanan RL, Edelson SG, Miller RL, Sapers GM (1999) Contamination of intact apples after immersion in an aqueous environment containing *Escherichia coli* O157:H7. J Food Prot 62: 444–450.
9. Eblen BS, Walderhaug MO, Edelson-Mammel S, Chirtel SJ, De Jesus A, et al. (2004) Potential for internalization, growth, and survival of Salmonella and *Escherichia coli* O157:H7 in oranges. J Food Prot 67: 1578–1584.
10. Islam M, Morgan J, Doyle MP, Jiang X (2004) Fate of *Escherichia coli* O157:H7 in manure compost-amended soil and on carrots and onions grown in an environmentally controlled growth chamber. J Food Prot 67: 574–578.
11. Deering AJ, Pruitt RE, Mauer LJ, Reuhs BL (2011) Identification of the cellular location of internalized *Escherichia coli* O157:H7 in mung bean, *Vigna radiata*, by immunocytochemical techniques. J Food Prot 74: 1224–1230.
12. Reinhold-Hurek B, Hurek T (2011) Living inside plants: bacterial endophytes. Curr Opin Plant Biol 14: 435–443.
13. Holden NJ, Pritchard L, Wright K, Toth IK (2013) Mechanisms of plant colonization by human pathogenic bacteria: An emphasis on the roots and rhizosphere, John Wiley & Sons, Inc., 1217.1226.
14. van Baarlen P, van Belkum A, Summerbell RC, Crous PW, Thomma BPHJ (2007) Molecular mechanisms of pathogenicity: how do pathogenic microorganisms develop cross-kingdom host jumps? FEMS Microbiol Rev 31: 239–277.
15. Wright KM, Chapman S, McGeachy K, Humphris S, Campbell E, et al. (2013) The endophytic lifestyle of *Escherichia coli* O157:H7: Quantification and internal localization in roots. Phytopathology 103: 333–340.
16. Erickson MC, Webb CC, Diaz-Perez JC, Phatak SC, Silvoy JJ, et al. (2010). Surface and internalized *Escherichia coli* O157:H7 on field-grown spinach and lettuce treated with spray-contaminated irrigation water. J Food Protect 7: 500–506.
17. Cooley MB, Miller WG, Mandrell RE (2003) Colonization of *Arabidopsis thaliana* with *Salmonella enterica* and enterohemorrhagic *Escherichia coli* O157:H7 and competition by *Enterobacter asburiae*. Appl Environ Microbiol 69: 4915–4926.
18. Monier JM, Lindow SE (2004) Frequency, size, and localization of bacterial aggregates on bean leaf surfaces. Appl Environ Microbiol 70: 346–355.
19. Zaragoza W, Noel J, Teplitski M (2012). Spontaneous non-rdar mutations increase fitness of *Salmonella* in plants. Environ Microbiol Rep 4: 453–458.
20. Queval G, Hager J, Gakiere B, Noctor G (2008) Why are literature data for H2O2 contents so variable? A discussion of potential difficulties in the quantitative assay of leaf extracts. J Exp Bot 59: 135–156.
21. Kyle JL, Parker CT, Goudeau D, Brandl MT (2010) Transcriptome analysis of *Escherichia coli* O157:H7 exposed to lysates of lettuce leaves. Appl Environ Microbiol 76: 1375–1387.
22. Bacilio-Jiménez M, Aguilar-Flores S, Ventura-Zapata E, Pérez-Campos E, Bouquelet S, et al. (2003) Chemical characterization of root exudates from rice (*Oryza sativa*) and their effects on the chemotactic response of endophytic bacteria. Plant and Soil 249: 271–277.
23. de Weert S, Vermeiren H, Mulders IH, Kuiper I, Hendrickx N, et al. (2002) Flagella-driven chemotaxis towards exudate components is an important trait for tomato root colonization by *Pseudomonas fluorescens*. Mol Plant Microbe Interact 11: 1173–1180.
24. Cormack BP, Valdivia RH, Falkow S (1996) FACS-optimized mutants of the green fluorescent protein (GFP). Gene 173: 33–38.
25. Parkinson JS, Houts SE (1982) Isolation and behavior of *Escherichia coli* deletion mutants lacking chemotaxis functions. J Bacteriol 151: 106–113.
26. Sanatinia H, Kofoid EC, Morrison TB, Parkinson JS (1995) The smaller of two overlapping cheA gene products is not essential for chemotaxis in *Escherichia coli*. J Bacteriol 177: 2713–2720.
27. Adler J (1973) A method for measuring chemotaxis and use of the method to determine optimum conditions for chemotaxis *by Escherichia coli*. J Gen Microbiol 74: 77–91.

Permissions

List of Contributors

Sompong Chankaew
Program in Plant Breeding, Faculty of Agriculture at Kamphaeng Saen, Kasetsart University, Kamphaeng Saen, Nakhon Pathom, Thailand

Takehisa Isemura, Akito Kaga, Norihiko Tomooka and Duncan A. Vaughan
Genetic Resources Center, National Institute of Agrobiological Sciences, Tsukuba, Ibaraki, Japan

Sachiko Isobe, Hideki Hirakawa and Kenta Shirasawa
Kazusa DNA Research Institute, Kisarazu, Chiba, Japan

Prakit Somta and Peerasak Srinives
Department of Agronomy, Faculty of Agriculture at Kamphaeng Saen, Kasetsart University, Kamphaeng Saen, Nakhon Pathom, Thailand

Ian P. Adams, Anna Skelton, Roy Macarthur, Neil Boonham, Adrian Fox and Laura Flint
Centre for Crop Protection, Food and Environment Research Agency, Sand Hutton, York, United Kingdom

Tobias Hodges
Centre for Crop Protection, Food and Environment Research Agency, Sand Hutton, York, United Kingdom
University of York, York, United Kingdom

Howard Hinds
RootCrop Ltd., Hoveringham, Nottinghamshire, United Kingdom

Palash Deb Nath
Department of Plant Pathology, Assam Agricultural University, Jorhat, India

Mathieu B. Brodeur
Douglas Mental Health University Institute and Department of Psychiatry, McGill University, Montréal (Québec), Canada

Katherine Guérard
Department of Psychology, Universitéde Moncton, Moncton (New Brunswick), Canada

Maria Bouras
Department of Education, University of Sheffield, Sheffield (South Yorkshire), United Kingdom

Alexander M. Muvea
Institute of Horticultural Production Systems, Section Phytomedicine, Leibniz Universität Hannover, Hannover, Germany
Plant Health Division, IPM cluster, International Centre of Insect Physiology and Ecology, Nairobi, Kenya

Rainer Meyhöfer and Hans-Michael Poehling
Institute of Horticultural Production Systems, Section Phytomedicine, Leibniz Universität Hannover, Hannover, Germany

Sevgan Subramanian, Sunday Ekesi and Nguya K. Maniania
Plant Health Division, IPM cluster, International Centre of Insect Physiology and Ecology, Nairobi, Kenya

Huihua Zhang, Junjian Chen, Guoyi Yang and Dingqiang Li
Guangdong Institute of Eco-environmental and Soil Sciences, Guangzhou, China

Li Zhu
Management School, Jinan University, Guangzhou, China

Qiaoying Zhang
Department of Botany, Institute of Ecology and Earth Sciences, University of Tartu, Tartu, Estonia
Department of Environmental Science and Engineering, Qilu University of Technology, Jinan, Shandong, China

Yunchun Zhang
Department of Environmental Science and Engineering, Qilu University of Technology, Jinan, Shandong, China

Shaolin Peng
State Key Laboratory of Biocontrol, School of Life Sciences, Sun Yat-sen University, Guangzhou, Guangdong, China

Kristjan Zobel
Department of Botany, Institute of Ecology and Earth Sciences, University of Tartu, Tartu, Estonia

Marzena Nowakowska, Marcin Nowicki, Urszula Kłosińska and Elżbieta U. Kozik
Department of Genetics, Breeding, and Biotechnology of Vegetable Crops, Research Institute of Horticulture, Skierniewice, Poland

Robert Maciorowski
Unit of Economics and Statistics, Research Institute of Horticulture, Skierniewice, Poland

Zining Cui
Guangdong Province Key Laboratory of Microbial Signals and Disease Control, Department of Plant Pathology, College of Natural Resource and Environment, South China Agricultural University, Guangzhou, China

Department of Nanobiology, Graduate School of Advanced Integration Science, Chiba University, Chiba, Japan
State Key Laboratory for Biology of Plant Diseases and Insect Pests, Institute of Plant Protection, Chinese Academy of Agricultural Sciences, Beijing, China

Jun Ito and Yoshimiki Amemiya
Department of Environment Science for Bioproduction, Graduate School of Horticulture, Chiba University, Chiba, Japan

Hirofumi Dohi and Yoshihiro Nishida
Department of Nanobiology, Graduate School of Advanced Integration Science, Chiba University, Chiba, Japan

Yigal Cohen, Avia E. Rubin and Mariana Galperin
Faculty of Life Sciences, Bar-Ilan University, Ramat Gan, Israel

Sebastian Ploch and Marco Thines
Biodiversity and Climate Research Centre, Frankfurt, Germany

Fabian Runge
Institute of Botany (210), University of Hohenheim, Stuttgart, Germany

Arthur Q. Villordon
Sweet Potato Research Station, Louisiana State University Agricultural Center, Chase, Louisiana, United States of America

Christopher A. Clark
Department of Plant Pathology and Crop Physiology, Louisiana State University Agricultural Center, Baton Rouge, Louisiana, United States of America

Thomas E. Simon, Ronan Le Cointe, Patrick Delarue, Stéphanie Morliére, Françoise Montfort, Maxime R. Hervé and Sylvain Poggi
INRA UMR 1349 IGEPP, Le Rheu, France

Bo Li
National Soil Fertility and Fertilizer Effects Long-term Monitoring Network, Institute of Agricultural Resources and Regional Planning, Chinese Academy of Agricultural Sciences, Beijing, P. R. China
Institute of Plant Nutrition and Environmental Resources, Liaoning Academy of Agricultural Sciences, Shenyang, P. R. China

Junxing Yang
National Soil Fertility and Fertilizer Effects Long-term Monitoring Network, Institute of Agricultural Resources and Regional Planning, Chinese Academy of Agricultural Sciences, Beijing, P. R. China
Centre forEnvironmental Remediation, Institute of Geographic Sciences and Natural Resources Research, Chinese Academy of Sciences, Beijing, P. R. China

Dongpu Wei, Shibao Chen, Jumei Li and Yibing Ma
National Soil Fertility and Fertilizer Effects Long-term Monitoring Network, Institute of Agricultural Resources and Regional Planning, Chinese Academy of Agricultural Sciences, Beijing, P. R. China

Artemis Giannakopoulou and Sophien Kamoun
The Sainsbury Laboratory, Norwich, United Kingdom

Sebastian Schornack
The Sainsbury Laboratory, Norwich, United Kingdom
Sainsbury Laboratory, Cambridge University, Cambridge, United Kingdom

Tolga O. Bozkurt
The Sainsbury Laboratory, Norwich, United Kingdom

Dave Haart
Institute of Food Research, Food & Health Programme, Norwich, United Kingdom

Dae-Kyun Ro
Department of Biological Sciences, University of Calgary, Calgary, Canada

Juan A. Faraldos
School of Chemistry, Cardiff University, Cardiff, United Kingdom

Paul E. O'Maille
Institute of Food Research, Food & Health Programme, Norwich, United Kingdom
John Innes Centre, Department of Metabolic Biology, Norwich, United Kingdom

Emile Faye
Institut de Recherche pour le Développement (IRD), UR 072, Laboratoire Evolution, Génomes et Spéciation, UPR 9034, Centre National de la Recherche Scientifique (CNRS), Gif sur Yvette, France et Université Paris-Sud 11, Orsay, France
UPMC Univ Paris06, Sorbonne Universités, Paris, France
Facultad de Ciencias Exactas y Naturales, Pontificia Universidad Católica del Ecuador, Quito, Ecuador

Mario Herrera
Facultad de Ciencias Exactas y Naturales, Pontificia Universidad Católica del Ecuador, Quito, Ecuador

Lucio Bellomo
Mediterranean Institute of Oceanography (MIO) CNRS/INSU, IRD, UM 110, Universitéde Toulon, La Garde, France

Jean-François Silvain
Institut de Recherche pour le Développement (IRD), UR 072, Laboratoire Evolution, Génomes et Spéciation, UPR 9034, Centre National de la Recherche Scientifique (CNRS), Gif sur Yvette, France et Universite´ Paris-Sud 11, Orsay, France

Olivier Dangles
Institut de Recherche pour le Développement (IRD), UR 072, Laboratoire Evolution, Génomes et Spéciation, UPR 9034, Centre National de la Recherche Scientifique (CNRS), Gif sur Yvette, France et Université Paris-Sud 11, Orsay, France
Facultad de Ciencias Exactas y Naturales, Pontificia Universidad Católica del Ecuador, Quito, Ecuador
Instituto de Ecología, Universidad Mayor San Andrés, Cotacota, La Paz, Bolivia

Yoko Kimata-Ariga and Toshiharu Hase
Institute for Protein Research, Osaka University, Suita, Osaka, Japan

Yanhong Qin, Zhenchen Zhang, Qi Qiao, Desheng Zhang, Yuting Tian, Shuang Wang and Yongjiang Wang
Key Laboratory of Crop Pest Control of Henan Province, Key Laboratory of Pest Management in South of North-China for Ministry of Agriculture of PRC, Institute of Plant Protection, Henan Academy of Agricultural Sciences, Zhengzhou, Henan, China

Li Wang
School of Life Sciences and technology, Nanyang Normal University, Nanyang, Henan, China

Zhaoling Yan
Institute of Agricultural Economics and Information, Henan Academy of Agricultural Sciences, Zhengzhou, Henan, China

Sean Chapman, Brian Harrower and Kara McGeachy
Cell and Molecular Sciences, James Hutton Institute, Invergowrie-Dundee, United Kingdom

Laura J. Stevens, Pauline S. M. Van Weymers and Paul R. J. Birch
Cell and Molecular Sciences, James Hutton Institute, Invergowrie-Dundee, United Kingdom
Division of Plant Sciences, University of Dundee at James Hutton Institute, Invergowrie-Dundee, United Kingdom
Dundee Effector Consortium, Invergowrie-Dundee, United Kingdom

Petra C. Boevink, Xinwei Chen and Ingo Hein
Cell and Molecular Sciences, James Hutton Institute, Invergowrie-Dundee, United Kingdom
Dundee Effector Consortium, Invergowrie-Dundee, United Kingdom

Stefan Engelhardt
Division of Plant Sciences, University of Dundee at James Hutton Institute, Invergowrie-Dundee, United Kingdom
Dundee Effector Consortium, Invergowrie-Dundee, United Kingdom

Colin J. Alexander
Biomathematics and Statistics Scotland, Invergowrie-Dundee, United Kingdom

Nicolas Champouret
J.R. Simplot Company, Simplot Plant Sciences, Boise, Idaho, United States of America

Qiang Sun and Lingbo Zhao
Center for Health Management and Policy, Key Lab of Health Economics and Policy Research of Ministry of Health in Shandong University, Shandong University, Jinan, Shandong, China

Maria Tärnberg, Maud Nilsson and Lennart E. Nilsson
Clinical Microbiology, Department of Clinical and Experimental Medicine, Linköping University, Linköping, Sweden

Cecilia Stålsby Lundborg
Global Health (IHCAR), Department of Public Health Sciences, Karolinska Institutet, Stockholm, Sweden

Yanyan Song
School of Public Health, Shandong University, Jinan, Shandong, China

Malin Grape
Antibiotics and Infection Control Unit, Public Health Agency of Sweden, Solna, Sweden

Göran Tomson
Global Health (IHCAR), Department of Public Health Sciences, Karolinska Institutet, Stockholm, Sweden
Medical Management Centre (MMC), Department of Learning, Informatics Management, Ethics, Karolinska Institutet, Stockholm, Sweden

Muhammad Tariq Rafiq
Ministry of Education Key Laboratory of Environmental Remediation and Ecological Health, College of Environmental and Resource Sciences, Zhejiang University, Hangzhou, China
Department of Environmental Science, International Islamic University, Islamabad, Pakistan

Rukhsanda Aziz, Xiaoe Yang, Wendan Xiao and Tingqiang Li
Ministry of Education Key Laboratory of Environmental Remediation and Ecological Health, College of Environmental and Resource Sciences, Zhejiang University, Hangzhou, China

Peter J. Stoffella
Indian River Research and Education Center, Institute of Food and Agricultural Sciences, University of Florida, Fort Pierce, Florida, United States of America

Aamir Saghir
Institute of Statistics, Zhejiang University, Hangzhou, China

Muhammad Azam
College of Agriculture and Biotechnology, Zhejiang University, Hangzhou, China

Gia Khuong Hoang Hua, Lien Bertier, Saman Soltaninejad and Monica Höfte
Laboratory of Phytopathology, Department of Crop Protection, Faculty of Bioscience Engineering, Ghent University, Gent, Belgium

Michio Murakami and Taikan Oki
Institute of Industrial Science, The University of Tokyo, Meguro, Tokyo, Japan
Japan Science and Technology Agency, Core Research for Evolutionary Science and Technology (CREST), Chiyoda, Tokyo, Japan

Maria de los Angeles Dublan
Instituto de Investigaciones en Biodiversidad y Biotecnología, Consejo Nacional de Investigaciones Científicas y Técnicas, Mar del Plata, Buenos Aires, Argentina
Fundación para Investigaciones Biológicas Aplicadas, Mar del Plata, Buenos Aires, Argentina
Laboratorio Integrado de Microbiología Agrícola y de Alimentos, Facultad de Agronomía, Universidad Nacional del Centro de la Provincia de Buenos Aires, Azul, Buenos Aires, Argentina

Juan Cesar Federico Ortiz-Marquez and Leonardo Curatti
Instituto de Investigaciones en Biodiversidad y Biotecnología, Consejo Nacional de Investigaciones Científicas y Técnicas, Mar del Plata, Buenos Aires, Argentina
Fundación para Investigaciones Biológicas Aplicadas, Mar del Plata, Buenos Aires, Argentina

Lina Lett
Laboratorio Integrado de Microbiología Agrícola y de Alimentos, Facultad de Agronomía, Universidad Nacional del Centro de la Provincia de Buenos Aires, Azul, Buenos Aires, Argentina

Index